Allele Mining for Genomic Designing of Cereal Crops

This book deliberates on the concept, strategies, tools, and techniques of allele mining in cereal crops and its application potential in genome elucidation and improvement, including studying allele evolution, discovery of superior alleles, discerning new haplotypes, assessment of intra- and interspecific similarity, and studies of gene expression and gene prediction. Available gene pools in global germplasm collections specifically consisting of wild allied species and local landraces for almost all major crops have facilitated allele mining. Development of advanced genomic techniques including PCR-based allele priming and Eco-TILLING-based allele mining are being widely used now for mining superior alleles. Allele's discovery has become more relevant now for employing molecular breeding to develop designed crop varieties matching consumer needs and with genome plasticity to adapt the climate change scenarios. All these concepts and strategies along with precise success stories are presented in the chapters dedicated to the major cereal crops.

The features of this book are as follows:

1. The first book on the novel strategy of allele mining in cereal crops for precise breeding.
2. Presents genomic strategies for mining superior alleles underlying agronomic traits from genomic resources.
3. Depicts case studies of PCR-based allele priming and Eco-TILLING-based allele mining.
4. Elaborates on gene discovery and gene prediction in major cereal crops.

This book will be useful to students and faculties in various plant science disciplines including genetics, genomics, molecular breeding, agronomy, and bioinformatics; the scientists in seed industries; and the policymakers and funding agencies interested in crop improvement.

Preface to the Series

Agriculture is going through a phase of transformation now. It has both challenges and opportunities. One of the major challenges it is facing is obviously food security. A population of about 9.3 billion projected by 2050 necessitates an increase of crop production by 60–70%. Another 10–15% increase would be needed to counter the reduction of crop production due to global warming and climate change. Nutrition security is another challenge with more than 800 million people suffering from malnutrition. Depletion of fossil fuel and increasing dependence on biofuel from crop residues is another challenge. Loss of biodiversity and land degradation also add to the list of challenges. The cultivable area is also decreasing fast due to housing, transport, and industrialization. Above all, climate change and global warming have a huge impact on agriculture, health, and environment.

Several novel concepts and strategies are emerging to address the above challenges. For example, practice of precision farming will lead to precise and timely use of inputs resulting in higher yield and better quality of produce on the one hand and reduction of loss of inputs on the other. Similarly, urban agriculture and vertical agriculture will add a lot of cultivable areas in future. Protected farming will facilitate growing crops under controlled conditions leading to reduction in biotic and abiotic stresses and precise utilization of inputs.

Agriculture requires maintenance of the philosophy farmers have been following since inception over thousands of years which are being sold today in new packages with sophisticatedly coined labels such as traditional agriculture, natural agriculture, sustainable agriculture, conservation agriculture, integrated agriculture, regenerative agriculture, organic agriculture, etc.

In the changing scenarios, next generation agriculture will need designed crop varieties and breeds produced through designed practices. This book series aims at deliberations on several novel concepts and innovative strategies developed for next generation agriculture and will hopefully benefit agricultural education, research, and outreach.

Chittaranjan Kole
Kolkata
January 10, 2024

NextGen Agriculture: Novel Concepts and Innovative Strategies
Chittaranjan Kole and Phullara Kole
Series Editor

Allele Mining for Genomic Designing of Fruit Crops
Chittaranjan Kole, Kenta Shirasawa, and Anil Kumar Singh

Allele Mining for Genomic Designing of Cereal Crops
Chittaranjan Kole, Sharat Kumar Pradhan, and Vijay K Tiwari

Allele Mining for Genomic Designing of Vegetable Crops
Chittaranjan Kole, Tusar Kanti Behera, and Prashant Kaushik

Allele Mining for Genomic Designing of Oilseed Crops
Chittaranjan Kole, Manish Kumar Pandey, and Naveen Puppala

Sustainable Urban Agriculture: New Frontiers
Kheir Al-Kodmany, Madhav Govind, Sharmin Khan, and Chittaranjan Kole

Allele Mining for Genomic Designing of Grain Legume Crops
Chittaranjan Kole, C. Bharadwaj, and Abhimanyu Sarkar

The Vertical Farm: Scientific Advances and Technological Developments
Kheir Al-Kodmany, Andrew Keong Ng, Abel Tablada, and Chittaranjan Kole

For more information, please visit our website: https://www.routledge.com/Nextgen-Agriculture/book-series/NGA

Allele Mining for Genomic Designing of Cereal Crops

Edited by
Chittaranjan Kole, Sharat Kumar Pradhan, and
Vijay K. Tiwari

CRC Press
Taylor & Francis Group
Boca Raton London New York

CRC Press is an imprint of the
Taylor & Francis Group, an **informa** business

First edition published 2024
by CRC Press
2385 NW Executive Center Drive, Suite 320, Boca Raton FL 33431

and by CRC Press
4 Park Square, Milton Park, Abingdon, Oxon, OX14 4RN

CRC Press is an imprint of Taylor & Francis Group, LLC

ISBN: 9781032471907 (hbk)
ISBN: 9781032471938 (pbk)
ISBN: 9781003385004 (ebk)

DOI: 10.1201/9781003385004

Typeset in Times
by Newgen Publishing UK

Dedicated to

Dr. Norman Ernest Borlaug

The Father of Green Revolution, Nobel Peace Prize winner and the Founder of the World Food Prize

Who inspired my research and academic works with his generous appreciations and advice.

Contents

Abbreviations

ABA	Abscisic acid
ABC	ATP-binding cassette
ABRE	Abscisic acid response element
AC	Amylose content
acdc1	*adult cyanide deficient 1* sorghum mutant
AFLP	Amplified fragment length polymorphism
AGPase	ADP glucose pyrophosphorylase
AM	Association mapping
AM	Axillary meristem
AP2	*Apetala2*
ARMS	Amplification refractory mutation system
asal	*Allium sativum*
ASI	Anthesis and silking interval
ASP	Allele-specific primers
AS-PCR	Allele-specific PCR
BAC	Bacterial artificial chromosome
BAP	Bioenergy Association Panel
BB	Bacterial blight
BC	Backcross population
BC3F1/ BC3F2	Backcross derived population
BETL	Basal endosperm transfer cell layer
bif2	*barren inflorescence 2*
BioDIG	Biological Database of Images and Genomes
BLB	Bacterial leaf blight
bmr	Sorghum *brown midrib* genes/mutants
BPH	Brown plant hopper
BS	Brown spot
bZIP	Basic region-leucine zipper protein
CAD	Cinnamyl alcohol dehydrogenase
CAP	Common Agricultural Policy
CAPS	Cleaved amplified polymorphic sequence
Cas	CRISPR associated protein
CDR	Ccarbon dioxide removal
CEP	C-terminal encoded peptide
CG	Candidate gene
CG	Comparative genomics
CGAM	Candidate gene-based association mapping
CIMMYT	International Maize and Wheat Improvement Center
CISP	Conserved intron specific primers
CK	Cytokinin
CLE	CLAVATA3/ESR-related
CLT	CACTA-like transposon element
CLV1/BAM	Clavata1/ barely any meristem
CLV-WUS	CLAVATA-WUSCHEL
CMS	Cytoplasmic male sterility
CNV	Copy number variant/variation
CORE	Collaborative Oat Research Enterprise

CRISPR	Clustered regularly interspaced short palindromic repeats
CRT	C-repeat element
CSL	Chromosome substitution line
CSSL	Chromosome segment substitution line
ct2	*compact plants 2*
CWR	Crop wild relatives
DAA	Days after anthesis
DArT	Diversity arrays Technology
DArTseq	DArT sequencing
DEGs	Differentially expressed genes
DH	Doubled haploid
Dhu1	Sorghum *Dhurrin 1* locus
DRE	DNA replication related element
DREB	Dehydration response element binding genes
DRO1	*Deeper Rooting 1* gene
DTY	*Yield under drought* gene
Dw1-4	Sorghum *Dwarfing 1-4* loci
EC	European Union
Eco-TILLING	Ecotype TILLING
ECQ	Eating and cooking quality
ED	Ear diameter
EFSA	European Food Safety Authority
EGW	Ear glume weight
EL	Ear length
EMS	Ethylmethane sulfonate
EREBP	Ethylene-responsive element binding protein
ERF	Ethylene responsive factor
EST	Expressed sequence tag
EW	Ear weight
FAO	Food and Agriculture Organization
FD	*FLOWERING LOCUS D*
fea2	*fascinated ear2*
fea3	*fascinated ear3*
fea4	*fascinated ear4*
FHB	Fusarium head blight
FIGS	Focused identification of germplasm strategy
FLO/LFY	*Floricaula and Leafy*
FLS2	*Flagellin-sensitive*
FMs	*Floral meristems*
FOAM	F-one association mapping
FRET	Fluorescence resonance energy transfer
FS	False smut
FT	*FLOWERING LOCUS T*
FTIR	Fourier-transform infrared spectroscopy
FWL	*Fw2.2-like*
GA	Gibberellic acid
GAB	Genomics-assisted breeding
GABA	Gamma-aminobutyric acid
GAB-ald	Gamma-aminobutyraldehyde
GBS	Genotyping-by-sequencing

GBSS	*granule bound starch synthase* gene
GBSSI	Granule-bound starch synthase I
GDD	Growing degree days.
GEBVs	Genome-assisted breeding values
GLW	Grain length and width
GM	Gall midge
GMO	Genetically modified organism
GMS	Gametic male sterile
GN	Grain number
gna	*Galanthus nivalis*
Gpc	Grain protein content
GRF	Growth-regulating factor
GRIN	Germplasm Resource Information Network
GS	Genomic selection
GS	Glutamine synthetase
GS	Grain shape
GSR	Green super rice
gSSRs	Genomic SSR
GT	Grain thickness/gelatinization temperature
GW	Grain weight
GWAS	Genome wide association study/studies
GxE	Genotype-by-environment
GxExM	Genotype-by-environment-by-management
GY	Grain yield
HCN	Hydrogen cyanide
HD-ZIP	Homeodomain-leucine zipper protein
HIF	Heterogeneous inbred family
HKW	Hundred kernel weight
HLT	Harbinger-like transposon
HMW	High molecular weight
HPC1	*PhosphatidylCholine 1*
HRW	Hard red winter
HTP	High-throughput phenotyping
HWW	Hard white winter
IAA	Indole acetic acid
IBD	Identity by descent
ICRISAT	International Crops Research Institute for Semi-Arid Tropics
id1	*indeterminate1 gene*
ids1	Indeterminate spikelet 1
IL	Introgression line
IM	Interval mapping
IM	Inflorescence meristem
InDel	Insertion-deletion
IPK	Institut für Pflanzengenetik und Kulturpflanzenforschung
Ipt	Isopentenyl transferase gene
IRD	Institut de Recherche Pour le Développement
JA	Jasmonic acid
KASP	Kompetitive allele specific PCR
KNPR	Kernel number per row
KRN	Kernel row number

KW	Kernel weight
LD	Linkage disequilibrium
LEA	Late embryogenesis abundant
LEA3	*LATE EMBRYOGENESIS ABUNDANT 3* locus
LMW	Low molecular weight
LOD	Logarithm of the odds
LRR	Leucine rich repeat
Ma1-6	Sorghum *Maturity 1-6* loci
MABB	Marker-assisted backcross breeding
MABC	Marker-assisted backcrossing
MAGIC	Multi-parent advanced generation intercross
MAI	Marker assisted introgression
maizezfl	Maize zeafloricaula
MALDI-TOF-MS	Matrix-assisted laser desorption ionization-time of flight mass spectrometry
MARS	Marker-assisted recurrent selection
MAS	Marker-assisted selection
MDH	Malondialdehyde
MEME	Multiple EM for Motif Elicitation
MFT-like	MOTHER OF FT-like
MIK	Myo-inositol kinase
MIPS	Myo-inositol-3-phosphate synthase
miRNA	MicroRNA
MPP	Multi-parental breeding population
MQTL	Meta-QTL
MTA	Marker-trait association
MYB	Myeloblastosis viral oncogene homolog
MYC	Myeloblastosis viral proto-oncogene homolog
NAC	Domain protein for NAM, ATAF1/2 and CUC2
NADH	Nicotinamide adenine dinucleotide (NAD) + hydrogen (H)
NAM	No apical meristem
NAM	Nested association mapping
NB-LRR	Nucleotide binding-LRR
NBPGR	National Bureau of Plant Genetic Resources
NBS	Nucleotide binding site
NCBI	National Center for Biotechnology Information
NDVI	Normalized difference vegetation index
NF-Y	Nuclear factor Y
NGS	Next-generation sequencing
NILs	Near isogenic lines
NPGS	National Plant Germplasm Systems
NSG	Non Stay-green
NSGC	National Society of Genetic Counselors
OFP	Ovate family protein
OPV	Open pollinated variety
OsSERK2	Rice Somatic Embryogenesis Receptor Kinase 2
P1	Maize *Pericarp color 1* gene
PAC	P1-derived artificial chromosome
PACE	PCR allele competitive extension
PAGE/RAGE	Promotion/removal of allele through genome editing
PAN	Parinthia

PASA	PCR-amplification of specific alleles
PBP	Pheromone-binding protein
PCR	Polymerase chain reaction
PGRC	Plant Gene Resources of Canada
PHYC	Phytochrome C
PID	PINOID protein kinase
PINII	Potato proteinase inhibitor
PLACE	Plant *cis*-acting regulatory DNA element
Pm3	Powdery mildew resistance
PMDTDb	Pearl Millet Drought Transcriptome Database
PMiGAP	Pearl millet inbred germplasm association panel
Ppd1	*Photoperiod 1*
PPR	P-type pentatricopeptide repeat
PSII	Photosystem II
pUBi	Maize ubiquitin promoter
QR-gene	Quantitative resistance gene
QTL	Quantitative trait locus
QTLs	Quantitative trait loci
QTN	Quantitative trait nucleotide
R allele	Resistant allele
ra	Ramose
RAD-seq	Restriction site-associated DNA sequencing
RAPD	Random amplified polymorphic DNA
rel2	*ramosa1 enhancer locus2*
RFLP	Restriction fragment length polymorphism
RFS	Renewable Fuel Standard
RGA	Rapid generation advancement
RGT	Rapid generation turn over
Rht	Reduced height
RIL	Recombinant inbred line
RLF	Rice leaf folder
RLK	Receptor-like kinase
RMES1	Sorghum Resistance to *Melanaphis sacchari 1* locus
ROS	Reactive oxygen species
RPG	Recurrent parent genome
RRL	Reduced representation library
RSA	Root system architecture
RSD	Rice stripe disease
RSV	Rice stripe virus
RTBV	Rice tungro bacilliform virus
RT-PCR	Reverse transcription-PCR
RTSV	Rice tungro spherical virus
RYMV	Rice yellow mottle virus
S allele	Susceptible allele
SAM	Shoot apical meristem
SAP	Stress associated protein
SAP	Sorghum association panel
SAVS	Senescence-associated vacuoles
SB	Stem borer
SBEIIb	*STARCH BRANCHING ENZYME IIb* gene

SBP	SQUAMOSA promoter-Binding protein
SCP	Sorghum Conversion Program
SES	Standard evaluation system
SG	Stay-green
ShB	Sheath Blight
SLAF-seq	Specific-locus amplified fragment sequencing
SM	Spikelet meristem
SNP	Single nucleotide polymorphism
Sor1	Sorghum *Sorgoleone 1* locus
SPDT	SULTR-like phosphorus distribution transporter
SPM	Spikelet-pair meristem
SRW	Soft red winter
SS	Starch synthase
SSCP	Single-stranded conformation polymorphism
SSIIa	*STARCH SYNTHASE IIa*
SSR	Simple sequence repeat
SSSII	Soluble starch synthase II
Stg1-5	Sorghum *Stay-green 1-5* loci
STS	Sequence-tagged site
SWDP	Sweet Sorghum Diversity Panel
SWEET	Sugars will eventually be exported transporters
SWW	Soft white winter
SYMD	SET and MYND
TALE	Transcription activation-like effector
TALEN	Transcription activator-like effector nuclease
TB1	*Teosinte branched 1*
tcd1	*totally cyanide deficient class 1* sorghum mutant
td1	*thick tassel dwarf1*
TF	Transcription factor
TFBS	Transcription factor binding site
TFIIA	Transcription Factor IIA
tga1	*teosinte glume architecture 1*
TGW	Thousand grain weight
TILLING	Targeting induced local lesions in genomes
ub2	*unbranched2*
UBN	Uniform blast nursery
UGT	UDP-glucosyltransferase
US FDA	United States Food and Drug Administration
USDA	United States Department of Agriculture
USDA-ARS	USDA-Agricultural Research Service
USDA-NPGS	USDA's National Plant Germplasm System
UTR	Untranslated region
UV	Ultraviolet
VI	Vegetation index
VIGS	Virus-induced gene silencing
Vip	Vegetative insecticidal protein
VIR	Vavilov All-Russian Institute of Plant Genetic Resources
VSP	Visual scale of plants
WAPO-A1	Wheat aberrant panicle organization-A1
WBPH	White backed plant hopper

WEW	Wild emmer wheat
WFZP	Wheat frizzy panicle
WGRS	Whole-genome resequence
WGS	Whole-genome sequencing
WUE	Water use efficiency
Wx	*Waxy* gene/locus
Wx-A1	Waxy gene and amylose biosynthesis
Xoo	*Xanthomonas oryzae* pv. *Oryzae*
XRD	X-ray diffraction
Y	Sorghum *YELLOW SEED* loci
YPP	Yield per plant
YSB	Yellow stem borer
ZEP	*Zeaxanthin epoxidase* gene
ZF-HD	Zinc finger homeodomain
Zfl	Zeafloricaula
ZFN	Zinc finger nuclease

Preface

The affordability, precision, and high-throughput nature of sequencing technologies in the last two decades have provided an unprecedented understanding of the genomes and target plant traits of economic importance. Further, the advances in the third-generation sequencing technologies including Hi-C especially in the last decade have made it possible for developing high-quality reference genomes, pangenomes, and gene annotations, including genome-wide structural variations. These developments have also increased the precision in gene and superior haplotype discovery followed by mining allelic diversity among breeding lines and gene bank collections. The high-throughput and cost-effective genotyping platforms have further boosted the adoption rate of genomic tools in the ongoing breeding programs across the world. Deployment of these advances in crops and making them well equipped to scale up the integration of genomic interventions in the breeding programs will facilitate development of future-ready crops.

The above advances have made gene/allele mining more precise and cost-effective for the target traits preferred by farmers, consumers, and industry as well as efforts towards mitigation of climate change impacts and achieving nutritional security prospective. It is important to accelerate the process of genomic designing to develop future-ready crops by bringing new allelic combination(s) at one or more loci underlying desired genes using modern breeding approaches, such as marker-assisted breeding, haplotype-based breeding, and gene/allele pyramiding; and also now through gene editing.

These new alleles originate through natural or induced mutations also at one or more genomic regions at a time. The traditional plant breeding methods practiced in the twentieth century depended solely on phenotypes for the development of new genotypes primarily through selection and hybridization. The advent of DNA-based molecular markers in the 1980s facilitated indirect identification of the alleles by using linked molecular markers using genetic linkage mapping. Molecular breeding as a supplement to traditional breeding has successfully been employed in almost all major crops for their genetic improvement in a short time. It is worthwhile to mention that transgenic breeding has also contributed significantly to crop improvement at least for some major input target traits, such as herbicide tolerance, insect resistance, etc., however, this technology is still facing challenges in societal adoption in different parts of the world, especially for food crops. Nevertheless, gene editing has been receiving positive vibes from the policymakers of different countries making a conducive environment for accelerating the use of this technology on large scale.

Despite the huge loss of diversity during domestication, several national and international gene banks have millions of diverse germplasms to explore variability for the traits that are needed to breed climate-resilient and future-ready crops. Genetic diversity played a pivotal role during the entire period of conventional plant breeding, traditional and molecular included. However, the plant breeders have been using mainly the primary gene pool constituting the indigenous germplasm lines and local landraces as a genetic resource of new alleles. In the meantime, a drastic dearth of new alleles in the primary gene pool happened due to genetic erosion caused naturally or due to domestication over time but aggravated because of the injudicious and irrational implementation of the so-called green revolution. Precisely, this highlighted the importance of allied wild crop relatives (WCRs) and thereby a mission of collection, conservation, characterization, and utilization of biodiversity. The emergence of advanced bioinformatics and the availability of cheaper and innovative genome sequencing technologies have empowered us to explore germplasm and identify a large number of donor alleles and allele mining has become a reality!

The strategy of association mapping, the most popular trait mapping approach in animal and human sciences, has received huge adoption in plant genomics studies. The availability of large sequenced diverse panels with multilocation phenotyping data has provided much-needed insights into trait genomics through genome-wide association studies (GWAS) leading to the discovery of genes and superior haplotypes in many crops. This has led to the possibility of mining a large number of genes in the available accessions in crops. Mutations occur in the genome as single nucleotide polymorphisms (SNPs) or insertion/deletions (In/Dels) resulting in the generation of new alleles or changes in existing alleles and allele combinations. Using exactly this principle, the approach of Eco-TILLING has facilitated large-scale allele mining in several major crops. On the other hand, sequence-based allele mining has led to the discovery of allelic diversity both at the target gene and the whole genome. Recent advances in modern genomic tools and technologies, such as genome engineering, speed breeding, etc., facilitate the deployment of these new alleles in breeding programs in a relatively short time to accelerate the rate of genetic gains. Truly speaking, allele mining has revolutionized the strategy of utilizing genetic as well as genomic resources and become an important component of next-generation precision breeding and biotechnology!

Allele discovery has become more relevant now for employing molecular breeding to develop designed crop varieties matching consumer needs and with genome plasticity to adapt the climate change scenarios using targeted and precision breeding tools validated by both genomics and phenomics studies. All these concepts and strategies along with precise success stories are presented in over 50 chapters allocated under five volumes dedicated to the five crop groups including cereals, oilseeds, grain legumes, fruits, and vegetables under this book series.

This volume entitled, "Allele Mining for Genomic Designing of Cereal Crops" includes 12 chapters dedicated to rice, wheat, maize, barley, oat, sorghum, pearl millet and finger millet contributed by 73 scientists from six countries including India, Italy, the Netherlands, Philippines, Spain, and the USA. We express our thanks for their excellent contributions and timely cooperation.

This book will hopefully be useful to students, research scholars, and teaching faculties in the academia to learn about fundamentals and applications of allele mining; the scientists practicing plant breeding in public and private institutes in improvising their programs; and also, the policymakers and funding agencies in prioritizing research fields.

Chittaranjan Kole, Kolkata, India
Sarat Kumar Pradhan, New Delhi, India
Vijay K Tiwari
College Park, USA

Contributors

Jauhar Ali
International Rice Research Institute, Los Banos, Philippines

Supriya Ambawat
ICAR-AICRP on Pearl Millet, Jodhpur 342 304, Rajasthan, India

Lokeshkumar B.M.
Division of Crop Improvement, ICAR-Central Soil Salinity Research Institute, Karnal, Haryana, India-132001

Kalyana Babu
ICAR-Indian Institute of Oilpalm Research, Pedavegi 534475, AP, India

Franz W. Badeck
Council for Agricultural Research and Economics (CREA), Research centre for Genomics & Bioinformatics, Via S. Protaso, 302, I -29017 Fiorenzuola d'Arda PC, Italy

Mary-Frances Behnke
Department of Plant & Environmental Sciences, Clemson University, Florence, SC, USA

Jayant S Bhat
ICAR-IARI-Regional Research Centre, Dharwad 580005, India

Zachary W. Brenton
Carolina Seed Systems, Inc., Darlington, SC, USA

Richard E. Boyles
Department of Plant & Environmental Sciences, Clemson University, Florence, SC, USA

Anilkumar C.
Division of Crop Improvement, ICAR-National Rice Research Institute, Cuttack, India-753006.

Ilaria Carrara
Council for Agricultural Research and Economics (CREA), Research centre for Genomics & Bioinformatics, Via S. Protaso, 302, I -29017 Fiorenzuola d'Arda PC, Italy

Luigi Cattivelli
Council for Agricultural Research and Economics (CREA), Research Centre for Genomics and Bioinformatics, Via San Protaso 302, 29017, Fiorenzuola d'Arda, Italy

Kasarla Chaithanya
Plant Breeding, ICAR-Indian Institute of Rice Research, Rajendranagar, Hyderabad-500030, India

Chandu singh
Division of Genetics, ICAR-Indian Agricultural Research Institute, New Delhi 110012, India

Hugo E. Cuevas
USDA-ARS, Tropical Agriculture Research Station, 2200 Pedro Albizu Campos Avenue, Mayagüez, Puerto Rico 00680, USA

Shivani D.
Plant Breeding, ICAR-Indian Institute of Rice Research, Rajendranagar, Hyderabad-500030, India

Stefano Delbono
Council for Agricultural Research and Economics (CREA), Research centre for Genomics & Bioinformatics, Via S. Protaso, 302, I -29017 Fiorenzuola d'Arda PC, Italy

Maroua Doghri
Department of Plant Biology, Faculty of Biology, Institute of Biotechnology and Biomedicine, University of Valencia, Valencia, Spain

Yves Emendack
Cropping Systems Research Laboratory, USDA-ARS, Lubbock, TX, USA

Franca Finocchiaro
Council for Agricultural Research and Economics (CREA), Research centre for Genomics & Bioinformatics, Via S. Protaso, 302, I -29017 Fiorenzuola d'Arda PC, Italy

Abdul Fiyaz R
Plant Breeding, ICAR-Indian Institute of Rice Research, Rajendranagar, Hyderabad-500030, India

RN Gadag
Division of Genetics, ICAR-Indian Agricultural Research Institute, New Delhi 110012, India

Vijay Gahlaut
Department of Biotechnology, Chandigarh University, Gharuan, Mohali, 140413, Punjab, India
University Center for Research and Development, Chandigarh University, Gharuan, Mohali, 140413, Punjab, India

Roberta Ghizzoni
Council for Agricultural Research and Economics (CREA), Research centre for Genomics & Bioinformatics, Via S. Protaso, 302, I -29017 Fiorenzuola d'Arda PC, Italy

R. Gobu
ICAR – Indian Institute of Spices Research, Kozhikode, Kerala, 673012, India

Davide Guerra
Council for Agricultural Research and Economics (CREA), Research Centre for Genomics and Bioinformatics, Via San Protaso 302, 29017, Fiorenzuola d'Arda, Italy

Chad M. Hayes
Cropping Systems Research Laboratory, USDA-ARS, Lubbock, TX, USA

Seyedali Hosseinirad
Department of Plant Science and Landscape Architecture, University of Maryland, College Park, MD-20783, USA

Vandana Jaiswal
Biotechnology Division, CSIR-Institute of Himalayan Bioresource Technology, Palampur, Himachal Pradesh 176061, India

Dinesh Chandra Joshi
ICAR-Vivekananda Parvatiya Krishi Anusandhan Sansthan, Almora 263 601, UK, India

Anmol Kajla
Department of Plant Science and Landscape Architecture, University of Maryland, College Park, MD-20783, USA

Ashvinkumar Katral
Division of Genetics, ICAR-Indian Agricultural Research Institute, New Delhi, India-110012

Chandan Kapoor
ICAR – Indian Agricultural Research Institute, Pusa campus, New Delhi- 110012, India

Dileep Kumar G.D
Plant Breeding, ICAR-Indian Institute of Rice Research, Rajendranagar, Hyderabad-500030, India.

Sachin Kumar
Department of Genetics and Plant Breeding, Ch. Charan Singh University, Meerut, India

Charu Lata
CSIR-National Institute of Science Communication and Policy Research, 14 Satsang Vihar Marg, New Delhi - 110067, India

Balram Marathi
Professor Jayashankar Telangana State Agricultural University, Hyderabad, India.

Kutubuddin A. Molla
Division of crop improvement, ICAR-National Rice Research Institute, Cuttack, India-753006

Caterina Morcia
Council for Agricultural Research and Economics (CREA), Research centre for Genomics & Bioinformatics, Via S. Protaso, 302, I -29017 Fiorenzuola d'Arda PC, Italy

Ganapati Mukri
Division of Genetics, ICAR-Indian Agricultural Research Institute, New Delhi 110012, India

Varunseelan Murugaiyan
International Rice Research Institute, Los Banos, Philippines

Sunitha N.C.
Department of Genetics and Plant Breeding, University of Agricultural Sciences, Bangalore, India-560065.

Christian John Robiso
International Rice Research Institute, Los Banos, Philippines

Chandra Nayaka Siddaiah
Pearl Millet Pathology Laboratory, University of Mysore, Manasagangotri, Mysore, India

Bernardo Ordas
Misión Biológica de Galicia, Spanish National Research Council (CSIC), Pontevedra, Spain

Arunava Pattanyak
ICAR-Indian Institute of Agricultural Biotechnology, Ranchi 834003, Jharkhand, India

Chandra Prabha
Division of Genetics, ICAR-Indian Agricultural Research Institute, New Delhi 110012, India

Fulvia Rizza
Council for Agricultural Research and Economics (CREA), Research centre for Genomics & Bioinformatics,Via S. Protaso, 302, I -29017 Fiorenzuola d'Arda PC, Italy

Davina H. Rhodes
Department Horticulture & Landscape Architecture, Colorado State University, Fort Collins, CO, USA

Krishnamurthy S.L.
Division of Crop Improvement, ICAR-Central Soil Salinity Research Institute, Karnal, Haryana, India-132001

Soumya SL
ICAR – Indian Agricultural Research Institute, Pusa campus, New Delhi- 110012, India

Rameswar Prasad Sah
Division of Crop Improvement, ICAR-National Rice Research Institute, Cuttack, India-753006

Jacobo Sanchez
Cropping Systems Research Laboratory, USDA-ARS, Lubbock, TX, USA

S. Mukesh Sankar
ICAR – Indian Agricultural Research Institute, Pusa campus, New Delhi- 110012, India

Gautam Saripalli
Department of Plant Science and Landscape Architecture, University of Maryland, College Park, MD-20783, USA

C Tara Satyavathi
ICAR-AICRP on Pearl Millet, Jodhpur 342 304, Rajasthan, India

Adam William Schoen
Department of Plant Science and Landscape Architecture, University of Maryland, College Park, MD-20783, USA

Hemant Sharma
Department of Genetics and Plant Breeding, Ch. Charan Singh University, Meerut, India

Samriti Sharma
University Institute of Biotechnology, Chandigarh University, Mohali, India, 140 103, India

Shilpa
Division of Genetics, ICAR-Indian Agricultural Research Institute, New Delhi 110012, India

Radha Shivhare
CSIR-National Botanical Research Institute, Rana Pratap Marg, Lucknow-226001, India

Rakesh. K. Srivastava
International Crops Research Institute for Semi Arid Tropics, Patancheru-502 324, Telangana, India

Salej Sood
ICAR-Central Potato Research Institute, Shimla 171001, HP 171 001, India

R.M. Sundaram
Biotechnology, ICAR-Indian Institute of Rice Research, Rajendranagar, Hyderabad-500030, India

Valeria Terzi
Council for Agricultural Research and Economics (CREA), Research centre for Genomics & Bioinformatics, Via S. Protaso, 302, I -29017 Fiorenzuola d'Arda PC, Italy

Vijay Kumar Tiwari
Department of Plant Science and Landscape Architecture, University of Maryland, College Park, MD-20783, USA

Alessandro Tondelli
Council for Agricultural Research and Economics (CREA), Research Centre for Genomics and Bioinformatics, Via San Protaso 302, 29017, Fiorenzuola d'Arda, Italy

Giorgio Tumino
Plant Breeding, Wageningen University and Research, Droevendaalsesteeg 1, 6708 PB Wageningen, the Netherlands

Wilfred Vermerris
Department of Microbiology & Cell Science, UF Genetics Institute, and Florida Center for Renewable Chemicals and Fuels, University of Florida, PO Box 103610, Gainesville, Florida 32610, USA

Zhanguo Xin
USDA-ARS, Plant Stress and Germplasm Development Unit, Crop Systems Research Laboratory, 3810, 4th Street, Lubbock, TX 79424, USA

Inderjit Singh Yadav
Department of Plant Science and Landscape Architecture, University of Maryland, College Park, MD-20783, USA

Venkata Rami Reddy Yannam
Sustainable Field Crops Program, Institute for Food and Agricultural Research and Technology (IRTA), 25198, Lleida, Spain

Melinda K. Yerka
Department of Agriculture, Veterinary & Rangeland Sciences, University of Nevada, Reno, NV, USA

Editor Biography

Chittaranjan Kole is an internationally reputed scientist with an illustrious professional career spanning over 40 years and globally recognized for his original contributions in the fields of plant genomics, biotechnology, and molecular breeding, leading to the publication of more than 150 quality research articles. He has edited over 180 books for the leading publishers of the world, including Springer, Wiley-Blackwell, and Taylor and Francis Group. His scientific contributions and editing acumen have been appreciated by seven Nobel Laureates including Profs. Norman Borlaug, Arthur Kornberg, Werner Arber, Phillip Sharp, Günter Blobel, Lee Hartwell, and Roger Kornberg. He has been honored with a number of fellowships, honorary fellowships, and national and international awards, including the 'Outstanding Crop Scientist Award' conferred by the International Crop Science Society. Prof. Kole has served in all prestigious positions in academia, including as Vice-Chancellor of the BC Agricultural university and Project Coordinator of Indo-Russian Center of Biotechnology in India, and Director of Research of the Institute of Nutraceutical Research of Clemson University in the USA. He worked also in the Pennsylvania State University and Clemson University as a Visiting Professor in the USA. He was awarded with the Raja Ramanna Fellow by the Department of Energy, Government of India. He is also heading the International Climate-Resilient Crop Genomics Consortium, International Phytomedomics and Nutriomics Consortium and Genome India International as their Founder President. Prof. Kole has recently established the Prof. Chittaranjan Kole Foundation for Science and Society and working as its Founding Chairman.

Sharat Kumar Pradhan, is Assistant Director General of the Indian Council of Agricultural Research (ICAR), New Delhi. At present, he is heading the Food and Fodder Crops (FFC) Section of the council. Before this assignment, he was Scientist at ICAR- NRRI, Cuttack for 25 years and 5 years under OUAT, Bhubaneswar. He has released 60 rice varieties for 18 states of the country. Dr. Pradhan is the lead developer of 35 and co-developer of 25 rice varieties. Many of his rice varieties namely CR Dhan409 (Pradhandhan), CR Dhan307 (Maudamani), CR Dhan800, CR Dhan801, CR Dhan802, CR Dhan507, CR Dhan508, etc., are very popular in the eastern region of the country. He has published about 200 peer reviewed research articles in various books and journals. He has reported more than 80 novel genes/QTL controlling various traits in rice through linkage and association mapping. He has guided 15 Ph.D and 12 M.Sc. students for their thesis research. Dr. Pradhan is a Fellow of National Academy of Agricultural Sciences (NAAS), New Delhi; Indian Society of Plant Breeding & Genetics (ISGPB), New Delhi, and Association of Rice Research Workers (ARRW), Cuttack. He is conferred with many prestigious Awards namely Dr. LK Sikka Endowment Award of NAAS; Professor Siddique Award & Hooker Award of IARI, New Delhi; Samanta Chandra Sekhar Award from DST, Govt. of Odisha; AB Joshi Award of ISGPB, New Delhi and many others. Dr. Pradhan is the former Editor-in-Chief, Oryza, and Vice President of the Indian Society of Genetics & Plant Breeding, New Delhi.

Vijay Kumar Tiwari is an assistant professor and small grain breeder at the University of Maryland College Park. His major research interest is focused on genetics and genomics of small grain crops with emphasis on wheat, barley, and triticale. His program integrates genomic tools and germplasm development to achieve the goals of sustainable agriculture. Identification of novel genes and alleles against abiotic and biotic stresses is a key step to perform genetic improvement of crop plants.

One of his research objectives involves genomics enabling gene discovery of important genes and alleles playing a vital role against abiotic and biotic stresses in small grain crops. His research is also focused on enhancing yield and yield contributing factors in bread wheat through the discovery and integration of some key genes playing an important role in spike morphology and grain number per spikelet. His program utilizes Speed Breeding and DH based approaches to deploy useful genes to develop high-yielding resilient MD wheat cultivars. He has published more than 60 peer reviewed research publications in high impact journals such as Nature, Science, Nature Biotechnology, Nature Genetics, Theoretical and Applied Genetics, Plant Journal, and Crop Science.

1 Allele Mining-Based Breeding Approaches to Enhance Tolerance to Various Biotic and Abiotic Stresses in Rice

Abdul Fiyaz R.[*$], *Jauhar Ali*[*$], *Balram Marathi*[*$],
Christian John Robiso, Varunseelan Murugaiyan,
Kasarla Chaithanya, Shivani D., Dileep Kumar G.D.,
and R.M. Sundaram
[$]Contributed equally.
*Corresponding authors: j.ali@irri.org; balumarathi@gmail.com; genefiyaz@gmail.com

1.1 INTRODUCTION

Rice is one of the most important staple crops in the world, providing food for billions of people globally. However, rice production is severely threatened by various abiotic and biotic factors, including drought, submergence, fungal, bacterial, and nematode-oriented biotic diseases, and pests such as brown plant hopper (BPH) and stem borer (SB). The principal biotic stresses, such as bacterial leaf blight (BLB), sheath blight (ShB), blast, brown spot (BS), false smut (FS), brown plant hopper (BPH), yellow stem borer (YSB), gall midge (GM), have a significant impact on the production and quality of rice grains. Among the different breeding techniques and control strategies available for reducing biotic stresses, the most effective, economical, and environmentally friendly is the host plant resistance, mainly acquired via conventional breeding methods. Traditional breeding approaches have successfully developed varieties with some level of resistance to these stresses. Still, the continuous evolution of new pathogen races and the emergence of new pests pose ongoing challenges to rice breeders. The progress made through these conventional breeding approaches has been slow due to the complexity of abiotic and biotic stress resistance. In addition, it is a polygenic trait influenced by multiple genes. Therefore, there is an urgent need to develop efficient and effective breeding approaches to enhance rice's tolerance to abiotic and biotic stresses. Agriculture's commercial significance depends on factors, such as crop yields, disease management techniques, and availability of disease-resistant plant sources. Disease resistance is vital in controlling various pathosystems, including viruses, nematodes, and fungal diseases, such as ergot in cereals (*Claviceps purpurea*), *Rhizoctonia solani* in maize, powdery scab in potato, or black leg in rapeseed. The genetic makeup of a plant determines its susceptibility to diseases, with homozygous and homogenous cultivars such as wheat, oats, barley, and peas being more prone to disease infections. Since most crops are homogenous, they are susceptible to disease outbreaks in different regions and countries.

Rice is a staple crop contributing to food and nutritional security (Fiyaz et al. 2022). Depending on the severity, rice disease incidences may lower yields by 20–100%. New minor diseases such as false smut, Bakanae, early seedling blight, sheath, and stem rot have developed as important issues. Disease losses vary on growth circumstances, susceptibility, etc. (Shivappa et al. 2021). Over 70 fungi, bacteria, viruses, and nematodes have infected rice (Zhang et al. 2009). Stem borer,

DOI: 10.1201/9781003385004-1

gall midge, plant hopper, Gundhi bug, and leaf folder are important rice pests (Jena et al. 2018). By 2030, rice demand will rise 40% to fulfill the needs of roughly 5.0 billion people (Khush 2005). Only increased production and productivity can meet this demand. Conventional plant breeding approaches were previously utilized for controlling abiotic and biotic stresses. Still, the current focus has shifted towards molecular breeding strategies due to environmental conditions, variability in phenotypic traits, and the labor-intensive nature of conventional approaches. Molecular breeding strategies help fill gaps in our understanding of abiotic and biotic stresses in different crops and aid in developing integrated management programs for controlling stresses. In recent decades, molecular markers, microarrays, and genetic alterations have been used to examine the genetic basis of stress tolerance and generate agricultural cultivars with better stress tolerance. DNA marker technology has helped create quantitative trait loci (QTLs) mapping, marker-assisted selection (MAS), genomic selection, and genetic transformation to generate stress-tolerant plants. Fine mapping may help breeders transfer resistance genes from donor cultivars into new, top rice cultivars via MAS. Connecting knowledge about genes and gene function is needed to generate highly productive, sustainable agricultural types (Ashkani et al. 2015).

This book chapter extensively covers the latest breakthroughs in enhancing rice's resistance to both biotic and abiotic stresses, specifically focusing on the use of allele mining-based breeding techniques. The chapter provides an inclusive overview of the successful implementation of these approaches, highlighting their potential impact in addressing food security concerns in the future. Alongside discussing the challenges, the chapter also identifies further research and development opportunities. This resourceful chapter will benefit policymakers, breeders, and researchers who aim to develop rice varieties with improved resistance to biotic and abiotic stresses. The innovative approaches highlighted in this chapter will undoubtedly play a crucial role in meeting future food security challenges.

1.2 MOLECULAR BREEDING FOR BACTERIAL BLIGHT RESISTANCE IN RICE

Over the last several decades, rice has developed a complex relationship with pathogens and pests. BLB is one of the most significant diseases (Vikal et al. 2017), substantially reducing rice productivity. Deployment of gene-conferred host plant resistance is an attractive technique that two critical approaches may enhance: marker-assisted breeding and genetic transformation for combating emerging diseases and deployment of resistance genes in plant breeding programs. However, the debate about food safety and restrictions in certain countries hindered the adoption of genetically modified crops, clearing the path for plant breeders to widely employ the marker-assisted breeding program (Jiang et al. 2020).

1.2.1 Disease Symptoms and Pathogenesis

BLB in rice is caused by the gram-negative bacterium *Xanthomonas oryzae pv. oryzae* (Xoo), depending on the development stage and climatic circumstances, may cause yield losses of up to 50 % (Liu et al. 2014). It is a vascular disease that may manifest throughout seedling, vegetative, and reproductive development. The symptoms are characterized by leaf dryness and yellowing that begins at the leaf tips and progresses downward. The disease generally prefers conducive temperatures between 25 to 34°C and relative humidity exceeding 70%. The presence of tiny droplets of bacterial ooze (pale amber in color) on the afflicted areas helps to visualize the disease. The most devastating phase of disease in the tropics is the "kresek" or wilt phase, which results from seed infection or early systemic infection in the nursery (Kumar and Rao 2014). In 1884, farmers first recognized the disease in the Fukuoka region of Kyushu prefecture, Japan (Mizukami and Wakimoto 1969). In India, the first report of bacterial blight disease on rice was made in Maharashtra (Srinivasan et al. 1959). Before its 1963 outbreak of the disease in the Shahabad area of Bihar, in India it was considered relatively mild (Srivastava 1967). However, damage caused by this disease was considerably exacerbated by the growth of semi-dwarf and hybrid rice types and extensive fertilizer use. It was prevalent in Asia, including India, Philippines, Nepal, Indonesia, Sri Lanka, Australia, and West Africa. In recent years, it has been reported in almost all rice-growing nations worldwide (Naqvi 2019).

1.2.2 Bacterial Blight R-Genes and Quantitative Trait Loci

According to reports, several wild species of cultivated rice, including *O. longistaminata*, *O. rufipogon*, *O. minuta*, *O. barthii*, *O. brachyantha*, *O. granulate*, *O. ridleyi*, and *O. nivara*, are resistant to BLB (Brar et al. 1997). In the last two decades, much genetic research has been undertaken on BLB resistance. So far, 47 resistance (R) genes giving resistance to distinct Xoo strains have been mapped from diverse sources on 10 of the 12 chromosomes (Rao et al. 2017, Kim et al. 2018; 2019, Neelam et al. 2020, Xing et al. 2021). Among the resistant genes, 17 are recessive (*xa5, xa8, xa13, xa15, xa19, xa20, xa24, xa25, xa26, xa28, xa31, xa32, xa33, xa34, xa41, xa42,* and *xa44*), while remaining 30 are dominant (*Xa1, Xa2, Xa3, Xa4, Xa6, Xa7, Xa9, Xa10, Xa12, Xa14, Xa16, Xa17, Xa18, Xa21, Xa23, Xa24, Xa27, Xa29, Xa30, Xa35, Xa36, Xa37, Xa38, Xa39, Xa40, Xa43, Xa45, Xa46,* and *Xa47*) (Biswas et al. 2021, Xing et al. 2021, Chen et al. 2020). Out of the 47 resistance genes, nine (*Xa1, Xa3/xa26, xa5, Xa10, xa13, Xa21, Xa23, xa25,* and *Xa27*) have been cloned, and twelve (*Xa2, Xa4, Xa7, Xa22, Xa30, xa31, xa33, xa34, Xa38, Xa39, Xa40,* and *Xa42*) have been physically mapped (Kim et al. 2015, Bhasin et al. 2012, Kumar et al. 2012, Song et al. 1995, Sun et al. 2003, Gu et al. 2005, Niño et al. 2005, Cheema et al. 2008, Zhang et al. 2015).

Xa23, a dominant resistant gene effective at all growth stages, was identified in the wild rice species *O. rufipogon* (Zhang et al. 1998; 2001) and was discovered to be highly resistant to ten Philippine races (P1–P10), seven Chinese pathotypes (C1–C7), and three Japanese races (TI–T3) at maximum tillering stage (Zhou et al. 2011) Jin et al. (2007) isolated the BB resistance gene *Xa*30 from the wild rice species *O. rufipogon* and transferred it to cultivated rice to generate near-isogenic lines. Tan et al. (2004) identified the *Xa29* gene in *O. officinalis* and localized it inside a 1.3 cM area on Chromosome 1 bordered by RFLP (restriction fragment length polymorphism) markers. Similarly, the *O. australiensis Xa32*(t) gene is resistant to Xoo strains. *P1, P4, P5, P6, P7, P8, P9, KXO85*, but resistant to *P2* and *P3*; mapped using two SSR markers on the long arm of chromosome 11 (Zheng et al. 2009). Guo et al. (2010) introduced *Xa35*(t), a unique source of BB resistance gene from *O. minuta* (Acc. No. 101133), into the *O. sativa* L. cultivar IR24. The discovered genes for bacterial blight resistance are listed in (Table 1.1). Zhang et al. (2021) carried out an elaborate 1D/2D GWAS strategy to investigate the genetic systems underlying the reciprocal adaptation of rice (*Oryza sativa*) and its bacterial pathogen, *Xanthomonas oryzae* pv. *oryzae* (*Xoo*) using the whole-genome sequencing and large-scale phenotyping data of 701 rice accessions and 23 diverse *Xoo* strains. Zhang et al. (2021) discovered 47 *Xoo* virulence-related genes and 318 rice quantitative resistance genes (QR-genes) located in 41 genomic regions. Genome-wide interactions between the detected virulence-related genes and QR genes were identified. This study helped understand the relationship between rice and *Xoo*, which was characterized by strong differentiation among *Xoo* races corresponding to the subspecific differentiation of rice. Further, genome-wide interactions between many rice QR genes and *Xoo* virulence genes, in a multiple-to-multiple manner, are likely to result from direct protein–protein interactions or genetic epistasis. Zhang et al. (2021) observed a complex genetic interaction system between rice and *Xoo* and in other crop–pathogen systems that would maintain high levels of diversity at their QR-loci/virulence-loci, resulting in dynamic coevolutionary consequences during their reciprocal adaptation.

1.2.3 BLB Genes Cloned

Categorizing nine cloned genes into four classes of resistance genes, including LRR-RLKs (leucine-rich repeat receptor-like kinases), NB-LRR, a wall-associated kinase, executor R proteins, SWEET

TABLE 1.1
List of Bacterial Blight Resistance Genes Identified, Origin, Source, and Linked Markers

R Gene	Chr	Position (bp)	Donor	Source	Inheritance	Cloned	Resistance to *Xoo* Race	Linked Markers	Marker Type	Reference
Xa1	4	31,638,099–31,644,795	Kogyoku, Java 14	Japan	Dominant	Yes	Japanese race-I	Npb235	RFLP	Sakaguchi, 1967; Yoshimura et al. 1998
Xa2	4	–	RantaiEmas II, Tetep	Vietnam	Dominant	No	Japanese race-II	HZR950-5	SSR	Sakaguchi, 1967; He et al. 2006; Kurata and Yamazaki (2006)
Xa3/*Xa26*	11	28,399,360–28,402,773	Wase Aikoku 3	Japan	Dominant	Yes	Chinese, Philippine, and Japanese races	C481S	RFLP	Ezuka et al. 1975; Yoshimura et al. 1992; Xiang et al. (2006); Gao et al. 2013
Xa4	11	–	TKM6, IR20, IR22, IR72	India	Dominant	Yes	Philippine race-I	Npb181 and RM224	RFLP and SSR	Petpisit et al. 1977; Wang et al. 2001
xa5	5	437,010–443,270	DZ192, IR1545-339	Bangladesh	Recessive	Yes	Philippine races I, II, and III	RG556 and RM122	CAPS and SSR	Petpisit et al. 1977; Blair et al. 2003
Xa6/xa3	11	–	MalagkitSungsong	Zenith USA	Dominant	No	Philippine race-I	Y68SSRA	RFLP	Sidhu et al. 1978
Xa7	6	–	DZ78, DV85	Bangladesh	Dominant	No	Philippine races	G1091, RM205S2	RFLP, SSR	Sidhu et al. 1978; Chen et al. 2008
xa8	7	–	PI231129	USA	Recessive	No	Philippine races	RM500, RM533	SSR	Singh et al. 2002; Vikal et al. 2014
Xa9	11	–	KhaolayNhay	Laos	Dominant	No	Philippine races	C4S1S	RFLP	Singh et al. 1983; Ogawa et al. 1988
Xa10	11	22,203,734–22,204,676	Cas 209	Philippines	Dominant	Yes	Philippine and Japanese races	M491/M419	RFLP, CAPS	Yoshimura et al. 1983; Kurata and Yamazaki (2006)
Xa11	3	–	RP9-3	Philippines	Dominant	No	Japanese races IB, II, IIIA, and V	–	–	Ogawa and Yamamoto, 1986; Goto et al. 2009; Kurata and Yamazaki (2006)
Xa12	4	–	Kogyoku, Java 14	Japan	Dominant	No	Indonesian race V	–	–	Ogawa et al. 1978
xa13	8	–	BJ1, ChinsurahEoro II	India	Recessive	Yes	Philippine race-6	RG136, xal3p	STS and SSR	Yoshimura et al. 1995; Zhang et al. 1996; Kurata and Yamazaki (2006)

Xa14	4	–	TN1	China	Dominant	No	Philippine race 5	VAZ190B/ RG163	RFLP	Taura et al. 1987; Kurata and Yamazaki (2006)
xa15	–	–	M41 Mutant	Japan	Recessive	No	Japanese races	–	-	Nakai et al. 1998; Ogawa 2008
Xa16	–	–	Tetep	Vietnam	Dominant	No	Japanese races	–	-	Sanchez et al. 1999; Kurata and Yamazaki (2006)
Xa17	–	–	Asominori	Japan	Dominant	No	Japanese races	–	–	Ogawa et al. 1989; Kurata and Yamazaki (2006)
Xa18	–	–	Toyonishiki, Miyang 23, IR24	Japan	Dominant	No	Burmese races	–	–	Ogawa et al. 1986; Kurata and Yamazaki (2006)
xa19	3	–	XM5 (Mutant of IR24)	Philippines	Recessive	No	Japanese races	–	–	Taura et al. 1992; Kurata and Yamazaki (2006)
xa20	–	–	XM6 (Mutant of IR24)	Philippines	Recessive	No	Japanese races	–	–	Taura et al. 1992; Kurata and Yamazaki (2006)
Xa21	11	20,802,924–20,806,518	*O. longistaminata, IRBB 21*	Africa, Mali	Dominant	Yes	Philippine and Japanese races	pTA248	STS	Khush et al. 1990; Song et al.1995
Xa22(t)	11	–	Zhachanglong	China	Dominant	No	Chinese races	L363B/P143	RFLP	Lin et al. 1996; Kurata and Yamazaki (2006)
Xa23	11	22,203,734–22,204,676	*O. rufipogon*(CBB 23)	China	Dominant	Yes	Indonesian races	–	–	Zhang et al. 1998; Zhang et al. 2001
xa24	2	–	DV86, DV85, Aus 295	Bangladesh	Recessive	No	Philippine and Chinese races	–	–	Mir and Khush, 1990; Khush and Angeles,1999
xa25(t)	12	17,302,073–17,305,326	HX-3, Minghui 63 (Somaclonal mutant of Minghui 63)	China	Recessive	Yes	Chinese and Philippine races	–	–	Lee et al. 2003; Liu et al. 2011
xa26(t)	11	–	Minghui 63, Nep Bha Bong	China	Recessive	No	Philippine races	C4S1S/ Y6855R	RFLP	Lee et al. 2003
Xa27(t)	6	–	*O. minuta* IRGC101141, IRBB27	Philippines	Dominant	Yes	Chinese strains and Philippine races 2–6	M10S1, M1095	RFLP	Amante-Bordeos et al. 1992; Lee et al. 2003; Gu et al. 2004
xa28(t)	11	–	Lota Sail	Bangladesh	Recessive	No	Philippine race 2	–	–	Lee et al. 2003

(continued)

TABLE 1.1 (Continued)
List of Bacterial Blight Resistance Genes Identified, Origin, Source, and Linked Markers

R Gene	Chr	Position (bp)	Donor	Source	Inheritance	Cloned	Resistance to *Xoo* Race	Linked Markers	Marker Type	Reference
Xa29(t)	1	–	*O. officinalis*, B5	Portugal, Spain	Dominant	No	Chinese races	–	–	Tan et al. 2004
Xa30(t)	11	–	*O. rufipogon* (Y238)	India	Dominant	No	Indonesian races	–	–	Jin et al. 2007; Cheema et al. 2008
xa31(t)	4	–	Zhachanglong	China	Recessive	No	Chinese races	–	–	Wang et al. 2009
xa32(t)	11	–	C4064 (Introgression line of *O. australiensis*)	Australia	Recessive	No	Philippine races	RM27256, 27274	SSRs	Zheng et al. 2009
Xa33	7	–	Ba 7 *O. nivara*	Thailand	Dominant	No	Thai races	RMWR7.1 and 7.6	SSRs	Kumar et al. 2012
Xa33(t)	6	–	Ba7 *O. nivara*	Thailand	Dominant	No	Thai races	RM20590	SSRs	Korinsak et al. 2009
xa34(t)	1	–	Pin Kaset*O.brachyantha*	Sri Lanka	Recessive	No	Thai races	RM493, RM446, RM10927, RM10591	SSRs	Ram et al. 2010; Chen et al. 2011b
Xa35(t)	11	–	*O. minuta (Acc No.*101133)	Philippines	Dominant	No	Philippine races	–	–	Guo et al. 2010
Xa36(t)	11	–	C4059	China	Dominant	No	Philippine races	–	–	Miao et al. 2010
Xa38	4	–	*O. nivara* IRGC81825	India	Dominant	No	Indian Punjab races	RM17499, RM459, RM317	STS/SSR	Kaur et al. 2005; Cheema et al. 2008;Bhasin et al.2012
Xa39	11	–	FF329	China, Philippines	Dominant	No	Chinese and Philippine races	RM21, RM206	STS/SSR	Zhang et al. 2015
Xa40(t)	11	–	IR65482-7-216-1-2	Korea	Dominant	No	Korean BB races	RM27320, ID55, WA18-5	STS/SSR	Kim et al. 2015
xa41(t)	11	–	Rice germplasm	-	Recessive	Yes	Various *Xoo* strains	–	–	Hutin et al. 2015
xa42	3	–	XM14, a mutant of IR24	Japan	Recessive	No	Japanese *Xoo* races	RM15189	SSR	Busungu et al. 2016

Xa43(t)	11	11,92,907–11,943,779	P8	Japan	Dominant	No	Korean BB races	–	–	Kim and Reinke, 2019
xa44(t)	11	11,964,077–11,985,463	IR73571-3B-11-3-K3 (P6)	Japan	Recessive	No	Korean BB races	–	–	Kim, 2018
xa45(t)	8	26,737,175–26,818,765	*O. glaberrima* IRGC102600B	Philippines	Recessive	No	Indian Punjab races	Xa13prom	STS	Neelam et al. 2020
Xa46(t)	11	–	Mutant H 120	Japan	Dominant	No	Chinese *Xoo* races	RM26981–RM26984	SSR	Chen et al. 2020
Xa47(t)	11	27,957,132–27,983,380	*G252 (IL of O. rufipogon Yuanjiang)*	–	Dominant	No	Various *Xoo*strains	R13I14 13rbq-71	InDel	Jiaxin Xing et al. 2021

(sugars will eventually be exported transporters) genes, and a transcription factor gamma subunit protein. Kinases are encoded by the resistance genes *Xa3/Xa26*, *Xa4*, and *Xa21*. *Xa4* is a wall-associated kinase that confers resistance to specific races of Xoo through cell-wall reinforcement (Ke et al. 2017, Jiang et al. 2020). *Xa3/Xa26* and *Xa21* are LRR-RLKs identifying Xoo-associated molecules *AvrXa*3 and sulfated *Rax X*, respectively. Resistance mediated by *Xa21* and *Xa3/Xa26* has been proven favorably controlled by *OsSERK2* (rice somatic embryogenesis receptor kinase 2). *Xa1*, an NB-LRR, identifies intact transcription activation-like effectors (TALEs) from Xoo [proteins used by *Xanthomonas oryzae pv. oryzae (Xoo)* targets and stimulates the expression of specific host genes to improve the susceptibility of the host and therefore induces resistance. SWEET genes, which code for sugar transporters, have been discovered as targets of several *Xoo* TALEs, serving as susceptibility factors. The promoters of three SWEET genes, *OsSWEET11/Os8N3/xa13*, *OsSWEET13/xa25, and OsSWEET14/Os11N3/xa41*, have natural polymorphisms that increase their induction by related TALEs, hence conferring recessive resistance. Executor R-genes include *Xa10, Xa23*, and *Xa27* since the expression of their respective resistance alleles is triggered by *Xoo* TALEs. Several potential transmembrane domains distinguish these genes, whose expression stimulation leads to HR and, consequently resistance to *Xoo*. Another recessive resistance gene is *xa5,* which encodes transcription factor IIA gamma subunit 5 (*TFIIA5*) with a valine-to-glutamine mutation at codon 39. *TALE*s hijack the susceptible allele, *Xa5*, to stimulate the expression of additional host susceptibility genes. The mutation impairs the capacity of *TALE*s to bind to *TFIIA5*, resulting in resistance to the disease (Ke et al. 2017, Jiang et al. 2020).

1.2.4 Molecular Breeding for Bacterial Blight Resistance

Several preventative measures, such as chemical and biological treatments, are employed to restrict the spread of disease and insect pests. Still, the use of pesticides is expensive and detrimental to the environment. In contrast to the ecologically damaging use of pesticides, host plant resistance is thus regarded as the most effective and eco-friendly method for controlling biotic stress (Khush et al. 1989, Suh et al. 2015). Globally, most disease-resistant rice cultivars were generated using traditional breeding techniques. Combining dominant and recessive genes into a single genetic background would ensure more durable and broad-spectrum resistance, but pyramiding of these genes is hindered by conventional plant breeding due to the masking of the dominant gene to recessive gene (Koide et al. 2010, Kumar and Rao 2014); whereas incorporation of a single resistance gene would frequently result in the breakdown of resistance in a short period (Fiyaz et al. 2022). However, gene pyramiding is challenging owing to the dominant nature of most genes, and it is difficult to discover transferred genes that respond similarly to several pathogen races (Sundaram et al. 2009, Rajpurohit et al. 2011). Advanced rice genomics offers tools, such as molecular markers, for deployment in breeding programs to generate resistant varieties without using agrochemicals in an ecologically responsible manner (Miah et al. 2013). Molecular breeding introduces one or more genes for biotic and abiotic stress tolerance from distinct gene pools (Suh et al. 2009). Marker-assisted selection cleared the path for indirect selection of plants containing two or more genes at any stage of the plant's development without conducting a bioassay (Sundaram et al. 2008, Shanti et al. 2010). In addition, new developments in DNA sequencing have facilitated the mapping and characterization of the targeted genes, hence facilitating the application of MAS. Recombination between the gene of interest and the associated marker has led to the likelihood of false positives in the marker-assisted selection procedure, which has been mitigated using gene-specific functional markers (Ingvardsen et al. 2008). A combination of foreground selection and background selection followed by continuous backcrossing may recover up to 99 percent of the recurrent parent genome (RPG) in marker-assisted backcross breeding (MABB) (Tanksley et al. 1989). In foreground selection, gene-linked markers or functional markers (SSRs, InDels, and SNPs) are used to screen the targeted resistance genes at any stage of plant development. Singh et al. (2001) employed background selection by utilizing polymorphism information, such as SSRs and SNPs, between the donor and recurrent parents. This technique was used to estimate the recovery of the recipient parent genome (RPG) in each backcross generation at different stages of plant development. Numerous functional indicators, including *xa5, xa13, Xa21* (Hajira et al. 2016), and *Xa38*, have been developed and employed in screening targeted BB-resistance genes (Yugander et al. 2018).

As a single gene or in combination with two, three, or four genes, several of these resistance genes have been effectively introduced into rice cultivars/hybrid rice parental lines that are widely farmed in many rice-growing nations (Huang et al. 1997, Singh et al. 2001, Rao et al. 2017). Rice varieties such as PR106 (Singh et al. 2001), Pusa Basmati 1 (Joseph et al. 2004), Improved Samba Mahsuri (Sundaram et al. 2008), and Samba Mahsuri (Yugander et al. 2018) were produced by the introgression of a single gene or as a combination. A list of varieties developed using marker-assisted breeding is presented in Table 1.2.

TABLE 1.2
List of Rice Varieties Developed Using Marker-Assisted Breeding for BLB Resistance

Genes/QTLs Introgressed	Donor Parent	Recurrent Parent	Reference
Xa7	IRBB7	MR219	Mi et al. 2018
Xa21	IRBB21	Minghui 63 (Restorer line)	Chen et al. 2000
Xa5	IRBB5	IR64	Toenniessen et al. 2003
	IR62266	RD 6	Pinta et al. 2013
Xa23	CBB23 (NIL of Jingan 30)	Guangzhan63S (photoperiodic and thermosensitive male sterile line)	Ni et al 2015
Xa38	PR114	Improved Samba Mahsuri	Yugander et al. 2018
Xa40	IR65482-7-126-1-2	Junam	Reinke et al. 2018
Xa21+Xa7	Huahui20	Yihui1577 (restorer line)	Xu et al. 2012
Xa21+Xa33	FBR1-15, ISM	DRR17B	Balachiranjeevi et al. 2018
Xa21+xa13	Pusa1460 (Improved Pusa Basmati 1)	Pusa 6B, PRR78	Basavaraj et al. 2010
	Improved Samba Mahsuri	Taraori Basmati, Basmati 386	Pandey et al. 2013
	IRBB55	PusaBasmathi 1	Joseph et al. 2004
	Improved Samba Mahsuri	MTU1010	Arunakumari et al 2016
Xa21+Xa38	Improved Samba Mahsuri and PR114	APMS6B (Maintainer line)	Yugander et al. 2018
Xa4+Xa21	WH421	Siputeh	Luo et al 2014
Xa33+Xa38	FBR1-15, PR114	CO43	Vennisa et al 2018
Xa21+xa13+xa5	IRBB62 (NIL of IR24)	PR106	Singh et al.2001
	SS1113	Samba Mahsuri	Sundaram et al. 2008
	Improved Samba Mahsuri	Safri 14, Dubraj	Yap et al. 2016
	IRBB 60	CB174 R (Restorer line)	
	IRBB59 (NIL of IR24)	IRS5441-2 (Aromatic line)	Salgotra et al. 2012
	IRBB 60	ASD 16, ADT 43	Ramalingam et al. 2020
	SS1113	Triguna	Sundaram et al. 2009
Xa21+Xa33 +xa5	RG-9 (RGD-07097-1-8M-9)	MK-75(Manawthukha rice line Yn 3248-2-128-76-4-3-75)	Win et al 2013
Xa4+Xa21+xa13+xa5	IRBB60	Putra 1	Chukwu et al. 2019
	IRBB60	KMR3R, PRR78 (Restorer lines), Mahsuri	Guvvala et al. 2013
	IRBB 60	KMR3R, PRR78, IR58025B, Pusa6B, Mahsuri	Shanti et al. 2010
Xa4+Xa7+Xa21+ xa13+xa5	IRBB66	Tainung 82	Hsu et al. 2020
	IRBB60	Tapaswini, Lalat	Dokku et al 2013a, b

1.3 MOLECULAR BREEDING IN RICE FOR BLAST RESISTANCE

Rice blast is a fungus that produces a plant disease that affects rice and other grains such as wheat, rye, barley, and pearl millet. The fungus is present in 85 countries and has been known to eradicate whole rice crops, resulting in substantial economic losses. Additionally, the fungus is believed to destroy enough rice annually to feed 60 million people. Blast disease is caused by the hemibiotrophic filamentous fungus *Magnaporthe oryzae* or *Pyricularia oryzae*, often known as rice rotting neck and the most damaging fungal diseases in Southern Africa. The disease was initially documented in West Africa (Sierra Leone) in 1930 but did not gain momentum until 1965, when additional types were imported (Chisi et al. 2019). It is a persistent issue for upland and lowland rice, which is highly sensitive to this fungus phytopathogen from seedling to maturity.

At any growth stage, blast symptoms may arise on any part of the rice plant, including the leaves, leaf collars, necks, panicles, pedicels, and seeds. Using local pathogen isolates, rice cultivars are routinely evaluated for sensitivity to rice blast by spraying conidial suspensions on the plant in greenhouse and field settings (Takahashi et al. 2009). The various terms such as nursery blast, leaf blast, node blast, neck blast, and panicle blast characterize the symptoms at various growth stages of its occurrence in rice crops. At different phases of the disease, neck, and panicle blast has been found to cause dramatic yield decreases, resulting in reduced grain weight, ripe spikelet percentages, and utterly mature grain percentages.

Until 20 days after heading, the panicle base was infected with the blast pathogen (Neck blast), leading to a yield loss of more than 50 percent. In India, blast-related production losses might reach up to 50 percent if the disease reaches pandemic proportions. Artificial leaf blast disease screening is often conducted in a uniform blast nursery (UBN) under field circumstances (Jamaloddin et al. 2020). Spraying a virulent blast race (fungal conidial suspension at a concentration of 1×10^5 spores/mL) over UBN beds 25–30 days after planting inoculates them with the disease (DAS). Later, throughout the night, the beds are covered with polythene sheeting to foster an environment favorable to disease development. 10–15 days after infection, based on the severity of the disease on the susceptibility check, the disease score is determined using the SES (IRRI 1996).

Currently, 109 significant genes for leaf blast resistance have been discovered, 25 of which have been molecularly characterized and listed in (Table 1.3) and with *Pi9* being the first cloned R gene. The vast majorities of these genes are allelic or closely linked genes that encode disease-resistant NBS-LRR proteins and zinc finger domains (Annegowda et al. 2021). *Pi54*, a significant blast resistance gene that provides durable resistance against Indian races of blast fungus, is one of the most critical genes mapped in India. *Pi37, Pit, Pi-sh, Pi64, Pi-b, Pi63, Pi9, Pi-2, Piz-t, Pid3, Pigm, Pi25, Pi36, Pi5, Pi-54, Pik-m, Pik, Pik-p, Pik-e, Pi-a, Pi1*, and *Pita* belong to the largest class of plant R genes that encode proteins with the nucleotide-binding site (NBS). *Pid2*, a serine–threonine–kinase membrane-spanning protein, has leucine-rich repeat (LRR) domains (Kalia et al. 2019).

Due to a lack of cultural strategies, such as nutrition and water management, planting time, spacing, and the administration of fungicides, the agricultural sector has yet to be able to combat blast disease effectively and efficiently. Traditional breeding techniques, marker-dependent breeding methods, breeding procedures that need genetic transformation, and genome editing are the four significant categories of rice breeding for blast resistance. Blast-resistant rice cultivars were produced using classic techniques, such as pedigree breeding, backcrossing, recurrent selection, and mutation breeding. The blast resistance genes *Pib, Pita, Pia, Pi1, Pikh, Pi2(t),* and *Pi4* have been used in several rice types to battle the blast pathogen by conventional breeding techniques. Through one of the most recent breeding techniques, marker-assisted introgression (MAI), Kumar et al. (2019) introduced the wide spectrum rice blast-resistant gene *Pi-2* coming from the isogenic line C101A51 into the highly vulnerable elite cultivar Samba Mahsuri (BPT 5204). Using MAS, Madhavi et al. (2012) introduced the blast resistance gene *Pi54* into the genetic background of the high-yielding bacterial blight rice variety Improved Samba Mahsuri. Using markers, Hari et al.

TABLE 1.3
List of Blast Resistance Genes Identified in Various Rice Cultivars

Sl. No	Gene	Chr. No.	Position Location (bp)	Source	Identification Procedure	Country	References
1	*Mpiz*	11	4073024–16730739	Zenith (J)	Linkage analysis using phenotypic markers	Japan	Goto et al. 1970
2	*Pb1*	11	21711437–21361768	Modan (I)	Cloned	Japan	Hayashi et al. 2010
3	*PBR*	11	20,125,000–30,075,000	St-No 1 (J)	Mapped within 22.9 cM	Japan	Fujii et al. 1995
4	*Pi(t)*	4	2,270,216–3,043,185	P167 (I)	Linkage analysis using phenotypic markers	–	Causse et al. 1994
5	*Pi1*	11	26498854–28374448	LAC23 (J)	Mapped within 11.4 cM	Philippines	Hua et al. 2012
6	*Pi10*	5	14521809–18854305	Tongil (I)	Mapped within 6.7 cM	India	Naqvi et al. 1995
7	*Pi11*	8	–	Zhai-Ya-Quing8 (I)	–	China	Causse et al. 1994
8	*Pi12*	12	6988220–15120464	K80-R-Hang (J) Moroberekan (J)	Linkage analysis using RFLP markers	Japan	Inukai et al. 1996
9	*Pi13(t)*	6	12456009–16303608	*O. minuta*(W), Kasalath (I), Maowangu	–	Philippines	Amante-Bordeos A et al. 1992, Pan et al. 1996
10	*Pi14(t)*	2	1–6725831	Maowangu	Linkage analysis using Isozyme marker	Japan	Pan et al. 1996
11	*Pi15*	9	9641358–9685993	GA25 (J)	Mapped within 0.7 cM	China	Pan et al. 1996
12	*Pi157*	12	8826555–18050447	Moroberekan (J)	–	India	Naqvi NI et al. 1996
13	*Pi16(t)*	2	1–6725831	Aus373 (I)	Linkage analysis using isozyme marker	Japan	Pan and Tanisaka 1997
14	*Pi17*	7	22250443–24995083	DJ123 (I)	Mapped within 1.8 cM	Philippines	Sallaud et al. 2003
15	*Pi18(t)*	11	26796917–28376959	Suweon365 (J)	Mapped using RFLP markers	Korea	Ahn et al. 1996
16	*Pi19(t)*	12	8826555–13417087	Aichi Asahi (J)	Linkage analysis to other resistance genes	Japan	Iwata et al. 1996
17	*Pi20*	12	6988220–10603823	IR24 (I)	–	Philippines	Imbe T et al. 1997
18	*pi21**	4	5242654–5556378	Owarihatamochi (J)	Cloned	Japan	Fukuoka et al. 2009
19	*Pi22(t)*	6	4897048–6023472	Suweon365 (J)	QTL mapping	Korea	Ahn et al. 1997
20	*Pi23*	5	10755867–19175845	Suweon365 (J)	QTL mapping	Korea	Ahn et al. 1997
21	*Pi24(t)*	1	5242654–5556378	Azuenca (J)	QTL mapping	France	Zhuang et al. 2002
22	*Pi25**	6	18080056–19257588	Gumei 2 (I)	Cloned	China	Zhuang JY et al. 2001
23	*Pi25(t)*	2	34360810–37725160	IR64 (I)	QTL mapping	France	Sallaud et al. 2003
24	*Pi26*	6	8751256–11676579	Gumei 2 (I)	QTL mapping	China	Wu and Tanksley et al. 1993

(continued)

TABLE 1.3 (Continued)
List of Blast Resistance Genes Identified in Various Rice Cultivars

Sl. No	Gene	Chr. No.	Position Location (bp)	Source	Identification Procedure	Country	References
25	*Pi26(t)*	5	2069318–2760202	Azucena (J)	–	France	Sallaud et al. 2003
26	*Pi27*	1	5556378–744329	Q14 (I)	–	France	Sallaud et al. 2003
27	*Pi27(t)*	6	6230045–6976491	IR64 (I)	Mapped within 21.6 cM	France	Sallaud et al. 2003
28	*Pi28(t)*	10	19565132–22667948	IR64 (I)	QTL mapping	France	Sallaud et al. 2003
29	*Pi29(t)*	8	9664057–16241105	IR64 (I)	Mapped within 0.7 cM	France	Sallaud et al. 2003
30	*Pi3(t)*	9	7,825,000–8,250,001	Pai-kan-tao (J)	Linkage analysis using RFLP markers	Philippines	Causse et al. 1994
31	*Pi30(t)*	11	441392–6578785	IR64 (I)	QTL mapping	France	Sallaud et al. 2003
32	*Pi31(t)*	12	7731471–11915469	IR64 (I)	QTL mapping	France	Sallaud et al. 2003
33	*Pi32(t)*	12	13103039–18867450	IR64 (I)	QTL mapping	France	Sallaud et al. 2003
34	*Pi33*	8	5915858–6152906	IR64 (I)	Mapped within 1.6 cM	France	Berruyer et al. 2003
35	*Pi34*	11	19423000–19490000	Chubu32 (J)	QTL mapping	Japan	Zenbayashi et al. 2002
36	*Pi35(t)**	1	33,000,000–34,150,000	Hokkai 188 (J)	Cloned	Japan	Xu et al. 2014
37	*Pi36**	8	2870061–2884353	Q61 (I)	Cloned	China	Liu et al. 2005
38	*Pi37**	1	33110281–33489931	St-No 1 (J)	Cloned	China	Lin et al. 2008
39	*Pi38*	11	19137900–21979485	Tadukan (I)	Mapped within 20 cM	India	Gowda et al. 2006
40	*Pi39(t)*	4, 12	–	Chubu 111 (J) Q15(I)	Mapped within 37 kb	China	Liu et al. 2007
41	*Pi40(t)*	6	16274830–17531111	*O. australiensis* (W)	Mapped within 1.8 cM	Philippines	Jeung et al. 2007
42	*Pi41*	12	33110281–34005652	93–11 (I)	–	China	Yang et al. 2009
43	*Pi42(t)*	12	19565132–22667948	DHR9 (I)	–	India	Kumar et al. 2010
44	*Pi44*	11	20549800–26004823	Moroberekan (J)	–	USA	Chen et al. 1999
45	*Pi47*	11	–	Xiangzi 3150 (I)	–	China	Huang et al. 2011
46	*Pi48*	12	–	Xiangzi 3150 (I)	–	China	Huang et al. 2011
47	*Pi5(t)*	9	–	Moroberekan (J)	–	Philippines	Jeon et al. 2003
48	*Pi6(t)*	12	4053339–18867450	Apura (I)	–	USA	McCouch et al. 1994
49	*Pi62(t)*	12	2426648–18050026	Yashiro-mochi (J)	Mapped within 1.9 cM	Japan	Wu et al. 2008
50	*Pi67*	–	–	Tsuyuake	–	Philippines	Sallaud et al. 2003
51	*Pi8*	6	6230045–8751256	Kasalath (I)	–	Japan	Pan et al. 1996
52	*Pi9**	6	10386510–10389466	*O. minuta* (W)	Cloned	China	Qu et al. 2006

53	*Pia**	11	4073024–8078510	Aichi Asahi (J)	–	Japan	Goto I et al. 1981; Okuyama Y et al. 2011
54	*Pib**	2	35107768–35112900	Tohoku IL9 (J)	Cloned	Japan	Wang et al. 1999; Yokoo et al. 1978
55	*Pib2*	11	26796917–28376959	Lemont (J)	–	Philippines	Tabien et al. 1996
56	*PiCO39(t)*	11	6304007–6888870	CO39 (I)	Cloned	USA	Chauhan et al. 2002
57	*Pid(t)1*	2	20143072–22595831	Digu (I)	–	China	Chen et al. 2004
58	*Pid2**	6	17159337–17163868	Digu (I)	–	China	Chen et al. 2004
59	*Pif*	11	24695583–28462103	Chugoku 31-1 (J)	QTL mapping	Japan	Shinoda et al. 1971
60	*Pig(t)*	2	34346727–35135783	Guangchangzhan (I)	Mapped within 11.8 cM	China	Zhou et al. 2004
61	*PiGD1*	8	–	Sanhuangzhan 2 (I)	–	China	Liu et al. 2004
62	*PiGD-2*	10	–	Sanhuangzhan 2 (I)	–	China	Deng et al. 2006
63	*PiGD3*	12	–	Sanhuangzhan 2 (I)	–	China	Deng et al. 2006
64	*Pigm(t)**	6	10367751–10421545	Gumei4 (I)	–	China	Ise et al. 1991
65	*Pii*	9	2291804–28431560	Ishikari Shiroke (J) Fujisaka 5 (J)	Cloned	Japan	Takagi et al. 2013
66	*Pii1*	6	2291804–28431560	Fujisaka 5 (J)	–	–	Pan et al.1996
67	*Pii2*	9	1022662–7222779	Ishikari Shiroke (J)	Linkage analysis using phenotypic markers	Japan	Kinoshita et al. 1997
68	*Piis1*	11	2840211–19029573	Imochi Shiraz (J)	–	Japan	Hayasaka et al. 1996
69	*Piis2*	–	–	Imochi Shiraz (J)	–	Japan	Hayasaka et al. 1996
70	*Piis3*	–	–	Imochi Shiraz (J)	–	–	Hayasaka et al. 1996
71	*Pik**	11	27314916–27532928	Kusabue (I)	–	China	Zhai et al. 2011; Wang et al. 2009
72	*Pikg*	11	27314916–27532928	GA20 (J)	–	Japan	Pan et al. 1996
73	*Pikh (Pi54)**	11	24761902–24762922	Tetep (I)	Cloned	India	Sharma et al. 2005
74	*Pikm**	11	27314916–27532928	Tsuyuake (J)	Cloned	China	Kaji et al.1996; Ashikawa et al. 2008
75	*Pikp**	11	27314916–27532928	HR22 (I)	–	China	Hayashi et al. 2006
76	*Piks*	11	27314916–27532928	Shin 2 (J)	–	Japan	Fjellstrom et al. 2004
77	*Pikur1*	4	24611955–33558479	Kuroka (J)	Linkage analysis using phenotypic markers	Japan	Goto et al.1988
78	*Pikur2*	11	2840211–18372685	Kuroka (J)	Linkage analysis using phenotypic markers	Japan	Goto et al. 1988
79	*Pilm2*	11	13635033–28377565	Lemont (J)	QTL mapping	Japan	Tabien et al. 2002
80	*Pir2-3(t)*	2	–	IR64 (I)	–	Indonesia	Utami et al. 2008
81	*Pirf2-1(t)*	2	–	*O. rufipogan* (W)	–	Indonesia	Utami et al. 2008

(continued)

TABLE 1.3 (Continued)
List of Blast Resistance Genes Identified in Various Rice Cultivars

Sl. No	Gene	Chr. No.	Position Location (bp)	Source	Identification Procedure	Country	References
82	*Pise*	11	5740642–16730739	Sensho (J)	Linkage analysis using phenotypic markers	Japan	Hayasaka et al. 1996
83	*Pise2*	–	–	Sensho (J)	–	Japan	Hayasaka et al. 1996
84	*Pise3*	–	–	Sensho (J)	–	Japan	Hayasaka et al. 1996
85	*Pish**	1	33381385–35283446	Shin 2 (J)	Cloned	Japan	Imbe et al. 1997
86	*Pi-sh*	11	33381385–35283446	Nipponbare (J)	–	Japan	Imbe et al. 1997
87	*Pit**	1	2270216–3043185	Tjahaja (I), K59 (I)	Cloned	Japan	Hayashi et al. 2006; Hayashi et al. 2010
88	*Pita**	12	10603772–10609330	Tadukan (I),	–	USA	Bryan et al. 2000
89	*Pita2*	12	10078620–13211331	Shimokita (J)	–	Japan	Kiyosawa et al. 1967; Nakamura et al. 1997
90	*Pitp(t)*	1	25135400–28667306	Tetep (I)	–	India	Barman et al. 2004
91	*Pitq1*	6	28599181–30327854	Tequing (I)	–	USA	Tabien et al. 2000
92	*Pitq2*	2	–	Teqing (I)	–	USA	Tabien et al. 1996
93	*Pitq3*	3	–	Teqing (I)	–	USA	Tabien et al. 1996
94	*Pitq4*	4	–	Teqing (I)	–	USA	Tabien et al. 1996
95	*Pi-tq5*	2	34614264–35662091	Tequing (I)	QTL mapping	USA	Tabien et al. 2002
96	*Pitq6*	12	5758663–7731471	Tequing (I)	QTL mapping	USA	Tabien et al. 2002
97	*Piy1(t)*	2	38,300,000–38,525,000	Yanxian No 1 (I)	Mapped within 1.6 cM	China	Lei et al.2005
98	*Piy2(t)*	2	38,300,000–38,525,001	Yanxian No1 (I)	Mapped within 3.0 cM	China	Lei et al. 2005
99	*Piz*	6	10155975–10517612	Zenith (J), Tadukan (I), Toride 1 (J), Fukunishiki (J)	Mapped within 0.43 cM	Japan	Ahn et al. 1996
100	*Pizh*	8	4372113–21012219	Zhai-Ya-Quing8 (I)	QTL mapping	China	Sallaud et al. 2003
101	*Pi2**	6	10076481–10204423	Jefferson	Cloned	China	Jian et al. 2012
102	*Pid3**	6	13055253–13058027	93-11(I),Nipponbare (J)	–	China	Shang et al. 2009
103	*Pi4*	12	–	Pai-kan-tao	–	Philippines	Joshi et al. 2009
104	*Pizt**	6	–	Wild rice	–	Japan	Hayashi et al. 2004

105	*Pi5**	9	S04G03-C1454 (Map position)	RIL260	–	Korea	Lee et al. 2009
106	*Pb1**	11	*Os11g0597700*	Kanto 2Kanto 209, Koshihikari Aichi	–	Japan	Inoue et al. 2017
107	*NLS1**	11	AC134922 (Accession number)	Wild rice	–	China	Yang et al. 2008
108	*Pi54rh**	11	–	*Oryza rhizomatis* (Wild rice)	–	India	Das et al. 2012
109	*Pi54of**	11	HE589448 (	*Oryza officinalis*	–	India	Devanna et al. 2014

* Cloned and characterized blast resistance genes in rice.

Note: Chr No. = Chromosome number; I = indica; J = japonica; – = Not Known.

Source: Annegowda et al. 2021; Jamaloddin et al. 2021

(2013) effectively inserted the blast resistance gene *Pi54* and the bacterial blight resistance gene *Xa21* into the genetic background of the elite maintenance line IR58025B. Balachiranjeevi et al. (2015) introduced two blast-resistant genes and the critical bacterial blight resistance gene, *Xa21*, into a select maintainer line, DRR17B. Abhilash Kumar et al. introduced the same combination of bacterial blight and blast-resistant genes into RPHR1005R, the male father of the commonly used rice hybrid DRRH3. Rambabu et al. 2016 integrate a significant blast-resistant gene, *Pi54*, into Swarna's genetic background by marker-assisted backcross breeding. Mutation breeding is the preferred strategy when the available alleles in the germplasm have been depleted, and new alleles are needed. Mutation breeding has shown to be an efficient supplement to conventional breeding techniques.

Marker-assisted selection is a technique where the presence of a molecular marker linked to an R gene or an R gene itself is used to select a resistant phenotype in a variety. This approach is highly reliable and saves time since it does not require disease-favoring environmental conditions, and it can identify resistant genotypes even when the pathogen or disease is absent. As a result, many breeders now use MAS to produce blast-resistant rice cultivars, effectively incorporating numerous blast-resistance genes into rice varieties in India (Koide et al. 2009, Ashkani et al. 2015). Molecular-assisted backcross breeding (MABB) is another technique that uses DNA markers like SSRs and SNPs. The primary difference between MAS and MABB is the type of parent variety being improved and the recurrent parent genome recovery. The MABB enhances blast resistance in high-yielding top cultivars or genotypes. The MAS introduces the blast-resistant gene into any genotype, while the MABB is utilized to increase blast resistance in otherwise high-yielding elite varieties or genotypes. Madhavi et al. (2016), and Rekha et al. (2018) increased the blast resistance of Improved Samba Mahsuri (ISM) by introgressing two essential blast resistance genes (*Pi2* and *Pi54*) by marker-assisted backcross breeding via two independent sets of backcrosses, respectively. The list of rice cultivars improved for blast resistance by following MABB is given in Table 1.4.

1.4 GALL MIDGE (*ORSEOLIA ORYZAE* – WOOD-MASON)

Rice is an important crop for global food security, and any pests or diseases that attack the crop can result in significant yield losses and economic impact. One such pest is the gall midge, which has recently emerged as a substantial threat to rice production. In several Asian countries, gall midge infestation has resulted in severe yield losses, reaching up to 70%. Adult rice gall midge flies resemble mosquitoes with pink (female) to brownish (male) bodies; they only survive for 1 to 2 days. The flies emerge in the early morning hours and mate before sunrise. On the succeeding day, females are less active and begin vigorous oviposition after dusk. Unmated females release a potent pheromone that attracts and assures mating with males. Females only mate once and deposit between 100 and 150 eggs on the host plant. On the fourth day, these eggs hatch, and freshly developed maggots move through the area between leaf sheaths to the apical meristem to begin eating. High humidity and a thin layer of water on the plant's surface enhance the formation of maggots.

With a pharyngeal spatula, maggots lacerate meristem tissue and feed on the flowing cell sap. Laceration and salivary secretion result in cell enlargement and hyperplasia and the formation of a gall chamber around the maggot. The maggot feeding spends 2 weeks on the leaking tissue content through the pharyngeal suction mechanism. During this vigorous feeding stage, maggots endure two molts and three instars. During the third instar, feeding ends, and the gall chamber protrudes quickly. Within the gall, the maggot undergoes a metamorphosis. The pupa wriggles up to the gall's tip, drills an escape hole using cephalic horns, and partly pushes out of the gall. The adult emerges from the puparium by eclosion. The formation of gall or silver shoots makes the tillers infertile, lowering grain yield (Bentur et al. 2003).

TABLE 1.4
Examples of Marker-Assisted Eelection (MAS) and Marker-Assisted Backcross Breeding (MABB) in Rice for Blast Resistance

Sl. No	Gene/QTL	Marker type	Technique	Trait	Application	Reference
1.	*Pi1, Piz-5, Pita*	RFLP	MAS	Blast resistance	Pyramiding of three near isogenic lines (C101LAC, C101A51 and C101PKT) for blast resistance into a single cultivar CO39, each carrying the major genes *Pi1, Piz-5* and *Pita*, respectively	Korinsak et al. 2011
2.	*Pi1*	SSR, ISSR	MABB	Blast resistance	Applied for backcross breeding of variety	Liu et al. 2002
3.	*Xa21, Piz*	SSR	MAS	Bacterial blight and blast resistance	Pyramiding of target traits	Narayanan et al. 2002
4.	*Pid1, Pib, Pita, Pi2*	SSR	MAS	Blast resistance	*Pid1, Pib* and *Pita* genes were introduced into G46B, while *Pi2* was introduced into Zhenshan 97B	Chen et al. 2004
5.	*Piz*	SSR	MAS	Blast resistance	Successfully used for selection of blast resistance in a wide array of rice germplasm	Fjellstrom et al. 2004
6.	*Xa13, Xa21, Pi54, qSBR11*	SSR	MAS	Blast, Bacterial blight and Sheath blight resistance	Transfer genes conferring the resistances toward three different diseases in rice	Singh et al. 2012
7.	*Pita*	Gene specific	MAS	Blast resistance	Existence of *Pita* gene in 141 rice germplasms was determined, but the results were more articulated when *Pita* gene was introduced through advanced breeding lines	Wang et al. 2007
8.	*Pi1, Pi2, Pi33*	SSR	MABB	Blast resistance	Introgressed into Jin23B	Chen et al. 2008
9.	*Pi1, Pi2, Xa23*	SSR	MABB	Blast resistance and bacterial blight	Successfully applied for breeding variety Rongfeng B	Fu et al. 2012
10.	*Piz-5, Pi54*	SSR	MABB	Blast resistance	Transfer blast resistance genes from donor lines (C101A51 and Tetep) into PRR78 to develop Pusa 1602 (PRR78 + *Piz-5*) and Pusa 1603(PRR78 + *Pi54*), respectively	Singh et al. 2013
11.	*Pi9*	Gene specific	MABB	Blast resistance	Applied to introgress the cultivar Luhui 17	Wen and Gao 2011
12.	*Pi1, Piz*	SSR	MABB	Blast resistance	Pyramiding of *Pi1* and *Piz-5* genes into introduced PRR78	Gouda et al. 2013
13.	*Pi39*	InDel	MABB	Blast resistance	Introgressed into Chinese cultivar Q15	Hua et al. 2015
14.	*Pi2, Xa21, Xa33*	SSR	MABB	Blast and Bacterial blight resistance	Introgressed into RPHR-1005	Kumar et al. 2016
15.	*Pi40*	SSR	MABB	Blast resistance	Introgressed into elite cultivars Turkish, Osmancik-97 and Halilbey	Beser et al. 2016

(continued)

TABLE 1.4 (Continued)
Examples of Marker-Assisted Eelection (MAS) and Marker-Assisted Backcross Breeding (MABB) in Rice for Blast Resistance

Sl. No	Gene/QTL	Marker type	Technique	Trait	Application	Reference
16.	*Pi1*, *Pi2*	SSR	MABB	Blast resistance	Introgressed into Intan variety and BPT5204	Hegde et al. 2016
17.	*Pi46*, *Pita*	SSR	MABB	Blast resistance	Introgressed into Hang hui 179 (HH179)	Xiao WM et al. 2016
18.	*Pi2*, *Pi9*	SNP	MABB	Blast resistance	Introgressed into R179	Luo WL et al. 2016
19.	*Pizt*, *Pi2*, *Pigm*, *Pi40*, *Pi9*, *Piz*	SSR	MAS	Blast resistance	Introgressed into Yangdao 6	Wu et al. 2016
20.	*Pi1*, *Pi2*, *Pi33*	SSR	MAS	Blast resistance	Introgressed into Russian rice varieties	Usatov et al. 2016
21.	*Pi9*, *Pizt*, *Pi54*	SNP	MABB	Blast resistance	Introgressed into japonica rice 07GY31	Xiao et al. 2016
22.	*Pi-b*, *Pik-h*	SSR	MAS	Blast resistance	Introgressed into MR219	Tanweer FA et al. 2015
23.	*Pi-ar*	RAPD	MAS	Blast resistance	Double haploid technique was used for the introgression of *Pi-ar* gene	Araujo et al. 2010
24.	*Pi2*, *Pi54*, *xa13* and *Xa21*	SSR	MABB	Blast and Bacterial blight resistance	Improvement of Basmati rice varieties	Ellur et al. 2016
25.	*Piz-5* and *Pi54*	SSR	MABB	Blast resistance	Incorporation of blast resistance into "PRR78", an elite Basmati rice restorer line	Singh et al. 2012
26.	*Pi1*, *Pi2* and *Pi33*	SSR	MABB	Blast resistance	Improve blast resistance in Indian rice (*Oryza sativa*) variety ADT43	Divya et al. 2014
27.	*Pi-2* and *Pi-54*	SSR	MABB	Blast resistance	Introgression of blast resistance genes into the genetic background of elite, bacterial blight resistant indica rice variety, Improved Samba Mahsuri	Madhavi et al. 2016; Rekha et al. 2018
28.	*Xa21*, *xa13* and *Pi54*	Gene specific	MABB	Blast and Bacterial blight resistance	Marker-Assisted Pyramiding of Genes Conferring Resistance Against Bacterial Blight and Blast Diseases into Indian Rice Variety MTU1010	Aruna Kumari et al. 2016
29.	*xa5* and 4 blast QTLs	SSR	MAS	Blast and Bacterial blight resistance	Pyramiding of blast and bacterial leaf blight resistance genes into rice cultivar RD6	Pinta et al. 2013
30.	*Pi54* and *Pi1*	SSR	MAS	Blast resistance	Introgression of blast resistance genes into cold tolerant variety Tellahamsa	Oddin et al. 2015
31.	*Pi46* and *Pita*	SSR	MABB	Blast resistance	Blast resistance genes were introgressed into an elite restorer line Hang-Hui-179 (HH179)	Xiao et al. 2016
32.	*Pi2* and *Xa23*	SSR	MAS	Blast and Bacterial blight resistance	Pyramiding of broad-spectrum disease resistance genes into GZ63-4S elite thermo-sensitive genic male-sterile line in rice	Jiefeng Jiang et al. 2015
33.	*Xa21*, *Xa33*, *Pi2*, *Rf3*, and *Rf4*	SSR	MAS	Blast and Bacterial blight resistance	Marker-assisted improvement of the elite restorer line of rice, RPHR-1005 for resistance against diseases	Kumar et al. 2016

34.	*Xa21, xa13,* and *Pi54*	SSR	MAS	Blast and Bacterial blight resistance	Introgression of Major Bacterial Blight And Blast Resistant Genes into Vallabh Basmati 22	Srikanth et al. 2016
35.	*Xa21* and *Pi54*	SSR	MABB	Blast and Bacterial blight resistance	Introgression of bacterial blight and blast resistance into DRR17B, an elite, fine-grain type maintainer line of rice	Balachiranjeevi et al. 2015
36.	*Pi54, qSBR11-1, qSBR11-2,* and *qSBR7-1*	SSR and QTLs	MABB	Blast and Sheath blight	Introgression of multiple disease resistance genes into a maintainer of Basmati rice CMS line	Singh et al. 2015
37.	*Pikh*	SSR	MABB	Blast	Introgression of Blast resistance gene into a Malaysian Cultivar, MR264	Hasan et al. 2015
38	*Pi2*	SSR	MABB	Blast	Introgression of broad spectrum Blast resistance gene into mega variety Swarna	Rambabu et al. 2016
39	*Pi1*	SSR	MABB	Blast	Introgression of Blast resistance gene into mega variety Swarna	Rambabu et al. 2016
40	*Pi2*	SSR	MABB	Blast	Introgression of broad spectrum Blast resistance gene into mega variety Samba Mahsuri	Kumar SV et al. 2019
41	*Pi54*	SSR	MABB	Blast	Introgression of Blast resistance gene into mega variety Swarna	Rambabu et al. 2016

Source: Ashkani et al. 2015; Annegowda et al. 2021.

1.4.1 Molecular Breeding in Rice for Gall Midge Resistance

Gall midge has several biotypes, where the word "biotype" refers to the diverse genotypes of the insect populations exhibiting distinct virulence resistance responses on a set of rice cultivars/ landraces. These biotypes were labeled as *GMB1, GMB2, GMB3*, etc., for gall midge. Vijaya et al. (2006) have classified seven unique gall midge biotypes based on the response pattern against various varieties, some of which have been chosen as a direct result of extended cultivation of these kinds (Table 1.5). During the 1950s and 1970s, controlled evaluations of rice germplasm for gall midge resistance were conducted in the hot spot locations of Cuttack, Sambalpur, and Warangal. This has led to the discovery of significant sources of resistance, such as Eswarakorra, Ptb18, Ptb21, Siam 29, Leuang 152, etc. (Bentur et al. 2003), which have been widely used in breeding resistant varieties.

Genetic investigations have so far identified 12 significant resistance genes (*Gm1, Gm2, Gm3, Gm4, Gm5, Gm6, Gm7, Gm8, Gm9, Gm10, Gm11,* and *Gm12*) (Table 1.6) (Kumar et al. 2005). *Gm1, Gm2*, and unidentified gene(s) from a landrace, Ptb21, have each provided resistance to different varieties (Bentur et al. 2003). In most rice-growing regions, improved rice cultivars containing Gm1 or Gm2 have lost their resistance to gall midge. Exceptionally, five of the seven biotypes resist variants deriving resistance from Ptb21. None of the discovered genes offer resistance against all gall midge biotypes, and none of the gall midge biotypes are virulent against all resistance genes.

MAS is a promising strategy for developing long-lasting gall midge resistance in rice through gene pyramiding. To achieve gene pyramiding objectives, molecular markers identify and map gall midge resistance genes in rice. Seven such genes have been mapped and identified using tightly connected and flanking co-dominant markers. The abundance of microsatellite or simple sequence repeat (SSR) markers in the rice genome and their known physical location make it possible to fine-map genes of interest and pyramiding them using MAS. CR57-MR1523 is a rice breeding line that inherits its resistance to gall midge from unidentified genes of the Ptb21 parent. Although the genetics of resistance in this breeding line is poorly understood, a novel gene that confers resistance to GMB4 has been identified and mapped using closely related flanking SSR markers suitable for MAS (Himabindu et al. 2009).

1.5 MOLECULAR BREEDING IN RICE FOR OTHER MINOR PESTS

This chapter discusses the use of several molecular techniques, such as genome editing and transgenics, to manage small pest infestations. It is well-established that molecular markers are

TABLE 1.5
Gall Midge Differential Groups and Their Genetic Diversity with Biotypes

			Reaction against biotype					
Group	**Differential**	**R gene**	**1**	**2**	**3**	**4**	**5**	**6**
I	Kavya	*Gm1*	R	S	R	S	R	R
II	Phalguna	*Gm2*	R	R	S	S	R	S
III	RP2068,	*gm3*	R	R	R	R	S	S
	Abhaya,	*Gm4*						
	Jhitpiti,	*Gm8*						
	MR 1523,	*Gm11*						
	Aganni	Unidentified						
IV	ARC 15831, INRC 3021	Unidentified	R	R	R	R	R	S
V	TN1, B95-1, IR64	None	S	S	S	S	S	S
GMB4	CR57-MR1523	*Gm11t*	Unknown					

TABLE 1.6
Gall Midge Resistance Genes

R gene	Chr. No	Source	Linked Marker	Reference
Gm1	9	Kavya	OPK7,OOPN15,OPE1,RM219 OPB2(470)	Katiyar *et al.* 2000
Gm2	4	Phalguna	RG 214,F8,F10 RM 317	Mohan *et al.* 1994 Sundaram *et al.*2003
gm3	4	RP 2068	OPQ12	Katiyar *et al.* 2000
Gm4	8	Abhaya	OPM12,RG 476	Nair*et al* 1996 and Mohan*et al.* 1997
Gm5	12	ARC5984	OOPB14	Katiyar *et al.* 2000
Gm6	4	Dukong	OPMO6(1400),RG124<>RG476	Katiyar *et al.* 2001
Gm7	4	RP 2333	SA 598,F8	Sardesi *et al.* 2002
Gm8	8	Jhitpiti	AR 257, AS 168	Jain *et al.* 2004
Gm9(t)	–	Madhuri Line 9	Un identified	–
Gm 10	–	BG 380-2	Un identified	–
Gm11t	12	CR57-MR1523		Himabindu et al. 2010

helpful in tagging, mapping, introgression, and cloning important plant resistance genes, such as gall midge, yellow stem borer, and brown plant hopper. However, recent research on transgenics has also concentrated on providing resistance against minor pests in rice. These unique methodologies have produced agricultural plants with an array of introduced insecticidal genes, demonstrating the economic viability of this technology. Unfortunately, even though modern molecular methods, such as "omics", are now being used to comprehend the connections between rice and major-minor pests, no insect-resistant cultivar is commercially accessible for cultivation. Nonetheless, the unrelenting efforts of the rice scientific community to address existing and future biotic challenges will lead to improved pest-insect management.

Reducing losses due to insect pests is one of the most challenging biotic stress problems in rice ecosystems on a global scale. Due to their coevolution with a single host, only 12 out of 200 insect species are significant (Ali and Wani 2021). According to Heinrichs (1994), the diverse types of rice pests include, among others, tissue borers, plant suckers, defoliators, and gall formers. Stem borers, plant hoppers, and gall midges are among the most significant and are classified as serious pests. According to Bentur (2003), the other insect groups, including stem flies, mealybugs, aphids, rice hispa, and grain bugs, are classified as minor pests and have regional significance. Future occurrences of higher rice output losses due to developing insect pests cannot be ruled out, given the current catastrophic global warming scenario (Savary et al. 2000, Deutsch et al. 2018). This suggests that the insects now classified as minor pests may obtain an evolutionary advantage in the future, resulting in a large-scale invasion, and subsequently being classified as significant pests. The striped rice stem borer (*Chilo suppresalis*), the pink stem borer (*Sesamiainferens walker*), the rice whorl maggot (*Hydrelliaphilippina ferino*), the rice thrips (*Stenchactothrips biformis Bagnall*), the rice mealy bug (*Brevenniarehi lindinger*), the mole cricket (*Gryllotalphus* or (*Pelopidas mathias Fabricius*; *Parnara guttata Bremer* and *Grey*). Classical and innovative molecular breeding techniques were used to develop resistant rice cultivars against the three significant pests: gall midge, yellow stem borer, and brown plant hopper. However, novel genomics and transgenic breeding techniques were equally effective in developing resistant rice cultivars against the stripped rice (stem) borer and pink (stem) borer. A survey of some of the most significant methodologies applied in different investigations.

Genome editing entails the precise substitution, insertion, or deletion of nucleotides in particular genes or genomes to produce the desired morphological alterations in an organism. Due to its simplicity, the method is extensively used in agriculture. As genome editing is ideally based on functional genomics leading to trait improvement in agricultural plants (Yin et al. 2017), biosafety issues still need to be raised. This has led to the evolution of crops with improved yields and abiotic-biotic stress resistance. Through mutation of the pheromone-binding protein 1 (PBP1) gene, it has been established that genome editing may be used to degrade insect reactions to pheromones of the sex. Through mutation of *PBP1* and *PBP3* genes in the striped stem borer, Dong et al. (2019) demonstrated the roles of these genes and changes in the perception of sex pheromones. The transgenic lines resistant to the stripped rice (stem) borer and pink (stem) borer has been generated using the transgenic method and genome editing. However, only some significant previous research on resistance to the striped stem borer is described briefly. Wunn et al. (1996) used the CaMV35S promoter to introduce the cry1Ab transgene into IR58 (indica rice) utilizing the cry1Ab transgene. Cheng et al. (1998) introduced the *cry1Ab, cry1Ac, hph*, and *gus* genes into *japonica* using the maize ubiquitin promoter. Using *ubi 1* or *mpi* promoter, *cry1B* and *cry1Aa* were introduced into Ariete and Senia (Breitler et al. 2004). Yang et al. (2011) used the maize ubiquitin promoter to introduce *cry1Ab, cry1Ac, cry1c*, and *cry2A* into Minghui 63. (elite *indica* restorer line). The pGreen promoter introduced the *cry11-like Cry1Ac* gene into rice (Yang et al. 2014). Using the *mpi* promoter, Quilis et al. (2014) introduced the *mpi-pci* fusion gene into the recipient rice variety Ariete. Zhao (2015) inserted *cry1Ac, cry1lg*, and *G10* (EPSPS gene) into Xiushui 134 using the maize ubiquitin promoter (*pUBi*) and the modified cauliflower 35S promoter.

Duan et al. (1996) used PINII (potato proteinase inhibitor) in Japonica rice to produce resistant rice variants against the pink stem borer. Also, Chen et al. (2010) implanted rice lines *G6H1, G6H2, G6H3, G6H4, G6H5,* and *G6H6* with the *cry1Ab* and *Vip3H* fusion gene.

Due to their evolutionary advantage, pest management is preferable to pest eradication because many pests are rapidly resistant to pesticides due to their evolutionary advantage (Ali and Wani 2021). Unquestionably, molecular techniques for pest control have led to the establishment of varieties, cultivars, and lines resistant to the vast majority of major and minor pests. However, commercial production on a broad scale of such transgenic lines has never occurred. According to Khajuria et al. (2018), each group of diverse pest kinds has its evolutionary method to circumvent various plant defensive systems. In addition, there is an urgent need to produce higher-yielding rice lines considering the dwindling natural resources available for agriculture and the enormous productivity losses caused by insect pest infestation. This will only be achieved via the use of multiple conventional-molecular breeding techniques and the constant evaluation of information from other agricultural disciplines, which may provide a long-lasting and efficient solution to rice insect pests.

1.6 MOLECULAR BREEDING IN RICE FOR MULTIPLE PEST RESISTANCE

Throughout the rice crop's growth cycle, several insect pests infest it. In the early phases of crop development, several common small insects, such as the leaf folder, whorl maggot, and armyworms, may create apparent damage signs; nevertheless, early damage seldom reduces production since the crop can compensate for early damage during the remainder of its growing season.

Major insect pests of rice provide a vital biotic stress factor and a substantial global production limitation. More than 100 kinds of insect pests feed on rice globally, but only around 20 species may cause significant economic harm (IRRI Rice Knowledge Bank). Stem borers, plant hoppers, leafhoppers, gall midges, and leaf folders are the most important. Mondal et al. (2017) examined yield losses due to several biotic stresses in a system of intensified rice production. In India, it is estimated that insect pests are responsible for 9.5% of the output losses attributed to biotic stress.

Due to climate change, the degree of damage and number of bug species that attack various crops have become unpredictable. Once upon a time, there was a great deal of optimism that chemical control tactics would successfully manage or eradicate insect pests. Ironically, researchers saw a rebound of pests, environmental pollution, increased agricultural production expenses, and the demise of a group of untargeted beneficial insects (predators and parasites). Therefore, varietal resistance may be the primary means of control (Painter 1951). To lower the amount of biotic stress in extremely sensitive agricultural plant varieties, molecular breeding has become an excellent and valuable instrument for resistance breeding on an urgent basis.

1.6.1 Advancements in Molecular Breeding for Multiple Insect Resistance

By crossing transgenic lines, Bharathi et al. (2011) acquired sustained resistance to BPH, green leaf-hopper, and white-backed plant hopper. The lectin genes of *Allium sativum (asal)* and *Galanthus nivalis (gna)* were pyramided by crossing single-gene-based transgenic strains. In all three hoppers, the resistance of the created lines has been determined to be superior. Chakraborty et al. (2016) revealed that transgenic rice expressing the *cry2AX1* gene gave resistance to many lepidopteron pests, including the rice leaf folder (RLF). Pradhan et al. (2016) also showed that transgenic rice expressing the vegetative insecticidal protein (Vip) of *Bacillus thuringiensis* exhibited extensive insecticidal activities and was safe for human consumption. Du et al. (2020) released a result of transgenic rice containing an exogenous Bt gene. Coding for an insecticidal toxin generated by *Bacillus thuringiensis* was used to develop insect-resistant rice. The *Bt* gene was effective against stemborers and leaf-folders but not against piercing-and-sucking insects like the brown plant hopper. Tan et al. (2004) identified *WBPH7* and *WBPH8* genes in *O. officinalis* that give resistance to WBPH, while Hu et al. (2015) mapped *qBPH3* and *qBPH4* genes in the same wild species *O. officinalis* in 2015. This wild rice species may be used for marker-assisted breeding of various plant hopper resistance in rice. Liu et al. (2015) performed map-based cloning and functional characterization of *BPH3*. They discovered *BPH3* genomic region consists of a cluster of three genes encoding the plasma membrane-localized proteins, lectin receptor kinases (*OsLecRK1*, *OsLecRK2*, and *OsLecRK3*) and reported that plants co-expressing all three genes displayed significantly enhanced, broad-spectrum resistance to multiple plant hoppers (BPH and WBPH). Maqbool et al. (2001) conducted gene pyramiding using the transgenic technique by co-transformation of *gna* gene for *BPH* and *cry1Ac* and *cry2A* for chewing insect pest in rice, achieving enhanced, broader resistance for the targeted insect pests.

1.7 MOLECULAR BREEDING IN RICE FOR MULTIPLE DISEASES AND PEST RESISTANCE

Multiple biotic stresses, mainly insect pests and diseases, devastate rice yield. Insect pests such as yellow stem borer (YSB), gall midge, brown plant hopper (BPH), white backed plant hopper (WBPH), and leaf folder are the most problematic biotic stresses. In contrast, blast, brown spot, BLB, and ShB are the most devastating diseases to the crop. In many rice-growing regions of the nation, insects and diseases, such as case worm, swarming caterpillars, mealybug, gundhi bug, false smut, bakanae, and sheath rot have recently emerged as the primary concern (Jena et al. 2018). Resistant sources/donors are crucial to the success of resistance breeding to generate pest-resistant cultivars. Recent advancements in breeding to transfer genes/QTLs with high accuracy have opened new avenues for combating problems posed by various biotic and abiotic stressess (Kumar et al. 2018). Utilizing molecular techniques for multiple and long-lasting insect resistance has led to several varietal enhancements. Presently, molecular markers are widely used in the tagging, mapping, introgressing, and cloning newly mapped gene(s)/QTL(s).

Wenzel (1985) anticipated that the pyramiding of essential genes (multigene approach) indicating a distinct mechanism of action with insecticidal and disease resistance genes against target species might be an effective technique. Datta et al. (2002) combined the *Xa21* gene (resistance to bacterial blight), the *Bt* fusion (cry) gene (for chewing insect resistance), and the chitinase gene (*RC7*) (for tolerance of sheath blight) in a single rice line by the reciprocal crossing of two transgenic homozygous IR72 lines. They confirmed successful introgression with a combined approach of bioassay analysis. Tang et al. (1999) effectively proved the advantages of multiple biotic stress-resistant types by co-transforming the *gna* gene (which provides resistance against BPH) and the *Xa21* gene (which offers resistance against BLB) into premier Chinese rice cultivars. Rani et al. (2011) showed that generating resistant varieties by focusing on one or two biotic stresses restricts the selection of variations for additional biotic stresses and leads to the co-evolution of virulent pest and disease biotypes. Such a circumstance necessitates improved breeding efforts to provide a broader resistance range.

Insect vectors mainly transmit numerous viral diseases to agricultural plants. As a result of breeding for insect-resistant cultivars, viral disease transmission may also be prevented. For example, rice stripe disease is one of the most dangerous viral diseases in temperate and subtropical Asia caused by the rice stripe virus (RSV). This virus is spread mainly by the little brown plant hopper (Hayano-Saito et al. 2000). Xu (2013) reported that marker-assisted introgression of two brown plant hopper resistance genes (*Bph14* and *Bph15*) and one rice stripe disease resistance gene (*Stv-bi*) into japonica rice cultivars effectively controlled the damage from BPH and RSD by providing durable and broad-spectrum resistance for rice stripe disease.

The lack of absolute resistance genes for single biotic stress occasionally necessitates searching for synergistic genes against numerous biotic pressures. Chen et al. (2014) observed that NILs (near isogenic lines) with the combination of these two genes, *qSB*-9TQ and *TAC1TQ*, regulating sheath blight resistance and tiller angle, respectively, exhibited more excellent resistance to ShB than NILs with just *qSB*-9TQ quantitative trait locus (QTL). Consequently, without novel R genes, a pyramiding of moderately resistant QTLs for ShB resistance and other QTLs regulating unrelated traits/stresses may be a viable alternative disease management method. Dixit et al. (2020) conducted research to produce a high-yielding, climate-resilient line from the Swarna variety employing genes giving resistance to blast (*Pi9*), bacterial leaf blight (*Xa4, xa5, xa13,* and *Xa21*), brown planthopper (*Bph3* and *Bph17*), and gall midge (*Gm4* and *Gm8*). ILs with seven to ten resistant genes were superior to the recurrent parent in terms of numerous biotic stresses and yield.

1.8 ABIOTIC STRESS TOLERANCE IN RICE

Rice is one of the most important staple crops in the world, and abiotic stresses are significant constraints to rice productivity and food security, particularly in regions that are susceptible to climate change. Over the years, several breeding strategies, including traditional and modern biotechnological approaches, have improved abiotic stress tolerance in rice. These strategies have focused on developing rice cultivars with enhanced drought, submergence, salinity, heat, and nutrient stress tolerance. Yu et al. (2022) reviewed the green super rice (GSR) breeding strategy in an elaborate manner that resulted in multiple stress-tolerant rice varieties. Yu et al. (2022) made efforts to connect germplasm research with genomics, typically by the large-scale resequencing of rice germplasms. Further utilizing the population genomics research had facilitated the mining of favorable alleles for agriculturally important traits resulting in the development of GSR varieties. Molecular breeding techniques, such as MAS and genome-wide association studies (GWAS), have been used to identify quantitative trait loci (QTLs) and candidate genes linked to abiotic stress tolerance in rice, which have helped to accelerate the development of stress-tolerant rice cultivars. Additionally, advances in understanding the molecular mechanisms of stress tolerance in rice have paved the way for developing new genetic engineering strategies to enhance the crop's stress tolerance.

1.8.1 Breeding for Drought Tolerance in Rice

Drought is one of the most significant abiotic stresses affecting worldwide rice production. Several genes have been identified in rice that contribute to drought tolerance, such as *OsDREB1A*, *OsMYB2*, and *OsNAC6*. These genes are vital in regulating various physiological and biochemical pathways that help plants cope with drought stress. Additionally, several QTLs have been identified for drought tolerance. One of rice's most well-known drought-tolerant QTLs is the *qDTY1*.1 QTL, which has been placed on chromosome 1. This QTL has been shown to enhance rice plants' root growth and drought tolerance. Another important QTL is the *qDTY2*.2 QTL, which has been mapped on chromosome 2. This QTL is associated with maintaining higher leaf water potential, which is crucial for drought tolerance. Other drought-tolerant QTLs in rice include *qDTY3*.1, *qDTY3.2, qDTY4.1, qDTY5.1, qDTY8.1,* and *qDTY12*.1. These QTLs have been found to affect various traits, such as root morphology, leaf rolling, water-use efficiency, and osmotic adjustment, which are all essential for drought tolerance. Incorporating these genes and QTLs into elite rice varieties through marker-assisted selection (MAS) has been shown to enhance drought tolerance and yield under water-limited conditions (Manavalan et al. 2012, Kumar et al. 2014, Sandhu et al. 2020)

1.8.2 Breeding for Submergence Tolerance in Rice

Submergence is a significant abiotic stress that severely affects rice yield worldwide. Several genes and quantitative trait loci (QTLs) have been identified and characterized for breeding rice varieties with tolerance to submergence stress. Some notable genes identified include *SUB1A, SUB1B, SUB1C, SUB1D, qAG-1-1, qAG-8-1,* and *qAG-9-2*. These genes have been used in MAS for developing submergence-tolerant varieties. *SUB1A* gene is incorporated into popular rice varieties, such as Swarna, Samba Mahsuri, CR1009, Ranjit, Bahadur, and IR64 has significantly improved their submergence tolerance. Similarly, the introgression of qAG-9-2 QTL into the elite variety IR64 has shown increased tolerance to submergence (Septiningsih et al. 2013, Singh et al. 2017).

1.8.3 Breeding for Salinity Tolerance in Rice

Salinity is another significant abiotic stress that limits the productivity of rice. Several genes and QTLs associated with salinity tolerance have been identified in rice, including Saltol, OsHKT1, 5, and SKC1. These genes and QTLs have been incorporated into elite rice varieties through conventional and marker-assisted backcross breeding approaches. The *Saltol* QTL has been introgressed into several select rice varieties, including IR64, BRRI dhan47, and Samba Mahsuri, which have shown improved salinity tolerance under field conditions. Similarly, the OsHKT1;5 and SKC1 genes have been used to improve the salinity tolerance of rice varieties. These efforts have resulted in the development of several salinity-tolerant rice varieties that can provide food security to millions of people living in saline-prone areas (Thomson et al. 2008, Rahman et al. 2016)

1.9 CONCLUSION

Cereal crops are recognized globally as the most vital food source. Therefore, crop improvement projects must identify genotypes and traits linked with resistance to various pressures, including biotic and abiotic stresses, which QTLs and genes regulate. Understanding the physiological and molecular basis of tolerance requires intensive efforts. Identifying candidate genes by assessing genetic resources, followed by their characterization, mapping, and cloning deserves greater attention. Molecular markers are increasingly used in QTL analyses in breeding programs to expedite the selection of complex quantitative traits. However, the practical application of MAS needs more consistency across different contexts. Advancements in discovering QTLs and candidate genes and gaining a deeper understanding of the underlying processes could lead to generating desirable plants with enhanced tolerance to biotic and abiotic stresses through genetic modification. Further improvements in discovering QTLs and candidate genes may result in crop varieties with increased stress tolerance.

REFERENCES

Aglawe SB, Barbadikar KM, Mangrauthia SK, Madhav MS (2018) New breeding technique "genome editing" for crop improvement: applications, potentials and challenges. *3 Biotech* 8:1–20.

Ahn SN, Kim YK, Han SS, Choi HC, Moon HP, McCouch SR (1996) Molecular mapping of a gene for resistance to a Korean isolate of rice blast. *Rice Genet Newsl* 13:74–6.

Ahn SN, Kim YK, Hong HC, Choi HC, Moon HP, Han SS, Mccouch SR (1997) Mapping of genes conferring resistance to Korean isolates of rice blast fungus using DNA markers. *Kor J Breed* 29:416–423.

Ali J, Wani SH (2021) *Rice Improvement: Physiological, Molecular Breeding and Genetic Perspectives.* Springer Nature, 498.

Amante-Bordeos A, Sitch LA, Nelson R, Dalmacio RD, Oliva NP, Aswidinnoor H, Leung H (1992) Transfer of bacterial blight and blast resistance from the tetraploid wild rice *Oryza minuta* to cultivated rice, *Oryza sativa. Theor Appl Genet* 84:345–54.

Annegowda DC, Prasannakumar MK, Mahesh HB, Siddabasappa CB, Devanna P, Banakar SN, Manojkumar HB, Prasad SR (2021) Rice blast disease in India: Present status and future challenges. *Integr Adv Rice Res* 21:1–42.

Araujo LG, Prabhu AS, Pereira PA, Silva GB (2010) Marker-assisted selection for the rice blast resistance gene *Pi-ar* in a backcross population. *Crop Breed Appl Biotechnol* 10:23–31.

Arunakumari K, Durgarani CV, Satturu V, Sarikonda KR, Chittoor PD, Vutukuri B, Laha GS, Nelli AP, Gattu S, Jamal M, Prasadbabu A (2016) Marker-assisted pyramiding of genes conferring resistance against bacterial blight and blast diseases into Indian rice variety MTU1010. *Rice Sci* 23(6):306–16.

Ashikawa I, Hayashi N, Yamane H, Kanamori H, Wu J, Matsumoto T, Ono K, Yano M (2008) Two adjacent nucleotide-binding site–leucine-rich repeat class genes are required to confer Pikm-specific rice blast resistance. *Genetics* 180(4):2267–76.

Ashkani S, Rafii MY, Shabanimofrad M, Miah G, Sahebi M, Azizi P, Tanweer FA, Akhtar MS, Nasehi A (2015) Molecular breeding strategy and challenges towards improvement of blast disease resistance in rice crop. *Front Plant Sci* 6:886.

Balachiranjeevi CH, Bhaskar NS, Abhilash V, Akanksha S, Viraktamath BC, Madhav MS, Hariprasad AS, Laha GS, Prasad MS, Balachandran SM, Neeraja CN (2015) Marker-assisted introgression of bacterial blight and blast resistance into DRR17B, an elite, fine-grain type maintainer line of rice. *Mol Breed* 35:1–2.

Balachiranjeevi CH, Abhilash Kumar V, Mahadev Swamy HK, Hajira SK, Kale RR, Pranathi K, Koushik MBVN (2018) Marker-assisted pyramiding of two major, broad-spectrum bacterial blight resistance genes, Xa21 and Xa33 into an elite maintainer line of rice, DRR17B. *PLoS One* 13(10):e0201271.

Barman SR, Gowda M, Venu RC, Chattoo BB (2004) Identification of a major blast resistance gene in the rice cultivar 'Tetep'. *Plant Breed* 123(3):300–2.

Basavaraj SH, Singh VK, Singh A, Singh A, Singh A, Anand D, Yadav S, Ellur RK, Singh D, Gopala Krishnan S, Nagarajan M (2010) Marker-assisted improvement of bacterial blight resistance in parental lines of Pusa RH10, a superfine grain aromatic rice hybrid. *Mol Breed* 26:293–305.

Bentur JS, Pasalu IC, Sarma NP, Prasada Rao U, Mishra B (2003). Gall midge resistance in rice. DRR Research paper Series 01/2003, 20pp. Directorate of Rice Research, Hyderabad India.

Berruyer R, Adreit H, Milazzo J, Gaillard S, Berger A, Dioh W, Lebrun MH, Tharreau D (2003) Identification and fine mapping of *Pi33*, the rice resistance gene corresponding to the *Magnaporthe grisea* avirulence gene ACE1. *Theor Appl Genet* 107:1139–47.

Beser N, Del Valle MM, Kim SM, Vinarao RB, Surek H, Jena KK (2016). Marker-assisted introgression of a broad-spectrum resistance gene, *Pi40* improved blast resistance of two elite rice (*Oryza sativa* L.) cultivars of Turkey. *Mol Plant Breed* 7(33).

Bharathi Y, Kumar SV, Pasalu IC, Balachandran SM, Reddy VD, Rao KV (2011) Pyramided rice lines harbouring *Allium sativum* (*asal*) and *Galanthus nivalis* (*gna*) lectin genes impart enhanced resistance against major sap-sucking pests. *J Biotechnol* 152(3):63–71.

Bhasin H, Bhatia D, Raghuvanshi S, Lore JS, Sahi GK, Kaur B, Vikal Y, Singh K (2012) New PCR-based sequence-tagged site marker for bacterial blight resistance gene *Xa38* of rice. *Mol Breed* 30:607–11.

Biswas PL, Nath UK, Ghosal S, Goswami G, Uddin MS, Ali OM, Latef AA, Laing AM, Gao YM, Hossain A (2021) Introgression of bacterial blight resistance genes in the rice cultivar Ciherang: Response against *Xanthomonas oryzae* pv. *oryzae* in the F6 generation. *Plant J* 10(10):2048.

Blair MW, Garris AJ, Iyer AS, Chapman B, Kresovich S, McCouch SR (2003) High resolution genetic mapping and candidate gene identification at the xa5 locus for bacterial blight resistance in rice (*Oryza sativa* L.). *Theor Appl Genet* 107:62–73.

Brar DS, Khush GS (1997) Alien introgression in rice. Oryza: from molecule to plant. *Plant Mol Biol* 35: 35–47.

Breitler JC, Vassal JM, Del Mar Catala M, Meynard D, Marfà V, Melé E, Royer M, Murillo I, San Segundo B, Guiderdoni E, Messeguer J (2004) Bt rice harbouring cry genes controlled by a constitutive or wound-inducible promoter: protection and transgene expression under Mediterranean field conditions. *Plant Biotechnol J* 2(5):417–30.

Bryan GT, Wu KS, Farrall L, Jia Y, Hershey HP, McAdams SA, Faulk KN, Donaldson GK, Tarchini R, Valent B (2000) A single amino acid difference distinguishes resistant and susceptible alleles of the rice blast resistance gene *Pi-ta*. *Plant Cell* 12(11):2033–45.

Busungu C, Taura S, Sakagami JI, Ichitani K (2016) Identification and linkage analysis of a new rice bacterial blight resistance gene from XM14, a mutant line from IR24. *Breed Sci* 66(4):636–45.

Causse MA, Fulton TM, Cho YG, Ahn SN, Chunwongse J, Wu K, Xiao J, Yu Z, Ronald PC, Harrington SE (1994) Saturated molecular map of the rice genome based on an interspecific backcross population. *Genetics* 138(4):1251–74.

Chakraborty M, Reddy PS, Mustafa G, Rajesh G, Narasu VL, Udayasuriyan V, Rana D (2016) Transgenic rice expressing the cry 2AX1 gene confers resistance to multiple lepidopteran pests. *Transgen Res* 25:665–78.

Chauhan RA, Farman MA, Zhang HB, Leong S (2002) Genetic and physical mapping of a rice blast resistance locus, *Pi-CO39 (t)*, that corresponds to the avirulence gene AVR1-CO39 of *Magnaporthe grisea*. *Mol Genet Genom* 267:603–12.

Cheema KK, Grewal NK, Vikal Y, Sharma R, Lore JS, Das A, Bhatia D, Mahajan R, Gupta V, Bharaj TS, Singh K (2008) A novel bacterial blight resistance gene from Oryza nivara mapped to 38 kb region on chromosome 4L and transferred to *Oryza sativa* L. *Genet Res* 90(5):397–407.

Chen DH, Dela Vina M, Inukai T, Mackill DJ, Ronald PC, Nelson RJ (1999) Molecular mapping of the blast resistance gene, *Pi44 (t)*, in a line derived from a durably resistant rice cultivar. *Theor Appl Genet* 98:1046–53.

Chen M, Shelton A, Ye GY (2011) Insect-resistant genetically modified rice in China: from research to commercialization. *Annu Rev Entomol* 56:81–101.

Chen S, Huang Z, Zeng L, Yang J, Liu Q, Zhu X (2008) High-resolution mapping and gene prediction of *Xanthomonas oryzae* pv. oryzae resistance gene Xa7. *Mol Breed* 22:433–41.

Chen S, Lin XH, Xu CG, Zhang Q (2000, Jan) Improvement of bacterial blight resistance of 'Minghui 63', an elite restorer line of hybrid rice, by molecular marker-assisted selection. *Crop Sci* 40(1):239–44.

Chen S, Wang C, Yang J, Chen B, Wang W, Su J, Feng A, Zeng L, Zhu X (2020) Identification of the novel bacterial blight resistance gene Xa46 (t) by mapping and expression analysis of the rice mutant H120. *Sci Rep* 10(1):12642.

Chen XW, Li SG, Ma YQ, Li HY, Zhou KD, Zhu LH (2004) Marker-assisted selection and pyramiding for three blast resistance genes, Pi-d (t) 1, Pi-b, Pi-ta2, in rice. *Chin J Biotechnol* 20(5):708–14.

Chen Y, Tian JC, Shen ZC, Peng YF, Hu C, Guo YY, Ye GY (2010) Transgenic rice plants expressing a fused protein of Cry1Ab/Vip3H has resistance to rice stem borers under laboratory and field conditions. *J Econ Entomol* Aug 1;103(4):1444–53.

Chen Z, Zuo S, Zhang Y, Chen H, Feng M, Jiang W, Feng F, Ma Y, Hua H, Li G, Pan X (2014) Breeding potential and interaction effects of qSB-9TQ and TAC1TQ in rice breeding against sheath blight disease. *Chin J Rice Sci* 28(5):479–86.

Cheng X, Sardana R, Kaplan H, Altosaar I (1998) Agrobacterium-transformed rice plants expressing synthetic cryIA (b) and cryIA (c) genes are highly toxic to striped stem borer and yellow stem borer. *Proc Natl Acad Sci USA* 95(6):2767–72.

Chisi M, Peterson G (2019) Breeding and agronomy. In: John R.N Taylor and Kwaku G (ed). *Duodu. Sorghum and Millets*. AACC International Press, pp 23–50.

Chukwu SC, Rafii MY, Ramlee SI, Ismail SI, Oladosu Y, Kolapo K, Musa I, Halidu J, Muhammad II, Ahmed M (2019) Marker-assisted introgression of multiple resistance genes confers broad spectrum resistance against bacterial leaf blight and blast diseases in Putra-1 rice variety. *Agronomy* 10(1):42.

Das A, Soubam D, Singh PK, Thakur S, Singh NK, Sharma TR (2012) A novel blast resistance gene, Pi54rh cloned from wild species of rice, *Oryza rhizomatis* confers broad spectrum resistance to *Magnaporthe oryzae*. *Funct Integr Genom* 12:215–28.

Datta K, Baisakh N, Maung Thet K, Tu J, Datta S (2002) Pyramiding transgenes for multiple resistance in rice against bacterial blight, yellow stem borer and sheath blight. *Theor Appl Genet* 106:1–8.

Deng Y, Zhu X, Shen Y, He Z (2006) Genetic characterization and fine mapping of the blast resistance locus Pigm (t) tightly linked to Pi2 and Pi9 in a broad-spectrum resistant Chinese variety. *Theor Appl Genet* 113:705–13.

Deutsch CA, Tewksbury JJ, Tigchelaar M, Battisti DS, Merrill SC, Huey RB, Naylor RL (2018) Increase in crop losses to insect pests in a warming climate. *Science* 361(6405):916–9.

Devanna NB, Vijayan J, Sharma TR (2014) The blast resistance gene Pi54of cloned from Oryza officinalis interacts with Avr-Pi54 through its novel non-LRR domains. *PLoS One* 9(8):e104840.

Divya B, Robin S, Rabindran R, Senthil S, Raveendran M, Joel AJ (2014) Marker assisted backcross breeding approach to improve blast resistance in Indian rice (*Oryza sativa*) variety ADT43. *Euphytica* 200:61–77.

Dixit S, Singh UM, Singh AK, Alam S, Venkateshwarlu C, Nachimuthu VV, Yadav S, Abbai R, Selvaraj R, Devi MN, Ramayya PJ (2020) Marker assisted forward breeding to combine multiple biotic-abiotic stress resistance/tolerance in rice. *Rice* 13:1–5.

Dokku P, Das KM, Rao GJ (2013) Pyramiding of four resistance genes of bacterial blight in Tapaswini, an elite rice cultivar, through marker-assisted selection. *Euphytica* 192:87–96.

Dong XT, Liao H, Zhu GH, Khuhro SA, Ye ZF, Yan Q, Dong SL (2019) CRISPR/Cas9-mediated PBP1 and PBP3 mutagenesis induced significant reduction in electrophysiological response to sex pheromones in male *Chilo suppressalis*. *Insect Sci* 26(3):388–99.

Du B, Chen R, Guo J, He G (2020) Current understanding of the genomic, genetic, and molecular control of insect resistance in rice. *Mol Breed* 40:1–25.

Duan X, Li X, Xue Q, Abo-El-Saad M, Xu D, Wu R (1996) Transgenic rice plants harboring an introduced potato proteinase inhibitor II gene are insect resistant. *Nat Biotechnol* 14(4):494–8.

Ellur RK, Khanna A, Yadav A, Pathania S, Rajashekara H, Singh VK, Krishnan SG, Bhowmick PK, Nagarajan M, Vinod KK, Prakash G (2016) Improvement of Basmati rice varieties for resistance to blast and bacterial blight diseases using marker assisted backcross breeding. *Plant Sci* 242:330–41.

Ezuka A (1975) Inheritance of resistance of rice variety Wase Aikoku 3 to Xanthomonas oryzae. *Bull Tokai-Kinki Nat Agric Exp Stn* 28:124–30.

Fiyaz RA, Shivani D, Chaithanya K, Mounika K, Chiranjeevi M, Laha GS, Viraktamath BC, Rao LS, Sundaram RM (2022) Genetic improvement of rice for bacterial blight resistance: Present status and future prospects. *Rice Sci* 29(2):118–32.

Fjellstrom R, Conaway-Bormans CA, McClung AM, Marchetti MA, Shank AR, Park WD (2004) Development of DNA markers suitable for marker assisted selection of three Pi genes conferring resistance to multiple Pyricularia grisea pathotypes. *Crop Sci* 44(5):1790–8.

Fu C, Wu T, Liu W, Wang F, Li J, Zhu X, Huang H, Liu ZR, Liao Y, Zhu M, Chen J (2012) Genetic improvement of resistance to blast and bacterial blight of the elite maintainer line Rongfeng B in hybrid rice (*Oryza sativa* L.) by using marker-assisted selection. *Afr J Biotechnol* 11(67):13104–14.

Fujii K, Hayano-Saito Y, Shumiya A, Inoue M (1995) Genetical mapping based on the RFLP analysis for the panicle blast resistance derived from a rice parental line St. No. 1. *Breed Sci* 45(Suppl 1):209.

Fukuoka S, Saka N, Koga H, Ono K, Shimizu T, Ebana K, Hayashi N, Takahashi A, Hirochika H, Okuno K, Yano M (2009) Loss of function of a proline-containing protein confers durable disease resistance in rice. *Science* 325(5943):998–1001.

Gao L, Cao Y, Xia Z, Jiang G, Liu G, Zhang W, Zhai W (2013) Do transgenesis and marker-assisted backcross breeding produce substantially equivalent plants?-A comparative study of transgenic and backcross rice carrying bacterial blight resistant gene Xa21. *BMC Genom* 14:1–2.

Goto I (1970) Genetic studies on the resistance of rice plant to the blast fungus 1. Inheritance of resistance in crosses Senshox H-79 and Imochi-shirazux H-79. *J Phytopathol* 36(5):304–12.

Goto I (1988) Genetic studies on resistance of rice plant to blast ungus (VII) blast resistance genes of Kuroka. *J Phytopathol* 54(4):460–5.

Goto I, Jaw YL, Baluch AA (1981) Genetic studies on resistance of rice plant to blast fungus IV. Linkage analysis of four genes, Pi-a, Pi-k, Pi-z and Pi-i. *J Phytopathol* 47(2):252–4.

Goto T, Matsumoto T, Furuya N, Tsuchiya K, Yoshimura A (2009) Mapping of bacterial blight resistance gene Xa11 on rice chromosome 3. *Jpn Agric Res Q: JARQ* 43(3):221–5.

Gouda PK, Saikumar S, Varma CM, Nagesh K, Thippeswamy S, Shenoy V, Ramesha MS, Shashidhar HE (2013) Marker-assisted breeding of Pi-1 and Piz-5 genes imparting resistance to rice blast in PRR 78, restorer line of P usa RH-10 B asmati rice hybrid. *Plant Breed* 132(1):61–9.

Gowda M, Roy-Barman S, Chattoo BB (2006) Molecular mapping of a novel blast resistance gene Pi38 in rice using SSLP and AFLP markers. *Plant Breed* 125(6):596–9.

Gu K, Tian D, Yang FW, Wu L, Sreekala C, Wang D, Wang GL, Yin Z (2004) High-resolution genetic mapping of Xa27 (t), a new bacterial blight resistance gene in rice, Oryza sativa L. *Theor Appl Genet* 108:800–7.

Gu K, Yang B, Tian D, Wu L, Wang D, Sreekala C, Yang F, Chu Z, Wang GL, White FF, Yin Z (2005) R gene expression induced by a type-III effector triggers disease resistance in rice. *Nature* 435(7045):1122–5.

Guo S, Zhang D, Lin X (2010) Identification and mapping of a novel bacterial blight resistance gene Xa35 (t) originated from Oryza minuta. *Scia Agric Sin* 43(13):2611–8.

Guvvala LD, Koradi P, Shenoy V, Marella LS (2013) Improvement of resistance to bacterial blight through marker assisted backcross breeding and field validation in rice (*Oryza sativa*). *Res J Biol* 1:52–66.

Hajira SK, Sundaram RM, Laha GS, Yugander A, Balachandran SM, Viraktamath BC, Sujatha K, Balachiranjeevi CH, Pranathi K, Anila M, Bhaskar S (2016) A single-tube, functional marker-based multiplex PCR assay for simultaneous detection of major bacterial blight resistance genes Xa21, xa13 and xa5 in rice. *Rice Sci* 23(3):144–51.

Hari Y, Srinivasarao K, Viraktamath BC, Hari Prasad AS, Laha GS, I. Ahmed M, Natarajkumar P, Sujatha K, Srinivas Prasad M, Pandey M, Ramesha MS (2013) Marker-assisted introgression of bacterial blight and blast resistance into IR 58025B, an elite maintainer line of rice. *Plant Breed* 132(6):586–94.

Hasan NA, Rafu MY, Rahim HA, Ali NS, Mazlan N, Abdullah S (2015) Introgression of Pi-kh Resistance Gene into a Malaysian Cultivar, MR264 using Marker-Assisted Backcrossing (MABC). *Int J Agri Biol* 31:17(6).

Hayano-Saito Y, Saito K, Nakamura S, Kawasaki S, Iwasaki M (2000) Fine physical mapping of the rice stripe resistance gene locus, Stvb-i. *Theor Appl Genet* 101:59–63.

Hayasaka H, Miyao A, Yano M, Matsunaga K, Sasaki T (1996) RFLP mapping of a rice blast resistance gene Pi-k. *Breed Sci* 46(Suppl 2):68.

Hayashi K, Hashimoto N, Daigen M, Ashikawa I (2004) Development of PCR-based SNP markers for rice blast resistance genes at the Piz locus. *Theor Appl Genet* 108:1212–20.

Hayashi K, Yoshida H, Ashikawa I (2006) Development of PCR-based allele-specific and InDel marker sets for nine rice blast resistance genes. *Theor Appl Genet* 113:251–60.

Hayashi N, Inoue H, Kato T, Funao T, Shirota M, Shimizu T, Kanamori H, Yamane H, Hayano-Saito Y, Matsumoto T, Yano M (2010) Durable panicle blast-resistance gene Pb1 encodes an atypical CC-NBS-LRR protein and was generated by acquiring a promoter through local genome duplication. *Plant J* 64(3):498–510.

He Q, Li D, Zhu Y, Tan M, Zhang D, Lin X (2006) Fine mapping of Xa2, a bacterial blight resistance gene in rice. *Mol Breed* 17:1–6.

Hegde SS, Prashanthi SK (2016) Identification of polymorphic markers and introgression of Pi1 and Pi2 genes for blast resistance in rice. *J Farm Sci* 29(3):327–31.

Heinrichs EA (1994) *Biology and management of rice insects*. Int Rice Res Inst, Manila, Philippines.

Himabindu K (2009) *Identification, tagging and mapping of rice gall midge resistance genes using microsatellite markers*. Thesis submitted to dept. of Biotechnology. Acharya Nagarjuna University, Guntur, p 58.

Himabindu K, Suneetha K, Sama VS, Bentur JS. (2010, July) A new rice gall midge resistance gene in the breeding line CR57-MR1523, mapping with flanking markers and development of NILs. *Euphytica* 174:179–87.

Hsu YC, Chiu CH, Yap R, Tseng YC, Wu YP (2020) Pyramiding bacterial blight resistance genes in Tainung82 for broad-spectrum resistance using marker-assisted selection. *Int J Mol Sci* 21(4):1281.

Hu J, Xiao C, Cheng MX, Gao GJ, Zhang QL, He YQ (2015). A new finely mapped Oryza australiensis-derived QTL in rice confers resistance to brown planthopper. *Gene* Apr 25 561(1):132–7.

Hua LX, Liang LQ, He XY, Wang L, Zhang WS, Liu W, Liu XQ, Lin F (2015) Development of a marker specific for the rice blast resistance gene Pi39 in the Chinese cultivar Q15 and its use in genetic improvement. *Biotechnol Biotechnol Equip* May 4, 29(3):448–56.

Huang H, Huang L, Feng G, Wang S, Wang Y, Liu J, Jiang N, Yan W, Xu L, Sun P, Li Z (2011) Molecular mapping of the new blast resistance genes Pi47 and Pi48 in the durably resistant local rice cultivar Xiangzi 3150. *Phytopathology* 101(5):620–6.

Huang N, Angeles ER, Domingo J, Magpantay G, Singh S, Zhang G, Kumaravadivel N, Bennett J, Khush GS (1997) Pyramiding of bacterial blight resistance genes in rice: marker-assisted selection using RFLP and PCR. *Theor Appl Genet* 95:313–20.

Hutin M, Sabot F, Ghesquière A, Koebnik R, Szurek B (2015) A knowledge-based molecular screen uncovers a broad-spectrum Os SWEET 14 resistance allele to bacterial blight from wild rice. *Plant J* 84(4):694–703.

Imbe T (1997) A new gene for blast resistance in rice cultivar IR24. *Rice Genet Newsl* 14:60–2.

Imbe T, Oba S, Yanoria MJT, Tsunematsu H (1997) A new gene for blast resistance in rice cultivar, IR24. *Rice Genet Newsl* 14:60–62.

Ingvardsen CR, Schejbel B, Lübberstedt T (2008) Functional markers in resistance breeding. *Prog Bot* 69:61–87.

Inoue H, Nakamura M, Mizubayashi T, Takahashi A, Sugano S, Fukuoka S, Hayashi N (2017) Panicle blast 1 (Pb1) resistance is dependent on at least four QTLs in the rice genome. *Rice* 10(1):36.

Inukai T, Nelson RJ, Zeigler RS, Sarkarung S, Mackill DJ, Bonman JM, Takamure I, Kinoshita T (1996) Genetic analysis of blast resistance in tropical rice cultivars using near-isogenic lines. *Rice Genetics III: (In 2 Parts)* 84(11):447–450.

IRRI (1996) Standard evaluation system for rice, 4th ed. IRRI.PO Box 933 Manila Philippines, 52 p.

Ise K (1991) Linkage analysis of some blast resistance gene in rice, *Oryza sativa* L. *Jpn J Breed* 42(Suppl 2):388–9.

Iwata N (1996) Registration of new gene symbols. *Rice Genet Newsl* 13:12–8.

Jain A, Ariyadasa R, Kumar A, Srivastava MN, Mohan M, Nair S. (2004, Nov) Tagging and mapping of a rice gall midge resistance gene, Gm8, and development of SCARs for use in marker-aided selection and gene pyramiding. *Theor Appl Genet* 109:1377–1384.

Jamaloddin M, Durga Rani CV, Swathi G, Anuradha C, Vanisri S, Rajan CP, Krishnam Raju S, Bhuvaneshwari V, Jagadeeswar R, Laha GS, Prasad MS (2020) Marker Assisted Gene Pyramiding (MAGP) for bacterial blight and blast resistance into mega rice variety "Tellahamsa". *PLoS One* 15(6):e0234088.

Jamaloddin M, Mahender A, Gokulan CG, Balachiranjeevi C, Maliha A, Patel HK, Ali J (2021) Molecular approaches for disease resistance in rice. In: Khush GS (ed). *Rice Improvement*. Springer, pp 315–378.

Jena M, Rath PC, Mukherjee AK, Raghu S, Prasanthi G, Yadav MK, Baite MS, Prabhukarthikeyan SR, Bag MK, Lenka S, Arvindan S (2018) Exploring new sources of resistance for insect pest and diseases of rice. ICAR-NRRI, Hyderabad, India.

Jeon JS, Chen D, Yi GH, Wang GL, Ronald PC (2003) Genetic and physical mapping of Pi5 (t), a locus associated with broad-spectrum resistance to rice blast. *Mol Genet Genom* 269:280.

Jeung JU, Kim BR, Cho YC, Han SS, Moon HP, Lee YT, Jena KK (2007) A novel gene, Pi40 (t), linked to the DNA markers derived from NBS-LRR motifs confers broad spectrum of blast resistance in rice. *Theor Appl Genet* 115:1163–77.

Jiang J, Yang D, Ali J, Mou T (2015) Molecular marker-assisted pyramiding of broad-spectrum disease resistance genes, Pi2 and Xa23, into GZ63-4S, an elite thermo-sensitive genic male-sterile line in rice. *Mol Breed* 35:1–2.

Jiang N, Li Z, Wu J, Wang Y, Wu L, Wang S, Wang D, Wen T, Liang Y, Sun P, Liu J (2012) Molecular mapping of the Pi2/9 allelic gene Pi2-2 conferring broad-spectrum resistance to Magnaporthe oryzae in the rice cultivar Jefferson. *Rice* 5:1–7.

Jiang N, Yan J, Liang Y, Shi Y, He Z, Wu Y, Zeng Q, Liu X, Peng J (2020) Resistance genes and their interactions with bacterial blight/leaf streak pathogens (*Xanthomonas oryzae*) in rice (*Oryza sativa* L.)—an updated review. *Rice* 13(1):1–2.

Jin XW, Wang CL, Yang Q, Jiang QX, Fan YL, Liu GC, Zhao KJ (2007) Breeding of near-isogenic line CBB30 and molecular mapping of Xa30(t), a new resistance gene to bacterial blight in rice. *Sci Agric Sin* 40(6):1094–1100. doi: 10.3321/j.issn:0578-1752.2007.06.002

Joseph M, Gopalakrishnan S, Sharma RK, Singh VP, Singh AK, Singh NK, Mohapatra T (2004) Combining bacterial blight resistance and Basmati quality characteristics by phenotypic and molecular marker-assisted selection in rice. *Mol Breed* 13:377–87.

Joshi BK, Bimb HP, Parajuli G, Chaudhary B (2009) Molecular tagging, allele mining and marker aided breeding for blast resistance in rice. *BSN e-Bulletin* 1:1–23.

Kaji R, Ogawa T (1996) RFLP mapping of blast resistance gene Pi-km in rice. Int Rice Research Notes, IRRI, Manila, Philippines.

Kalia S, Rathour R (2019) Current status on mapping of genes for resistance to leaf-and neck-blast disease in rice. *3 Biotech* 9:1–4.

Katiyar SK, Tan Y, Nagaliyadde L, Chandel G, Bentur JS, Inthawong S, Zhang Y, Huang B, Constantino S, Bennett J (2000) Biodiversity of Asian rice gall midge (Orseolia oryzae Wood Mason) from five countries examined by AFLP analysis. *Genome* 43:322–332

Katiyar SK, Tan Y, Huang B, Chandel G, Xu Y, Zhang Y, Xie Z, Bennett J. Molecular mapping of gene Gm-6 (t) which confers resistance against four biotypes of Asian rice gall midge in China. Theoretical and Applied Genetics. 2001 Nov;103:953–61.

Kaur R, Grewal N, Das A, Vikal Y, Singh J, Bharaj TS, Sidhu JS, Singh K (2005) Inheritance of bacterial blight resistance in two accessions of wild rice, *Oryza nivara. Rice Genet Newsl* 22:78–82.

Ke Y, Deng H, Wang S (2017) Advances in understanding broad-spectrum resistance to pathogens in rice. *Plant J* 90(4):738–48.

Khajuria C, Ivashuta S, Wiggins E, Flagel L, Moar W, Pleau M, et al. (2018) Development and characterization of the first dsRNA-resistant insect population from western corn rootworm, Diabrotica virgifera virgifera LeConte. *PLoS One* 13:e0197059. doi: 10.1371/journal.pone.0197059

Khush GS (2005) What it will take to feed 5.0 billion rice consumers in 2030. *Plant Mol Biol* 59:1–6.

Khush GS, Angeles ER (1999) A new gene for resistance to race 6 of bacterial blight in rice, *Oryza sativa* L. *Rice Genet Newsl* 16:92–3.

Khush GS, Bacalangco E, Ogawa T (1990) A new gene for resistance to bacterial blight from O. longistaminata. *Rice Genet. News Lett* 7:121–2.

Khush GS, Mackill DJ, Sidhu GS (1989) Breeding rice for resistance to bacterial blight. Bacterial blight of rice. *Proceedings of the International Workshop on Bacterial Blight of Rice* 207–17.

Kim SM (2018) Identification of novel recessive gene xa44 (t) conferring resistance to bacterial blight races in rice by QTL linkage analysis using an SNP chip. *Theor Appl Genet* 131:2733–43.

Kim SM, Reinke RF (2019) A novel resistance gene for bacterial blight in rice, Xa43 (t) identified by GWAS, confirmed by QTL mapping using a bi-parental population. *PLoS One* 14(2):e0211775.

Kim SM, Suh JP, Qin Y, Noh TH, Reinke RF, Jena KK (2015) Identification and fine-mapping of a new resistance gene, Xa40, conferring resistance to bacterial blight races in rice (*Oryza sativa* L.). *Theor Appl Genet* 128:1933–43.

Kinoshita T, Kiyosawa S (1997) Some considerations on linkage relationships between Pii and Piz in the blast resistance of rice. *Rice Genet Newsl* 14:57–9.

Kiyosawa S (1967) Genetic studies on host-pathogen relationship in the rice blast disease. In Proc. Symp. Rice Diseases and their Control by Growing resistant Varieties and Other Measures. *Tokyo Agr Forest Fish Res Counc Trop Agr Res Ser* 1:137–153.

Koide Y, Kawasaki A, Telebanco-Yanoria MJ, Hairmansis A, Nguyet NT, Bigirimana J, Fujita D, Kobayashi N, Fukuta Y (2010) Development of pyramided lines with two resistance genes, Pish and Pib, for blast disease (*Magnaporthe oryzae* B. Couch) in rice (*Oryza sativa* L.). *Plant Breed* 129(6):670–5.

Koide Y, Kobayashi N, Xu D, Fukuta Y (2009) Resistance genes and selection DNA markers for blast disease in rice (*Oryza sativa* L.). *Jpn Agric Res Quart* 43(4):255–80.

Korinsak S, Sirithunya P, Meakwatanakarn P, Sarkarung S, Vanavichit A, Toojinda T (2011) Changing allele frequencies associated with specific resistance genes to leaf blast in backcross introgression lines of Khao Dawk Mali 105 developed from a conventional selection program. *Field Crops Res* 122(1):32–9.

Korinsak S, Sriprakhon S, Sirithanya P, Jairin J, Vanavichit A, Toojinda T (2009) Identification of microsatellite markers (SSR) linked to a new bacterial blight resistance gene xa33 (t) in rice cultivar'Ba7'. *Maejo Int J Sci Technol* 3(2):235–47.

Kumar A, Jain A, Sahu RK, Shrivastava MN, Nair S, Mohan M (2005). Genetic analysis of resistance genes for the rice gall midge in two rice genotypes. *Crop Sci* 45:1631–5.,

Kumar K, Sarao PS, Bhatia D, Neelam K, Kaur A, Mangat GS, Brar DS, Singh K (2018) High-resolution genetic mapping of a novel brown planthopper resistance locus, Bph34 in *Oryza sativa* L. X *Oryza nivara* (Sharma & Shastry) derived interspecific F2 population. *Theor Appl Genet* 131:1163–71.

Kumar PN, Sujatha K, Laha GS, Rao KS, Mishra B, Viraktamath BC, Hari Y, Reddy CS, Balachandran SM, Ram T, Madhav MS (2012) Identification and fine-mapping of Xa33, a novel gene for resistance to Xanthomonas oryzae pv. oryzae. *Phytopathology* 102(2):222–8.

Kumar P, Pathania S, Katoch P, Sharma TR, Plaha P, Rathour R (2010) Genetic and physical mapping of blast resistance gene Pi-42 (t) on the short arm of rice chromosome 12. *Mol Breed* 25:217–28.

Kumar SV, Prasad MS, Rambabu R, Bhaskar B, Sundaram RM, Prakasam V, Ladhalakshmi D, Laha GS, Madhav MS (2019) Marker assisted introgression of broad spectrum blast resistance gene Pi-2 into an elite rice cultivar, Samba Mahsuri. *Indian J Plant Protec* 47(3&4):154–63.

Kumar S, Rao M (2014) Conventional and molecular breeding for bacterial leaf blight and blast resistance in rice. *Res Rev J Ecol* 3:1–3.

Kumar VA, Balachiranjeevi CH, Naik SB, Rambabu R, Rekha G, Madhavi KR, Harika G, Vijay S, Pranathi K, Hajira SK, Srivastava A (2016) Marker-assisted introgression of the major bacterial blight resistance gene, Xa21 and blast resistance gene, Pi54 into RPHR-1005, the restorer line of the popular rice hybrid, DRRH3. *J Plant Biochem Biotechnol* 25:400–9.

Kumar SV, Prasad MS, Rambabu R, Bhaskar B, Sundaram RM, Prakasam V, Ladhalakshmi D, Laha GS, Madhav MS (2019) Marker assisted introgression of broad spectrum blast resistance gene Pi-2 into an elite rice cultivar, Samba Mahsuri. *Indian J Plant Protec* 47(3&4):154–63.

Kurata N, Yamazaki Y (2006) Oryzabase. An integrated biological and genome information database for rice. *Plant Physiol* 140(1):12–7.

Lin XH, Zhang DP, Xie YF, Gao HP, Zhang Q (1996) Identifying and mapping a new gene for bacterial blight resistance in rice based on RFLP markers. *Phytopathology* Nov 1, 86(11):1156–9.

Lee KS, Rasabandith S, Angeles ER, Khush GS (2003) Inheritance of resistance to bacterial blight in 21 cultivars of rice. *Phytopathology* 93(2):147–52.

Lee SK, Song MY, Seo YS, Kim HK, Ko S, Cao PJ, Suh JP, Yi G, Roh JH, Lee S, An G (2009) Rice Pi5-mediated resistance to Magnaporthe oryzae requires the presence of two coiled-coil–nucleotide-binding–leucine-rich repeat genes. *Genetics* 181(4):1627–38.

Lei CL, Huang DY, Li W, Wang JL, Liu ZL, Wang XT, Shi K, Cheng ZJ, Zhang X, Ling ZZ, Wan JM (2005) Molecular mapping of a blast resistance gene in an indica rice cultivar Yanxian No. 1. *Rice Genet Newsl* 22:76–7.

Lin F, Chen S, Que Z, Wang L, Liu X, Pan Q (2008) The blast resistance gene Pi37 encodes an NBS-LRR protein and is a member of a resistance gene cluster on rice chromosome 1. *Genetics* 98(6):S90–S90.

Lin XH, Zhang DP, Xie YF, Gao HP, Zhang Q (1996) Identifying and mapping a new gene for bacterial blight resistance in rice based on RFLP markers. *Phytopathology* 86(11):1156–9.

Liu B, Zhang S, Zhu X, Yang Q, Wu S, Mei M, Mauleon R, Leach J, Mew T, Leung H (2004) Candidate defense genes as predictors of quantitative blast resistance in rice. *Mol Plant Microbe Interact* 17(10):1146–52.

Liu Q, Yuan M, Zhou YA, Li X, Xiao J, Wang S (2011) A paralog of the MtN3/saliva family recessively confers race-specific resistance to Xanthomonas oryzae in rice. *Plant Cell Environ* 34(11):1958–69.

Liu SP, Li X, Wang CY, Li XH, He YQ (2002) Improvement of resistance to rice blast in Zhenshan 97 by molecular marker aided selection. *J Integ Plant Biol* 45:1346–1350.

Liu W, Liu J, Triplett L, Leach JE, Wang GL (2014) Novel insights into rice innate immunity against bacterial and fungal pathogens. *Annu Rev Phytopathol* 52:213–41.

Liu XQ, Wang L, Chen S, Lin F, Pan QH (2005) Genetic and physical mapping of Pi36 (t), a novel rice blast resistance gene located on rice chromosome 8. *Mol Genet Genom* 274:394–401.

Liu X, Yang Q, Lin F, Hua L, Wang C, Wang L, Pan Q (2007) Identification and fine mapping of Pi39 (t), a major gene conferring the broad-spectrum resistance to Magnaporthe oryzae. *Mol Genet Genom* 278:403–10.

Liu Y, Wu H, Chen H, Liu Y, He J, Kang H, Sun Z, Pan G, Wang Q, Hu J, Zhou F (2015) A gene cluster encoding lectin receptor kinases confers broad-spectrum and durable insect resistance in rice. *Nat Biotechnol* 33(3):301–305.

Luo W, Huang M, Guo T, Xiao W, Wang J, Yang G, Liu Y, Wang H, Chen Z, Zhuang C (2017, Feb). Marker-assisted selection for rice blast resistance genes Pi2 and Pi9 through high-resolution melting of a gene-targeted amplicon. *Plant Breed* 136(1):67–73.

Luo Y, Zakaria S, Basyah B, Ma T, Li Z, Yang J, Yin Z (2014) Marker-assisted breeding of Indonesia local rice variety Siputeh for semi-dwarf phonetype, good grain quality and disease resistance to bacterial blight. *Rice* 7(1):1–8.

Madhavi KR, Prasad MS, Madhav MS, Laha GS, Mohan KM, Sundaram RM, Viraktamath BC (2012) Introgression of blast resistance gene Pi-k^h into elite indica rice variety improved samba mahsuri. *Indian J Plant Protec* 40(1):52–56.

Madhavi KR, Rambabu R, Abhilash Kumar V, Vijay Kumar S, Aruna J, Ramesh S, Prasad MS (2016) Marker assisted introgression of blast (*Pi-2* and *Pi-54*) genes in to the genetic background of elite, bacterial blight resistant indica rice variety, improved Samba Mahsuri. *Euphytica* 212(2):331–342.

Manavalan LP, Guttikonda SK, Phan Tran LS, Nguyen HT (2012) Physiological and molecular approaches to improve drought resistance in rice. *Plant Cell Rep* 31(4):627–638.

Maqbool SB, Riazuddin S, Loc NT, Gatehouse AM, Gatehouse JA, Christou P (2001) Expression of multiple insecticidal genes confers broad resistance against a range of different rice pests. *Mol Breed* 7(1):85–93.

McCouch SR, Nelson RJ, Tohme J, Zeigler RS (1994) Mapping of blast resistance genes in rice. Adaptation to climate change through genomic breeding in sesame. In: Zeigler RS, Leong SA, Teng PS (eds). *Rice blast disease*. CAB International, Wallingford, UK, pp 167.

Mi J, Yang D, Chen Y, Jiang J, Mou H, Huang J, andMou T (2018) Accelerated molecular breeding of a novel P/TGMS line with broad-spectrum resistance to rice blast and bacterial blight in two-line hybrid rice. *Rice* 11(1):1–12.

Miah G, Rafii MY, Ismail MR, Puteh AB, Rahim HA, Asfaliza R, Latif MA (2013) Blast resistance in rice: a review of conventional breeding to molecular approaches. *Mol Biol Rep* 40(3):2369–2388.

Miao L, Wang C, Zheng C, Che J, Gao Y, Wen Y, Zhao K (2010) Molecular mapping of a new gene for resistance to rice bacterial blight. *Sci Agric Sin* 43(15):3051–3058.

Mir GN, Khush GS (1990) Genetics of resistance to bacterial blight in rice cultivar DV86. *Crop Res* 3(2):194–198.

Mizukami T, Wakimoto S (1969) Epidemiology and control of bacterial leaf blight of rice. *Annu Rev Phytopathol* 7(1):51–72.

Mohan M, Nair S, Bentur JS, Rao UP, Bennett J. (1994, Feb) RFLP and RAPD mapping of the rice Gm2 gene that confers resistance to biotype 1 of gall midge (Orseolia oryzae). *Theor Appl Genet* 87:782–788.

Mondal D, Ghosh A, Roy D, Kumar A, Shamurailatpam D, Bera S, Majumder A (2017) Yield loss assessment of rice (*Oryza Sativa* L.) due to different biotic stresses under system of rice intensification (SRI). *J Entomol Zool Stud* 5(4): 1974–1980.

Nakai H, Nakamura K, Kuwahara S, Saito M (1988) Genetic studies of an induced rice mutant resistant to multiple races of bacterial leaf blight. *Rice Genet Newsl* 5:101–3.

Nakamura S, Asakawa S, Ohmido N, Fukui K, Shimizu N, Kawasaki S (1997) Construction of an 800-kb contig in the near-centromeric region of the rice blast resistance gene *Pi-ta* 2 using a highly representative rice BAC library. *Mol Genet Genom* 254(6):611–620.

Naqvi NI, Chattoo BB (1996) Development of a sequence characterized amplified region (SCAR) based indirect selection method for a dominant blast-resistance gene in rice. *Genome* 39(1):26–30.

Naqvi NI, Bonman JM, Mackill DJ, Nelson RJ, Chattoo BB (1995) Identification of RAPD markers linked to a major blast resistance gene in rice. *Mol Breed* 1(4):341–348.

Naqvi SAH (2019) Bacterial leaf blight of rice: An overview of epidemiology and management with special reference to Indian sub-continent. *Pak J Agri Sci* 32(2):359.

Narayanan NN, Baisakh N Vera, Cruz CM, Gnanamanickam SS, Datta KB, Datta SK (2002) Molecular breeding for the development of blast and bacterial blight resistance in rice cv IR50. *Crop Sci* 42(6):2072–2079.

Neelam K, Mahajan R, Gupta V, Bhatia D, Gill BK, Komal R, Singh K (2020) High-resolution genetic mapping of a novel bacterial blight resistance gene xa-45 (t) identified from *Oryza glaberrima* and transferred to *Oryza sativa*. *Theor Appl Genet* 133(3):689–705.

Ni D, Song F, Ni J, Zhang A, Wang C, Zhao K, Li, L (2015) Marker-assisted selection of two-line hybrid rice for disease resistance to rice blast and bacterial blight. *Field Crops Res* 184:1–8.

Niño-Liu DO, Darnielle L, Bogdanove AJ (2005) A simple method of mass inoculation of rice effective for both pathovars of *Xanthomonas oryzae*, and the construction of comparable sets of host cDNA libraries spanning early stages of bacterial leaf blight and bacterial leaf streak. *J Phytopathol* 153(7–8):500–504.

Oddin J, Durga Rani HV, Swathi G, Anuradha CH, Vanisri S, Sheshu Madhav M, Sundaram RM, Arun Prem Kumar N (2015) Introgression of blast resistance genes *Pi-54* and *Pi1* into cold tolerant variety Tellahamsa, by marker assisted selection. *Int J Curr Res* 7(11):22348–22353.

Ogawa T, Kaku H, Yamamoto T (1989) Resistance gene of rice cultivar, Asaminori to bacterial blight of rice. *Jpn J Breed* 39:196–197.

Ogawa T, Morinaka T, Fujii K, Kimura T (1978) Inheritance of resistance of rice varieties Kogyoku and Java 14 to bacterial group V of *Xanthomonas oryzae*. *Jpn J Phytopathol* 44(2):137–141.

Ogawa T, Yamamoto T (1986) Inheritance of resistance to bacterial blight in rice. *Rice Genetics I: (In 2 Parts)*: 471–479.

Okuyama Y, Kanzaki H, Abe A, Yoshida K, Tamiru M, Saitoh H, Terauchi R (2011) A multifaceted genomics approach allows the isolation of the rice Pia-blast resistance gene consisting of two adjacent NBS-LRR protein genes. *Plant J* 66(3):467–479.

Painter RH (1955). Insect resistance in crop plants. LWW Dec 1.

Pan Q, Tanisaka T (1997) Studies on the genetics and breeding of blast resistance in rice. VII. Gene analysis for the blast resistance of Indian native cultivar, Aus 373. *Breed Sci* 47:35.

Pan QH, Tanisaka T, Ikehashi H (1996) Studies on the genetics and breeding of blast resistance in rice VI Gene analysis for the blast resistance of two Yunnan native cultivars GA20 and GA25. *Breed Sci* 46(2):70.

Pandey MK, Shobha Rani N, Sundaram RM, Laha GS, Madhav MS, Srinivasa Rao K, Viraktamath BC (2013) Improvement of two traditional Basmati rice varieties for bacterial blight resistance and plant stature through morphological and marker-assisted selection. *Mol Breed* 31(1):239–246.

Petpisit V, Khush GS, Kauffman HE (1977) Inheritance of resistance to bacterial blight in rice. *Crop Sci* 17(4):551–554.

Pinta W, Toojinda T, Thummabenjapone P, Sanitchon J (2013) Pyramiding of blast and bacterial leaf blight resistance genes into rice cultivar RD6 using marker assisted selection. *Afr J Biotechnol* 12(28):4432–4438.

Pradhan S, Chakraborty A, Sikdar N, Chakraborty S, Bhattacharyya J, Mitra J, Sen SK (2016) Marker-free transgenic rice expressing the vegetative insecticidal protein (Vip) of *Bacillus thuringiensis* shows broad insecticidal properties. *Planta* 244(4):789–804.

Qu S, Liu G, Zhou B, Bellizzi M, Zeng L, Dai L, Wang GL (2006) The broad-spectrum blast resistance gene *Pi9* encodes a nucleotide-binding site-leucine-rich repeat protein and is a member of a multigene family in rice. *Genetics* 172(3):1901–1914.

Quilis J, López-García B, Meynard D, Guiderdoni E, San Segundo B (2014) Inducible expression of a fusion gene encoding two proteinase inhibitors leads to insect and pathogen resistance in transgenic rice. *Plant Biotechnol J* 12(3):367–377.

Rahman MA, Thomson MJ, Shah-E-Alam M (2016) Marker-assisted selection for salinity tolerance in rice: recent progress and future perspectives. *Plant Breed Biotechnol* 4:1–18.

Rajpurohit D, Kumar R, Kumar M, Paul P, Awasthi A, Osman Basha P, Dhaliwal HS (2011) Pyramiding of two bacterial blight resistance and a semidwarfing gene in Type 3 Basmati using marker-assisted selection. *Euphytica* 178(1):111–126.

Ram T, Laha GS, Gautam SK, Deen R, Madhav MS, Brar DS, Viraktamath BC (2010) Identification of new gene introgressed from *Oryza brachyantha* with broad-spectrum resistance to bacterial blight of rice in India. *Rice Genet Newsl* 25:57.

Ramalingam J, Raveendra C, Savitha P, Vidya V, Chaithra TL, Velprabakaran S, Vanniarajan C (2020) Gene pyramiding for achieving enhanced resistance to bacterial blight, blast, and sheath blight diseases in rice. *Front Plant Sci* 11:1662.

Rambabu R, Vijay Kumar S, Abhilash Kumar V, Aruna J, Madhavi KR, Madhav MS, Ravindrababu V, Srinivas Prasad M (2016) Enhancement of leaf blast resistance in rice cultivar 'Swarna' by marker assisted back-cross breeding. *Int J Dev Res* 06(8):9194–9198.

Rani NS, Varaprasad GS, Rao SLV, Sudharshan I, Hariprasad AS, Prasad ASR, Ram T, Babu VR, Padmavathi G, Vijaipal B, Suneetha K, Viraktamath BC (2011) *High Yielding rice varieties of India DRR technical bulletin 55/2011 Directorate of Rice Research*, Hyderabad, India 95 p.

Rao SR, Priyanka M, Kumar MA, Ramanaiah C, Yashwanth B, Mohan KM, Shanti ML (2017) Marker-assisted breeding for bacterial blight resistance in parental lines of hybrid rice. *J Plant Pathol* 99(3):691–701.

Reinke R, Kim SM, Kim BK (2018) Developing japonica rice introgression lines with multiple resistance genes for brown planthopper, bacterial blight, rice blast, and rice stripe virus using molecular breeding. *Mol Genet Genom* 293(6):1565–1575.

Rekha G, Abhilash Kumar V, Viraktamath BC, Pranathi K, Kousik MBVN, Laxmi Prasanna B, Sundaram RM (2018) Improvement of blast resistance of the popular high-yielding, medium slender-grain type, bacterial blight resistant rice variety, Improved Samba Mahsuri by marker-assisted breeding. *J Plant Biochem Biotechnol* 27(4):463–472.

Sakaguchi S (1967) Linkage studies on the resistance to bacterial leaf blight, *Xanthomonas oryzae* (Uyeda et Ishiyama) DOWSON, in rice. *Inst Agric Sci Ser* 16:1–18.

Salgotra RK, Gupta BB, Millwood RJ, Balasubramaniam M, Stewart CN (2012) Introgression of bacterial leaf blight resistance and aroma genes using functional marker-assisted selection in rice (*Oryza sativa L*). *Euphytica* 187(3):313–323.

Sallaud C, Lorieux M, Roumen E, Tharreau D, Berruyer R, Svestasrani P, Notteghem JL (2003) Identification of five new blast resistance genes in the highly blast-resistant rice variety IR64 using a *QTL* mapping strategy. *Theor Appl Genet* 106(5):794–803.

Sanchez AC, Ilag LL, Yang D, Brar DS, Ausubel F, Khush GS, Yano M, Sasaki T, Li Z, Huang N. (1999) Genetic and physical mapping of xa13, a recessive bacterial blight resistance gene in rice. *Theor Appl Genet* 98:1022–1028.

Sandhu N, Yadav S, Kumar A (2020) Recent efforts in developing high-yield, drought-tolerant rice varieties. In *Advancement in crop improvement techniques*, Jan 1 pp 111–128.

Sardesai N, Kumar A, Rajyashri K, Nair S, Mohan M (2002, Oct) Identification and mapping of an AFLP marker linked to Gm7, a gall midge resistance gene and its conversion to a SCAR marker for its utility in marker aided selection in rice. *Theor Appl Genet* 105:691–698.

Savary S, Willocquet L, Elazegui FA, Castilla NP, Teng PS (2000) Rice pest constraints in tropical Asia: quantification of yield losses due to rice pests in a range of production situations. *Plant Dis* 84(3):357–369.

Septiningsih EM, Collard BCY, Heuer S, Bailey-Serres J, Ismail AM (2013) Improving drought tolerance in rice: advances and challenges. *Plant Cell Envt* 36: 381–397.

Shang J, Tao Y, Chen X, Zou Y, Lei C, Wang J, Zhu L (2009) Identification of a new rice blast resistance gene, Pid3, by genomewide comparison of paired nucleotide-binding site–leucine-rich repeat genes and their pseudogene alleles between the two sequenced rice genomes. *Genetics* 182(4):1303–1311.

Shanti ML, Shenoy VV, Devi GL, Kumar VM, Premalatha P, Kumar GN, Freeman WH (2010) Marker-assisted breeding for resistance to bacterial leaf blight in popular cultivar and parental lines of hybrid rice. *J Plant Pathol* 92(2):495–501.

Sharma TR, Shanker P, Singh BK, Jana TK, Madhav MS, Gaikwad K, Rathour R (2005) Molecular mapping of rice blast resistance gene *Pi-kh* in the rice variety Tetep. *J Plant Biochem Biotechnol* 14(2):127–133.

Shinoda H (1971) Studies on the varietal resistance of rice to blast 6 Linkage relationship of blast resistance genes. *Annu Rev Chugoku Natl Agric Exp Stn* 20:1–25.

Shivappa R, Navadagi DB, Baite MS, Yadav MK, Rathinam PS, Umapathy K, Rath PC (2021) Emerging Minor Diseases of Rice in India: Losses and Management Strategies. In Min Huang (ed). *Integrative Advances in Rice Research*. IntechOpen, pp 1–22.

Sidhu GS, Khush GS (1978) Dominance reversal of a bacterial blight resistance gene in some rice cultivars. *Phytopathology* 68:461–463.

Singh A, Singh VK, Singh SP, Pandian RTP, Ellur RK, Singh D, Singh AK (2012) Molecular breeding for the development of multiple disease resistance in Basmati rice. *AoB Plants* 29:1–13.

Singh AK, Singh VK, Singh A, Ellur RK, Pandian RTP, Gopala Krishnan S, Prabhu KV (2015) Introgression of multiple disease resistance into a maintainer of Basmati rice CMS line by marker assisted backcross breeding. *Euphytica* 203(1):97–107.

Singh K, Vikal Y, Singh S, Leung H, Dhaliwal HS, Khush GS (2002) 40 Mapping of bacterial blight resistance gene xa8 using microsatellite markers. *Rice Genet Newsl* 19:94–97.

Singh N, Dang T, Vergara GV, Pandey DM, Sanchez DL, Neeraja CN, Septiningsih EM, Mendioro MS, Tecson-Mendoza EM, Ismail AM (2017) Molecular marker-assisted breeding for submergence tolerance in rice: recent progress and future directions. *Plant Sci* 256:45–54.

Singh RJ, Khush GS, Mew TW (1983) A new gene for resistance to bacterial blight in rice. *Crop Sci* 23(3):558–560.

Singh S, Sidhu JS, Huang N, Vikal Y, Li Z, Brar DS, Khush, GS (2001) Pyramiding three bacterial blight resistance genes (*xa5*, *xa13* and *Xa21*) using marker-assisted selection into indica rice cultivar PR106. *Theor Appl Genet* 102(6):1011–1015.

Singh VK, Singh A, Singh SP, Ellur RK, Singh D, Gopala Krishnan S, Singh AK (2013) Marker-assisted simultaneous but stepwise backcross breeding for pyramiding blast resistance genes *Piz5* and *Pi54* into an elite Basmati rice restorer line 'PRR 78'. *Plant Breed* 132(5):486–495.

Song WY, Wang GL, Chen LL, Kim HS, Pi LY, Holsten T, Ronald P (1995) A receptor kinase-like protein encoded by the rice disease resistance gene, *Xa21*. *Science* 270(5243):1804–1806.

Srikanth S, Manish K, Pandey BC, Hajira SK, Kumar VS, Bhadana VP, Suneetha K (2016) Introgression of major bacterial blight and blast resistant genes into Vallabh Basmati 22 an elite Basmati variety. *Int J Dev Res* 6:8366–8370.

Srinivasan MC, Thirumalachar MJ, Patel MK (1959) Bacterial blight disease of rice. *Curr Sci* 28:469–470.

Srivastava DN (1967) Epidemiology and control of bacterial blight of rice in India. In: *Proceedings of a symposium of rice diseases and their control by growing resistant varieties and other measures.* 25th Sep, Tokyo, Japan, pp 11–18.

Suh JP, Cho YC, Won YJ, Ahn EK, Baek MK, Kim MK, Jena KK (2015) Development of resistant gene-pyramided japonica rice for multiple biotic stresses using molecular marker-assisted selection. *Plant Breed Biotech* 3:333–345. https://doi.org/10.9787/PBB.2015.3.4.333

Suh JP, Noh TH, Kim KY, Kim JJ, Kim YG, Jena KK (2009) Expression levels of three bacterial blight resistance genes against K3a race of Korea by molecular and phenotype analysis in japonica rice (*O sativa* L) *J Crop Sci Biotechnol* 12(3):103–108.

Sun X, Yang Z, Wang S, Zhang Q (2003) Identification of a 47-kb DNA fragment containing *Xa4*, a locus for bacterial blight resistance in rice *Theor Appl Genet* 106(4):683–687.

Sundaram RM, Bentur JS, Sarma NP (2003) Molecular tagging and mapping of rice gall midge resistance gene *Gm1* and validation of markers for gall midge resistance gene *Gm2*. In Abstracts 4th Gall midge mini-network meeting. Directorate of Rice Research, Hyderabad, 1–4 December 2003.

Sundaram RM, Vishnupriya MR, Biradar SK, Laha GS, Reddy GA, Rani NS, Sonti RV (2008) Marker assisted introgression of bacterial blight resistance in Samba Mahsuri, an elite indica rice variety. *Euphytica* 160(3):411–422.

Sundaram RM, Vishnupriya MR, Laha GS, Rani NS, Rao PS, Balachandran SM, Sonti RV (2009) Introduction of bacterial blight resistance into Triguna, a high yielding, mid-early duration rice variety. *Healthcare Nutr Technol* 4(3):400–407.

Tabien RE, Li Z, Paterson AH, Marchetti MA, Stansel JW, Pinson SRM, Park WD (2000) Mapping of four major rice blast resistance genes from' Lemont' and 'Teqing' and evaluation of their combinatorial effect for field resistance. *Theor Appl Genet* 101(8):1215–1225.

Tabien RE, Pinson SRM, Marchetti MA, Li Z, Park WD, Paterson AH, Stansel JW (1996) Blast resistance genes from Teqing and Lemont In: Khush GS (ed). *Rice genetics III proceedings IRRI.* Los Banos, Laguna, Philippines, 3:451–455.

Tabien R, Li Z, Paterson A, Marchetti M, Stansel J, Pinson S (2002) Mapping *QTLs* for field resistance to the rice blast pathogen and evaluating their individual and combined utility in improved varieties. *Theor Appl Genet* 105(2):313–324.

Takagi H, Abe A, Yoshida K, Kosugi S, Natsume S, Mitsuoka C, Terauchi R (2013) *QTL*-seq: rapid mapping of quantitative trait loci in rice by whole genome resequencing of DNA from two bulked populations. *Plant J* 74(1):174–183.

Takahashi W, Miura Y, Sasaki T (2009) A novel inoculation method for evaluation of grey leaf spot resistance in Italian ryegrass. *J Plant Pathol* 91(1):171–176.

Tang K, Tinjuangjun P, Xu Y, Sun X, Gatehouse JA, Ronald PC, Kohli A (1999) Particle-bombardment-mediated co-transformation of elite Chinese rice cultivars with genes conferring resistance to bacterial blight and sap-sucking insect pests. *Planta* 208(4): 552–563.

Tanksley SD, Young ND, Paterson AH, Bonierbale MW (1989) RFLP mapping in plant breeding: new tools for an old science. *Biotechnology* 7(3):257–264.

Tanweer FA, Rafii MY, Sijam K, Rahim HA, Ahmed F, Ashkani S, Latif MA (2015) Introgression of blast resistance genes (putative Pi-b and Pi-kh) into elite rice cultivar MR219 through marker-assisted selection. *Front plant Sci* Dec 17, 6:1002.

Taura S, Ogawa T, Tabien RE, Khush GS, Yoshimura A, Omura T (1987) The specific reaction of Taichung Native 1 to Philippine races of bacterial blight and inheritance of resistance to race 5 (Pxo112). *Rice Genet Newsl* 4:101–102.

Taura S, Ogawa T, Yoshimura A, Ikeda R (1992) ldentification of a recessive resistance gene to rice bacterial blight of mutant line XM6, Oryza sativa L. *Jap J Breed* 42(1):7–13.

Thomson MJ, de Ocampo M, Egdane J (2008) Characterizing the Saltol quantitative trait locus for salinity tolerance in rice. *Rice* 1:37–42.

Toenniessen GH, Toole JC, DeVries J (2003) Advances in plant biotechnology and its adoption in developing countries. *Curr Opinion Plant Biol* 6(2):191–198.

Usatov AV, Kostylev PI, Azarin KV, Markin NV, Makarenko MS, Khachumov VA, and Bibov MY (2016) Introgression of the rice blast resistance genes *Pi1, Pi2* and *Pi33* into Russian rice varieties by marker-assisted selection. *Indian J Genet* 76(1):55–61.

Utami DW, Moeljopawiro S, Hanarida I, Tharreau D (2008) Fine mapping of rice blast QTL from Oryza rufipogon and IR64 by SNP markers. *SABRAO J Breed Genet* Dec 1;40(2).

Vennisa R (2018) Marker-assisted selection of bacterial blight broad spectrum resistance genes *Xa33* and *Xa38* into CO43 in ICF3 generation. *Elect J Plant Breed* 9(3):978–984.

Vijaya LP, Amudhan S, Himabindu K, Cheralu C, Bentur JS (2006) A new biotype of the Asian rice gall midge *Orseolia oryzae* (Diptera: Cecidomyiidae) characterized from the Warangal population in Andhra Pradesh. *India Int J Insect Sci* 26:207–211.

Vikal Y, Bhatia D (2017) Genetics and genomics of bacterial blight resistance in rice. In: Min Huang (ed). *Integrative Advances in Rice Research.* IntechOpen, pp 175–213.

Vikal Y, Chawla H, Sharma R, Lore JS, Singh K (2014) Mapping of bacterial blight resistance gene *xa8* in rice (*Oryza sativa* L). *Indian J Genet* 74:589–595.

Wang C, Wen G, Lin X, Liu X, Zhang D (2009) Identification and fine mapping of the new bacterial blight resistance gene, *Xa31* (t), in rice. *Eur J Plant Pathol* 123(2):235–240.

Wang W, Zhai W, Luo M, Jiang G, Chen X, Li X, Zhu L (2001) Chromosome landing at the bacterial blight resistance gene *Xa4* locus using a deep coverage rice BAC library. *Mol Genet Genom* 265(1):118–125.

Wang A, Jia Y, Rutger JN, Xia U (2007) Rapid survey for presence of a blast resistance gene *Pi-ta* in rice cultivars using the dominant DNA markers derived from portions of the *Pi-ta* gene. *Plant Breed* 126(1):36–42.

Wang ZX, Yano M, Yamanouchi U, Iwamoto M, Monna L, Hayasaka H, Sasaki T (1999) The *Pib* gene for rice blast resistance belongs to the nucleotide binding and leucine-rich repeat class of plant disease resistance genes. *Plant J* 19(1):55–64.

Wenzel G (1985) Strategies in unconventional breeding for disease resistance. *Annu Rev Phytopathol* 23(1):149–172.

Win KM, Korinsak S, Sirithunya P, Lanceras-Siangliw J, Jamboonsri W, Da T, Toojinda T (2013) Marker assisted introgression of multiple genes for bacterial blight resistance into aromatic Myanmar rice MK-75. *Field Crops Res* 154:164–171.

Wu Y, Yu L, Pan C, Dai Z, Li Y, Xiao N, Li A (2016) Development of near-isogenic lines with different alleles of *Piz* locus and analysis of their breeding effect under Yangdao 6 background. *Mol Breed* 36(2):1–12.

Wünn J, Klöti A, Burkhardt PK, Biswas GC, Launis K, Iglesias VA, Potrykus I (1996) Transgenic Indica rice breeding line IR58 expressing a synthetic cryIA (b) Gene from *Bacillus thuringiensis* provides effective insect pest control. Bio/technology Feb 1, 14(2):171–176. doi: 10.1038/nbt0296-171. PMID: 9636319.

Xiang Y, Cao Y, Xu C, Li X Wang S (2006) *Xa3*, conferring resistance for rice bacterial blight and encoding a receptor kinase-like protein, is the same as *Xa26. Theor Appl Genet* 113(7):1347–1355.

Xiao WM, Luo LX, Hui WA, Tao Guo, Liu YZ, Zhou JY, Chen ZQ (2016) Pyramiding of *Pi46* and *Pita* to improve blast resistance and to evaluate the resistance effect of the two R genes. *J Integr Agric* 15(10):2290–2298.

Xing J, Zhang D, Yin F, Zhong Q, Wang B, Xiao S, Chen L (2021) Identification and fine-mapping of a new bacterial blight resistance gene, *Xa47* (t), in G252, an introgression line of yuanjiang common wild rice (*Oryza rufipogon*). *Plant Dis* 105(12):4106–4112.

Xu J (2013) Pyramiding of two BPH resistance genes and Stv-b i gene using marker-assisted selection in Japonica rice. *Crop Breed Appl Biotechnol* 13(2):99–106.

Xu J, Jiang J, Dong X, Ali J, Mou T (2012) Introgression of bacterial blight (BB) resistance genes *Xa7* and *Xa21* into popular restorer line and their hybrids by molecular marker-assisted backcross (MABC) selection scheme. *Afr J Biotechnol* 11(33):8225–8233.

Yang Q, Lin F, Wang L, Pan Q (2009) Identification and mapping of *Pi41*, a major gene conferring resistance to rice blast in the *Oryza sativa* subsp indica reference cultivar, 93–11. *Theor Appl Genet* 118(6):1027–1034.

Yang S, Gu T, Pan C, Feng Z, Ding J, Hang Y, Tian D (2008) Genetic variation of NBS-LRR class resistance genes in rice lines. *Theor Appl Genet* 116(2):165–177.

Yang YY, Mei F, Zhang W, Shen Z, Fang J (2014) Creation of Bt rice expressing a fusion protein of *Cry1Ac* and *Cry1I*-like using a green tissue-specific promoter. *J Econ Entomol* 107(4):1674–1679.

Yang Z, Chen H, Tang W, Hua H, Lin Y (2011) Development and characterisation of transgenic rice expressing two *Bacillus thuringiensis* genes. *Pest Manag Sci* 67(4):414–422.

Yap R, Hsu YC, Wu YP, Lin YR, Kuo CW (2016) Multiplex PCR genotyping for five bacterial blight resistance genes applied to marker-assisted selection in rice (*Oryza sativa*). *Plant Breed* 135(3):309–317.

Yin K, Gao C, Qiu JL (2017) Progress and prospects in plant genome editing. *Nat Plants* 3(8):1–6.

Yokoo M, Kikuchi F, Fujimaki H, Nagai K (1978) Breeding of blast resistant lines (BLI to 7) from Indica-Japonica crosses of rice. *Jpn J Breed* 28(4):359–365.

Yoshimura A, Mew TW, Khush GS, Omura T (1983) Inheritance of resistance to bacterial blight in rice cultivar Cas 209. *Genetics* 73(10):1409–1412.

Yoshimura S, Yamanouchi U, Katayose Y, Toki S, Wang ZX, Kono I, Sasaki T (1998) Expression of *Xa1*, a bacterial blight-resistance gene in rice, is induced by bacterial inoculation. *Proc Natl Acad Sci USA* 95(4):1663–1668.

Yoshimura S, Yoshimura A, Iwata N, McCouch SR, Abenes ML, Baraoidan MR, Nelson RJ (1995) Tagging and combining bacterial blight resistance genes in rice using RAPD and RFLP markers. *Mol Breed* 1(4):375–387.

Yoshimura S, Yoshimura A, Saito A, Kishimoto N, Kawase M, Yano M, Iwata N (1992) RFLP analysis of introgressed chromosomal segments in three near-isogenic lines of rice for bacterial blight resistance genes, *Xa-1, Xa-3* and *Xa-4*. *Jpn J Genet* 67(1):29–37.

Yu S, Ali J, Zhou S, Ren G, Xie H, Xu J, Yu X, Zhou F, Peng S, Ma L, Yuan D, Li Z, Chen D, Zheng R, Zhao Z, Chu C, You A, Wei Y, Zhu S, Gu Q, He G, Li S, Liu G, Liu C, Zhang C, Xiao J, Luo L, Li Z, Zhang Q (2022) From Green Super Rice to green agriculture: Reaping the promise of functional genomics research. *Mol Plant* 15:9–26.

Yugander A, Sundaram RM, Singh K, Senguttuvel P, Ladhalakshmi D, Kemparaju KB, Laha GS (2018) Improved versions of rice maintainer line, APMS 6B, possessing two resistance genes, *Xa21* and *Xa38*, exhibit high level of resistance to bacterial blight disease. *Mol Breed* 38(8):1–14.

Zenbayashi K, Ashizawa T, Tani T, Koizumi S (2002) Mapping of the *QTL* (quantitative trait locus) conferring partial resistance to leaf blast in rice cultivar Chubu 32. *Theor Appl Genet* 104(4):547–552.

Zhai C, Lin F, Dong Z, He X, Yuan B, Zeng X, Pan Q (2011) The isolation and characterization of *Pik*, a rice blast resistance gene which emerged after rice domestication. *New Phytol* 189(1):321–334.

Zhang CQ, Liu YH, Ma XY, Feng Z, Ma ZH (2009) Characterization of sensitivity of *Rhizoctonia solani*, causing rice sheath blight, to mepronil and boscalid. *Crop Protec* 28(5):381–386.

Zhang F, Wang MM, Shi XR, Cui YR, Cruz CV, Zhuo DL, Hua DD, Li M, Wang WS, Zhao, XQ, Zheng TQ, Fu BY, Ali J, Zhou YL, Li ZK (2021) Reciprocal adaptation of rice and *Xanthomonas oryzae pv oryzae*: cross-species two-dimensional GWAS reveals the underlying genetics. *Plant Cell* 33(8): 2538–2561.

Zhang GQ, Angeles ER, Abenes MLP, Khush GS, Huang N (1996) RAPD and RFLP mapping of the bacterial blight resistance gene *xa-13* in rice. *Theor Appl Genet* 93(1):65–70.

Zhang Q, Lin SC, Zhao BY, Wang CL, Yang WC, Zhou YL, Zhu LH (1998) Identification and tagging a new gene for resistance to bacterial blight (*Xanthomonas oryzaepv oryzae*) from *O rufipogon*. *Rice Genet Newsl* 15:138–142.

Zhang Q, Wang CL, Zhao KJ, Zhao YL, Caslana VC, Zhu XD, Jiang QX (2001) 31 The effectiveness of advanced rice lines with new resistance gene *Xa23* to rice bacterial blight. *Rice Genet Newsl* 12:238–240.

Zhang SY, Tan GL, Ren GM, Li MR, Li YY, Lan PX, Li F (2015) Investigation of rice virus diseases and analysis of the molecular variation of RSV isolates in the main rice-growing areas of Yunnan Province from 2013 to 2014. *Chin J Rice Sci* 29(5):535.

Zhao QC, Liu MH, Zhang XW, Lin CY, Zhang Q, Shen ZC (2015, Oct) Generation of insect-resistant and glyphosate-tolerant rice by introduction of a T-DNA containing two Bt insecticidal genes and an EPSPS gene. *J Zhejiang Univ Sci B* 16(10):824.

Zheng CK, Chun-Lian, W, Yuan-Jie Y, Liang Y, Kai-Jun Z (2009) Identification and molecular mapping of *Xa32* (t), a novel resistance gene for bacterial blight (*Xanthomonas oryzaepv oryzae*) in rice. *Acta Agron Sin* 35(7):1173–1180.

Zhou YL, Uzokwe VN, Zhang CH, Cheng LR, Wang L, Chen K, Li ZK (2011) Improvement of bacterial blight resistance of hybrid rice in China using the *Xa23* gene derived from wild rice (*Oryza rufipogon*). *Crop Protec* 30(6):637–644.

Zhuang JY, Wu JL, Fan YY, Rao ZM, Zheng KL (2001) 30 Genetic drag between a blast resistance gene and *QTL* conditioning yield trait detected in a recombinant inbred line population in rice. *Rice Genet Newsl* 18:69–70.

2 Allele Mining in Rice

Importance, Progress, and Prospects

*Lokeshkumar B.M., Ashvinkumar Katral, Sunitha N.C., Rameswar Prasad Sah, Krishnamurthy S.L., Kutubuddin A. Molla, and Anilkumar C.**

*Corresponding author: anilcgpb@gmail.com

2.1 INTRODUCTION

In the era of climate change, it is important to improve rice varieties for varying climatic conditions over locations and years (Gobu et al. 2020). The change in climatic factors like rain fall, temperatures, and CO_2 concentrations in crop growing environments alters the plant's circadian rhythms, leading to unanticipated crop losses. Rice, an important cereal crop feeding the stomachs of more than half of the world's population, is known to be more sensitive to climate change (Beena et al. 2022). However, rice breeders face a difficult task in improving the genetic base of new rice varieties so that they can adapt well to changing climates. The conventional breeding approaches followed to develop new varieties contributed significantly to the human race. However, the resources utilized and time consumed in varietal development via conventional approaches are quite high by virtue of the selection methodologies involved (Patra et al. 2020; Chakraborti et al. 2021). In order to accelerate the varietal improvement process to develop climate smart varieties, adopting new breeding strategies is gaining importance (Verma et al. 2021; Azharudheen et al. 2022). Advances in biotechnological tools and their importance in plant breeding have added a new essence to plant breeding programs (Katara et al. 2021).

Many of the economic traits, including yield, are known to be inherited in a complex pattern in rice, which makes selection only based on phenotypes difficult. With the development and advances in different marker systems, several genomic regions or quantitative trait loci (QTLs) for numerous traits have been identified and utilized in many breeding programs through marker-assisted breeding programs (Pang et al. 2017). Further, developments in data science provided the foundation for breeding science's decision making in the selection of appropriate progenies for developing elite varieties (Varshney et al. 2016). In rice, along with the identification of QTLs, a few consensus QTLs or meta-QTLs are also identified for economically important traits like yield (Khahani et al. 2021), thousand grain weight (Anilkumar et al. 2022a), salinity tolerance (Prakash et al. 2022), nitrogen use efficiency (Sandhu et al. 2021), etc. These are better choices for breeders to introgress into an elite background. However, for a quantitative trait, marker-assisted selection programs were not successful owing to the multi-allelic contribution to the expression of the trait (Kumar et al. 2018). Genome-wide association study (GWAS), a tool that can detect the contributing alleles of a trait across the genome, found promising results in the identification of genomic regions responsible for multiple stress tolerance in rice (Anilkumar et al. 2022b). Further, selection of progeny is also possible based on the genotypic values estimated from the genome-wide distributed markers' effects on a specific trait. Since selection of progeny is based on its own genotypic value, the approach is called genomic selection (Anilkumar et al. 2022c). Genomic selection has greater scope in rice

DOI: 10.1201/9781003385004-2

breeding, however, only limited studies have been performed and reported for a few major traits (Anilkumar et al. 2022d). However, it is difficult to dissect the individual gene and study the effect and variation of the gene on the trait's expression. To comprehend the quantitative trait, it is necessary to dissect all of the variant alleles that contribute to trait expression as well as diversity.

Allele mining is one promising approach to dissecting natural variant alleles of a candidate gene that are known to contribute to the expression of a trait (Figure 2.1). The major objective of allele mining is to identify a small number of accessions that carry diverse alleles in a large population for a quantitative trait (Leung et al. 2015). It is well understood that phenotype varies with different alleles of a gene and that interactions between genes governing the same trait exist. At the same time, it is also true that allele diversity for a trait is most important in crop improvement programs. As a result, in the era of genomics-assisted breeding, the identification or mining of alleles for important economic traits in rice is gaining traction (Ogawa et al. 2018). Conventionally, bi-parent populations were used to detect the QTLs and alleles responsible for trait expression. However, allelic richness in bi-parental populations is limited to only two parent genomes. Further, attention was shifted towards developing multi-parent populations to harness more alleles for a trait (Bandillo et al. 2013). The availability of low-cost single nucleotide polymorphism (SNP) genotyping eased the process of candidate gene identification and allele mining in multi-parent populations (Pascual et al. 2016). GWAS on natural populations, on the other hand, was found to be equally effective for the purpose. However, the composition of multi-parent advanced generation intercross (MAGIC)

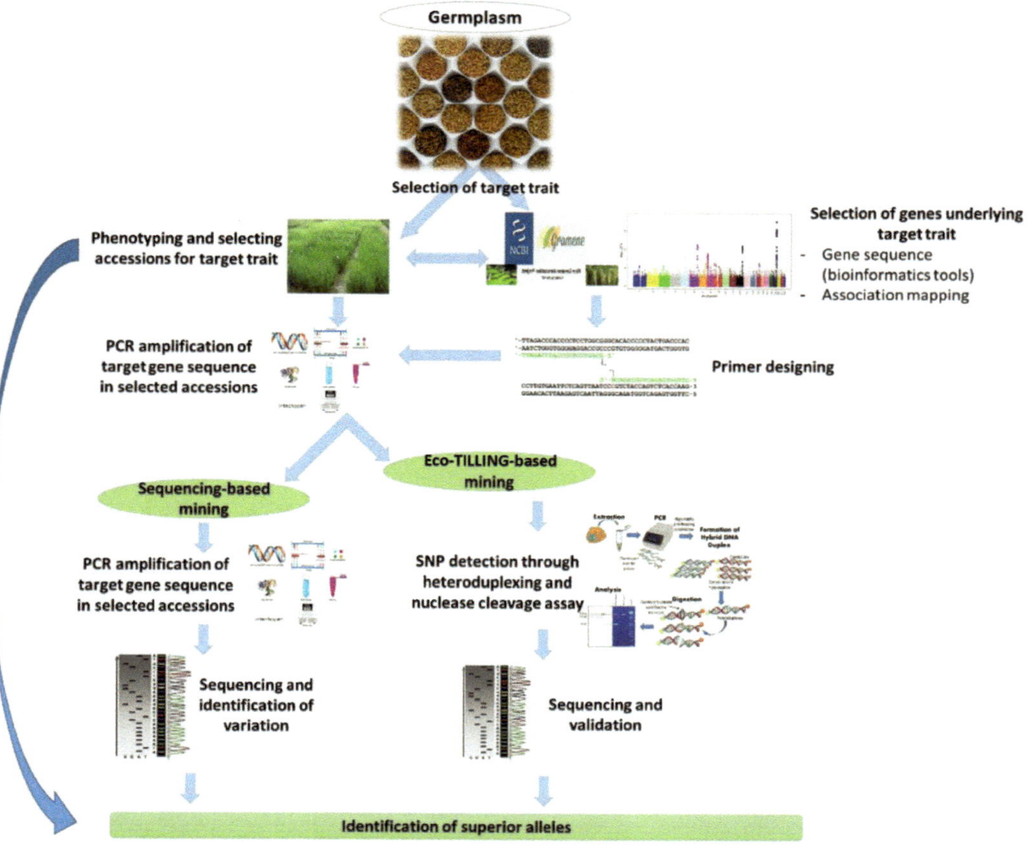

FIGURE 2.1 General procedure of allele mining in crop plants.

populations was able to identify only those alleles that exist in the parental populations, while GWAS on natural populations identified a wide array of alleles for economic traits (Meng et al. 2016). In the early phase of GWAS studies, candidate genes and alleles have rarely been identified considering the low depth of genotyping. With the emergence of deep sequencing platforms, it is possible to detect nucleotide variations in all the genes and identify the putative functions of all the alleles (Zhou et al. 2016).

In rice, several studies have been reported to identify candidate genes with putative alleles for many economic traits. The availability of genome data provides new avenues to shrink the number of accessions and identify relevant genes for a phenotype. Also, known gene sequences can be searched in the databases available to identify the unique accessions carrying targeted alleles for a phenotype. The present chapter focuses on the current status and future prospects of allele mining for various traits in rice, including allele mining for yield-related traits, quality traits, biotic stress tolerance, abiotic stress tolerance, and other adoptive traits contributing to rice improvement programs.

2.2 ALLELE MINING UNRAVELED THE DIVERSITY OF BIOTIC STRESS TOLERANCE GENES

In rice, resistance genes for major diseases have been identified with greater allelic diversity. However, host–pathogen interactions are considered to be dynamic, which leads to the evolution of new races of pathogens. Hence, a continued search for new resistant alleles or genes in natural populations is most vital to combat the emergence of disease epidemics. Several alleles of major disease resistance genes in rice have been reported. Identified superior alleles from known genes are being exploited to develop resistant cultivars in modern rice breeding programs. Allele mining also helps in the identification of conserved resistance gene sequences among different species. With the availability of low-cost sequencing facilities, it is possible to identify new sources of resistance in the natural population using approaches like GWAS.

Rice blast disease, caused by *Magnaporthe oryzae*, is the most dangerous and devastating disease in the rice crop. Despite numerous efforts to dissect the resistance for this disease, the pathogen's rapid evolution has hampered the consistency of resistance sources (Devanna et al. 2022). The blast fungus *Magnaporthe oryzae* was the first phyto-pathogenic fungus to have its genome sequenced (Dean et al. 2005). The availability of sequence information facilitated the rapid mining of alleles for resistance. Exceptionally high genomic plasticity due to high repeat elements in the genome leads to high genome instability, resulting in the evolution of new races in populations (Singh et al. 2019). As of now, more than 100 resistance genes have been documented over all the rice chromosomes, and there are still many more to document (Devanna et al. 2022). Most of the identified genes encode protein variants containing nucleotide binding sites and leucine rich (NBS-LR) domains. Most of the *Pi* allele variants identified from different donor sources are located in the same genomic region either as orthologs or as paralogs (Leung et al. 2015). For example, *Pish* and *Pi35* are orthologs to each other as they are located in the same genomic position (Fukuoka et al. 2014). Most of the *Pi* genes are known to have several variant alleles that impart resistance reactions; for example, Zhou et al. (2020) identified 13 novel alleles for the *Pi9* gene from a diverse population. Similarly, Vasudevan et al. (2015) reported nine unique alleles for *Pi54* from 885 Indian rice genotypes, and Ying et al. (2022) reported seven novel alleles for the *Pik* gene from 2000 rice accessions. The sequence similarities between a few R and S alleles were purely due to localized mutation, and in some cases, they were due to small insertions/deletions. However, in some cases, R alleles and S alleles show no sequence similarity, which helps in the development of functional markers. Next-generation sequencing facilities eased the process of allele mining for blast resistance. Functional SNPs identified in the genic regions or tightly linked SNPs can be best utilized in the identification of resistance haplotypes, which are more useful in breeding programs to incorporate precise genomic regions resistant to blast disease.

Bacterial blight (BB) is another major disease in rice caused by *Xanthomonas oryzae* pv. *oryzae* (*Xoo*). The *Xoo* pathogen resistance reaction has been shown to be strongly race-specific, and several reports on the identification of resistance genes are publicly available. A minimum of 39 genes (*Xa*) have been reported from cultivated and wild rice accessions. Amplification of homologs of BB resistance genes helped in the identification of several superior alleles that are useful in stress resistance breeding programs (Utami et al. 2013; Bimolata et al. 2013; 2015). Like the blast pathogen, the *Xoo* pathogen also exhibits a high level of plasticity that triggers evolution and the fast adaptation of new races of pathogen. In this context, searching for new genic/allelic variations of resistance in landraces, cultivated genotypes, and wild relatives is most crucial for imparting durable resistance against BB (Borba et al. 2009). Some *Xa7* resistance alleles were isolated and used in the development of BB-resistant varieties using an allele mining approach in Indonesian local landraces (Utami et al. 2013). With the BB-resistant gene *Xa27*, which provides resistance against *Xoo* strains with the *AVrXa27* gene. The sequence differences in the promoter regions impart the resistance reaction; otherwise, both alleles code for the same protein (Gu et al. 2005). Many of the *Xa27* resistance alleles identified in *O. nivara* and *O. sativa* exhibited sequence differences in both the 5' UTR region and the promotor regions, which enhanced the functional diversification of BB alleles. Among all the genes identified, *Xa26* is reported to have the highest nucleotide diversity, whereas *Xa21* is known to have the highest number of single nucleotide polymorphisms (Bimolata et al. 2015). Functional characterization of newly identified resistance alleles helps to dissect their role in disease reactions and to develop new resistant cultivars.

Rice sheath blight is yet another serious disease that causes higher levels of yield losses to rice farmers globally. Resistance to sheath blight is known to be regulated by several genes with minor effects. Several sheath blight resistance QTLs have been identified on all 12 rice chromosomes (Richa et al. 2016; Molla et al. 2020). Among all the QTLs identified for sheath blight resistance, *qShB9*-2 (Liu et al. 2009) and *qSBR11*-1 (Channamallikarjuna et al. 2010) are considered major QTLs, contributing 25% and 14% of the phenotypic variance to sheath blight resistance, respectively. It was also reported that *qShB9*-2 has ten candidate genes for defense against sheath blight (Silva et al. 2012). Further, Al-Bader et al. (2019) identified four allelic variations that exist in these QTL regions between susceptible and resistant genotypes. Similarly, from *qSBR11*-1, 26 genes were reported as defense responsive genes (Channamallikarjuna et al. 2010). Apart from major QTLs, several minor QTLs have been identified for sheath blight resistance responses across different genetic backgrounds and environments. These minor QTLs can play a significant role in fine-tuning the major QTL and hence should not be ignored in mining alleles for resistance against sheath blight (Molla et al. 2020).

Apart from major diseases, several minor diseases, including viruses known to infect rice plants, cause a range of losses to rice growers (Zhou et al. 2013). In recent years, the brown spot disease caused by *Bipolaris oryzae* has spread rapidly in Southeast Asian countries. The endemic occurrence of the disease causes chronic losses to farmers. Efforts have been made to delineate the genetic control of pathogens, and a few QTLs were identified through bi-parental mapping by Banu et al. (2008). Further, QTLs were confirmed in multi-parent populations, and tolerance alleles have been reported. Similar efforts were made to find resistant sources and alleles for viral diseases infecting rice. *Rice yellow mottle virus* (RYMV), *Rice stripe virus* (RSV), and *Rice tungro spherical virus* (RTSV) are some of the most common viruses that infect rice crops. The genetic diversity in the rice germplasm collection proved the availability of resistance sources against RTSV, however, resistance sources to another variant of RTSV, *Rice tungro bacilliform virus* (RTBV), are very limited (Hibino 1990). Later, it was found that the resistance to RTSV is a recessive trait and regulated by a gene encoding the translation initiation factor (eIF4G) (Lee et al. 2010). Similarly, the QTLs linked to RSV resistance was found in Kasalath, an indica rice (Wang et al. 2014). RSV resistance in indica rice was proven to be a dominant trait controlled by the *STV11* gene, which is predominantly found in indica rice (Wang et al. 2014). While, the resistant allele for RYMV is more frequently found in *O. glaberrima* than *O. sativa* (indica rice). Screening results of 3000 accessions of *O. sativa* indicated the RYMV-resistant allele is extremely rare in *O. sativa* (Leung et al. 2015). Allele mining using 3000 accessions suggests that there is considerable diversity available for breeding virus-resistant varieties of rice.

Apart from diseases, rice crop suffer from the infestation of insects that threaten the farmers, causing significant crop loss. Among the array of insects that infest the rice crop, the brown planthopper (BPH) stands out, causing severe crop loss upon infestation. Host-plant resistance to BPH was first reported in 1969 with the resistant gene *BPH1* (Pathak et al. 1969). In the last few decades, more than 30 BPH-resistant genes and their associated alleles have been identified (Liu et al. 2015; Zhao et al. 2016; Zhou et al. 2021). Similarly, more than ten genes have been identified for gall midge resistance in rice (Leelagud et al. 2020) and molecular markers have been developed and validated to identify resistant sources (Sahu et al. 2022). These molecular markers are significantly useful in the identification of resistant alleles in the population and introgression into elite genetic backgrounds in resistant breeding programs.

2.3 WIDE VARIATION OF ALLELES FOR ABIOTIC STRESS TOLERANCE IN RICE

Developing stress tolerance in plants requires a thorough understanding of the abiotic stress response. Traditional plant breeding has significantly increased crop plants' genetic productive capacity and improved their ability to withstand biotic and abiotic stresses (Takeda and Matsuoka 2008). A number of important genes, quantitative trait loci, and molecular networks that control how plants react to drought, heat, salinity, and other abiotic stresses have been recently identified (Krasensky and Jonak 2012). Allele mining is one technique for acquiring access to the genetic diversity that is still unexplored for a number of agronomic characteristics. By associating the function of an allele with a phenotype through allele mining, it is feasible to develop allele-specific markers that will be useful in the molecular breeding programs. Although, Targeting Induced Local Lesions in Genomes (TILLING) populations are typically tested for phenotypic or genotypic differences, it has been shown that some TILLING mutants can be used to better understand stress responses (Lorenzo et al. 2009). Similarly, Eco-TILLING has been used to find several kinds of polymorphisms in natural populations (Comai et al. 2004).

A crop cannot complete its life cycle without access to water during a drought, which is a meteorological feature characterized by low or no rainfall. The main danger to future agricultural food production is drought, which is predicted to damage more than 50% of the world's arable land by the year 2050 (Singhal et al. 2016). Rice is known to be susceptible to drought, and severe yield loss can be seen up to around 100% (Sahebi et al. 2018). Therefore, knowing how plants react to drought stress becomes crucial and the reference point for creating rice varieties that can withstand it. Numerous QTLs or genes linked to drought resistance characteristics, such as maintaining leaf water status, regulating root shape and controlling stomatal closure, have been discovered (Courtois et al. 2000; Yue et al. 2006; Li et al. 2017). The genes involved in a variety of plant developmental processes, such as cell membrane integrity (LEA proteins) (Xu et al. 1996), signaling (kinase) (Saijo et al. 2001; Liu et al. 2003), water uptake and its transportation (aquaporins) (Martre et al. 2002), carbohydrate metabolism (Jang et al. 2003) and the regulation of oxidative stress management (Prakash et al. 2016) are mainly responsible for drought tolerance in crop plants. Drought tolerance is primarily influenced by four root characteristics: root length, root growth angle, root volume, and root thickness (Uga et al. 2011). *Deeper Rooting 1 (DRO1)* is the most important of the three significant QTLs identified for rapid generation advancement (RGA) system in rice (Uga et al. 2011, 2013, 2015). It has also been demonstrated that the *DRO1* and the other two QTLs, *DRO2* and *DRO3*, interact and work together. The early auxin responsive factor that underlies the *DRO1* QTL has been cloned using the cultivars Kinandang Patong (KP), which has deep roots, and IR64, which has shallow roots. For allele mining of the *DRO1* (3058 bp) gene, Singh et al. (2021) used 24 significant Indian upland and lowland rice genotypes along with *yield under drought* (*DTY*) QTL donors. Based on single-nucleotide (SNP) variations and respective amino acid changes, 24 genotypes were classified into 16 primary structure variants using allele mining, ranging from fully functional alleles (Satti, IR36, and DTY 3.1 donor, IR81896-B-B-195) to truncated non-functional alleles (PMK2, IR64, IR20, and Swarna). Plants express a wide range of proteins in response to drought stress, including transcription factors, a crucial family of regulatory elements (Singh et al. 2002). Drought stress can activate transcription factor families such MYB, MYC, AP2/ERF, NAC, bZIP, HD-ZIP, and NF-Y (nuclear factor Y) and changes in the expression of certain of genes may affect drought tolerance (Hu et al. 2006; Li et al. 2008). Dehydration response element binding (*DREB*) gene-encoded plant transcription factors (TFs) are critical for controlling the expression of several genes induced by drought stress in a largely ABA-independent manner. They accomplish this by interacting with a *DRE/CRT* cis-element that is present in the promoter region of numerous genes, which respond to abiotic stress (Dubouzet et al. 2003; Lata et al. 2011). The gene, *OsDREB1F*, a DREB transcription factor, was cloned and characterized from upland rice Nipponbare by Wang et al. (2008). An analysis of its expression patterns showed that it was induced by drought, salt, and cold stresses, as well as by the application of ABA. The *OsDREB1F* protein can only bind to DRE/CRT elements and not ABRE elements. The *OsDREB1F* gene, which is related with drought resistance, was examined for natural allelic diversity by Singh et al. (2015) in the Indian wild rice germplasm and found six protein variants for the gene. Rice transcription factors linked to drought stress were found using Eco-TILLING, where promoters of eight genes associated with drought resistance in five cultivars, including three genes associated with a drought tolerant index and five genes with significant levels of drought tolerance (Table 2.1). The ability of plants to survive osmotic stress and develop was influenced by *OsALF11*, *OsAE128*, and *OsZIM14* genes, which belong to the Alfin-like, AP2/EREBP, CCAAT-HAP3, GRF, and Tify families (Park et al. 2004; Svensson et al. 2006). The allelic variation with different tolerant mechanisms were made available in the form of diverse germplasm accessions reported as tolerant, which can be used in marker-aided breeding programs.

Rice has been shown to have a large number of genes that are triggered by low temperature (Fowler and Thomashow 2002; Seki et al. 2002; Rabbani et al. 2003), some of which code for regulatory proteins such as transcription factors and protein kinases. There are four *DREB2*-type genes and at least ten *DREB1*-type genes in the rice genome, of which *OsDREB1A* and *OsDREB1B* are triggered by cold stress (Dubouzet et al. 2003). Two transcription factors known as *DREB1A* and

TABLE 2.1
The QTLs, Genes, and Transcription Factor Promoters Associated with Drought Tolerance in Rice

Drought Tolerance Parameter	QTL-Associated Traits	QTL Name	Underlying Gene Name	Gene Family
Drought-Tolerant Index	Biomass yield in aerobic land	AQAL037	*OsALF11*	Alfin-like
	Root number of upland rice	AQO089	*OsGRF5*	GRF
	Leaf rolling	AQA047	*OsAE115*	AP2/EREBP
Drought-Tolerant Level	Osmotic tolerance	CQG4	*OsGRF8*	GRF
	Leaf yellowing tolerance	CQAA15		
	Relative water content	DQE53	*OsNFYB12*	CCAAT
	Penetrated root number of upland rice	DQC5	*OsAE25*	AP2/EREBP
	Root thickness of low moisture regime	CQAI39		
	Root thickness of low moisture regime	CQAI39		
	Tiller number of low moisture regime	AQA010	*OsAE128*	AP2/EREBP
	Deep root dry weight in aerobic land	AQAL056	*OsZIM14*	Tify

DREB1B have been implicated in the regulation of low-temperature tolerance. The allelic variations of these genes translated to tolerant breeding lines exhibited significant tolerance response to low temperature (Challam et al. 2015). However, accumulation of alleles of temperature-tolerant genes distributed among different germplasm resources in rice required to develop a multi-gene combination of tolerance source for marker-assisted breeding programs. The diverse alleles in distant sources should be brought into few elite breeding lines before introgressing into elite cultivar in order to reduce the linkage drag in marker-assisted back cross breeding programs.

Soil salinity is yet another abiotic stress, which has a detrimental effect on the growth and production of rice. Numerous genes and QTLs with high allelic diversity that are related to salt tolerance have been reported in rice. These include *qSKC*-1 (for shoot K+ level), *qSNC*-7 (for shoot Na+ accumulation), and *saltol* QTL (for Na+/K+ ratio and seedling stage salt-tolerance). The precise mapping of QTLs has shown molecular changes in *OsHKT1*;5, which directly affect Na+/K+ homeostasis. The candidate gene-based method and GWAS have highlighted the significance of key SNPs in genes linked with salt tolerance, including the HKT1;1/1;5 gene, MYB and MADS-box family transcription factors, the OST3/OST6 family of proteins, and the vascular cation/proton exchanger. For the purpose of improving salinity stress tolerance, several researchers have investigated the isolation, characterization, and over expression of several Na+/H+ antiporter genes (Horie et al. 2011; Ganie et al. 2019). Of them, the *HKT* gene family in rice makes a considerable contribution to Na+ exclusion (Ali et al. 2019), and in particular, *OsHKT1*;5 is a key contributor found in the region beneath *Saltol* QTL (Platten et al. 2006; Aliyu et al. 2010). For allele mining of stress-tolerance rice genes, including late embryogenesis-associated protein coding gene, calmodulin, and salt-inducible rice gene, wild relatives with AA-genomes and non-AA genomes were reported to have a large number of allelic variations (Latha et al. 2004). Studies have been done using the Eco-TILLING technique to find gene variations in salt responsive genes associated to the Na+/K+ ratio, signaling pathways, and osmo-protection. Utilizing the Eco-TILLING method, alleles linked to salinity tolerance were reported from 375 rice germplasm (Negrao et al. 2011). Ten *OsCPK17* haplotypes and six *SalT* haplotypes with significant impact on salt tolerance discovered (Negrão et al. 2013). Two of the three non-synonymous SNPs that were discovered in *OsHKT1*;5 with significant relationships had two residual variations between the "Nipponbare genotype" (haplotype A) and IR29 (haplotype B). A total of seven major alleles and several minor allelic variations for *OsHKT1*;5 gene were reported among and between japonica aromatic and indica rice genotypes (Platten et al. 2013).

Around 15 Mha of lowland rice cultivation is regularly impacted by flooding in the river basins of South and South-East Asia. Flooding causes hypoxic conditions that hinder seed germination and seedling establishment. Water stagnation also contributes to severe yield losses when larger proportion of rice aerial tissues are submerged for weeks to months. The flash flooding rice variety "FR13A" and the *Submergence 1A* (*SUB1A*) gene were discovered and then introgressed into high-yielding rice cultivars. Similar to this, genes for rapid stem elongation during deep-water flooding (*SNORKEL1* and *SNORKEL2*) and anaerobic germination ability (*TPP7*) have been discovered in tolerant landraces of rice. In the *SUB1* QTL region of chromosome 9, three ERF genes, *SUB1A*, *SUB1B*, and *SUB1C*, were discovered (Xu et al. 2006). Under flash flooding, *SUB1A* was found to be the main factor determining submergence tolerance, with *SUB1C* playing a smaller but still substantial supporting role and *SUB1B* playing no discernible part (Xu et al. 2006; Septiningsih 2009; Singh et al. 2010). The genes *SUB1A*-1 and *SUB1A*-2 are two alleles that were found as a result of studies on sequence variation in the *SUB1A* protein-coding regions. Typically, submergence-tolerant cultivars have the *SUB1A*-1 allele, whereas submergence-sensitive types either have the *SUB1A*-2 allele or lack the *SUB1A* gene (null allele). Pro186 in the intolerant accessions is changed to Ser186 in the tolerant accessions at the protein level (Xu et al. 2006). Studies have been carried out to investigate allelic variation at this locus among rice cultivars and wild relatives following the map-based cloning of *SUB1A* and correlate this with variations in submergence tolerance. Among the different alleles of *SUB1* locus, *SUB1A* was found common in indica and aus rice accessions whereas, *SUB1B* and *SUB1C* in japonica accessions (Fukao et al. 2009). Re-sequence information of SUB1 locus revealed that allelic variation within the locus are functions of gene duplications during the domestication (Singh 2015; Singh et al. 2020). The predominant tolerant alleles of submergence tolerance and respective haplotypes identified in different genetic background have been utilized in submergence tolerance breeding programs.

2.4 ALLELE MINING FOR GRAIN YIELD AND COMPONENT TRAITS IN RICE

The interplay of the environment has a significant impact on grain yield, a complex quantitative trait. The hypothesis that many of these genes/QTLs may not have been utilized in current varieties of rice during the course of its evolution and may still be present in the wild gene pool led to the beginning of the search for genes/QTLs governing polygenically regulated yield and its key components (Xiao et al. 1996). With significant efforts following different approaches, several QTLs/genes have been identified with favorable alleles for yield traits in rice (Table 2.2). Through marker-assisted breeding, new QTLs or alleles might be discovered using recent advancements in plant biotechnological tools, and the potential ones could be stacked in existing high-yielding varieties to further increase genetic gains. Two yield QTLs, *yld* 1.1 and *yld* 2.1 in *Oryza rufpogan*, were successfully mapped and validated by Xiao et al. (1996) using a marker-associated QTL method. The quest for more and more yield genes/QTLs in wild/weedy relatives and primitive cultivars was prompted by the increasing demands for higher production across the globe, and this led to the discovery of numerous potential yield QTLs in landraces, *O. rufpogan*, and *O. nivara* (Marri et al. 2005; Kaladhar et al. 2008; Swamy and Sarla 2008; Sudhakar et al. 2012; Swamy et al. 2011, 2012, 2014). This ground-breaking study culminated in the creation of higher-yielding varieties as Jefferson employing *yld2*.1 from *O. rufpogan* (Imai et al. 2013), DRR Dhan 40 utilizing yield genes/QTLs obtained from *O. nivara* (Haritha et al. 2017), and the high-yielding salinity-tolerant IET21943 based on a yield QTL from *O. rufpogon* (Ganeshan et al. 2016).

TABLE 2.2
Gene(s)/QTL(s) Governing Yield and Attributing Traits in Rice

Trait	Gene(s)/QTL(s)	Reference
Grain yield	*yld 1.1* and *yld 2.1*	Xiao et al. (1996); Imai et al. (2013)
	yldp3.1, yldp3.2, yldp4.1, yldp8.1 and *yldp11.1*	Kaladhar et al. (2008); Swamy and Sarla (2008)
Grain weight	*gw2.1* and *gw9.1*	Marri et al. (2005)
	tgw11	Oh et al. 2011
Grain weight and shape	*qTGW3.1, qTGW9, qTGW11, qGL4/qRLW4, qGL10* and *qRLW1*	Niu et al. (2021)
	qGRL7.1	Singh et al. (2012)
Grain shape	*GS9*	Zhao et al. (2018)
Grain length and grain width	*GS3*	Fan et al. (2006); Mao et al. (2010)
	qSS7	Qiu et al. 2012
Grain length	*qGL4b*	Kato et al. (2011)
Grain length and filling	*qGL3/qGL3.1*	Qi et al. (2012)
Grain filling	*GIF1*	Wang et al. (2008)
Grain size and filling	*TGW6*	Izawa et al. (2010)
Grain width and filling	*GW2*	Song et al. (2007)
	GS5	Li et al. (2011)
Grain width	*GW5/qSW5*	Weng et al. (2008)
Grain width	*GW8*	Wang et al. (2012)

Grain yield is a complex variable, thus understanding genetics and the relationships between traits that affect grain yield are essential (Katral et al. 2022). It is controlled by factors that are associated with yield, such as grain shape (GS), grain thickness (GT), 1000-grain weight (TGW), and grain length/width (GLW) (Tan et al. 2000; Gong et al. 2017). In rice, grain yield has been mostly influenced by grain weight, or TGW, while grain quality is influenced by grain shape (Niu et al. 2021). Over ten significant QTLs that influence grain weight and grain shape have been identified and cloned to date (Zuo et al. 2014; Guo et al. 2018; Zhao et al. 2018; Zhang et al. 2021). Notably, *GW2*, *GW7/GL7*, *GW8/OsSPL16*, and *GS9* modify cell division to regulate grain form (Zhao et al. 2018; Niu et al. 2021). In addition, a number of QTL for grain weight or grain shape have been discovered, including qGRL7.1 (Singh et al. 2012), qGL4b (Kato et al. 2011), qSS7 (Qiu et al. 2012), and tgw11 (Oh et al. 2011), however it is not yet understood how these genes function. Additionally, a number of well-known genes, including *An1, Gn1a, DEP1, DEP3, APO1, GS3, GW5, PROG1, LRK1, OsSPL14, EP3,* and *SP1*, influence the amount of grains per panicle (Gouda et al. 2019). *An-1* gene, for example, has a pleiotropic function that controls rice grain size, grain number, and awn development. Additionally, it has been discovered that several genes, including *Gn1a, GS312, DEP114, Ghd713,* and *sd115*, influence grain yield (Vemireddy et al. 2019). Through genome-wide association studies, Niu et al. (2021) recently discovered seven candidate genes linked to traits relevant to grain weight and shape in rice (GWAS).

2.5 ALLELE MINING EXPLORED THE ALLELIC DIVERSITY FOR NUTRITIVE TRAITS IN RICE

Due to its high carbohydrate content, rice has relatively low levels of important micronutrients as zinc, iron, protein, and phosphorus. As a result, great efforts have been made to enhance the rice's cooking quality, flavor, and nutritional value. In order to identify the molecular process, several unique genes affecting the vital minerals and cooking characteristics of rice have been found and thoroughly characterized. To investigate the use of allelic differences in breeding programs to improve grain nutritional status, it is critical to have a thorough grasp of genetic regulation as well as knowledge of their impacts.

Granule-bound starch synthase I (GBSSI), which is encoded by the *Wx* locus on chromosome 6, plays a major role in controlling the amylose production in rice endosperm (Wang et al. 1990). Allelic variation at the *Wx* locus is substantially responsible for the genetic diversity of amylose content (AC) (Zhang et al. 2019a). Eight *Wx* alleles, including Wx^{lv}, Wx^{a}, Wx^{in}, Wx^{b}, $Wx^{op/hp}$, Wx^{mq}, Wx^{mp}, and *wx,* have been identified as being related with the five amylose types that have been detected in rice cultivars (Yang et al. 2013; Zhang et al. 2019b). The majority of these alleles have been effectively inserted into contemporary rice cultivars, where $Wx^{op/hp}$, Wx^{mq} and Wx^{mp} (managing low to very low AC) have been bred to produce eating and cooking quality (ECQ) variants of rice known as soft rice. Since the grain appearance of these soft rice cultivars with AC values typically lower than 13% is dull or opaque, they might not match the commercial appearance standards set by consumers (Li et al. 2018). As a result, the novel rare *Wx* allele Wx^{mw}, which was created by recombination between two significant natural Wx alleles (Wx^{in} and Wx^{b}), has been discovered. It combines a favorable AC with increased ECQ and grain transparency (Zhang et al. 2021).

On chromosome 6, the gene known as *Alkali degeneration gene (ALK)*, also known as *SSII-3*, controls rice gelatinization temperature (GT) (Zhang et al. 2011). According to Gao et al. (2003), the *ALK* gene encodes for soluble starch synthase II (SSSII), which is essential for the endosperm's short A and B1 chains of amylopectin to be lengthened (Gao et al. 2011; Yu et al. 2011). Three functional single nucleotide polymorphisms (SNPs), the GC/TT SNP at 4329/4330 bp and the G/A SNP at 4198 bp, have been reported to be intimately related to GT and amylopectin structure (Chen et al. 2020). In accordance with these functional nucleotide polymorphisms, three ALK haplotypes in rice cultivars, A-GC, G-TT, and G-GC, have been identified (Zhou et al. 2016). Based on the level of GT, they were divided into two types: active indica type *ALK*^{i} (G-GC), which controls a high GT and inactive japonica type *ALK*^{j} (A-GC) or (G-TT), which controls a low GT (Wang et al. 2015; Fan et al. 2017).

Betaine aldehyde dehydrogenase 2 (OsBadh2), also known as *fgr/badh2/osbadh2/os2AP*, was discovered on chromosome 8 and is the primary gene controlling aroma in rice. The betaine aldehyde dehydrogenease enzyme was truncated due to an 8-bp deletion in exon 7 of this gene, and the accumulation of a significant aromatic compound (2AP) in fragrant rice was caused by this loss of function. In fragrant rice, the *osbadh2* gene's nonfunctional *badh2* enzyme prevents the conversion of gamma-aminobutyraldehyde (GAB-ald) to gamma-aminobutyric acid (GABA), permitting the synthesis of 1-pyrroline and eventually the 2AP (Bradbury et al. 2008). On rice chromosome 4, a different rice fragrance locus known as *aro4-1* has been discovered (Amarawathi et al. 2008). *OsBadh1*, a putative gene at this location, performs similar molecular functions to *OsBad2*. Additionally, a new region of the rice genome harbors the fragrance QTL *aro3-1*, which may be unique to the Basmati rice variants (Amarawathi et al. 2008). In addition, the CHR3_22 and CHR3_24 region of chromosome 3 harbors one gene that is differently expressed and is known to be a minor quantitative trait locus (QTL) for grain aroma that plays a complementary role to the other QTLs on chromosome 4 (*aro4-1*) (Pachauri et al. 2014). Through bulk segregant analysis, a single gene (*LOC Os03g21040*) from the family of hydroxyproline-rich glycoproteins that is expressed under abiotic stress and is also in charge of giving rice its distinctive flavor was discovered (Pachauri et al. 2014). Furthermore, a significant QTL on chromosome 5, which also contains the *OsGlyI* gene, has been found, suggesting that methylglyoxal contributes to the scent of rice (Talukdar et al. 2017).

There were 14 QTL reported for grain Fe and Zn contents, and candidate genes that were close to QTLs were found, including *OsYSL1* and *OsMTP1* for Fe; *OsARD2*, *OsIRT1*, *OsNAS1*, and *OsNAS2* for Zn; and *OsNAS3*, *OsNRAMP1*, and *APRT* for Fe and Zn (Anuradha et al. 2012). Fe homeostasis genes including nicotianamine (NA) synthase *OsNAS3*, vacuolar Fe transport *OsVIT1*, and also *OsMTP6*, *OsMT2D*, and *OsNRAMP7* were discovered using GWAS together with the grain Fe content, and they were co-located with QTLs for Fe and Zn (Descalsota et al. 2018). Based on genotypic and multi-season phenotypic data, Dixit et al. (2019) identified two significant QTLs for Fe and three for Zn. Based on genotypic and multi-season phenotypic data, Dixit et al. (2019) identified two significant QTLs for Fe and three for Zn. The comparison study of the two seasons also revealed two QTLs for Zn content and two QTLs for Fe content, as well as prospective candidate genes, including *OsPOT, OsZIP4, OsFDR3*, and *OsIAA5* that were previously characterized to affect grain Fe and Zn content. On chromosome 2, two frequently occurring QTLs for grain Zn polished, designated *qZnPR*.2.1 (PVE 19.42%), and grain Zn brown, designated *qZnBR*.2.1 (PVE 20.80%), have recently been discovered (Uttam et al. 2022).

In brown rice, three QTLs, *qPC1*.1, *qPC11*.1, and *qPC11*.2, were linked to the amount of protein (Qin et al. 2009). On chromosome 11, the QTLs known as *qPC11*.1 and *qPC11*.2 were discovered. Five QTLs, designated as *qPC-3, qPC-4, qPC-5, qPC-6*, and *qPC-10* by Yu et al. (2009), were located on chromosomes 3, 4, 5, 6, and 10, respectively. On the short arm of rice chromosome 6, the QTL, *qPC*-6, was shown to be closest to the *Wx* marker between RM190 and RZ516. Later, Zhong et al. (2011) reported finding two reliable quantitative trait loci (QTLs) for protein concentration in milled rice, *qPr1* and *qPr7*, which were consistently found over 2 years and mapped on chromosomes 1 and 7, respectively. Additionally, Yang et al. (2019) discovered two significant consistent QTLs, *qGPC*-1 and *qGPC*-10, the latter of which harbors *OsGluA2*, a gene that regulates variation in both japonica and indica rice. Furthermore, by association mapping, Pradhan et al. (2019) discovered three novel QTLs, namely *qPC3*.1, *qPC5*.1, and *qPC9*. Three QTLs, known as *qPC1*.1, *qPC1*.2, and *qPC7*.1, have recently been found to co-localize with QTLs for serine, histidine, and threonine amino acid concentration in rice (Jang et al. 2020).

The rice bran layer's color is controlled by the *Rc* gene, and selection for the *rc* allele, which results in a white pericarp, took place during crop domestication. It was claimed that two loci, *Rc* and *Rd*, control how the color of the pericarp develops (Sweeney et al. 2006; Furukawa et al. 2007). The locus *Rc* causes the development of brown pericarp and seed coat; when *Rd* is added, however, red pericarp and seed coat are produced, although *Rd* by itself has no phenotypic impact. According to Sweeney et al. (2006), the locus Rd encodes the enzyme DFR (dehydroflavonolreductase), which is implicated in the anthocyanin and proanthocyanidins pathway, whereas *Rc* encodes a regulatory protein (Basic Helix-Loop-Helix Protein) that permits the accumulation of proanthocyanidins (Furukawa et al. 2007). Two loci, *Pb* (*Prp-b*) and *Pp* (*Prp-a*), which control the pericarp coloration of black rice with anthocyanin, were discovered by Yoshimura et al. (1997). Furthermore, Wang and Shu (2007) located the gene *Pb*, which controls the purple pericarp, on chromosome 4, and hypothesized that it might be a mutant of the *Ra* gene brought on by a two base deletion (GT) in exon 7.

Myo-inositol-3-phosphate synthase (MIPS), an enzyme that catalyzes the conversion of D-glucose 6-phosphate to 1D-myo-inositol 3-phosphate [Ins(3)P1], was mutated in the first step of the phytic acid biosynthesis pathway, resulting in the development of the low phytic acid mutant in rice known as *lpa1*. The locus for this mutation was located on chromosome 3S (Larson et al. 2000). The nature of the mutations causing the decrease in phytic acid was unknown in two *lpa* rice mutants, *Os-lpa-XS110-1* and *Os-lpa-XS110-2*, which were created using gamma irradiation (Liu et al. 2007). Kim et al. (2008) used chemical mutagenesis to establish a novel *lpa* line in rice called lpa N15-186, which is orthologous to the maize gene *MIK*. The gene mapped on to the chromosome 3, which encodes for myo-inositol kinase (MIK), is an enzyme, which catalyzes the synthesis of Ins(3)P1. A new rice transporter for phosphorus (P) delivery to the grain is called the SULTR-like phosphorus distribution transporter (SPDT). The *SPDT* transporter gene underwent a single nucleotide alteration, which decreased phytic acid and enhanced the bioavailability of minerals from rice grain (Kumar et al. 2021). The allelic variations for *SPDT* gene proved to regulate various levels of phytic acid content in rice genotypes (Azharudheen et al. 2022).

2.6 FUTURE PROSPECTS

The breeding programs should be designed to develop varieties resilient to changing climatic conditions. Extreme climatic fluctuations will become more common in the coming decades around the world. The effects of climate change will certainly impose new biotic and abiotic stresses on rice grain production. It is ideal to create and utilize the new genetic variation along with available information for more important traits in rice breeding programs. Exploring the allelic and genetic diversity to impart necessary traits into new rice varieties to combat climatic changes and, at the same time, increase productivity and nutritional value. It is more important to have a breeding system that can explore allelic diversity and deliver it to rice breeding programs.

The availability of high-quality genome sequence information for rice and low-cost genotyping facilities can aid in mining new alleles for major productivity and stress-resilient traits in rice. The existence of enormous genetic diversity among genotypes in gene bank collections can serve as base genetic material to explore favorable alleles and use as donors in breeding programs. Improvements in data science and the availability of bioinformatics tools may assist in mining the most valuable alleles for nutritive and stress-tolerant traits. Identification of favorable resistance allele haplotypes is critical for use in haplotype introgression and haplo-near-isogenic line development. Haplo-near-isogenic lines may serve as elite breeding material with favorable alleles for improving many economically important traits. The diverse array of lines with favorable alleles may be further deployed for combining alleles using inter-mating designs like marker-assisted recurrent selection or multi-parent advanced generation intercross approaches. The combination of genomic diversity with mined alleles for major traits and mating designs that accumulate the alleles can benefit the development of new genomic resources and elite varieties for cultivation.

REFERENCES

Al-Bader N, Meier A, Geniza M, Gongora YS, Oard J, Jaiswal P (2019, Jan) Loss of premature stop codon in the Wall-Associated Kinase 91 (OsWAK91) gene confers sheath blight disease resistance in rice. *bioRxiv*. 1:625509. https://doi.org/10.1101/625509

Ali A, Maggio A, Bressan RA, Yun DJ (2019) Role and functional differences of HKT1-type transporters in plants under salt stress. *Int J Mol Sci* 20(5):1059. https://doi.org/10.3390/ijms20051059

Aliyu R, Adamu AK, Muazu S, Alonge SO, Gregorio GB (2010) Tagging and validation of SSR markers to salinity tolerance QTLs in rice *(Oryza spp)*. In: *International Conference on Biology, Environment and Chemistry IPCBEE* IACSIT press; Singapore. pp. 328–332.

Amarawathi Y, Singh R, Singh AK, Singh VP, Mohapatra T, Sharma TR, Singh NK (2008) Mapping of quantitative trait loci for Basmati quality traits in rice (*Oryza sativa* L.). *Mol Breed* 21(1): 49–56. https://doi.org/10.1007/s11032-007-9108-8

Anilkumar C, Sah RP, Muhammed Azharudheen TP, Behera S, Singh N, Prakash NR, Sunitha NC, Devanna BN, Marndi BC, Patra BC, Nair SK (2022a) Understanding complex genetic architecture of rice grain weight through QTL-meta analysis and candidate gene identification. *Sci Rep* 12(1):1–3. https://doi.org/10.1038/s41598-022-17402-w

Anilkumar C, Lokeshkumar BM, Sunitha NC, Basavaraj PS, Hanjagi PS (2022b) Genome-wide association mapping and genomic selection approaches for stress resilience in rice. In: Gowdra Mallikarjuna M, Nayaka SC, Kaul T (eds) *Next-Generation Plant Breeding Approaches for Stress Resilience in Cereal Crops*. Springer, Singapore, pp 45–94. https://doi.org/10.1007/978-981-19-1445-4_2

Anilkumar C, Sunitha NC, Devate NB, Ramesh S (2022c) Advances in integrated genomic selection for rapid genetic gain in crop improvement: a review. *Planta* 256(5):1–20. https://doi.org/10.1007/s00425-022-03996-y

Anilkumar C, Sah RP, Muhammed AT, Sunitha NC, Behera S, Marndi BC, Sharma TR, Singh AK (2022d) Genomic selection in rice: current status and future prospects. In: Elias AA, Goel S (eds) Genomic Selection in Plants A Guide for Breeders. CRC Press, pp 68–82. https://doi.org/10.1201/9781003214991

Anuradha K, Agarwal S, Rao YV, Rao KV, Viraktamath BC, Sarla N (2012) Mapping QTLs and candidate genes for iron and zinc concentrations in unpolished rice of Madhukar× Swarna RILs. *Gene* 508(2):233–240. https://doi.org/10.1016/j.gene.2012.07.054

Azharudheen MTP, Kumar A, Anilkumar C, Sah RP, Behera S, Marndi BC (2022) Understanding natural genetic variation for grain phytic acid content and functional marker development for phytic acid-related genes in rice. *BMC Plant Biol* 22(1):446. https://doi.org/10.1186/s12870-022-03831-2

Bandillo N, Raghavan C, Muyco PA, Sevilla MA, Lobina IT, Dilla-Ermita CJ, Tung CW, McCouch S, Thomson M, Mauleon R, Singh RK (2013) Multi-parent advanced generation inter-cross (MAGIC) populations in rice: progress and potential for genetics research and breeding. *Rice* 6(1):1–5. https://doi.org/10.1186/1939-8433-6-11

Banu SP, Meah B, Ali A, Brar DS, Leung H, Cruz CV (2008) Inheritance and molecular mapping for brown spot disease resistance in rice. *J Plant Pathol* 90(S2):295.

Beena R, Sunitha NC, Sah RP, Md Azharudheen TP, Krishna GK, Deepika Kumar Umesh, Sini Thomas, Anilkumar C, Sameer Upadhyaya, Awadhesh Kumar, Manikanta Ch LN, Behera S, Marndi BC, Siddique KHM (2022) Physiological and molecular implications of multiple abiotic stresses on yield and quality of rice. *Front Plant Sci* 13:996514. https://doi.org/10.3389/fpls.2022.996514

Bimolata W, Kumar A, Sundaram RM, Laha GS, Qureshi IA, Reddy GA, Ghazi IA (2013) Analysis of nucleotide diversity among alleles of the major bacterial blight resistance gene Xa27 in cultivars of rice (*Oryza sativa*) and its wild relatives. *Planta* 238(2):293–305. https://doi.org/10.1007/s00425-013-1891-3

Bimolata W, Kumar A, Sundaram RM, Laha GS, Qureshi IA, Ghazi IA (2015) Nucleotide diversity analysis of three major bacterial blight resistance genes in rice. *PLoS One* 10(3):e0120186. https://doi.org/10.1371/journal.pone.0120186

Borba TC, Mendes CD, Guimarães ÉP, Brunes TO, Fonseca JR, Brondani RV, Brondani C (2009) Genetic variability of Brazilian rice landraces determined by SSR markers. *Pesq Agropec Bras Brasılia* 44:706–712. https://doi.org/10.1590/S0100-204X2009000700009

Bradbury LMT, Gillies SA, Brushett DJ, Waters DLE, Henry RJ (2008) Inactivation of an aminoaldehyde dehydrogenase is responsible for fragrance in rice. *Plant Mol Biol* 68(4):439–449. https://doi.org/10.1007/s11103-008-9381-x

Chakraborti M, Anilkumar C, Ram Lakhan Verma, Abdul Fiyaz R, Reshmi Raj KR, Patra BC, Divya Balakrishnan, Sutapa Sarkar, Nimai P Mondal, Meera K Kar, Jitendriya Meher, Sundaram RM, Subba Rao LV (2021) Rice breeding in India: eight decades of journey towards enhancing the genetic gain for yield, nutritional quality, and commodity value. *Oryza*, 58(spl iss):69–88. https://doi.org/10.35709/ory.2021.58.spl.2

Challam C, Ghosh T, Rai M, Tyagi W (2015) Allele mining across DREB1A and DREB1B in diverse rice genotypes suggest a highly conserved pathway inducible by low temperature. *J Genet* 94(2):231–238. https://doi.org/10.1007/s12041-015-0507-z

Channamallikarjuna V, Sonah H, Prasad M, Rao GJ, Chand S, Upreti HC, Singh NK, Sharma TR (2010) Identification of major quantitative trait loci qSBR11-1 for sheath blight resistance in rice. *Mol Breed* 25(1):155–166. https://doi.org/10.1007/s11032-009-9316-5

Chen Z, Lu Y, Feng L, Hao W, Li C, Yang Y, Fan X, Li Q, Zhang C, Liu Q (2020) Genetic dissection and functional differentiation of *ALK(a)* and *ALK(b),* two natural alleles of the *ALK/SSIIa* gene, responding to low gelatinization temperature in rice. *Rice (NY)* 13(1):39. https://doi.org/10.1186/s12284-020-00393-5

Comai L, Young K, Till BJ, Reynolds SH, Greene EA, Codomo CA, Enns LC, Johnson JE, Burtner C, Odden AR, Henikoff S (2004) Efficient discovery of DNA polymorphisms in natural populations by Ecotilling. *Plant J* 37(5):778–786. https://doi.org/10.1111/j.0960-7412.2003.01999.x

Courtois B, McLaren G, Sinha PK, Prasad K, Yadav R, Shen L (2000) Mapping QTL associated with drought avoidance in upland rice. *Mol Breed* 6:55–66. https://doi.org/10.1023/A:1009652326121

Dean RA, Talbot NJ, Ebbole DJ, Farman ML, Mitchell TK, Orbach MJ, Thon M, Kulkarni R, Xu JR, Pan H, Read ND (2005) The genome sequence of the rice blast fungus *Magnaporthe grisea. Nature* 434(7036):980–986. https://doi.org/10.1038/nature03449

Descalsota GI, Swamy BM, Zaw H, Inabangan-Asilo MA, Amparado A, Mauleon R, Chadha-Mohanty P, Arocena EC, Raghavan C, Leung H, Hernandez JE (2018) Genome-wide association mapping in a rice magic plus population detects QTLs and genes useful for biofortification. *Front Plant Sci* 9:1347. https://doi.org/10.3389/fpls.2018.01347

Devanna BN, Jain P, Solanke AU, Das A, Thakur S, Singh PK, Kumari M, Dubey H, Jaswal R, Pawar D, Kapoor R, Singh J, Arora K, Saklani BK, Anilkumar C, Maganti SM, Sonah H, Deshmukh R, Rathour R, Sharma TR (2022) Understanding the dynamics of blast resistance in rice-*Magnaporthe oryzae* interactions. *J Fungi* 2022 (8):584. https://doi.org/10.3390/jof8060584

Dixit S, Singh UM, Abbai R, Ram T, Singh VK, Paul A, Virk PS, Kumar A (2019) Identification of genomic region(s) responsible for high iron and zinc content in rice. *Sci Rep* 9:8136. https://doi.org/10.1038/s41598-019-43888-y

Dubouzet JG, Sakuma Y, Ito Y, Kasuga M, Dubouzet EG, Miura S, Seki M, Shinozaki K, Yamaguchi-Shinozaki K (2003) OsDREB genes in rice, Oryza sativa L., encode transcription activators that function in drought-, high-salt-and cold-responsive gene expression. *Plant J* 33(4):751–763. https://doi.org/10.1046/j.1365-313X.2003.01661.x

Fan C, Xing Y, Mao H, Lu T, Han B, Xu C, Li X, Zhang Q (2006) *GS3*, a major QTL for grain length and weight and minor QTL for grain width and thickness in rice, encodes a putative transmembrane protein. *Theor Appl Genet* 112(6):1164–1171. https://doi.org/10.1007/s00122-006-0218-1

Fan XY, Guo M, Li RD, Yang YH, Liu M, Zhu Q, Tang SZ, Gu MH, Xu RG, Yan CJ (2017) Allelic variations in the soluble starch synthase II gene family result in changes of grain quality and starch properties in rice (*Oryza sativa* L.). *J Agric Sci* 155(1):129–140. https://doi.org/10.1017/S0021859615001331

Fowler S, Thomashow MF (2002) Arabidopsis transcriptome profiling indicates that multiple regulatory pathways are activated during cold acclimation in addition to the CBF cold response pathway. *Plant Cell* 14(8):1675–1690. https://doi.org/10.1105/tpc.003483

Fukao T, Harris T, Bailey-Serres J (2009) Evolutionary analysis of the Sub1 gene cluster that confers submergence tolerance to domesticated rice. *Ann Bot* 103(2):143–150. https://doi.org/10.1093/aob/mcn172

Fukuoka S, Yamamoto SI, Mizobuchi R, Yamanouchi U, Ono K, Kitazawa N, Yasuda N, Fujita Y, Thi Thanh Nguyen T, Koizumi S, Sugimoto K (2014) Multiple functional polymorphisms in a single disease resistance gene in rice enhance durable resistance to blast. *Sci Rep* 4(1):1–7. https://doi.org/10.1038/srep04550

Furukawa T, Maekawa M, Oki T, Suda I, Iida S, Shimada H, Takamure I, Kadowaki K-i (2007) The *Rc* and *Rd* genes are involved in proanthocyanidin synthesis in rice pericarp. *Plant J* 49:91–102. https://doi.org/10.1111/j.1365-313X.2006.02958.x

Ganeshan P, Jain A, Parmar B, Rao AR, Sreenu K, Mishra P, Mesapogu S, Subrahmanyam D, Ram T, Sarla N, Rai V (2016) Identifcation of salt tolerant rice lines among interspecifc BILs developed by crossing *Oryza sativa* × *O. rufpogon* and *O. sativa* × *O. nivara*. *Aust J Crop Sci* 10(2):220–228.

Ganie SA, Molla KA, Henry RJ, Bhat KV, Mondal TK (2019) Advances in understanding salt tolerance in rice *Theor Appl Genet* 132(4):851–870. https://doi.org/10.1007/s00122-019-03301-8

Gao Z, Zeng D, Cui X, Zhou Y, Yan M, Huang D, Li J, Qian Q (2003) Map-based cloning of the ALK gene which controls the gelatinization temperature of rice. *Sci China* 46:661–668. https://doi.org/10.1360/03yc0099

Gao ZY, Zeng DL, Cheng FM, Tian ZX, Guo LB, Su Y, Yan MX, Jiang H, Dong GJ, Huang YC, Han B, Li JY Qian Q (2011) *ALK,* the key gene for gelatinization temperature, is a modifier gene for gel consistency in rice. *J Integr Plant Biol* 53(9): 756–765. https://doi.org/10.1111/j.1744-7909.2011.01065.x

Gobu R, Shiv A, Anilkumar C, Basavaraj PS, Harish D, Adhikari S, Ramtekey V, Hudedamani U, Mulpuri S (2020) Accelerated Crop Breeding Towards Development of Climate Resilient Varieties. In: Srinivasarao Ch., Srinivas T, Rao RVS, Srinivasarao N, Vinayagam SS, Krishnan P (eds) *Climate Change and Indian Agriculture: Challenges and Adaptation Strategies*, ICAR-National Academy of Agricultural Research Management, Hyderabad, Telangana, India, pp 59–79.

Gong J, Miao J, Zhao Y, Zhao Q, Feng Q, Zhan Q, Cheng B, Xia J, Huang X, Yang S, Han B (2017) Dissecting the genetic basis of grain shape and chalkiness traits in hybrid rice using multiple collaborative populations. *Mol Plant* 10(10):1353–1356. https://doi.org/10.1016/j.molp.2017.07.014

Gouda G, Gupta MK, Donde R, Kumar J, Vadde R, Mohapatra T, Behera L (2019) Computational approach towards understanding structural and functional role of cytokinin oxidase/dehydrogenase 2 (CKX2) in enhancing grain yield in rice plant. *J Biomol Struct Dyn* 38(4):1158–1167. https://doi.org/10.1080/07391102.2019.1597771

Gu K, Yang B, Tian D, Wu L, Wang D, Sreekala C, Yang F, Chu Z, Wang GL, White FF, Yin Z (2005) R gene expression induced by a type-III effector triggers disease resistance in rice. *Nature* 435(7045):1122–1125. https://doi.org/10.1038/nature03630

Guo T, Chen K, Dong N-Q, Shi C-L, Ye W-W, Gao J-P, Shan J-X, Lin H-X (2018) *GRAIN SIZE AND NUMBER1* negatively regulates the OsMKKK10-OsMKK4- OsMPK6 cascade to coordinate the trade-off between grain number per panicle and grain size in rice. *Plant Cell* 30(4):871. https://doi.org/10.1105/tpc.17.00959

Haritha G, Vishnukiran T, Yugandhar P, Sarla N, Subrahmanyam D (2017) Introgressions from *Oryza rufpogon* increase photosynthetic efficiency of KMR3 rice lines. *Rice Sci* 24(2):85–96. https://doi.org/10.1016/j.rsci.2016.07.006

Hibino H (1990) Resistances in rice to tungro-associated viruses. *Plant Dis* 74:923.

Horie T, Sugawara M, Okada T, Taira K, Kaothien-Nakayama P, Katsuhara M, Shinmyo A, Nakayama H (2011) Rice sodium-insensitive potassium transporter, OsHAK5, confers increased salt tolerance in tobacco BY2 cells. *J Biosci Bioeng* 111(3):346–356. https://doi.org/10.1016/j.jbiosc.2010.10.014

Hu H, Dai M, Yao J, Xiao B, Li X, Zhang Q, Xiong L (2006) Overexpressing a NAM, ATAF, and CUC (NAC) transcription factor enhances drought resistance and salt tolerance in rice. *Proc Natl Acad Sci USA* 103(35):12987–12992. https://doi.org/10.1073/pnas.0604882103

Imai I, Kimball JA, Conway B, Yeater KM, McCouch SR, McClung A (2013) Validation of yield-enhancing quantitative trait loci from a low-yielding wild ancestor of rice. *Mol Breed* 32:101–120. https://doi.org/10.1007/s11032-013-9855-7

Izawa Y, Takayanagi Y, Inaba N, Abe Y, Minami M, Fujisawa Y, Kato H, Ohki S, Kitano H, Iwasaki Y (2010) Function and expression pattern of the α subunit of the heterotrimeric G protein in rice. *Plant Cell Physiol* 51(2):271–281. https://doi.org/10.1093/pcp/pcp186

Jang IC, Oh SJ, Seo JS, Choi WB, Song SI, Kim CH, Kim YS, SeoHS, DoChoi Y, Nahm BH, Kim JK (2003) Expression of abifunctional fusion of the Escherichia coli genes for trehalose-6-phosphatesynthase and trehalose-6-phosphate phosphatase intransgenic rice plants increases trehalose accumulation andabiotic stress tolerance without stunting growth. *Plant Physiol* 131:516–524. https://doi.org/10.1104/pp.007237

Jang S, Han J-H, Lee YK, Shin N-H, Kang YJ, Kim C-K, Chin JH (2020) Mapping and validation of QTLs for the amino acid and total protein content in brown rice. *Front Genet* 11:240. https://doi.org/10.3389/fgene.2020.00240

Kaladhar K, Swamy BPM, Babu AP, Reddy CS, Sarla N (2008) Mapping quantitative trait loci for yield traits in BC_2F_2 population derived from Swarna x *O. nivara* cross. *Rice Genet Newsl* 24:34–36.

Katara JL, Parameswaran C, Devanna BN, Verma RL, Anilkumar C, Patra BC, Samantaray S (2021) Genomics assisted breeding: The need and current perspective for rice improvement in India. *ORYZA Int J Rice* 58(1): 61–68. https://doi.org/10.35709/ory.2021.58.spl.1

Kato T, Segami S, Toriyama M, Kono I, Ando T, Yano M, Kitano H, Miura K, Iwasaki Y (2011) Detection of QTLs for grain length from large grain rice (*Oryza sativa* L.). *Breed Sci* 61(3):269–74. https://doi.org/10.1270/jsbbs.61.269

Katral A, Biradar H, Harijan Y, Aruna YR, Hadimani J and Hittalmani S (2022) Genetic analysis and traits association study in marker-assisted multi-drought-traits pyramided genotypes under RMS in rice. *Euphytica* 218(3):1–13. https://doi.org/10.1007/s10681-022-02974-5

Khahani B, Tavakol E, Shariati V, Rossini L (2021) Meta-QTL and ortho-MQTL analyses identified genomic regions controlling rice yield, yield-related traits and root architecture under water deficit conditions. *Sci Rep* 11(1):1–18. https://doi.org/10.1038/s41598-021-86259-2

Kim SI, Andaya CB, Newman JW, Goyal SS, Tai TH (2008) Isolation and characterization of a low phytic acid rice mutant reveals a mutation in the rice orthologue of maize MIK. *Theor Appl Genet* 117(8): 1291–1301. https://doi.org/10.1007/s00122-008-0863-7

Krasensky J, Jonak C (2012) Drought, salt, and temperature stress-induced metabolic rearrangements and regulatory networks. *J Exp Bot* 63(4):1593–1608. https://doi.org/10.1093/jxb/err460

Kumar A, Nayak S, Ngangkham U, Sah RP, Lal MK, Tp A, Behera S, Swain P, Behera L, Sharma S (2021) A single nucleotide substitution in the SPDT transporter gene reduced phytic acid and increased mineral bioavailability from rice grain (*Oryza sativa* L.). *J Food Biochem* 45(7):e13822. https://doi.org/10.1111/jfbc.13822

Kumar A, Sandhu N, Dixit S, Yadav S, Swamy BP, Shamsudin NA (2018) Marker-assisted selection strategy to pyramid two or more QTLs for quantitative trait-grain yield under drought. *Rice* 11(1):35. https://doi.org/10.1186/s12284-018-0227-0

Larson SR, Rutger JN, Young KA, Raboy V (2000) Isolation and genetic mapping of a non-lethal rice (*Oryza sativa* L.) *low phytic acid 1* mutation. *Crop Sci* 40(5):1397–1405. https://doi.org/10.2135/cropsci2000.4051397x

Lata C, Bhutty S, Bahadur RP, Majee M, Prasad M (2011) Association of an SNP in a novel DREB2-like gene SiDREB2 with stress tolerance in foxtail millet [*Setaria italica* (L.)]. *J Exp Bot* 62(10):3387–3401. https://doi.org/10.1093/jxb/err016

Latha R, Rubia L, Bennett J, Swaminathan MS (2004) Allele mining for stress tolerance genes in Oryza species and related germplasm. *Mol Biotechnol* 27(2):101–108. https://doi.org/10.1385/MB:27:2:101

Lee JH, Muhsin M, Atienza GA, Kwak DY, Kim SM, De Leon TB, Angeles ER, Coloquio E, Kondoh H, Satoh K, Cabunagan RC (2010) Single nucleotide polymorphisms in a gene for translation initiation factor (eIF4G) of rice (*Oryza sativa*) associated with resistance to Rice tungro spherical virus. *Mol Plant-Microbe Interact* 23(1):29–38. https://doi.org/10.1094/MPMI-23-1-0029

Leelagud P, Kongsila S, Vejchasarn P, Darwell K, Phansenee Y, Suthanthangjai A, Uparang C, Kawichai R, Yajai P, Boonsa-Nga K, Chamarerk V (2020) Genetic diversity of Asian rice gall midge based on mtCOI gene sequences and identification of a novel resistance locus *gm12* in rice cultivar MN62M. *Mol Biol Rep* 47(6):4273–4283. https://doi.org/10.1007/s11033-020-05546-9

Leung H, Raghavan C, Zhou B, Oliva R, Choi IR, Lacorte V, Jubay ML, Cruz CV, Gregorio G, Singh RK, Ulat VJ (2015) Allele mining and enhanced genetic recombination for rice breeding. *Rice* 8(1):1–11. https://doi.org/10.1186/s12284-015-0069-y

Li QF, Huang LC, Chu R, Li J, Jiang MY, Zhang CQ, Fan XL, Yu HX, Gu MH, Liu QQ (2018) Down-regulation of *SSSII-2* gene expression results in novel low-amylose rice with soft, transparent grains. *J Agric Food Chem* 66(37):9750–9760. https://doi.org/10.1021/acs.jafc.8b02913

Li WX, Oono Y, Zhu J, He XJ, Wu JM, et al. (2008) The Arabidopsis NFYA5 transcription factor is regulated transcriptionally and post-transcriptionally to promote drought resistance. *Plant Cell* 20(8):2238–2251. https://doi.org/10.1105/tpc.108.059444

Li X, Guo Z, Lv Y, Cen X, Ding X, Wu H, Li X, Huang J, Xiong L (2017) Genetic control of the root system in rice under normal and drought stress conditions by genome-wide association study. *PLoS Genet* 13(7):e1006889 https://doi.org/10.1371/journal.pgen.1006889

Li Y, Fan C, Xing Y, Jiang Y, Luo L, Sun L, Shao D, Xu C, Li X, Xiao J, He Y (2011) Natural variation in GS5 plays an important role in regulating grain size and yield in rice. *Nat Genet* 43(12):1266–1269. https:// doi.org/10.1038/ng.977

Liu GU, Jia Y, Correa-Victoria FJ, Prado GA, Yeater KM, McClung A, Correll JC (2009) Mapping quantitative trait loci responsible for resistance to sheath blight in rice. *Phytopathology* 99(9):1078–1084.

Liu QL, Xu XH, Ren XL, Fu HW, Wu DX, Shu QY (2007) Generation and characterization of low phytic acid germplasm in rice (*Oryza sativa* L.). *Theor Appl Genet* 114(5):803–814. https://doi.org/10.1007/s00 122-006-0478-9

Liu W, Xu ZH, Luo D, Xue HW (2003) Roles of OsCKI1, a ricecasein kinase i, in root development and plan hormonesensitivity. *Plant J* 36:189–202. https://doi.org/10.1046/j.1365-313x.2003.01866.x

Liu Y, Wu H, Chen H, Liu Y, He J, Kang H, Sun Z, Pan G, Wang Q, Hu J, Zhou F (2015) A gene cluster encoding lectin receptor kinases confers broad-spectrum and durable insect resistance in rice. *Nat Biotechnol* 33(3):301–305. https://doi.org/10.1038/nbt.3069

Lorenzo L, Merchan F, Laporte P, Thompson R, Clarke J, Sousa C, Crespi M (2009) A novel plant leucine-rich repeat receptor kinase regulates the response of *Medicago truncatula* roots to salt stress. *Plant Cell* 21(2):668–680. https://doi.org/10.1105/tpc.108.059576

Mao H, Sun S, Yao J, Wang C, Yu S, Xu C, Li X, Zhang Q (2010) Linking differential domain functions of the GS3 protein to natural variation of grain size in rice. *Proc Natl Acad Sci USA* 107(45):19579–1984 https://doi.org/10.1073/pnas.1014419107

Marri PR, Sarla N, Vemireddy LR, Siddiq EA (2005) Identifcation of yield enhancing quantitative trait loc (QTLs) from *Oryza rufpogon* of Indian origin. *BMC Genet* 6:33. https://doi.org/10.1186/1471-2156-6-3

Martre P, Morillon R, Barrieu F, North GB, Nobel PS, Chrispeels MJ (2002) Plasma membrane aquaporins play a significant role during recovery from water deficit. *Plant Physiol* 130:2101–2110. https://doi.org/ 10.1104/pp.009019

Meng L, Guo L, Ponce K, Zhao X, Ye G (2016) Characterization of three indica rice multiparent advanced generation intercross (MAGIC) populations for quantitative trait loci identification. *Plant Genome* 9(2):1–14. https://doi.org/10.3835/plantgenome2015.10.0109

Molla KA, Karmakar S, Molla J, Bajaj P, Varshney RK, Datta SK, Datta K (2020) Understanding sheath bligh resistance in rice: the road behind and the road ahead. *Plant Biotechnol J* 18(4):895–915. https://doi.org/ 10.1111/pbi.13312

Muhammed Azharudheen TP, Molla KA, Anilkumar C, Sah RP (2022) Advanced Technologies for Climate-Smart Breeding. In: Bhattacharyya P, Chakraborty K, Molla KA, Poonam A, Bhaduri D, Sah RP, Paul S Hanjagi PS, Basana-Gowda G, Swain P (eds) *Climate Resilient Technologies for Rice based Production Systems in Eastern India*. ICAR-National Rice Research Institute, Cuttack, Odisha, India, pp 408.

Negrao S, Almadanim C, Pires I, McNally KL, Oliveira MM (2011) Use of EcoTILLING to identify natural allelic variants of rice candidate genes involved in salinity tolerance. *Plant Genet Resour* 9(2):300–304 https://doi.org/10.1017/S1479262111000566

Negrão S, Cecília Almadanim M, Pires IS, Abreu IA, Maroco J, Courtois B, Gregorio GB, McNally KL Margarida Oliveira M (2013) New allelic variants found in key rice salt-tolerance genes: an association study. *Plant Biotechnol J* 11(1):87–100. https://doi.org/10.1111/pbi.12010

Niu Y, Chen T, Wang C, Chen K, Shen C, Chen H, Zhu S, Wu Z, Zheng T, Zhang F, Xu J. (2021) Identification and allele mining of new candidate genes underlying rice grain weight and grain shape by genome-wide association study. *BMC Genom* 22(1):1–4. https://doi.org/10.1186/s12864-021-07901-x

Ogawa D, Yamamoto E, Ohtani T, Kanno N, Tsunematsu H, Nonoue Y, Yano M, Yamamoto T, Yonemaru J (2018) Haplotype-based allele mining in the Japan-MAGIC rice population. *Sci Rep* 8(1):1–11. https:// doi.org/10.1038/s41598-018-22657-3

Oh J-M, Balkunde S, Yang P, Yoon D-B, Ahn S-N (2011) Fine mapping of grain weight QTL, *tgw11* using near isogenic lines from a cross between *Oryza sativa* and *O. grandiglumis*. *Genes Genom* 33(3):259–265 https://doi.org/10.1007/s13258-011-0038-9

Pachauri V, Mishra V, Mishra P, Singh AK, Singh S, Singh R, Singh NK (2014) Identification of candidate genes for rice grain aroma by combining QTL mapping and transcriptome profiling approaches. *Cereal Res Commun* 42(3):376–388. https://doi.org/10.1556/CRC.42.2014.3.2

Pang Y, Chen K, Wang X, Wang W, Xu J, Ali J, Li Z (2017) Simultaneous improvement and genetic dissection of salt tolerance of rice (*Oryza sativa* L.) by designed QTL pyramiding. *Front Plant Sci* 8:1275. https:// doi.org/10.3389/fpls.2017.01275

Park HC, Kim ML, Kang YH, Jeon JM, Yoo JH, Kim MC, Park CY, Jeong JC, Moon BC, Lee JH, Yoon HW (2004) Pathogen-and NaCl-induced expression of the SCaM-4 promoter is mediated in part by a GT-1 box that interacts with a GT-1-like transcription factor. *Plant Physiol* 135(4):2150–2161. https://doi.org/10.1104/pp.104.041442

Pascual L, Albert E, Sauvage C, Duangjit J, Bouchet JP, Bitton F, Desplat N, Brunel D, Le Paslier MC, Ranc N, Bruguier L (2016) Dissecting quantitative trait variation in the resequencing era: complementarity of bi-parental, multi-parental and association panels. *Plant Sci* 242:120–130. https://doi.org/10.1016/j.plantsci.2015.06.017

Pathak MD, Cheng CH, Fortuno ME (1969) Resistance to Nephotettix impicticeps and *Nilaparvata lugens* in varieties of rice. *Nature* 223(5205):502–504. https://doi.org/10.1038/223502a0

Patra BC, Anilkumar C, Chakraborti M (2020) Rice breeding in India: A journey from phenotype based pure-line selection to genomics assisted breeding. *Agril Res J* 57(6): 816–825.

Platten JD, Cotsaftis O, Berthomieu P, Bohnert H, Davenport RJ, Fairbairn DJ, Horie T, Leigh RA, Lin HX, Luan S, Mäser P (2006) Nomenclature for HKT transporters, key determinants of plant salinity tolerance. *Trends Plant Sci* 11(8):372–374. https://doi.org/10.1016/j.tplants.2006.06.001

Platten JD, Egdane JA, Ismail AM (2013) Salinity tolerance, Na+ exclusion and allele mining of HKT1; 5 in *Oryza sativa* and *O. glaberrima*: many sources, many genes, one mechanism?. *BMC Plant Biol* 13(1):32 https://doi.org/10.1186/1471-2229-13-32

Pradhan SK, Pandit E, Pawar S, Barsha B, Chatopadhyay K, Singh S, Dash P, Reddy JN (2019) Association mapping reveals multiple QTLs for grain protein content in rice useful for biofortification. *Mol Genet Genom* 294(4):963–983. https://doi.org/10.1007/s00438-019-01556-w

Prakash C, Mithra SVA, Singh PK, Mohapatra T, Singh NK (2016) Unraveling the molecular basis of oxidative stress management in a drought tolerant rice genotype nagina 22. *BMC Genom* 17:774–787. https://doi.org/10.1186/s12864-016-3131-2

Prakash NR, Lokeshkumar BM, Rathor S, Warraich AS, Yadav S, Vinaykumar NM, Dushynthkumar BM, Krishnamurthy SL, Sharma PC (2022) Meta-analysis and validation of genomic loci governing seedling and reproductive stage salinity tolerance in rice. *Physiol Planta* 174(1):e13629. https://doi.org/10.1111/ppl.13629

Qi P, Lin YS, Song XJ, Shen JB, Huang W, Shan JX, Zhu MZ, Jiang L, Gao JP, Lin HX (2012) The novel quantitative trait locus *GL3.1* controls rice grain size and yield by regulating Cyclin-T1; 3. *Cell Res* 22(12):1666–1680. https://doi.org/10.1038/cr.2012.151

Qin Y, Kim SM, Sohn JK (2009) QTL analysis of protein content in double-haploid lines of rice. *Kor J Crop Sci* 54(2):165–171.

Qiu X, Gong R, Tan Y, Yu S (2012) Mapping and characterization of the major quantitative trait locus qSS7 associated with increased length and decreased width of rice seeds. *Theor Appl Genet* 125(8):1717–1726. https://doi.org/10.1007/s00122-012-1948-x

Rabbani MA, Maruyama K, Abe H, Khan MA, Katsura K, Ito Y, Yoshiwara K, Seki M, Shinozaki K, Yamaguchi-Shinozaki K (2003) Monitoring expression profiles of rice genes under cold, drought, and high-salinity stresses and abscisic acid application using cDNA microarray and RNA gel-blot analyses. *Plant Physiol* 133(4):1755–1767. https://doi.org/10.1104/pp.103.025742

Richa K, Tiwari IM, Kumari M, Devanna BN, Sonah H, Kumari A, Nagar R, Sharma V, Botella JR, Sharma TR (2016) Functional characterization of novel chitinase genes present in the sheath blight resistance QTL: qSBR11-1 in rice line tetep. *Front Plant Sci* 7:244. https://doi.org/10.3389/fpls.2016.00244

Sahebi M, Hanafi MM, Rafii MY, Mahmud TM, Azizi P, Osman M, Abiri R, Taheri S, Kalhori N, Shabanimofrad M, Miah G (2018) Improvement of drought tolerance in rice (*Oryza sativa* L.): genetics, genomic tools, and the WRKY gene family. *BioMed Res Int* 2018:3158474. https://doi.org/10.1155/2018/3158474

Sahu N, Gowda BG, Rath LK, Anilkumar C, Sah RP, Mandal L, Pandi GP, Patil NB, Adak T, Mahendiran A, Rath PC (2022) Marker-trait association analysis for gall midge (*Orseolia oryzae*) resistance in a diverse rice population. *Ann Appl Biol* 182(3):361–370.

Saijo Y, Kinoshita N, Ishiyama K, Hata S, Kyozuka J, Hayakawa T, Nakamura T, Shimamoto K, YamayaT IK (2001) A Ca2?-dependent protein kinase that endows rice plants with cold-andsaltstresstolerance functions in vascular bundles. *Plant Cell Physiol* 42:1228–1233. https://doi.org/10.1093/pcp/pce158

Sandhu N, Pruthi G, Prakash Raigar O, Singh MP, Phagna K, Kumar A, Sethi M, Singh J, Ade PA, Saini DK (2021) Meta-QTL analysis in rice and cross-genome talk of the genomic regions controlling nitrogen use efficiency in cereal crops revealing phylogenetic relationship. *Front Plant Sci* 2021:2609. https://doi.org/10.3389/fgene.2021.807210

Seki M, Narusaka M, Ishida J, Nanjo T, Fujita M, Oono Y, Kamiya A, Nakajima M, Enju A, Sakurai T, Satou M (2002) Monitoring the expression profiles of 7000 Arabidopsis genes under drought, cold and high-salinity stresses using a full-length cDNA microarray. *Plant J* 31(3):279–292. https://doi.org/10.1046/j.1365-313X.2002.01359.x

Septiningsih EM, Pamplona AM, Sanchez DL, Neeraja CN, Vergara GV, Heuer S, Ismail AM, Mackill DJ (2009) Development of submergence-tolerant rice cultivars: the Sub1 locus and beyond. *Ann Bot* 103(2):151–160. https://doi.org/10.1093/aob/mcn206

Silva J, Scheffler B, Sanabria Y, De Guzman C, Galam D, Farmer A, Woodward J, May G, Oard J (2012) Identification of candidate genes in rice for resistance to sheath blight disease by whole genome sequencing. *Theor Appl Genet* 124(1):63–74. https://doi.org/10.1007/s00122-011-1687-4

Singh A (2015) *Sequence variation in the Sub1 gene and its relation to flooding tolerance in the land races of rice (Oryza sativa L.)*. PhD thesis, Ch. Charan Singh University, Meerut, India.

Singh A, Singh Y, Mahato AK, Jayaswal PK, Singh S, Singh R, Yadav N, Singh AK, Singh PK, Singh R, Kumar R (2020) Allelic sequence variation in the Sub1A, Sub1B and Sub1C genes among diverse rice cultivars and its association with submergence tolerance. *Sci Rep* 10(1):8621 https://doi.org/10.1038/s41598-020-65588-8

Singh BK, Ramkumar MK, Dalal M, Singh A, Solanke AU, Singh NK, Sevanthi AM (2021) Allele mining for a drought responsive gene DRO1 determining root growth angle in donors of drought tolerance in rice (*Oryza sativa* L.). *Physiol Mol Biol Plants* 27(3):523–534. https://doi.org/10.1007/s12298-021-00950-2

Singh BP, Jayaswal PK, Singh B, Singh PK, Kumar V, Mishra S, Singh N, Panda K, Singh NK (2015) Natural allelic diversity in OsDREB1F gene in the Indian wild rice germplasm led to ascertain its association with drought tolerance. *Plant Cell Rep* 34(6):993–1004. https://doi.org/10.1007/s00299-015-1760-6

Singh K, Foley RC, Onate-Sanchez L (2002) Transcription factors in plant defense and stress responses. *Curr Opin Plant Biol* 5(5):430–436. https://doi.org/10.1016/S1369-5266(02)00289-3

Singh N, Dang T, Vergara GV, Pandey DM, Sanchez D, Neeraja CN, Septiningsih EM, Mendioro M, Tecson-Mendoza EM, Ismail AM, Mackill DJ (2010) Molecular marker survey and expression analyses of the rice submergence-tolerance gene SUB1A. *Theor Appl Genet* 121(8):1441–1453. https://doi.org/10.1007/s00122-010-1400-z

Singh PK, Mahato AK, Jain P, Rathour R, Sharma V, Sharma TR (2019) Comparative genomics reveals the high copy number variation of a retro transposon in different Magnaporthe isolates. *Front Microbiol* 10:966. https://doi.org/10.3389/fmicb.2019.00966

Singh R, Singh AK, Sharma TR, Singh A, Singh NK (2012) Fine mapping of grain length QTLs on chromosomes 1 and 7 in Basmati rice (*Oryza sativa* L.). *J Plant Biochem Biotechnol* 21(2):157–166. https://doi.org/10.1007/s13562-011-0080-3

Singhal P, Jan AT, Azam M, Haq QM (2016) Plant abiotic stress: a prospective strategy of exploiting promoters as alternative to overcome the escalating burden. *Front Life Sci* 9(1):52–63. https://doi.org/10.1080/21553769.2015.1077478

Song X, Huang W, Shi M, Zhu M, Lin H (2007) A QTL for rice grain width and weight encodes a previously unknown RING-type E3 ubiquitin ligase. *Nat Genet* 39:623–630. https://doi.org/10.1038/ng2014

Sudhakar T, Panigrahy M, Lakshmanaik M, Babu AP, Reddy CS, Anuradha K, Swamy BPM, Sarla N (2012) Variation and correlation of phenotypic traits contributing to high yield in KMR3: *Oryza rufpogon* introgression lines. *Int J Plant Breed Genet* 6:69–82.

Svensson JT, Crosatti C, Campoli C, Bassi R, Stanca AM, Close TJ, Cattivelli L (2006) Transcriptome analysis of cold acclimation in barley Albina and Xantha mutants. *Plant Physiol* 141(1):257–270. https://doi.org/10.1104/pp.105.072645

Swamy BPM, Kaladhar K, Ramesha MS, Viraktamath BC, Sarla N (2011) Molecular mapping of QTLs for yield and yield-related traits in *Oryza sativa* cv. Swarna × *O. nivara* (IRGC81848) backcross population. *Rice Sci* 18(3):178–186. https://doi.org/10.1016/S1672-6308(11)60025-5

Swamy BPM, Kaladhar K, Reddy GA, Viraktamath BC, Sarla N (2014) Mapping and introgression of QTL for yield and related traits in two backcross populations derived from *Oryza sativa* cv. Swarna and two accessions of O. nivara. *J Genet* 93(3):643–654. https://doi.org/10.1007/s12041-014-0420-x

Swamy BPM, Kaladhar K, Shobharani N, Prasad GSV, Viraktamath BC, Reddy GA, Sarla N (2012) QTL analysis for grain quality traits in 2 BC_2F_2 populations derived from crosses between *Oryza sativa* cv. Swarna and 2 accessions of *O. nivara*. *J Hered* 103(3):442–452. https://doi.org/10.1093/jhered/esr145

Swamy BPM, Sarla N (2008) Yield-enhancing quantitative trait loci (QTLs) from wild species. *Biotechnol Adv* 26(1):106–120. https://doi.org/10.1016/j.biotechadv.2007.09.005

Sweeney MT, Thomson MJ, Pfeil BE, McCouch S (2006) Caught redhanded: Rc encodes a basic helix-loop-helix protein conditioning red pericarp in rice. *Plant Cell* 18(2):283–294. https://doi.org/10.1105/tpc.105.038430

Takeda S, Matsuoka M (2008) Genetic approaches to crop improvement: responding to environmental and population changes. *Nat Rev Genet* 9(6):444–457. https://doi.org/10.1038/nrg2342

Talukdar PR, Rathi S, Pathak K, Chetia SK, Sarma RN (2017) Population structure and marker-trait association in indigenous aromatic rice. *Rice Sci* 24(3):145–154. https://doi.org/10.1016/j.rsci.2016.08.009

Tan Y, Xing Y, Li J, Yu S, Xu C, Zhang Q (2000) Genetic bases of appearance quality of rice grains in Shanyou 63, an elite rice hybrid. *Theor Appl Genet* 101(5):823–829. https://doi.org/10.1007/s001220051549

Uga Y, Kitomi Y, Yamamoto E, Kanno N, Kawai S, Mizubayashi T, Fukuoka S (2015) A QTL for root growth angle on ricechromosome 7 is involved in the genetic pathway of DEEPERROOTING 1. *Rice* 8:8. https://doi.org/10.1186/s12284-015-0044-7

Uga Y, Okuno K, Yano M (2011) Dro1, a major QTL involved indeep rooting of rice under upland field conditions. *J Exp Bot* 62(8):2485–2494. https://doi.org/10.1093/jxb/erq429

Uga Y, Sugimoto K, Ogawa S, Rane S, Ishitani M, Hara N, Kitomi Y, Inukai Y, Ono K, Kanno N, Inoue H, Takehisa H, Motoyama R, Nagamura Y, Wu J, Matsumoto T, Takai T, Okuno K, Yano M (2013) Control of root system architecture by DEEPERROOTING 1 increases rice yield under drought conditions. *Nat Genet* 45(9):1097–1102. https://doi.org/10.1038/ng.2725

Utami DW, Lestari P, Koerniati S (2013) A relative expression of *Xa7* gene controlling bacterial leaf blight resistance in indonesian local rice population (*Oryza sativa* L.). *J Crop Sci Biotechnol* 16(1):1–7. https://doi.org/10.1007/s12892-012-0091-1

Uttam GA, Suman K, Jaldhani V, Babu PM, Rao DS, Sundaram RM, Neeraja CN (2022) Identification of genomic regions associated with high grain Zn content in polished rice using genotyping-by-sequencing (GBS). *Plants* 12(1):144. https://doi.org/10.3390/plants12010144

Varshney RK, Singh VK, Hickey JM, Xun X, Marshall DF, Wang J, Edwards D, Ribaut JM (2016) Analytical and decision support tools for genomics-assisted breeding. *Trends Plant Sci* 21(4):354–363. https://doi.org/10.1016/j.tplants.2015.10.018

Vasudevan K, Gruissem W, Bhullar NK (2015) Identification of novel alleles of the rice blast resistance gene Pi54. *Sci Rep* 5(1):1–12. https://doi.org/10.1038/srep15678

Vemireddy LR, Kadambari G, Reddy GE, Kola VS, Ramireddy E, Puram VR, Badri J, Eslavath SN, Bollineni SN, Naik BJ, Chintala S (2019) Uncovering of natural allelic variants of key yield contributing genes by targeted resequencing in rice (*Oryza sativa* L.). *Sci Rep* 9(1):1–3. https://doi.org/10.1038/s41598-019-44708-z

Verma RL, Katara JL, Anilkumar C, Devanna BN, Parameswaran C, Dash B, Samantaray S (2021) Advanced breeding strategies for rice improvement. In: Nayak AK, Samantaray S, Baig MJ, Tripathi R, Kumar U, Devanna BN, Maiti D (eds) *Rice Research: Recent Advances and Perspective*. ICAR-National Rice Research Institute, Cuttack, Odisha, India

Wang CX, Shu QY (2007) Fine mapping and candidate gene analysis of purple pericarp gene *Pb* in rice (*Oryza sativa* L.). *Chin Sci Bull* 52:3097–3104. https://doi.org/10.1007/s11434-007-0472-x

Wang E, Wang J, Zhu X, Hao W, Wang L, Li Q, Zhang L, He W, Lu B, Lin H, Ma H (2008) Control of rice grain-filling and yield by a gene with a potential signature of domestication. *Nat Genet* 40:1370–1374. https://doi.org/10.1038/ng.220

Wang K, Hasjim J, Wu AC, Li E, Henry RJ, Gilbert RG (2015) Roles of *GBSSI* and *SSIIa* in determining amylose fine structure. *Carbohyd Polym* 127:264–274. https://doi.org/10.1016/j.carbpol.2015.03.081

Wang Q, Liu Y, He J, Zheng X, Hu J, Liu Y, Dai H, Zhang Y, Wang B, Wu W, Gao H (2014) STV11 encodes a sulphotransferase and confers durable resistance to rice stripe virus. *Nat Commun* 5(1): 4768. https://doi.org/10.1038/ncomms5768.

Wang S, Wu K, Yuan Q, Liu X, Liu Z, Lin X, Zeng R, Zhu H, Dong G, Qian Q, Zhang G (2012) Control of grain size, shape and quality by *OsSPL16* in rice. *Nat Genet* 44(8):950–954. https://doi.org/10.1038/ng.2327

Wang ZY, Wu ZL, Xing YY, Zheng FG, Guo XL, Zhang WG, Hong MM (1990) Nucleotide sequence of rice waxy gene. *Nucleic Acids Res* 18(19):5898. https://doi.org/10.1093/nar/18.19.5898

Weng J, Gu S, Wan X, Gao H, Guo T, Su N, Lei C, Zhang X, Cheng Z, Guo X, Wang J (2008) Isolation and initial characterization of *GW5*, a major QTL associated with rice grain width and weight. *Cell Res* 18(12):1199–1209. https://doi.org/10.1038/cr.2008.307

Xiao J, Grandillo S, Ahn SN, McCouch SR, Tanksley SD, Li J, Yuan L (1996) Genes from wild rice improve yield. *Nature* 384:223–224. https://doi.org/10.1038/384223a0

Xu D, Duan X, Wang B, Hong B, Ho TH, Wu R (1996) Expression ofa late embryogenesis abundant (LEA) protein gene, HvA1, frombarley confers tolerance to drought and salinity in transgenicrice. *Plant Physiol* 110:249–257. https://doi.org/10.1104/pp.110.1.249

Xu K, Xu X, Fukao T, Canlas P, Maghirang-Rodriguez R, Heuer S, Ismail AM, Bailey-Serres J, Ronald PC, Mackill DJ (2006) Sub1A is an ethylene-response-factor-like gene that confers submergence tolerance to rice. *Nature* 442(7103):705–708. https://doi.org/10.1038/nature04920

Yang J, Wang J, Fan FJ, Zhu JY, Chen T, Wang CL, Zheng TQ, Zhang J, Zhong WG, Xu JL (2013) Development of AS-PCR marker based on a key mutation confirmed by resequencing of Wx-mp in Milky Princess and its application in japonica soft rice (*Oryza sativa* L.) breeding. *Plant Breed* 132(6):595–603. https://doi.org/10.1111/pbr.12088

Yang Y, Guo M, Sun S, Zou Y, Yin S, Liu Y, Tang S, Minghong G, Yang Z, Yan C (2019) Natural variation of *OsGluA2* is involved in grain protein content regulation in rice. *Nat Commun* 10(1):1–2. https://doi.org/10.1038/s41467-019-09919-y

Ying Z, Tao W, Bin Y, Fang L, Meijuan C, Qiong W, Ping H, Shuyan K, Wenxiu Q, Li L (2022) Improving rice blast resistance by mining broad-spectrum resistance genes at Pik locus. *Rice Sci* 29(2):133–142. https://doi.org/10.1016/j.rsci.2022.01.002

Yoshimura A, Ideta O, Iwana N (1997) Linkage map of phenotype and RFLP markers in rice. *Plant Mol Biol* 35:49–60. https://doi.org/10.1007/978-94-011-5794-0_5

Yu G, Olsen KM, Schaal BA (2011) Association between nonsynonymous mutations of *starch synthase IIa* and starch quality in rice (*Oryza sativa*). *New Phytol* 189(2):593–601. https://doi.org/10.1111/j.1469-8137.2010.03495.x

Yu YH, Li G, Fan YY, Zhang KQ, Min J, Zhu ZW, Zhuang JY (2009) Genetic relationship between grain yield and the contents of protein and fat in a recombinant inbred population of rice. *J Cereal Sci* 50(1):121–125. https://doi.org/10.1016/j.jcs.2009.03.008

Yue B, Xue WY, Xiong LZ, Yu XQ, Luo LJ, Cui KH, Jin DM, Xing YZ, Zhang QF (2006) Genetic basis of drought resistance at reproductive stage in rice: separation of drought tolerance from drought avoidance. *Genetics* 172(2):1213–1228. https://doi.org/10.1534/genetics.105.045062

Zhang C, Yang Y, Chen S, Liu X, Zhu J, Zhou L, Lu Y, Li Q, Fan X, Tang S, Gu M (2021) A rare Waxy allele coordinately improves rice eating and cooking quality and grain transparency. *J Integr Plant Biol* 63(5):889–901. https://doi.org/10.1111/jipb.13010

Zhang C, Zhu J, Chen S, Fan X, Li Q, Lu Y, Wang M, Yu H, Yi C, Tang S, Gu M (2019a) Wxlv, the ancestral allele of rice Waxy gene. *Mol Plant* 12(8):1157–1166. https://doi.org/10.1016/j.molp.2019.05.011

Zhang G, Cheng Z, Zhang X, Guo X, Su N, Jiang L, Mao L, Wan J (2011) Double repression of soluble starch synthase genes *SSIIa* and *SSIIIa* in rice (*Oryza sativa* L.) uncovers interactive effects on the physico-chemical properties of starch. *Genome* 54(6):448–459. https://doi.org/10.1139/g11-010

Zhang L, Zhao L, Zhang J, Cai X, Liu Q, Wei C (2019b) Relationships between transparency, amylose content, starch cavity, and moisture of brown rice kernels. *J Cereal Sci* 90:102854. https://doi.org/10.1016/j.jcs.2019.102854

Zhao D-S, Li Q-F, Zhang C-Q, Zhang C, Yang Q-Q, Pan L-X, Ren X-Y, Lu J, Gu M-H, Liu Q-Q (2018) *GS9* acts as a transcriptional activator to regulate rice grain shape and appearance quality. *Nat Commun* 9(1):1240. https://doi.org/10.1038/s41467-018-03616-y

Zhao Y, Huang J, Wang Z, Jing S, Wang Y, Ouyang Y, Cai B, Xin XF, Liu X, Zhang C, Pan Y (2016) Allelic diversity in an NLR gene BPH9 enables rice to combat planthopper variation. *Proc Natl Acad Sci USA* 113(45):12850–12855. https://doi.org/10.1073/pnas.1614862113

Zhong M, Wang L, Yuan J, Luo L, Xu C, He YQ (2011) Identification of QTL affecting protein and amino acid contents in rice. *Rice Sci* 18(3):187–195. https://doi.org/10.1016/S1672-6308(11)60026-7

Zhou C, Zhang Q, Chen Y, Huang J, Guo Q, Li Y, Wang W, Qiu Y, Guan W, Zhang J, Guo J (2021) Balancing selection and wild gene pool contribute to resistance in global rice germplasm against planthopper. *Integr Plant Biol* 63(10):1695–1711. https://doi.org/10.1111/jipb.13157

Zhou G, Xu D, Xu D, Zhang M (2013) Southern rice black-streaked dwarf virus: a white-backed planthopper-transmitted fijivirus threatening rice production in Asia. *Front Microbiol* 4:270. https://doi.org/10.3389/fmicb.2013.00270

Zhou L, Liu S, Wu W, Chen D, Zhan X, Zhu A, Zhang Y, Cheng S, Cao L, Lou X, Xu H (2016) Dissection of genetic architecture of rice plant height and heading date by multiple-strategy-based association studies. *Sci Rep* (6):29718. https://doi.org/10.1038/srep29718

Zhou Y, Lei F, Wang Q, He W, Yuan B, Yuan W (2020) Identification of novel alleles of the rice blast-resistance gene *Pi9* through sequence-based allele mining. *Rice* 13(1):1–15. https://doi.org/10.1186/s12284-020-00442-z

Zuo J, Li J (2014) Molecular genetic dissection of quantitative trait loci regulating rice grain size. *Annu Rev Genet* 48(1):99–118. https://doi.org/10.1146/annurev-genet-120213-092138

3 Allele Mining for Improvement of Plant Architecture, Grain Quality and Abiotic Stress Tolerance Traits in Wheat

*Gautam Saripalli, Seyedali Hosseinirad, Inderjit Singh Yadav, Anmol Kajla, Adam William Schoen, and Vijay Kumar Tiwari**
*Corresponding author: vktiwari@umd.edu

3.1 INTRODUCTION

Wheat is the world's most widely grown crop, after maize and rice and as per the statistics of 2020, the total production worldwide was 760 million tons with China, India, and Russia accounting for 41% of total wheat production (FAO, 2020: www.fao.org/faostat/en/#data/QCL). Importantly, wheat also provides 20% of the total protein and calories in human nutrition (Shiferaw et al. 2013), and dietary intake of essential micronutrients, including zinc, iron, manganese, magnesium, and vitamins B and E (Shewry and Hey 2015)

Among the different wheat species, bread wheat (*Triticum aestivum*, $2n=6x=42$; AABBDD) accounts for 95%, whereas the remaining include tetraploid durum wheat (*T. turgidum* L. *durum*, $2n=4x=28$; AABB) and other wheat types of smaller economic importance (Dubcovsky and Dvorak 2007). Global wheat production faces stiff challenges due to changing climatic conditions, biotic and abiotic stresses, and also due to reducing agricultural lands resulting from increasing human population. Sustainable wheat improvement is needed to cope with these challenges, and it requires integration of novel genes and alleles in the crop breeding programs worldwide.

Continuous efforts are being made to develop wheat varieties, which are tolerant to different biotic and abiotic stresses using different conventional and molecular approaches (Thudi et al. 2021). With the advent of next generation sequencing technology, a breakthrough was achieved in 2018 with the availability of reference genome sequence of wheat in the genetic background of cv. Chinese Spring (IWGSC 2018) increasing the power of the genomic toolbox associated with bread wheat. However, it is a known fact that there is a significant variation in genome of different genotypes of a particular crop species, like wheat, and this genomic variation may also be associated with functional trait(s) (Fernandez et al. 2022). Therefore, since the release of the reference genome, efforts have been made to create a pan-genome of wheat with the 10+ genome project in which 13 reference quality genomes of cultivars from across the globe have been released for public use (Walkowiak et al. 2020). Moreover, the 1000 exome capture project has been a powerful tool into the insight of wild-relative introgressions, as well as giving a glimpse of the genetic variation that is present in the wheat pan-genome (He et al. 2019). This compilation of multiple genome sequences into a pangenome instead of a single reference genome may facilitate the discovery of important alleles that are involved in controlling different agronomically important traits, and this allelic information can in turn be used to develop functional markers for trait discovery.

DOI: 10.1201/9781003385004-3

Recent advances in positional cloning of several agronomically important genes have allowed us to integrate key molecular markers in marker-assisted breeding programs. Some recent examples of positional cloning include (i) a gene *TaCol-B5* (encoding a constans like protein), which is involved in controlling the number of spikelet nodes in wheat (Zhang et al. 2022). (ii) A leaf rust-resistant gene *Lr42* transferred from *Aegilops tauschii* (Lin et al. 2022) and (iii) powdery mildew-resistant gene originating from wild emmer wheat; both encoded NLR-type resistant genes (Wu et al. 2022). Other important candidate genes, which have been cloned using map-based cloning approach, include *VRN1* locus, which contained two MADS box (*AP1* and *AGL1*; Agamous like gene from grasses) in the interval (Yan et al. 2004), *Gpc-B1* (grain protein content), which encoded for NAC transcription factor (Uauy et al. 2006), *Q* locus, which encoded for AP2 transcription factor influencing threshing character and spike phenotype (Faris et al. 2003) and *Ph1* locus, which is a multigene family, heterochromatin or both (Griffiths et al. 2006).

Allele mining is one of the promising approaches for genetic dissection of agronomically important traits in crop plants like wheat (Kumar et al. 2010). With the advent of next-generation sequencing technologies and availability of tremendous genomic resources in wheat, it has become possible to develop sequence-based functional markers, which are associated with traits of interest. This has also enabled us to capture the phenotypes at the genome level and thereby correlating the sequence variation with the phenotype (for details, see Hussain et al. 2022). This helps in developing allele-specific markers, which can in turn be used in breeding wheat for further improvement for different agronomically important traits, including biotic and abiotic stresses. In a review (Salgotra et al. 2020), development and use of functional markers for precision breeding has been described in important crop plants. This review highlights the single nucleotide polymorphism (SNP)-based markers as functional markers and their reliability as compared to random markers like simple sequence repeats (SSRs). This is due to the fact that SNPs are generally linked functionally to different phenotypic traits and hence better as compared to other markers (Salgotra et al. 2020).

The present article is focused on allele mining for important traits involved in abiotic stress tolerance, grain quality, and plant architecture in wheat. In wheat, to date, more than 10,000 quantitative trait loci (QTLs) have been reported for different agronomically important traits including for biotic and abiotic stress tolerance (Singh et al. 2021, 2022). Out of these QTLs, ~800 QTLs have been specifically identified for heat stress, which were also recently used for conducting metaQTL analysis in wheat (Kumar et al. 2021). Similarly, 2852 major QTLs for yield-related traits were also used for meta-analysis (Saini et al. 2022). Several of these QTLs may be possessing favorable alleles, which may be actively involved in controlling a particular trait. Similarly, a number of candidate genes (CGs) have also been predicted using different *in silico* approaches from the regions underlying the QTLs (using interval mapping) or linkage disequilibrium (LD)-based genome-wide association studies (GWAS) (for details of CGs, refer to review by Hussain et al. 2022). This information reveals a tremendous wealth of important genomic regions in wheat, which can be used for identification of important alleles and development of functional markers for important traits.

Candidate gene-based association mapping (CGAM) has also been used to identify the alleles in the candidate genes that are involved in controlling a specific trait. For example, in one of the studies involving CGAM, novel loci were identified for the candidate genes involved in grain weight in wheat (Sehgal et al. 2019). Using *in silico* approaches, some of the important genes like ADP glucose pyrophosphorylase (AGPase) and starch synthase (SS), which are involved in heat stress tolerance have been characterized (Batra et al. 2017; Mishra et al. 2017); however, the important alleles for these genes need to be explored and identified. Several such examples are described in detail with their potential application for future studies in wheat, which may aid in development of improved wheat cultivars.

Furthermore, in the later section of the chapter, modern mutagenesis-based approaches have also been described some of which (like Targeting Induced Local Lesions IN Genomes or TILLING; Schoen et al. 2021; Sakuma et al. 2019) have already been used for allele mining in wheat. These approaches mainly include mutagenesis-based approaches such as TILLING (McCallum et al. 2000), as well as other approaches involving mutagenesis-bulk segregant analysis (BSA) like Shoremap (Schneeberger et al. 2009), MutMap (Abe et al. 2012), MutMap+ (Fekih et al. 2013), and MutMap-Gap (Takagi et al. 2013a).

3.2 ALLELE MINING FOR PLANT ARCHITECTURE

3.2.1 Plant Height and Rht Genes

Plant height has been a major agronomic trait, which affects the wheat yield and relates to plant architecture and lodging resistance (Wang et al. 2017). Compared to tall plants, dwarf plants have several advantages, which help in improving harvest index and yield under different environmental conditions. Plants with short stems pose less competition for limited resources, meaning that a greater proportion of the plant's photosynthate is partitioned into spike and grain development (Youssefian et al. 1992). Therefore, the dwarfing genes (*Reduced height* or *Rht*) in wheat have played a significant role in yield improvement.

To date, 25 major Rht genes (*Rht1-Rht25*) [including gibberellic acid (GA) insensitive and GA responsive] are known in wheat, out of which the alleles for *Rht1* (*Rht-B1b* and *Rht-D1b*) are the most widely used dwarfing genes, which have resulted in increased yield during green revolution (Hedden 2003; Van De Velde et al. 2017, 2021). These two alleles influence the intensity of the stem length characteristic resulting in the improvement of harvest index and resistance against lodging (Jobson et al. 2019). Unfortunately, they also induce a reduction in stem length as well as the length of the coleoptile and roots, resulting in a loss in the efficiency of water utilization from the soil and a reduction in drought tolerance. Therefore, presence of these alleles in plants restricts their usefulness in arid climates where seeds must be planted deeply. *Rht18*, however, which has a similar effect on height reduction as *Rht-D1b* but has no effect on coleoptile length (Pearce 2021) can be used in diverse environmental conditions. Additionally, *Rht-B1c* mutation causes a greater reduction in plant stem length, resulting in a dwarf phenotype (Flintham and Gale 1982; Flintham et al. 1997). This allele is also reported to be responsible for reduced flag leaf photosynthesis rate in wheat under field conditions (Jobson et al. 2019)

Other than the two important alleles for *Rht1* described above, several more mutant alleles have been identified for *Rht1* (Thomas 2017) as well for other genes like *Rht8* (Asplund et al. 2012). *Rht8*, in fact, is the most recent addition in the list of positionally cloned dwarfing gene on 2D chromosome of wheat and is known to provide sensitivity against gibberellic acid (Xiong et al. 2022). Another important allele *Rht24b*, which was found in wild emmer wheat initially, is also known to reduce the plant height without a significant effect on the yield attributes. The above two genes (*Rht24* and *Rht18*) along with *Rht14* are linked to increased *TaGA2ox-A9* expression (Pearce 2021). This enzyme removes GA precursors from the stem, lowering bioactive GA levels and consequently cell elongation.

QTLs have also been identified, which modify the expression of *Rht* alleles. For instance, a modifier QTL was identified on the long arm of chromosome 2B, which showed a reduction of about 60% in plant height due to the expected presence of enhancers and suppressors modifying the expression of *Rht4c* (Agarwal et al. 2020). Other than the *Rht* genes, several other genes have also been reported, which possess favorable alleles for plant height. For instance, a gene associated with plant height, *TaIAA15*, was identified on chromosome 3B. This gene had favorable effects, which impact some of the plant architecture and yield components (Fu et al. 2022).

3.2.2 Spike Morphology

Studies on the heritability of inflorescence development could lead to the development of improved spike architectures with the goal of increasing grain production. A spike with a primary axis (the spike rachis) and lateral sessile spikelet that are linked to the rachis, as well as a terminal spikelet, make up the complete wheat inflorescence. Wheat yield is partially governed by spike architecture, which is constituted by spike number per unit area, seed/kernel weight, and seed number per spike (Hawkesford et al. 2013). Seed/kernel number per spike in turn is controlled by two more factors, which are spikelet number per spike and number of seeds/kernels per spikelet. Further, wheat spikelets have more than one grain, which, in contrast to other cereal crops, which only have one grain, makes them the most important yield component (Wolde et al. 2019).

A number of alleles have been identified that control the above underlying factors to increase the yield of the crop. (i) *WFZP (wheat frizzy panicle)* is one of the important alleles, which control the supernumerary spike (SS trait) in wheat thereby increasing the grain numbers (Dobrovolskaya et al. 2015). (ii) *WAPO-A1* (*wheat aberrant panicle organization-A1*) is another candidate gene for a 7AL locus controlling spikelet quantity per spike, which is a wheat homolog of the rice gene *APO* (*aberrant panicle organization*). This gene is responsible for more branches and spikelet per panicle (Kuzay et al. 2022). (iii) A photoperiod insensitive allele *Ppd-1* responsible for reducing flowering time is known to have association with *Rht* genes and is also responsible for reducing tiller and spikelet numbers (Okada et al. 2020). (iv) A recent study also identified an important gene encoding MADS-box gene *VRT2* whose high expression was found to increase the basal rudimentary spikelets in wheat (Backhaus et al. 2022). Some more examples are mentioned in Table 3.1.

3.2.3 Flowering Time

Flowering time is another important factor that affects the spike architecture as well. Winter wheats need to be exposed to low temperatures for an extended period in order to develop the ability to effectively flower, and thus produce grain. Alleles, such as *VRN1* and *VRN2*, are known to be involved in the vernalization response in diploid wheat. Another allele for VRN gene, *VRN4* has been identified by scientists but is not homoeologous to any other known common wheat Vrn gene therefore, has not been mapped using DNA markers (Goncharov et al. 2003).

3.2.4 Grain Yield/Weight

Grain yield (GY) is one of the most complex polygenic traits, often determined by the number of grains per unit area and grain weight. According to the current human population increase trends, it is predicted that by 2050, wheat production must increase by at least 60% to fulfill the increased demand associated with the anticipated population increase (Ray et al. 2013). Grain weight (GW) is a fairly stable yield component with a reasonably high heritability that is controlled by a complex genetic network expressed at different developmental stages. Furthermore, there is a significant impact of different climatic conditions on grain yield. GW and grain number (GN) are often found to be negatively associated to each other (Gegas et al. 2010), which is one of the major reasons for low potential grain weight in high yielding varieties. Therefore, the alleles responsible for high grain weight usually decrease grain number as shown by an allele *TaGW2-A1* (Zhang et al. 2018). Identification of favorable alleles, which increase grain weight without any penalty on grain number will be certainly beneficial in improving the grain yield in wheat and therefore, continuous efforts are being in this direction. A study in tetraploid wheat identified one such allele *GNI-1*, which mediated the trade off. Some more examples of alleles reported for GY are listed in Table 3.1.

3.3 ALLELE MINING FOR GRAIN QUALITY

Wheat in the United States is classified as hard red winter (HRW), hard red spring (HRS), soft red winter (SRW), durum, white [soft white winter (SWW) and club], hard white winter (HWW), mixed, and unclassified (Delwiche and Norris 1993). Generally, the breeding programs

TABLE 3.1
Important Genes/Alleles for Plant Architecture-Related Traits

Trait	S. No.	Allele	Chromosome	Effect of Gene/Allele on Plant	Reference
Plant height	1.	*Rht24b*	6AL	Reduced plant height, increased nitrogen use efficiency, and photosynthetic rate; GA insensitive	Tian et al. (2022)
	2.	*RhtB4c*	2BL	"Dwarf mutant-30" with plant height 44 cm.	Agarwal et al. (2020)
	4.	*Rht-B1b*	4BS	Semi-dwarf plant phenotype; increased seed output by 5–10% while reducing plant height by 20%.	Peng et al. (1999)
	5.	*Rht-D1b*	4DS	Semi-dwarf plant phenotype	Peng et al. (1999)
	6.	*TaCOLD1*	Homologs on 2A, 2B, and 2D	Regulation of plant height through heterotrimeric G protein	Dong et al. (2019)
	8.	*Rht13*	7BS	Higher grain number per main spike, grain number per spikelet and higher number of productive tillers	Rebetzke et al. (2011)
	9.	*TaSTE*	3A	The ortholog of AtSTE1 (Arabidopsis) for height reduction	Zhang et al. (2013)
	10.	*Rht8*	2DS	Gibberllin sensitive gene; height reduction with no significant effect on coleoptile length	Xiong et al. (2022)
	11.	*Rht18*	6A	Height reduction with no significant effect on coleoptile length	Pearce et al. (2021); Grant et al. (2018)
Spikelet number per spike	1.	*Wheat FRIZZY PANICLE (WFZP)*	2DS	Supernumerary spikelets in wheat	Dobrovolskaya et al. (2015)
	2.	*TaAPO* (wheat ortholog ABERRANT PANICLE ORGANIZATION)	7AL	Increased spikelet number per spike	Kuzay et al. (2022); Muqaddasi et al. (2019);
	3.	*FLOWERING LOCUS T* (*FT-B1*)	7B	Increases the number of fertile spikelets	Boden et al. (2015)
	4.	*Photoperiod-1* (*Ppd-1*)	2D	Reduction in spikelet number and tiller number	Addisu et al. (2009)
	5.	*TaTFL*	2D	Increases the number of spikelets and florets	Wang et al. (2017b)
Inflorescence	1.	*Vrn1*	5AL	Vernalization-mediated flowering; due to mutations in the promoter region of gene AP1	Yan et al. (2003, 2004, 2006)
	2.	*Vrn2/ZCCT1*	5A	Vernalization-mediated flowering	Yan et al. (2003, 2004, 2006), Distelfeld et al. (2009)
	3.	*Vrn3*	5D	Vernalization-mediated flowering	Yan et al. (2003, 2004, 2006)

Grain number and grain weight	1.	*GNI1*	2A	Increased floret fertility	Sakuma et al. (2019)
	2.	*GNI-A1*	2AL	Increase grain number	Golan et al. (2019)
	3.	*TaGW2-A1*	6A	Increases Thousand Grain Weight, grain width and grain length	Zhai et al. (2018)
	4.	*TaAGPL.B1*	1B	Alleles involved in starch content improvement identified which ultimately improve grain weight	Guo et al. (2017)
Spike Architecture	1.	*TaTFL1-2D*	2D	Increases the numbers of spikelets, florets, and seeds per spike	Wang et al. (2017b)
	2.	*TB1 TEOSINTE BRANCHED1 (TB1)*	4DS	Regulates inflorescence architecture	Dixon et al. (2018)
	3.	*TaVRS1*	2B	Overexpression negatively regulates spike branching by inhibiting SM initiation	Wang et al. (2017b)
	4.	*TaPAP2*	5A	Increase wheat spike complexity and thus grain yield	Wang et al. (2017b)
	5.	*TaIAA15*	3B	Reduced plant height, increased spike length and 1000-grain weight; increased the leaf angle	Fu et al. (2022)
		TaGRF3	2A	Grain filling and earlier heading	Bazhenov et al. (2021)
	7.	*gene Q*	5A	Free-threshing character and square spike phenotype	Faris et al. (2003), Sormacheva et al. (2015)
	8.	*B1 (Tipped 1), B2 (Tipped 2) and Hd (Hooded)*		Suppressors for awn development	Niu et al. (2020), Yoshioka et al. (2017)

for genetic improvement of bread wheat have three primary goals: (i) increasing yield, (ii) overcoming environmental stresses, and (iii) improving end-use quality traits (Mann et al. 2009, Hosseinirad 2022).

Wheat is known for its high protein content and excellent quality traits for human consumption and is one of the most important sources of carbohydrates and plant protein in human diets (Fatiukha et al. 2020). Carbohydrates (simple starches) make up 55–75% of the total dry grain weight, while storage proteins provide 10–20% of the starchy endosperm of mature wheat grains, which is a crucial source of nutrition (Gillies et al. 2012). Grain protein content (GPC), milling, flour, dough, and bread baking quality traits are complex traits determined by different genetics and environmental factors (Rousset et al. 1992; Peterson et al. 1998). However, grain protein content often shows negative correlation with yield traits (Naraghi et al. 2019) and therefore, many breeders worldwide are looking for genes or QTLs, which have a more negligible effect on yield while still increasing protein content.

One of the key issues that wheat breeders are facing is the dramatic decrease in genetic diversity as a result of allelic selection during the evolution and breeding process (Guo et al. 2019). The low genetic diversity in wheat could also be due to the replacement of local landraces by modern elite cultivars in the last century (Winfield et al. 2017; Balfourier et al. 2019). In the most modern breeding programs, a base population with narrow genetic variability (e.g., a few landraces or introduced breeding lines dedicated to a particular environment) is used. Research has shown that intensive breeding selection can reduce genetic diversity and evolution. The availability of landraces and other genomic resources could bring a more diverse germplasm in wheat (Salsman et al. 2021). More favorable and superior alleles in wild relatives and landraces could be used to improve the quality of wheat traits, especially GPC. Thus, understanding the genetic background of GPC and other protein-related traits and identification of favorable alleles could assist breeders to improve wheat quality traits (Salsman et al. 2021).

3.3.1 Grain Protein Content

Protein is one of the essential components of grain and GPC is one of the main quality traits in wheat (Fatiukha et al. 2020). Biotic and abiotic environmental conditions and their interactions with diverse genetic backgrounds (genotypes) have a considerable impact on the total amount and composition of grain proteins (Triboi-Blondel et al. 2003). Several studies have been conducted on GPC as an end-use quality trait because of the significant role of GPC in making products like bread, cake, noodles, and pasta (Zhao et al. 2010). Furthermore, wheat markets are determined by the GPC (Naraghi et al. 2019). Grain protein content determines the nutritional value of a diet, as well as the rheological properties of bread wheat and the pasta-making qualities of durum wheat (Autran et al. 1986). The negative correlation between GPC and wheat yield must be accounted for to properly enhance cultivars for high GPC (Blanco et al. 2012).

With reference genome assemblies being available for tetraploid (Avni et al. 2017; Maccaferri et al. 2019) and hexaploid wheat (IWGSC 2018), comparison of the genetic and physical locations of markers and identification of the genes responsible for high GPC have become more common. In order to identify the genes/alleles for high GPC and the markers that are closely linked to GPC, the chromosomal regions that increase GPC need to be identified. This may be followed by introducing beneficial alleles into elite wheat cultivars, cloning valuable genes, and discovering the underlying molecular pathways (Fatiukha et al. 2020).

Several studies have indicated that genes/QTLs linked to GPC are present on the majority of wheat chromosomes (Table 3.1) (Prasad et al. 2003; Groos et al. 2003; Sourdille et al. 2003; Kulwal et al. 2005; Blanco et al. 2006; Huang et al. 2006; Kunert et al. 2007; Tiwari et al. 2016; Boehm et al. 2017; Quraishi et al. 2017, Naraghi et al. 2019, Fatiukha et al. 2020; Leonova et al.

2022). For instance, (i) in a study, nine QTLs on chromosomes 1B, 3B, 4A, 5A, 5B, 6B, 7A, 7B, and 7D were discovered in a RIL population generated from a hybrid between the winter wheat varieties *T. aestivum* L. var. "Forno" and *T. spelta* L. variety "Oberkulmer" (Zanetti et al. 2001). (ii) In another study, 15 genomic regions linked to GPC were identified across nine chromosomes using an association mapping panel of 196 common wheat cultivars from varied geographic origins (Bordes et al. 2013). (iii) Other studies have identified significant QTL associated with GPC on chromosomes 1A, 7A, and 7B using double-haploid and RIL populations (Tiwari et al. 2016; Boehm et al. 2017). (iv) Thirteen QTLs on chromosomes 2A, 2B, 2D, 3D, 4A, 6B, 7A, and 7D were identified as related to GPC, and each explained 3 to 32% of phenotypic diversity (Prasad et al. 2003). The favorable alleles for the above QTLs/genomic regions could be potential targets for GPC improvement in wheat.

3.3.2 Importance of Wild Emmer Wheat (WEW) EW Allele for GPC

Wild emmer wheat (WEW) (Krugman et al. 2018; Klymiuk et al. 2019b), *T. turgidum* ssp. *dicoccum* and other wild relatives (Alvarez and Guzman 2018) have been shown to be viable sources for wheat genetic diversity. WEW, a tetraploid relative of modern tetraploid durum and hexaploid common wheat, is one of the most important wild relatives of modern cultivated wheat. The WEW germplasm contains high-quality trait variation, especially GPC and therefore possesses the excellent potential for improving cultivated wheat nutritional value (Huang et al. 2016; Klymiuk et al. 2019a). One of the most important genes for improving GPC is the *Gpc-B1*, found on chromosome 6BS, derived initially from WEW variety FA15-3 (Joppa et al. 1997; Olmos et al. 2003). The *Gpc-B1* in different environments has shown consistent GPC increases in tetraploid and hexaploid wheat (Mesfin et al. 1999; Chee et al. 2001). The WEW allele of *Gpc-B1* demonstrated positive effects on GPC and other quality parameters and minor effects on yield-related traits (Carlos Brevis and Dubcovsky, 2010; Tabbita et al. 2017). Hence, wheat breeders worldwide could use WEW as a viable source of beneficial alleles for wheat breeding programs (Fatiukha et al. 2020).

3.3.3 *Gpc-B1* Renamed to *NAM-B1* Having Three Allelic Variations

The *Gpc-B1* gene was renamed to *NAM-B1* by Uauy et al. (2006); they cloned *Gpc-B1* and discovered that it encodes for No Apical Meristem protein (NAM) that belongs to the NAC transcription factor family, and so changed the name to *NAM-B1*. The *NAM-B1* has three allelic variations that have been identified (wild-type, 1-bp insertion, and deletion). In addition to the functioning wild-type allele seen in wild and cultivated emmer wheat, the low GPC allele of *NAM-B1* in 57 durum wheat lines was discovered. Also, another four significant GPC QTLs with favorable alleles from WEW were found on chromosomes 4BS, 5AS, 6BS, and 7BL; where the 6BS GPC QTL coincided with the physical position of the NAC transcription factor *TtNAM-B1*, underlying the cloned QTL, *Gpc-B1* (Uauy et al. 2006).

The absence of wild-type *NAM-B1* alleles in durum and bread wheat suggested that the allele had been lost during wheat domestication (Uauy et al. 2006; Dubcovsky and Dvorak 2007). A study was conducted to find the *NAMB-B1* allele in 365 winter bread wheat varieties and advanced lines cultivated in China; however, researchers were not able to find a functional *NAM-B1* allele in those varieties (Chen et al. 2017). Asplund et al. (2010) concentrated on outdated nineteenth-century wheat cultivars from before the age of modern plant breeding, demonstrating that the wild-type gene was still present in certain widely farmed types at the time using historical materials. They hypothesized that if the wild-type allele had been lost, it happened during the previous century of plant improvement rather than domestication. It is also likely that the wild-type allele exists in the wheat genome and in gene banks (Hagenblad et al. 2012).

3.3.4 *Gpc-B1* Allele improves Zn Content

It has been demonstrated that the functional *Gpc-B1* allele is also associated with high Zn content in wheat genotypes. In one of the earlier studies, it was suggested that *Gpc-B1* locus is involved in efficient remobilization of protein, Zn, Fe, and Mn from leaves to grains in addition to its effect on earlier senescence of green tissues (Distelfeld et al. 2006). Similarly, in a recent study (Kaznina et al. 2021), conducted on wheat genotypes with a functional and non-functional *Gpc-B1* alleles, wheat with a functional allele of the *Gpc-B1* gene growing under Zn deficiency was capable of grain production with a sufficient Zn concentration without a decrease in yield.

3.3.5 Other Protein-Related Traits

The fundamental factor driving the differences in baking and nutritional qualities of flours obtained from different wheat types is variation in the amount and type of wheat grain proteins. Gluten is created when glutenin and gliadin (storage protein) interact after water is added to flour (Stone and Savin 1999). Detailed knowledge on the genes that code for the endosperm storage proteins is becoming available. This understanding will eventually lead to the creation of more precise and direct breeding strategies for improving flour quality in different ways (Law and Payne 1983).

Wheat has nine complicated loci that determine the endosperm proteins. The genes for the principal storage proteins, gliadins and glutenins, have been found on the chromosomes of homoeologous groups 1 and 6. Glutenins contain high molecular weight (HMW) and low molecular weight (LMW) subunits that give the dough its strength. High molecular weight glutenin subunits are encoded by the genes *Glu-Al*, *Glu-Bl*, and *Glu-Dl*, found on the long arms of chromosomes lA, 1B, and 1D, respectively. On the other hand, LMW glutenins are controlled by the genes *Glu-A3*, *Glu-B3*, and *Glu-D3* (Payne, 1987). In one study (Henkrar et al. 2017), allelic variations in HMW and LMW glutenin alleles were studied in Moroccan durum and bread wheat cultivars, and it was found that, compared to durum, more allelic variations were observed in bread wheat cultivars. It was suggested that for improving the durum wheat cultivars, the alleles *Glu-A1c* and *Glu-B3D* should be replaced with more favorable alleles for improving the durum glutenin content.

The HMW glutenin subunits are determined by a single complex locus located about nine cM from the centromere on each of the long arms of chromosomes lA, IB, and ID (Wrigley et al. 1973). *Gli-Al*, *Gli-Bl*, and *Gli-Dl*, which code for ω- and y-gliadins, are positioned distally on each of these chromosomes' short arms. The remaining three loci, which mostly code for ex- and p-gliadins, are located on the short arms of 6A, 6B, and 6D, respectively, and are named *Gli-A2*, *Gli-B2*, and *Gli-D2*. The loci are assumed to be formed of closely connected genes. Variation occurs at each locus and results in a vast diversity of proteins. This diversity has been linked to bread-making quality, particularly among the HMW glutenin subunits (Law and Payne 1983).

Wheat relatives have even more variation than modern cultivated wheat itself for HMW and LMW. A study by the Plant Breeding Institute in Cambridge showed that modern cultivated wheat does not have many HMW glutenin subunits. The same study reported that *Aegilops sp.* is a suitable source for HMW subunits since it has several gliadin subunit variations (Law and Payne 1983). Chromosome engineering is used to transfer genes for new protein components from *Ae. umbellulata* to wheat. The 1U chromosome from *Ae. umbellulata* (2n = 14, UU), for example, has been swapped for each of the chromosomes of group 1 of the Chinese Spring variation of *T. aestivum*. Interestingly, this chromosome contains genes for novel HMW glutenin subunits and ω- and y-gliadins (Chapman et al. 1975).

Also, endosperm proteins regulated by chromosomes from related species with group 1 have recently been studied. Chromosome 1R from rye, chromosome 5 from barley, and chromosome 1 from *Agropyron elongatum* are all comparable to the wheat group 1 chromosome as they carry genes for HMW proteins on their long and short arms (Lawrence et al. 1981). For example, some wheat

TABLE 3.2
Summary of Quantitative Trait Loci Associated with Grain Protein Content in Wheat

Trait	Chromosome/Arm	PVE%$	References
GPC*	2AS, 2BL, 2DL, 3DS, 4AL, 6BS, 7AS, 7DS	0.6–35.8	Prasad et al. (2003)
	1A, 2AS, 3AL, 3BS, 4AS, 4DL, 5BL, 6AL, 7AS, 7DL	4.2–10.4	Groos et al. (2003)
	1BL, 6AS	6.5–17.1	Sourdille et al. (2003)
	1AS, 1BL, 1DL, 2AS, 2AL, 2BL, 2DS, 2DL, 3BS, 4AS, 5BL, 5DL, 6DL, 7AL, 7DS	8.4–16.6	Kulwal et al. (2005)
	2AS, 6AS, 7BL	4.2–21.7	Blanco et al. (2006)
	4DS, 7BL	12.6–32.7	Huang et al. (2006)
	3AL, 4AL, 4BL, 5DL, 7BS, 7DS	4.5–22.4	Kunert et al. (2007)
	2A, 2B, 2D, 3A, 4A, 6B, 7A	NA	Quraishi et al. (2017)
	1A, 1B, 2A, 2B, 3A, 3B, 4B, 5B, 7A	6.6–20.0	Naraghi et al. (2019)
	1A, 2A, 3A, 4A, 4B, 5A, 6A, 6B, 7B	0.6–15.9	Fatiukha et al. (2020)
	4A, 6A	9.7–20.0	Leonova et al. (2022)

*: Grain protein content, $: Phenotypic variation explanation

loci are connected to the barley hordein loci *Hor-1*, *Hor-2*, and *Hor-3*, as well as possibly analogous loci in the rye. The *Hor-3* is homologous to *Glu-1*, while *Hor-l* and may be *Hor-2* are homologous to *Gli-l*. In barley or rye, there is no homologous to wheat's *Gli-2* loci. Hence, there are valuable sources for raising protein levels in wheat relatives (Law and Payne 1983).

Overall, allele mining provides insight into the genetic diversity improvements and identification of favorable genomic regions to enhance protein content in modern wheat cultivars; wild relatives of modern tetra- and hexaploid wheat being a potential source of desirable alleles for the improvement of GPC and other quality traits. Further studies are warranted to improve our knowledge in the application of allele mining to characterize the novel genetic sources for quality traits.

3.4 ALLELE MINING FOR HEAT AND DROUGHT TOLERANCE

Heat and drought are the two major abiotic stresses, which affect the yield of wheat crop drastically worldwide. Both are highly complex traits often controlled by a number of QTLs/genes; therefore, identification of beneficial alleles will certainly help in development of markers, which can be used for breeding wheat cultivars for improved abiotic stress tolerance. Some attempts have already been made in this direction, both at the genome wide level as well as in individual genes. At the genome wide level, a number of favorable alleles, which have been associated to QTLs for important agronomic traits under heat or drought conditions have been identified either using biparental mapping populations or using LD based GWAS (Singh et al. 2021, 2022). The use of these approaches for identification of favorable alleles is also summarized in Figure 3.1. These favorable alleles of QTLs have also been transferred into elite wheat cultivars using MAS. For instance, (i) Rai et al. (2018) used marker-assisted backcross breeding to improve wheat for drought tolerance. Three different QTLs were introgressed into sensitive wheat line, which contained positive alleles for physiological traits like canopy temperature, chlorophyll content, normalized difference vegetation index (NDVI) and yield contributing traits. (ii) In another study, Gautam et al. (2021), attempted to introgress a major QTL for yield under drought stress in four Indian wheat cultivars. (iii) Attempts were also made to use the QTL alleles from wheat wild relatives like wild emmer wheat. Three QTL alleles were validated in either durum wheat or bread wheat, which enhanced both yield as well as biomass under drought stress (Merchuk-Ovnat et al. 2016).

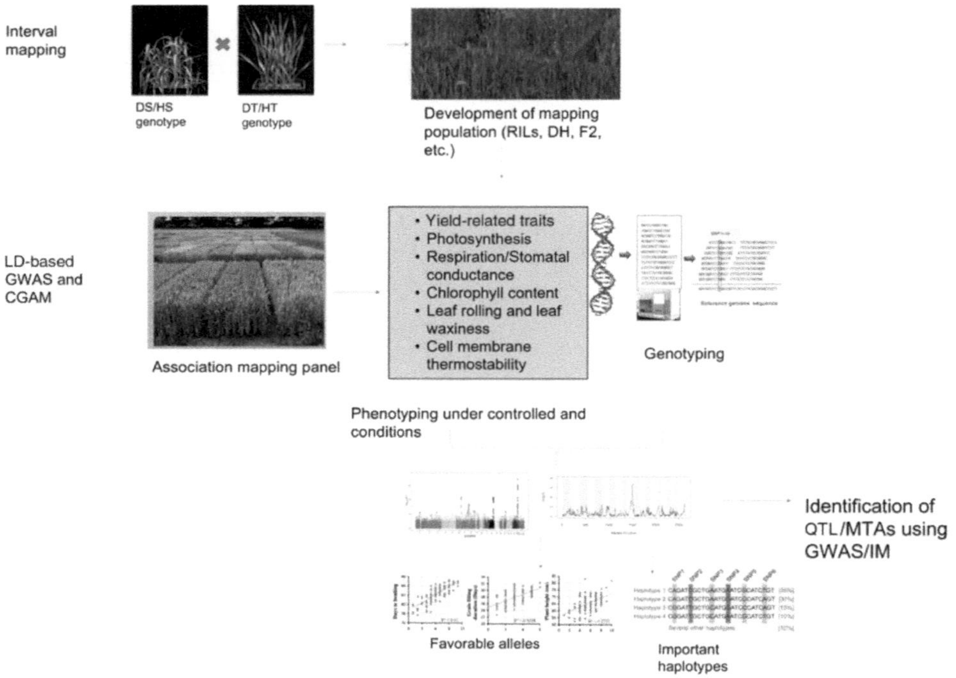

FIGURE 3.1 Use of linkage-based IM (interval mapping), LD-based GWAS and candidate gene-based association mapping (CGAM) for identification of favorable alleles/haplotypes for yield and other traits that are associated to abiotic stress tolerance.

Allele mining has also been conducted to identify useful alleles in individual genes (Table 3.3) For instance, for drought stress, alleles have been identified in genes encoding for MYB, DREB HKT, and WRKY transcription factor, which could be potential targets for wheat improvement after further validation. Similarly, for heat stress, alleles have been identified for different heat shock proteins. Apart from the above, several putative candidate genes have been identified for both the abiotic stresses, mostly using *in silico* approaches. These include ADP glucosepyrophosphorylase soluble starch synthase, sucrose synthase, heat shock proteins, as well as others for heat stress and genes like SWEET (Gautam et al. 2021), calcium-dependent protein encoding gene (Li et al. 2020). *TaARGOS* (Auxin regulated gene involved in organ size; Ahmed et al. 2022), *TaOFP* (Ovate family protein, Wang et al. 2020), etc., for drought stress. Therefore, it would be interesting to further explore these genes for identifying the alleles, which are involved in providing tolerance against heat/drought, either using the CGAM approach or mutagenesis-based approach like TILLING, which has not been explored to date for allele mining for abiotic stress tolerance in wheat. This is further discussed in the later section on TILLING.

In cereal crops, only few studies have been reported where TILLING has been used to identify alleles involved in abiotic stress tolerance. For instance, in wheat, TILLING has been used to screen a set of durum wheat mutants and four new Hsp alleles were identified, which can be suitably used in wheat for improving heat tolerance. Similarly, in rice, two Hsp genes were screened for allelic variants using TILLING and the variants identified may be potential candidates for enhanced heat tolerance. One of the major advantages of using TILLING-based methods is that it is non-GMO and therefore, free of any ethical issues. Therefore, efforts need to be made to identify important allelic variants for different genes using this approach. The Small Grains Breeding and Genetics Lab at University of Maryland possesses genomic resources for TILLING mutants in wheat, including

TABLE 3.3
List of Some Important Candidate Genes Involved in Heat/Drought Tolerance and Their Important Alleles

Trait	Gene	Allele(s)	Study	Reference
Drought	*TaSAP1* (member of stress associated protein)	SNP2606, SNP2606A, InDel39-1637	Sequence variations identified in promoter region	Chang et al. (2013)
	TaMYB2	SNP at 458 bp (T/A transition)	Validation in core set and development of allele specific marker	Garg B et al. (2012)
	QTLs/alleles identified on chromosomes 3D, 5D and 7A for grain yield related traits		GWAS analysis	Itam et al. (2022)
	DREB1, HKT1 and *WRKY1*	Alleles identified		
	L-galactone 1,4, lactone dehydrogenase	Ta-GLDHA1b	This allele enhances drought tolerance by affecting stomatal aperture	Zhang et al. (2016)
	DREB1	SNPS770	Genetic mapping of DREB1	Wei et al. (2009)
	sHSP26	Different alleles identified	Used tilling approach to identify alleles	Comastri et al. (2018)
Heat	*HSP16.9*	One SNP found between tolerant and sensitive genotype	A marker developed for heat tolerance	Garg D et al. (2012)

wild wheat species like *T. monococcum* and *T. durum*. *T. monococcum*, a close relative to diploid A-genome wheat progenitor *T. urartu*, and can serve as an excellent model to simplify wheat genomics due to its much less complex genome. Therefore, some of the important putative candidate genes identified for heat and drought tolerance can be validated using TILLING resources, which may ultimately lead us to favorable alleles for wheat breeding.

3.5 ALLELE MINING FOR SALT TOLERANCE

Like heat and drought, salt stress also affects wheat yield and yield of wheat starts to decline at a salt concentration of 6–8 dS m^{-1} (Royo and Abio 2003). Salt stress often leads to hyperosmotic and hyperionic (ion toxicity) stresses due to low water potential of soil and excess sodium ion accumulation takes place within the plant. To date, ~1500 QTLs have been identified for salt stress tolerance (Singh et al. 2021, 2022) in wheat; many of which may contain favorable alleles that can be used for MAS for improvement of salt stress tolerance in wheat. In some of these studies, putative candidate genes (CGs) have also been identified, which underly these important QTLs. These putative CGs can be potential targets for gene-specific allele mining and development of allele-specific markers like Kompetetive allele-specific polymerase chain reaction (KASP). For instance, using GWAS analysis (Oyiga et al. 2019), genes *NRAMP2* and *OPAQUE1* were identified on the chromosomes 4BL and 6AL; these two genes showing contrasting expression patterns in sensitive and tolerant wheat genotypes and these genes also showed variations in coding sequences.

Important alleles have also been earlier identified for some CGs. For instance, four important alleles have been identified for transporter like *HKT1*, which encodes high affinity potassium transporter. This gene was sequenced in diverse wheat collections including the wild species and four important alleles were identified, which could be distinguished by the number of SNPs (Thiyagarajan et al. 2022). In another study performed in durum wheat background (Makhloufi et al. 2015), a candidate gene (*TdERF1*) encoding ethylene responsive transcription factor was isolated and allele mining and gene expression analysis was conducted in salt-tolerant and sensitive genotypes. Four allelic variants of this gene were identified, all being transcriptionally active and, interestingly, the expression of one allelic form was specific to tolerant genotype, which indicated its importance as a functional marker to distinguish salt-tolerant and sensitive durum wheat genotypes.

3.6 DIFFERENT APPROACHES FOR ALLELE MINING

The conventional gene/allele mapping approaches during the nineteenth century mainly involved the use of gel-based molecular markers like SSRs, RFLPs, and RAPDs, which are now becoming obsolete. Due to the advent of next-generation sequencing (NGS) technology, reference genome sequences are now available for a number of food crops like rice, barley, maize, wheat, etc. This has helped in identification of a large number of single nucleotide polymorphisms (SNPs) that are involved in controlling different traits. The NGS-based markers exhibit cross-species validation and are very effective in mapping genes in other crops. But still, mining trait-specific genes and their respective allele is not a trivial task, especially in a complex crop species like bread wheat, which has a hexaploid genome. Further, most of the agronomically important traits are highly complex and therefore governed by multiple QTLs and candidate genes. Therefore, QTL-seq and mutagenesis-based approaches, like TILLING, SHOREmap, MutMap, MutMap+, MutMap-Gap, which were earlier developed in *Arabidopsis thaliana* and/or *Oryza sativa* may play an important role in candidate gene discovery with an application in wheat breeding. These different approaches are described in some more detail below as well as summarized in Figure 3.2.

3.6.1 QTL-seq

QTL identification constitutes an important aspect of candidate region identification for marker-assisted selection. It is a time-consuming and intensive process involving mapping populations developed from contrasting phenotypes and the development of markers, spanning multiple years. Takagi et al. (2013b) demonstrated a new methodology based on whole-genome sequencing of bulks with extreme phenotypic values. The method involves the generation of a segregating F_2 progeny by the cross between two parents with distinct phenotypes, for example, plant height. Since plant height is a multigenic trait, a normal distribution of height will be observed in the progeny. Two extreme bulks are generated by selecting the group of plants with the higher and lower height. DNA from these bulks is sequenced and aligned to the reference genome to calculate the SNP index. The QTL can be identified as peaks or valleys of the SNP-index plot or Δ(SNP-index). QTL-seq was used in the identification of the QTL region contributing resistance for rice blast in rice cultivar Nortai. An F_2 progeny was developed by crossing resistant cultivar Nortai to the susceptible cultivar Hitimebore. Two DNA bulks were sequenced for the resistant and susceptible progeny. SNP index plot revealed a contrasting region from 2.39 to 4.39 Mb on chromosome 6, indicating the presence of major QTL differentiating the Nortai and Hitomebore partial resistance.

3.6.2 Tilling

TILLING (targeting induced local lesions in genomes) is a powerful genetics tool that utilizes mutagenesis via chemicals or radiation of seeds, gametophytes, or in the case of animal systems, males or hermaphrodites in order to create F_2 populations with induced genetic variations (Colbert et al. 2001; Winkler et al. 2005; Gilchrist et al. 2006; Uauy et al. 2009; Zulfiqar et al. 2021). Once these populations are established, they can be utilized for forward screens for traits of interest, or for TILLING. In the

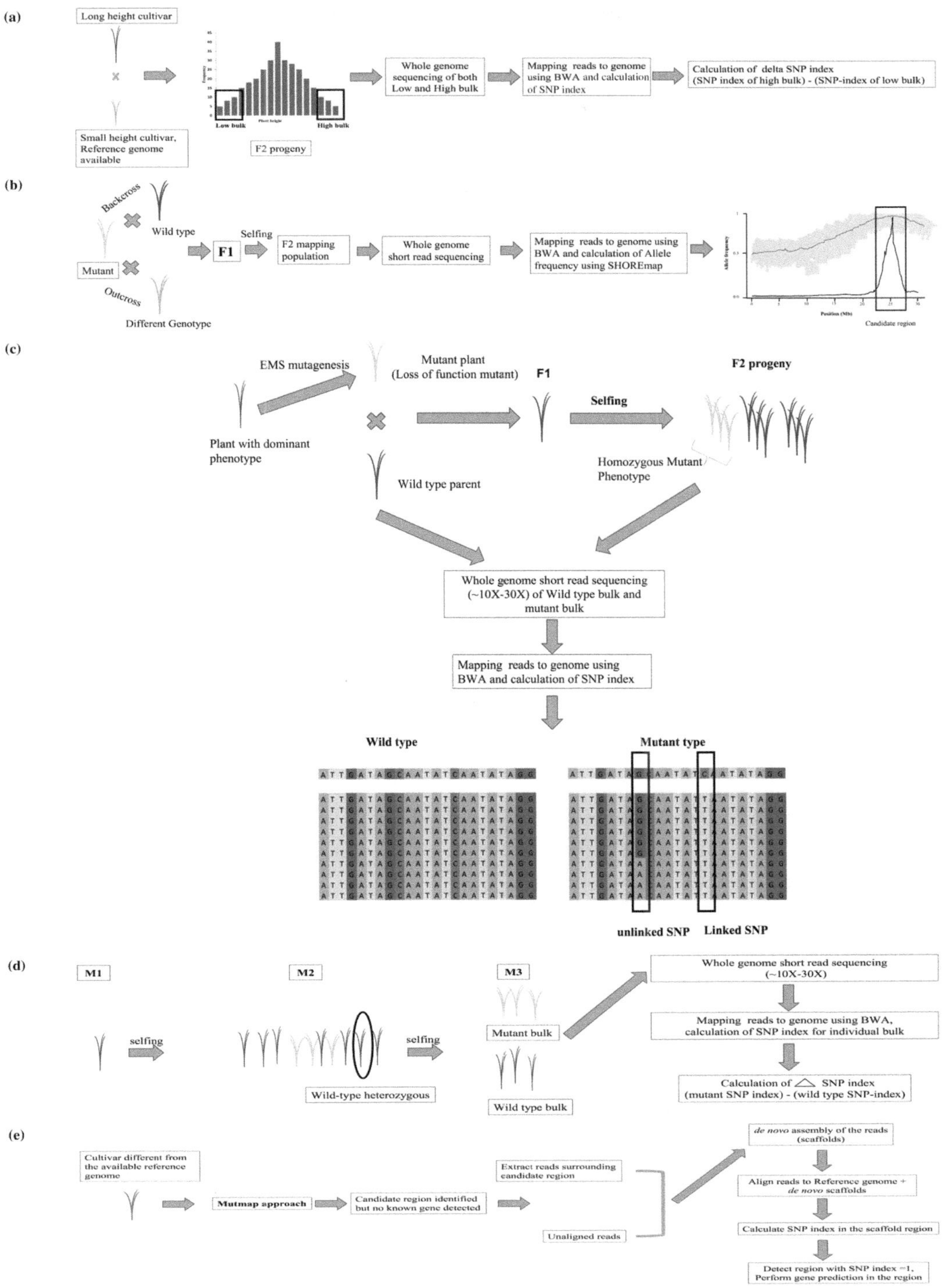

FIGURE 3.2 Allele mining approaches: (a) QTL-seq, (b) SHOREmap, (c) Mutmap, (d) Mutmap+, (e) Mutmap-Gap.

latter, a reverse genetics approach is used in which markers developed from candidate genes are used to screen the population for deleterious SNPs or other induced variations in order to discover knockouts for the gene of interest (Colbert et al. 2001; Singh et al. 2019). Briefly, when a mutant is found using CEL-1 assay, NGS sequencing, or through electrophoresis, the mutant sequence is analyzed for the effect on the peptide sequence. Nonsense, deleterious missense mutants, and/or mis-splice are then phenotyped to determine if the trait of interest is changed by the induced variation.

This technique has been shown to be incredibly beneficial for allele discovery in several systems, including small grains, such as rice, barley, and wheat (Till et al. 2007; Talamè et al. 2008; Gottwald et al. 2009; Schreiber et al. 2019). Specifically in wheat, several populations have been developed not only in hexaploid bread wheat backgrounds, but also in tetraploid wheat, as well as diploid wheat species (Rawat et al. 2012, 2018, 2019; Krasileva et al. 2017; Singh et al. 2019; Zulfiqar et al. 2021). Forward and reverse genetic approaches in these resources have been paramount for discovery and validation of genes that relate to agronomically important traits ranging from disease resistance, abiotic stress resistance, as well as yield related traits, such as spike morphology, shoot architecture, tiller number, and height (Worland et al. 1980; Kuraparthy et al. 2007; Sakuma et al. 2019; Nilsen et al. 2020; Lethin et al. 2020; Chhabra et al. 2021).

3.6.3 SHOREmap

Mapping by sequencing relies on the comparison of polymorphic markers between the parental lines. The process involves crossing the plants with the mutant phenotype (plant height or spike architecture) to a non-mutant genotype. Plant in the first filial generation (F_1 progeny) are self-pollinated or crossed with other F_1 plants to generate an F_2 population. F_2 population shows segregation for the mutant phenotype. The plants with the mutant phenotype are then separated and their DNA is extracted, pooled, and sequenced. Allele frequency is measured for each position by comparing the parental alleles, which helps in the identification of genomic regions controlling the phenotype. The technique is implemented in a user-friendly manner using the SHOREmap pipeline. The method supports genome-wide genotyping and candidate-gene sequencing in a single step through analysis of deep sequencing data. This technique was demonstrated for the first time in *Arabidopsis thaliana* (Schneeberger et al. 2009) for sluggish growth and pale-green leaf mutants.

3.6.4 MutMap

Mutational cloning is gaining momentum, in which an elite cultivar with a distinct dominant trait is treated with EMS (ethyl methanesulfonate) or other mutagens, causing alteration in its genome. Mutagenized plants with the loss of function mutation are characterized in the M_2 generation obtained by selfing of M_1 plants. M_2 plants with a mutant phenotype are then crossed with the wild-type parent and the resulting F_1 progenies are again allowed to self to create a segregating F_2 population. In the F_2 plants, most of the causal mutations get fixed in a homogeneous state. Non-linked SNPs usually segregate in 1:1 ratio, while the causal SNP demonstrates a homozygous state. DNA is collected from the mutant and wild-type bulks for short read sequencing. Sequence data is used to calculate the SNP index (ratio of mutant reads/wild-type reads) at each position. The causal SNP will demonstrate an SNP-index of 1, in which all reads contain the SNP, and an elevated peak region while the non-linked SNPs show mutant and wild-type reads in a ratio of 1:1.

MutMap was first demonstrated using the pale-green leaves and semi-dwarf mutants of rice cultivar Hitomebore (Abe et al. 2012). In this study, an EMS mutagenized population of about 12000 lines in M_3 to M_4 stage was generated for pale-green leaf mutants in elite background (Hitomebore). The two mutants (Hit1917-pl1 and Hit0813-pl2), were crossed with the wild-type and which shows the segregation ratio of 3:1 in F_2 generation, demonstrating the role of recessive allele. DNA was extracted from 20 F_2 plants and bulked in an equimolar ratio for sequencing. Using MAQ software,

TABLE 3.4
List of Some Studies Showing the Use of Tilling Approach for Identification of Important Alleles in Wheat

Species	Trait	Gene	Chromosome	Effect of Mutation	Reference
Triticum aestivum	Floret Fertility	*GNI-A1*	2A	Reduced Floret Fertility, Reduced Grain Weight per Plant	Sakuma et al. (2019)
	Gluten Content	*WPBF*	5A, 5B, 5D	Lowered gliadin and low-molecular-weight glutenin	Moehs et al. (2019)
	Grain Hardiness	*PINa, PINb*	5D	Increased Grain Hardiness	Feiz et al. 2009
	Starch Content	*Sgp-1*	7A, 7B, 7D	Reduced Starch	Sestili et al. 2010; Yamamori et al. 2000
	Grain Weight	*TaGW2*	5A, 5B, 5D	Increased Grain Width	Zhang et al. 2018
	Resistant Starch Content	*SSIIa*	7A, 7B, 7D	Increased Resistant Starch	Schoen et al. 2021, Botticella et al. 2018
	Starch Quality	*SBEIIa*	2A, 2B, 2D	Increased amylose content, reduced grain weight	Slade et al. 2012
Triticum turgidum	Height	*RhtB1b*	4B	Increased Height	Mo et al. 2018
	Starch Quality	*SBEIIa*	2A, 2B	Increased amylose content, reduced grain weight	Slade et al. 2012
	Heading Time	*VRN1*	5A, 5B	Increased Heading time	Chen and Dubcovsky, 2012
	Shoot architecture	*TdDof1*	3B	Loss of Pith	Nilsen et al. 2020
Triticum monococcum	Tiller Number	*TIN3*	3A	Decreased Tillers	Kuraparthy et al. 2007
Triticum durum	Heat stress	*Hsp26*	4AS and 4BL		Comastri et al. 2018

high-quality filtered reads were aligned to the Hitomebore reference sequence (370 Mb) leading to the detection of 1,001 SNPs in Hit1917-pl1 and 1,339 SNPs in Hit0813-pl2 after a stringent filtering based on EMS type transition mutations (GA and CT). The ratio of mutant read/wild-type reads was used to calculate chromosome-wise SNP index for each position, leading to the discovery of cluster of SNPs on chromosome 10 for Hit1917-pl1 and chromosome 1 for Hit0813-pl2. The mapped causal mutation was mapped to the gene chlorophyllide, an oxygenase (OsCAO1) in Hit1917-pl1.

This approach suffers from two major drawbacks: (i) mutagenesis sometimes leads to lethal recessive mutations causing sterility or early development lethality. In this scenario, F_2 mapping populations are challenging to achieve; (ii) varietal and cultivar differences from the available reference genome may result in non-detection of causative SNPs using the classical MutMap approach. Therefore, two derivatives of MutMap, i.e., MutMap+ and MutMap-Gap were introduced, which are described below. While MutMap+ offers a solution for the first problem, MutMap-Gap is helpful in the second scenario.

3.6.4.1 MutMap+

Mutmap+ is useful for detecting mutations that result in infertility, early development lethality, or overall impaired crossing. In MutMap+, heterozygous mutant plants are allowed to self in order to create an M_2 segregating population. From this population, two bulks representing the mutant and wild-type phenotypes are prepared. After following the Mutmap protocol, program calculates ΔSNP-index by subtracting SNP-index values of the wild-type bulk from the mutant bulk. The genomic region harboring the causal mutation should display positive ΔSNP-index values. Fekih et al. (2013) demonstrated the application using rice mutant Hit9188 of Hitomebore. The mutant was characterized by dwarfism and pale-green leaves followed by premature death within 3 weeks after germination. Due to lethality, mutants cannot be crossed with the wild-type Hitomebore. Heterozygous M_1 plants were self-crossed and M_2 progeny was grown to maturity and self-fertilize. From M_3 plants two bulks were developed with mutant and wild-type phenotypes and were sequenced separately. A unique region was identified by comparing the SNP index peak specific to mutant bulk using the Δ(SNP-index) on the chromosome 1. The mutation was mapped to the gene encoding to SufBD domain.

3.6.4.2 MutMap-gap

MutMap-gap combines MutMap and *de novo* assembly of reads to identify the causative mutations (Takagi et al. 2013b). Although reference genome sequences are available in almost all of the major crops, but the regional and varietal differences make it difficult to use a single reference for all the grown cultivars. The method involves the generation of reference for your cultivar of interest by aligning short reads to the available reference and substituting the SNP positions with the new sequenced base. SNP index is calculated using the MutMap approach for detecting the candidate region. In the absence of mutations in the coding genes, reads surrounding the candidate region and unmapped reads are combined to generate a *de novo* assembly. SNP index is calculated for the genome and the new assembled scaffolds. If the gene is absent in whole-genome assembly, assembled scaffolds with a higher SNP index are most likely to contain the causal nucleotide change responsible for the mutant phenotype.

The approach was first demonstrated in rice to locate a gene (*Pii*) providing resistance against rice blast. SNP index peaks were found in the regions 7.88 to 11.98 Mb and 7.18 to 13.05 Mb, respectively, on chromosome 9 in the mutant Hit5948 and Hit6780. *De novo* assembly of the reads in the mapping interval and unmapped reads led to the identification of mutation causing mis-splicing of mRNA encoding an NBS-LRR protein. The occurrence of a single candidate gene was confirmed by crossing Hit5948 with Hit6780, where F1 heterozygous plants showed susceptibility.

3.7 CONCLUDING REMARKS AND FUTURE PROSPECTS

Overall, the present chapter highlights the recent advances in the identification of alleles for yield, quality, and abiotic stress tolerance traits in wheat. Modern bread wheat has a complex hexaploid genome, which possess several challenges for the people working in the area of wheat genomics. However, the availability of genomic resources has simplified this to some extent making it possible to identify important alleles for different agronomic traits, which are described in this chapter. We believe that to date, only limited efforts have been made in wheat to introgress the identified alleles into elite wheat cultivars. Genomic selection (GS) technology certainly offers a new hope in this direction, which is superior to traditional MAS in several aspects (Merrick et al. 2022). For instance, the number of markers used per trait is generally low in MAS whereas in GS, all the possible high-quality markers are used for modeling the performance of a particular trait. Furthermore, the multiple small effect genes/markers contributing for the overall trait performance are often ignored in MAS, which is not the case in GS. Some attempts have already been made in wheat where GS has been used to predict the performance of a particular trait (Sandhu et al. 2021; Juliana et al. 2020; Lozada and Carter 2020; Michel et al. 2016). However, still efforts need to be made to use GS for development of improved wheat cultivars for different agronomic traits.

Furthermore, recent technologies like CRISPR-Cas9 also need to be explored for development of improved wheat cultivars. Protocols have already been developed using which multiplex genome editing may be possible in wheat (Li et al. 2020). Using this protocol, three genes were simultaneously targeted in wheat and eight genomic sites could be successfully knocked down. Similarly, attempts have been made in wheat where important alleles/genes have been knocked down leading to improved wheat lines for certain traits like thousand kernel weight, high amylose content, low gluten wheat (for details see review by Li et al. 2021). Therefore, the above examples certainly offer a hope for using the modern tools for developing improved wheat cultivars, which are resistant to different abiotic and biotic stresses.

REFERENCES

Abe A, Kosugi S, Yoshida K, Natsume S, Takagi H, et al. (2012) Genome sequencing reveals agronomically important loci in rice using MutMap. *Nat Biotechnol* 30(2): 174–178.

Addisu M, Snape JW, Simmonds JR, Gooding MJ (2009) Reduced height (Rht) and photoperiod insensitivity (Ppd) allele associations with establishment and early growth of wheat in contrasting production systems. *Euphytica* 166(2): 249–267.

Agarwal P, Balyan HS, Gupta PK (2020) Identification of modifiers of the plant height in wheat using an induced dwarf mutant controlled by *RhtB4c* allele. *Physiol Mol Biol Plants* 26(11): 2283–2289.

Ahmed K, Shabbir G, Ahmed M, Noor S, Mohi Ud Din A, Qamar M, Rehman N (2022) Expression profiling of *TaARGOS* homoeologous drought responsive genes in bread wheat. *Sci Rep* 12: 3595.

Alvarez JB, Guzman C (2018) Interspecific and intergeneric hybridization as a source of variation for wheat grain quality improvement. *Theor Appl Genet* 131: 225–251.

Asplund L, Hagenblad J, Leino MW (2010) Re-evaluating the history of the wheat domestication gene *NAM-B1* using historical plant material. *J Archaeol Sci* 37: 2303–2307

Asplund L, Leino MW, Hagenblad J (2012) Allelic variation at the Rht8 locus in a 19th century wheat collection. *Sci World J* 2012: 385.

Autran, JG, Abecassis J, Feillet P (1986) Statistical evaluation of different technological and biochemical tests for quality assessment in durum wheats. *Cereal Chem* 63: 390–394.

Avni R, Nave M, Barad O, Baruch K, Twardziok SO et al (2017) Wild emmer genome architecture and diversity elucidate wheat evolution and domestication. *Science* 357: 93–97.

Backhaus AE, Lister A, Tomkins M, Adamski NM, Simmonds J, et al. (2022) High expression of the MADS-box gene *VRT2* increases the number of rudimentary basal spikelets in wheat. *Plant Physiol* 189(3): 1536–1552.

Balfourier F, Bouchet S, Robert S, De Oliveira R, Rimbert H, et al. (2019) Worldwide phylogeography and history of wheat genetic diversity. *Sci Adv* 5(5): eaav0536.

Batra R, Saripalli G, Mohan A, et al. (2017) Comparative analysis of AGPase genes and encoded proteins in eight monocots and three dicots with emphasis on wheat. *Front Plant Sci* 8: 19.

Bazhenov MS, Chernook AG, Bespalova LA, Gritsay TI, Polevikova NA, Karlov GI, Divashuk MG (2021) Alleles of the *GRF3-2A* gene in wheat and their agronomic value. *Int J Mol Sci* 22(22): 12376.

Blanco A, Mangini G, Giancaspro A, Gioove S, Colasuonno P, et al. (2012) Relationships between grain protein content and grain yield components through quantitative trait locus analyses in a recombinant inbred line population derived from two elite durum wheat cultivars. *Mol Breed* 30: 79–92.

Blanco A, Simeone R, Gadaleta A (2006) Detection of QTLs for grain protein content in durum wheat. *Theor Appl Genet* 112: 1195–1204.

Boden SA, Cavanagh C, Cullis BR, Ramm K, Greenwood J, et al. (2015) *Ppd-1* is a key regulator of inflorescence architecture and paired spikelet development in wheat. *Nat Plants* 1: 14016.

Boehm JD, Ibba MI, Kiszonas AM, See DR, Skinner DZ, et al. (2017) Identification of genotyping-by-sequencing sequence tags associated with milling performance and end-use quality traits in hard red spring wheat (*Triticum aestivum* L.). *J Cereal Sci* 77: 73–83.

Bordes J, Ravel C, Jaubertie JP, Duperrier B, Gardet O, et al. (2013) Genomic regions associated with the nitrogen limitation response revealed in a global wheat core collection. *Theor Appl Genet* 126: 805–822.

Botticella E, Sestili F, Sparla F, Moscatello S, Marri L, et al. (2018) Combining mutations at genes encoding key enzymes involved in starch synthesis affects the amylose content, carbohydrate allocation and hardness in the wheat grain. *Plant Biotechnol J* 16(10): 1723–1734.

Carlos Brevis J, Dubcovsky J (2010) Effects of the chromosome region including the *Gpc-B*1 locus on wheat grain and protein. *Crop Sci* 50:93–104.
Chang J, Zhang J, Mao X, Li A, Jia J, Jing R (2013) Polymorphism of *TaSAP1-A*1 and its association with agronomic traits in wheat. *Planta* 237(6):1495–508.
Chapman V, Riley R, Miller TE (1975) 'Annual Report Plant Breeding Institute 1974', Plant Breeding Institute, p125.
Chee PW, Elias EM, Anderson JA, Kianian SF (2001) Evaluation of a high grain protein QTL from *Triticum turgidum* L. var. *dicoccoides* in an adapted durum wheat background. *Crop Sci* 41: 295–301.
Chen A, Dubcovsky J (2012) Wheat TILLING mutants show that the vernalization gene *VRN1* down-regulates the flowering repressor *VRN2* in leaves but is not essential for flowering. *PLoS Genet* 8(12): e1003134
Chen X, Song G, Zhang S, Li Y, Gao J, et al. (2017) The allelic distribution and variation analysis of the *NAM-B*1 gene in Chinese wheat cultivars. *J Integr Agri* 16: 1294–1303.
Chhabra B, Singh L, Wallace S, Schoen A, Dong Y, Tiwari V, Rawat N (2021) Screening of an ethyl methane sulfonate mutagenized population of a wheat cultivar susceptible to fusarium head blight identifies resistant variants. *Plant Dis* 105(11): 3669–3676.
Colbert T, Till BJ, Tompa R, Reynolds S, Steine MN, et al. (2001) High-throughput screening for induced point mutations. *Plant Physiol* 126(2):480–4.
Comastri A, Janni M, Simmonds J, Uauy C, Pignone D, Nguyen HT, Marmiroli N (2018). Heat in wheat: Exploit reverse genetic techniques to discover new alleles within the *Triticum durum sHsp*26 family. *Front Plant Sci* 9:1337.
Delwiche SR, Norris KH (1993) Classification of hard red wheat by near-infrared diffuse reflectance spectroscopy. *Cereal Chem* 70: 29–29.
Distelfeld A, Tranquilli G, Li C, Yan L, Dubcovsky J (2009) Genetic and molecular characterization of the *VRN2* loci in tetraploid wheat. *Plant Physiol* 149(1): 245–257.
Distelfeld A, Uauy C, Fahima T, Dubcovsky J (2006) Physical map of the wheat high-grain protein content gene *Gpc-B1* and development of a high-throughput molecular marker. *New Phytol* 169: 753–763.
Dixon LE, Greenwood JR, Bencivenga S, Zhang P Cockram J, et al. (2018) *Teosinte Branched1* regulates inflorescence architecture and development in bread wheat (*Triticum aestivum*). *Plant Cell* 30: 563–581.
Dobrovolskaya O, Pont C, Sibout R, Martinek P, Badaeva E, et al. (2015) *Frizzy Panicle* drives supernumerary spikelets in bread wheat. *Plant Physiol* 167(1):189–99.
Dong H, Yan S, Liu J, Liu P, Sun J (2019) *TaCOLD1* defines a new regulator of plant height in bread wheat *Plant Biotechnol J* 17(3): 687–699.
Dubcovsky J, Dvorak J (2007) Genome plasticity a key factor in the success of polyploid wheat under domestication. *Science* 316: 1862–1866.
FAO (2020) Food and Agriculture Organization. www.fao.org/faostat/en/#data
Faris JD, Fellers JP, Brooks SA, Gill BS (2003) A bacterial artificial chromosome contig spanning the major domestication locus *Q* in wheat and identification of a candidate gene. *Genetics* 164(1): 311–321.
Fatiukha A, Filler N, Lupo I, Lidzbarsky G, Klymiuk V, et al. (2020) Grain protein content QTLs identified in a Durum × Wild Emmer wheat mapping population tested in five environments. *Theor Appl Genet* 133: 119–131.
Fekih R, Takagi H, Tamiru M, Abe A, Natsume S, et al. (2013) Mutmap+: genetic mapping and mutant identification without crossing in rice. *PLoS One* 8(7): e68529.
Feiz L, Martin JM, Giroux MJ (2009) Creation and functional analysis of new Puroindoline alleles in *Triticum aestivum*. *Theor Appl Genet* 118: 247–257.
Fernandez CGT, Nestor BJ, Danilevicz MF, Marsh JI, Petereit J, Bayer PE, Batley J, Edwards D (2022) Expanding gene-editing potential in crop improvement with pangenomes. *Int J Mol Sci* 23(4): 2276.
Flintham J, Borner A, Worland A, Gale M (1997) Optimizing wheat grain yield: Effects of Rht (gibberellin-insensitive) dwarfing genes. *J Agric Sci* 128(1): 11–25.
Flintham JE, Gale MD (1982) The Tom Thumb dwarfing gene, *Rht3* in wheat, I. Reduced pre-harvest damage to breadmaking quality. *Theor Appl Genet* 62(2):121–6.
Fu L, Yan D, Gao L, Liu P, Zhao G, et al. (2022) *TaIAA15* genes regulate plant architecture in wheat. *J Integr Agri* 21(5): 1243–1252.
Garg B, Lata C, Prasad M (2012) A study of the role of gene *TaMYB2* and an associated SNP in dehydration tolerance in common wheat. *Mol Biol Rep* 39(12):10865–71.

Garg D, Sareen S, Dalal S, Tiwari R, Singh R (2012) Heat shock protein based SNP marker for terminal heat stress in wheat (*Triticum aestivum* L.). *Aus J Crop Sci* 6(11): 1516–1521.

Gautam T, Amardeep, Saripalli G, Rakhi, Kumar A, et al. (2021) Introgression of a drought insensitive grain yield QTL for improvement of four Indian bread wheat cultivars using marker assisted breeding without background selection. *J Plant Biochem Biotechnol* 30: 172–183.

Gautam T, Saripalli G, Gahlaut V, Kumar A, Sharma PK, et al. (2019) Further studies on sugar transporter (SWEET) genes in wheat (*Triticum aestivum* L.). *Mol Biol Rep* 46(2):2327–2353.

Gegas VC, Nazari A, Griffiths S, Simmonds J, Fish L, et al. (2010) A genetic framework for grain size and shape variation in wheat. *Plant Cell* 22(4): 1046–1056.

Gilchrist EJ, O'Neil NJ, Rose AM, Zetka MC, Haughn GW (2006) TILLING is an effective reverse genetics technique for *Caenorhabditis elegans*. *BMC Genomics* 7: 262.

Gillies SA, Futardo A, Henry RJ (2012) Gene expression in the developing aleurone and starchy endosperm of wheat. *Plant Biotechnol J* 10: 668—679

Golan G, Ayalon I, Perry A, Zimran G, Ade-Ajayi T, Mosquna A, Peleg Z (2019) *GNI-A1* mediates trade-off between grain number and grain weight in tetraploid wheat. *Theor Appl Genet* 132(8): 2353–2365.

Goncharov NP (2003) Genetics of growth habit (spring vs winter) in common wheat: confirmation of the existence of dominant gene *Vrn4*. *Theor Appl Genet* 107(4): 768–772.

Gottwald S, Bauer P, Komatsuda T, Lundqvist U, Stein N (2009) TILLING in the two-rowed barley cultivar "Barke" reveals preferred sites of functional diversity in the gene *HvHox1*. *BMC Res Notes* 2: 258.

Grant NP, Mohan A, Sandhu D, Gill KS (2018) Inheritance and genetic mapping of the reduced height (*Rht18*) gene in wheat. *Plants* 7(3): 58.

Griffiths S, Sharp R, Foote TN, Bertin I, Wanous M, Reader S, Colas I, Moore G (2006) Molecular characterization of *Ph*1 as a major chromosome pairing locus in polyploid wheat. *Nature* 439(7077): 749–52.

Groos C, Robert N, Bervas E, Charmet G (2003) Genetic analysis of grain protein-content, grain yield and thousand-kernel weight in bread wheat. *Theor Appl Genet* 106: 1032–1040.

Guo H, Xiong H, Xie Y, Zhao L, Gu J, et al. (2019) Functional mutation allele mining of plant architecture and yield-related agronomic traits and characterization of their effects in wheat. *BMC Genet* 20:102.

Hagenblad J, Asplund L, Balfourier F, Ravel C, Leino MW (2012) Strong presence of the high grain protein content allele of *NAM-B1* in Fennoscandian wheat. *Theor Appl Genet* 125: 1677–1686.

Hawkesford MJ, Araus JL, Park R, Calderini D, Miralles D, et al. (2013) Prospects of doubling global wheat yields. *Food Energy Secur* 2(1): 34–48.

He F, Pasam R, Shi F, Kant S, Keeble-Gagnere G, et al. (2019) Exome sequencing highlights the role of wild-relative introgression in shaping the adaptive landscape of the wheat genome. *Nat Genet* 51: 896–904.

Hedden P (2003) The genes of the green revolution. *Trends Genet* 19(1): 5–9.

Henkrar F, El-Haddoury J, Iraqi D, Bendaou N, Udupa SM (2017) Allelic variation at high-molecular weight and low-molecular weight glutenin subunit genes in Moroccan bread wheat and durum wheat cultivars. *3 Biotech* 7(5):287.

Hosseinirad SA (2022) *Identification of Genomic Regions Associated with Color-related Traits in Durum Wheat*. [Doctoral dissertation, North Dakota State University]. ProQuest Dissertation Publishing.

Huang L, Raats D, Sela H, Klymiuk H, Lidzbarsky G, et al. (2016) Evolution and adaptation of wild emmer wheat populations to biotic and abiotic stresses. *Annu Rev Phytopathol* 54: 279–301.

Huang XQ, Cloutier S, Lycar L, Radovanovic N, Humphreys DG, et al. (2006) Molecular detection of QTLs for agronomic and quality traits in a doubled haploid population derived from two Canadian wheats (*Triticum aestivum* L.). *Theor Appl Genet* 113: 753–766.

Hussain B, Akpınar BA, Alaux M, Algharib AM, Sehgal D, et al. (2022) Capturing wheat phenotypes at the genome level. *Front Plant Sci* 13: 851079.

International Wheat Genome Sequencing Consortium (IWGSC) (2018) Shifting the limits in wheat research and breeding using a fully annotated reference genome. *Science* 361:eaar7191.

Itam M, Mega R, Tadano S, Abdelrahman M, Matsunaga S, et al. (2022) Metabolic and physiological responses to progressive drought stress in bread wheat. *Sci Rep* 10: 17189.

Jobson EM, Johnston RE, Oiestad AJ, Martin JM, Giroux MJ (2019) The impact of the wheat *Rht-B1b* semi-dwarfing allele on photosynthesis and seed development under field conditions. *Front Plant Sci* 51.

Joppa LR, Du C, Hart GE, Hareland GA (1997) Mapping gene(s) for grain protein in tetraploid wheat (*Triticum turgidum* L.) using a population of recombinant inbred chromosome lines. *Crop Sci* 37:1586–1589.

Juliana P, Singh RP, Braun HJ, Huerta-Espino J, Crespo-Herrera L, Govindan V, Mondal S, Poland J, Shrestha S (2020). Genomic selection for grain yield in the CIMMYT wheat breeding program-status and perspectives. *Front Plant Sci* 11: 564183.

Kaznina N, Dubovets N, Batova Y, Ignatenko A, Orlovskaya O, Repnika N (2021) The response of wheat with different allele statuses of the *Gpc-B1* gene under zinc deficiency. *Agronomy* 11(6): 1057.

Klymiuk V, Fatiukha A, Fahima T (2019a) Wheat tandem kinases provide insights on disease resistance gene flow and host-parasite co-evolution. *Plant J* 98 (4): 667–679.

Klymiuk V, Fatiukha A, Huang L, Wei ZZ, Kis-Papo T, et al. (2019b) Durum wheat as a bridge between wild emmer wheat genetic resources and bread wheat. In: Miedaner T, Korzun V (eds) *Application of genetic and genomic research in cereals, Woodhead publishing series in food science, technology and nutrition* Woodhead Publishing, Cambridge, pp 201–230.

Krasileva KV, Vasquez-Gross HA, Howell T, Bailey P, Paraiso F, et al. (2017) Uncovering hidden variation in polyploid wheat. *Proc Natl Acad Sci USA* 114(6): E913–E921.

Krugman T, Nevo E, Beharav A, Sela H, Fahima T, et al. (2018) The institute of evolution wild cereal gene bank at the University of Haifa. *Isr J Plant Sci* 65: 129–146.

Kulwal P, Kumar N, Kumar A, Gupta R, Balyan H, et al. (2005) Gene networks in hexaploid wheat: interacting quantitative trait loci for grain protein content. *Funct Integr Genomics* 5: 254–259.

Kumar GR, Sakthivel K, Sundaram RM, Neeraja CN, Balachandran SM, Rani NS, Viraktamath BC, Madhav MS (2010) Allele mining in crops: prospects and potentials. *Biotechnol Adv* 28(4):451–61.

Kumar S, Singh VP, Saini DK, Sharma H, Saripalli G, et al. (2021) Meta-QTLs, ortho-MQTLs, and candidate genes for thermotolerance in wheat (*Triticum aestivum* L.). *Mol Breed* 41: 69.

Kunert A, Naz AA, Dedeck O, Pillen K, Leon J (2007) AB-QTL analysis in winter wheat: I. Synthetic hexaploid wheat (*T turgidum* ssp *dicoccoides* x *T tauschii*) as a source of favourable alleles for milling and baking quality traits. *Theor Appl Genet* 115: 683–695.

Kuraparthy V, Sood S, Dhaliwal HS, Chhuneja P, Gill BS (2007) Identification and mapping of a tiller inhibition gene (*Tin3*) in wheat. *Theor Appl Genet* 114(2): 285–294.

Kuzay S, Lin H, Li C, Chen S, Woods DP, Zhang J, Dubcovsky J (2022) *WAPO-A1* is the causal gene of the 7AL QTL for spikelet number per spike in wheat. *PLoS Genet* 18(1): e1009747.

Law CN, Payne PI (1983) Genetical aspects of breeding for improved grain protein content and type in wheat. *J Cereal Sci* 1: 79–93.

Lawrence GJ, Shepherd KW (1981) Chromosomal location of genes controlling seed proteins in species related to wheat. *Theor Appl Genet* 59: 25–31.

Leonova IN, Kiseleva AA, Berezhnaya AA, Stasyuk AL, Likhenko IE, et al. (2022) Identification of QTLs for grain protein content in Russian spring wheat varieties. *Plants* 11(3): 437.

Lethin J, Shakil SSM, Hassan S, Sirijovski N, Töpel M, Olsson O, Aronsson H (2020) Development and characterization of an EMS-mutagenized wheat population and identification of salt-tolerant wheat lines. *BMC Plant Biol* 20: 18.

Li GZ, Li HX, Xu MJ, Wang PF, Xiao XH, Kang GZ (2020) Functional characterization and regulatory mechanism of wheat *CPK34* kinase in response to drought stress. *BMC Genom* 21: 577.

Li Y, Li L, Zhao M, Guo L, Guo X, et al. (2021) Wheat *Frizzy Panicle* activates *Vernalization1-A* and *Homeobox4-A* to regulate spike development in wheat. *Plant Biotechnol J* 19: 1141–1154.

Lin G, Chen H, Tian B, Sehgal SK, Singh L, et al. (2022) Cloning of the broadly effective wheat leaf rust resistance gene *Lr42* transferred from *Aegilops tauschii*. *Nat Commun* 13: 3044.

Lozada DN, Ward BP, Carter AH (2020) Gains through selection for grain yield in a winter wheat breeding program. *PLoS One* 15(4): e0221603.

Maccaferri M, Harris NS, Twardziok SO, Pasam RK, Gundalch H, et al. (2019) Durum wheat genome highlights past domestication signatures and future improvement targets. *Nat Genet* 51:885–895.

Makhloufi E, Yousfi FE, Pirrello J, Bernadac A, Ghorbel A, Bouzayen M (2015) *TdERF1*, an ethylene response factor associated with dehydration responses in durum wheat (*Triticum turgidum* L. subsp. *durum*). *Plant Signal Behav* 10(10):e1065366.

Mann G, Diffey S, Cullis B, Azanza F, Martin D, et al. (2009) Genetic control of wheat quality: interactions between chromosomal regions determining protein content and composition, dough rheology, and sponge and dough baking properties. *Theor Appl Genet* 118: 1519–1537.

McCallum CM, Comai L, Greene EA, Henikoff S (2000) Targeting induced local lesions IN genomes (TILLING) for plant functional genomics. *Plant Physiol* 123(2):439–442.

Merchuk-Ovnat L, Barak V, Fahima T, Ordon F, Lidzbarsky GA, Krugman T, Saranga Y (2016) Ancestral QTL alleles from wild emmer wheat improve drought resistance and productivity in modern wheat cultivars. *Front Plant Sci* 7:452.

Merrick LF, Herr AW, Sandhu KS, Lozada DN, Carter AH (2022) Utilizing genomic selection for wheat population development and improvement. *Agronomy* 12(2): 522.

Mesfin A, Frohberg RC, Anderson JA (1999) RFLP markers associated with high grain protein from *Triticum turgidum* L. var. *dicoccoides* introgressed into hard red spring wheat. *Crop Sci* 39: 508–513.

Michel S, Ametz C, Gungor H, Epure D, Grausgruber H, et al. (2016) Genomic selection across multiple breeding cycles in applied bread wheat breeding. *Theor Appl Genet* 129, 1179–1189.

Mishra BP, Kumar R, Mohan A, Gill KS (2017) Conservation and divergence of Starch Synthase III genes of monocots and dicots. *PLoS One* 12(12):e0189303.

Mo Y, Howell T, Vasquez-Gross H, de Haro LA, Dubcovsky J, Pearce S (2018) Mapping causal mutations by exome sequencing in a wheat TILLING population: a tall mutant case study. *Mol Genet Genom* 293(2):463–477.

Moehs CP, Austill WJ, Holm A, Large TAG, Loeffler D, et al. (2019) Development of decreased-gluten wheat enabled by determination of the genetic basis of lys3a barley. *Plant Physiol* 179(4):1692–1703.

Muqaddasi QH, Brassac J, Koppolu R, Plieske J, Ganal MW, Röder MS (2019) *TaAPO-A1*, an ortholog of rice *ABERRANT PANICLE ORGANIZATION 1* is associated with total spikelet number per spike in elite European hexaploid winter wheat (*Triticum aestivum* L) varieties. *Sci Rep* 9(1): 1–12.

Naraghi SM, Simsek S, Kumar A, Al Rabbi SH, Alamri MS, et al. (2019) Deciphering the genetics of major end-use quality traits in wheat. *G3: Genes Genome Genet* 9(5): 1405–1427.

Nilsen KT, Walkowiaka S, Xiang D, Gao P, Quilichini TD, et al. (2020) Copy number variation of *TdDof* controls solid-stemmed architecture in wheat. *Proc Natl Acad Sci USA* 117(46): 28708–28718.

Niu J, Zheng S, Shi X, Si Y Tian S, He Y, Ling HQ (2020) Fine mapping and characterization of the awn inhibitor B1 locus in common wheat (*Triticum aestivum* L). *Crop J* 8: 613–622.

Okada T, Jayasinghe JEARM, Eckermann P, Watson-Haigh NS, Warner P, et al. (2020) Genetic factors associated with favourable pollinator traits in the wheat cultivar Piko. *Funct Plat Biol* 48: 434–447.

Olmos S, Distelfeld A, Chicaiza O, Schlatter AR, Fahima T, et al. (2003) Precise mapping of a locus affecting grain protein content in durum wheat. *Theor Appl Genet* 107: 1243–1251.

Oyiga BC, Ogbonnaya FC, Sharma RC, Baum M, Léon J, Ballvora A (2019) Genetic and transcriptional variations in *NRAMP-2* and *OPAQUE1* genes are associated with salt stress response in wheat. *Theor Appl Genet* 132(2):323–346.

Pearce S (2021) Towards the replacement of wheat 'Green Revolution'genes. *J Exp Bot* 72(2): 157–160.

Payne PI, Nightingale MA, Krattiger AF, Holt LM (1987) The relationship between HMW glutenin subunit composition and the bread-making quality of British-grown wheat varieties. *J Sci Food Agric* 40(1): 51–65.

Peng J, Richards DE, Hartley NM, Murphy GP, Devos KM, Flintham JE, et al. (1999) 'Green revolution' genes encode mutant gibberellin response modulators. *Nature* 400: 256–261.

Peterson CJ, Graybosch RA, Shelton DR, Baenziger PS (1998) Baking quality of hard winter wheat: Response of cultivars to environment in the great plains. *Euphytica* 100(1): 157–162.

Prasad MN, Kumar PL, Kulwal MS, Roder HS, Balyan, et al. (2003) QTL analysis for grain protein content using SSR markers and validation studies using NILs in bread wheat. *Theor Appl Genet* 106: 659–667.

Quraishi UM, Pont C, Ain Q-u, Flores R, Burlot L, et al. (2017) Combined genomic and genetic data integration of major agronomical traits in bread wheat (*Triticum aestivum* L.). *Front Plant Sci* 8:1843.

Rai N, Bellundagi A, Kumar PSC, Thimmappa RK, Rani S, et al. (2018) Marker-assisted backcross breeding for improvement of drought tolerance in bread wheat (*Triticum aestivum* L. em Thell). *Plant Breed* 137(4): 514–526.

Rawat N, Joshi A, Pumphrey M, Singh L, Mahlandt A, et al. (2019) A TILLING resource for hard red winter wheat variety jagger. *Crop Sci* 59(4): 1666–1671.

Rawat N, Schoen A, Singh L, Mahlandt A, Wilson DL, et al. (2018). TILL-D: An *Aegilops tauschii* TILLING resource for wheat improvement. *Front Plant Sci* 9: 1665.

Rawat N, Sehgal SK, Joshi A, Rothe N, Wilson DL, et al. (2012). A diploid wheat TILLING resource for wheat functional genomics. *BMC Plant Biol* 12: 205.

Ray DK, Mueller ND, West PC, Foley JA (2013) Yield trends are insufficient to double global crop production by 2050. *PLoS One* 8(6): e66428.

Rebetzke GJ, Ellis MH, Bonnett DG, Condon AG, Falk D, Richards RA (2011) The *Rht13* dwarfing gene reduces peduncle length and plant height to increase grain number and yield of wheat. *Field Crops Res* 124(3): 323–331.
Rousset M, Carrillo JM, Qualset CO, Kasarda DD (1992) Use of recombinant inbred lines of wheat for study of associations of high-molecular-weight glutenin subunit alleles to quantitative traits. *Theor Appl Genet* 83(4): 403–412.
Royo A, Abio D (2003) Salt tolerance in durum wheat cultivars. *Span J Agric Res* 1: 3.
Saini DK, Srivastava P, Pal N, Gupta PK (2022) Meta-QTLs, ortho-meta-QTLs and candidate genes for grain yield and associated traits in wheat (*Triticum aestivum* L.). *Theor Appl Genet* 135(3): 1049–1081.
Sakuma S, Golan G, Guo Z, Ogawa T, Tagiri A, Sugimoto K, Komatsuda T (2019) Unleashing floret fertility in wheat through the mutation of a homeobox gene. *Proc Natl Acad Sci* 116(11): 5182–5187.
Salgotra RK, Stewart CN Jr (2020) Functional markers for precision plant breeding. *Int J Mol Sci* 21(13): 4792.
Salsman E, Liu Y, Hosseinirad SA, Kumar A, Manthey F, et al. (2021) Assessment of genetic diversity and agronomic traits of durum wheat germplasm under drought environment of the northern Great Plains. *Crop Sci* 61(2): 1194–1206.
Sandhu K, Patil SS, Pumphrey M, Carter A (2021). Multitrait machine- and deep-learning models for genomic selection using spectral information in a wheat breeding program. *Plant Genome* 14(3): e20119.
Schneeberger K, Ossowski S, Lanz C, Juul T, Petersen AH, et al. (2009) SHOREmap: simultaneous mapping and mutation identification by deep sequencing. *Nat Methods* 6(8): 550–1.
Schoen A, Joshi A, Tiwari V, Gill BS, Rawat N (2021) Triple null mutations in starch synthase *SSIIa* gene homoeologs lead to high amylose and resistant starch in hexaploid wheat. *BMC Plant Biol* 21(1): 74.
Schreiber M, Barakate A, Uzrek N (2019) A highly mutagenised barley (cv. Golden Promise) TILLING population coupled with strategies for screening-by-sequencing. *Plant Methods* 15: 99.
Sehgal D, Mondal S, Guzman C, Barrios GG, Franco C, et al. (2019) Validation of candidate gene-based markers and identification of novel loci for thousand-grain weight in spring bread wheat. *Front Plant Sci* 10: 1189.
Sestili F, Botticella E, Bedo Z et al. (2010) Production of novel allelic variation for genes involved in starch biosynthesis through mutagenesis. *Mol Breed* 25: 145–154.
Shewry PR, Hey SJ (2015) The contribution of wheat to human diet and health. *Food Energy Secur* 4: 178–202.
Shiferaw B, Smale M, Braun HJ, Duveiller E, Reynolds M, et al. (2013) Crops that feed the world 10. Past successes and future challenges to the role played by wheat in global food security. *Food Sec* 5: 291–317.
Singh K, Batra R, Sharma S, Saripalli G, Gautam T, Singh R, et al. (2021) WheatQTLdb: a QTL database for wheat. Mol Genet Genom 296(5): 1051–1056. Epub 2021 Jun 11. Erratum in: Mol Genet Genomics.
Singh K, Saini DK, Saripalli G, Gautam T, Rakhi, et al. (2022) WheatQTLdb V2.0: a supplement to the database for wheat QTL. *Mol Breed* 42: 56 (2022).
Singh L, Schoen A, Mahlandt A, Chhabra B, Steadham J, et al. (2019). Development of targeting induced local lesions in genomes (tilling) populations in small grain crops by ethyl methanesulfonate mutagenesis. *J Vis Exp* 149: e59743.
Slade AJ, McGuire C, Loeffler D, Mullenberg J, Skinner W, et al. (2012). Development of high amylose wheat through TILLING. *BMC Plant Biol* 12: 69.
Sormacheva I, Golovnina K, Vavilova V, Kosuge K, Watanabe N, Blinov A, Goncharov NP (2015) Q gene variability in wheat species with different spike morphology. *Genet Resour Crop Evol* 62(6): 837–852.
Sourdille P, Cadalen T, Guyomarch TH, Snape JW, Perretant MR, et al. (2003) An update of the Courtot × Chinese Spring intervarietal molecular marker linkage map for the QTL detection of agronomic traits in wheat. *Theor Appl Genet* 106: 530—538.
Stone PJ, Savin R (1999) Grain quality and its physiological determinants. In: Satorre EH, Slafer GA (eds) *Wheat: A Physiological Ecological Approach to Understand Yield and its Determining Process at the Crop Level of Organization.* Food Product Press, New York, pp 85–120.
Tabbita F, Pearce S, Barneix AJ (2017) Breeding for increased grain protein and micronutrient content in wheat: ten years of the *GPCB*1 gene. *J Cereal Sci* 73: 183–191.
Takagi H, Abe A, Yoshida K, Kosugi S, Natsume S, et al. (2013a) QTL-seq: rapid mapping of quantitative trait loci in rice by whole genome resequencing of DNA from two bulked populations. *Plant J* 74(1): 174–83.
Takagi H, Uemura A, Yaegashi H, Tamiru M, Abe A, Mitsuoka C, et al. (2013b) MutMap-Gap: whole-genome resequencing of mutant F2 progeny bulk combined with de novo assembly of gap regions identifies the rice blast resistance gene *Pii*. *New Phytol* 200(1): 276–283.

Talamè V, Bovina R, Sanguineti MC, Tuberosa R, Lundqvist U, Salvi S (2008). TILLMore, a resource for the discovery of chemically induced mutants in barley. *Plant Biotechnol J* 6(5): 477–485.

Thiyagarajan K, Mathur PN, Vikram P, Sehgal D, Valluru R, et al. (2022) Allele mining, evolutionary genetic analysis of *TaHKT1.5* gene and evaluation of salinity stress in selected lines of wheat. *bioRxiv* doi: https://doi.org/10.1101/2022.04.01.486792

Thomas SG (2017) Novel *Rht-1* dwarfing genes: tools for wheat breeding and dissecting the function of DELLA proteins. *J Exp Bot* 68(3): 354–358.

Thudi M, Palakurthi R, Schnable JC, Chitkineni A, Dreisigacker S, et al. (2021) Genomic resources in plant breeding for sustainable agriculture. *J Plant Physiol* 257: 153351.

Tian X, Xia X, Xu D, Liu Y, Xie L, Hassan MA, Cao S (2022) *Rht24b*, an ancient variation of *TaGA2ox-A9*, reduces plant height without yield penalty in wheat. *New Phytol* 233(2): 738–750.

Till BJ, Cooper J, Tai TH, Colowit P, Greene EA, Henikoff S, Comai L (2007). Discovery of chemically induced mutations in rice by TILLING. *BMC Plant Biol* 7: 19.

Tiwari C, Wallwork H, Arun B, Mishra VK, Velu G, et al. (2016) Molecular mapping of quantitative trait loci for zinc, iron and protein content in the grains of hexaploid wheat. *Euphytica* 207: 563–570.

Triboï-Blondel A, Triboï E, Martre P (2003) Environmentally induced changes in protein composition in developing grains of wheat are related to changes in total protein content. *J Exp Bot* 54:1731–1742.

Uauy C, Distelfeld A, Fahima T, Blechl A, Dubcovsky J (2006) A NAC gene regulating senescence improves grain protein, zinc, and iron content in wheat. *Science* 314(5803):1298–301.

Uauy C, Paraiso F, Colasuonno P, Tran RK, Tsai H, Berardi S, Comai L, Dubcovsky J (2009) A modified TILLING approach to detect induced mutations in tetraploid and hexaploid wheat. *BMC Plant Biol* 9: 115.

Van De Velde K, Chandler PM, Van Der Straeten D, Rohde A (2017) Differential coupling of gibberellin responses by Rht-B1c suppressor alleles and *Rht-B1b* in wheat highlights a unique role for the DELLA N-terminus in dormancy. *J Exp Bot* 68(3): 443–455.

Van De Velde K, Thomas SG, Heyse F, Kaspar R, Van Der Straeten D, Rohde A (2021) N-terminal truncated *RHT-1* proteins generated by translational reinitiation cause semi-dwarfing of wheat Green Revolution alleles. *Mol Plant* 14(4): 679–687.

Walkowiak S, Gao L, Monat C, Haberer G, Kassa MT, et al. (2020) Multiple wheat genomes reveal global variation in modern breeding. *Nature* 588: 277–283.

Wang D, Cao Z, Wang W, Zhu W, Hao X, et al. (2020) Genome-wide characterization of OFP family genes in wheat (*Triticum aestivum* L.) reveals that *TaOPF29a-A* promotes drought tolerance. *Biomed Res Int* 7: 9708324.

Wang Y, Yu H, Tian C, Sajjad M, Gao C, Tong Y, Jiao Y (2017b) Transcriptome association identifies regulators of wheat spike architecture. *Plant Physiol* 175(2): 746–757.

Wei B, Jing R, Wang C, Mao X, Chang X, Jia Z (2009) *Dreb1* genes in wheat (*Triticum aestivum* L.): development of functional markers and gene mapping based on SNPs. *Mol Breed* 23: 13–22.

Winfield M, Allen A, Wilkinson P, Burridge A, Barker G, et al. (2017) High-density genotyping of the A. E. Watkins Collection of hexaploid landraces identifies a large molecular diversity compared to elite bread wheat. *Plant Biotechnol J* 16(1):165–175.

Winkler S, Schwabedissen A, Backasch D, Bökel C, Seidel C, et al. (2005) Target-selected mutant screen by TILLING in Drosophila. *Genome Res* 15(5): 718–723.

Wolde GM, Mascher M, Schnurbusch T (2019) Genetic modification of spikelet arrangement in wheat increases grain number without significantly affecting grain weight. *Mol Genet Genom* 294: 457–468.

Worland AJ, Law CN, Shakoor A (1980) The genetical analysis of an induced height mutant in wheat. *Heredity* 45: 61–71.

Wrigley CW, Shepherd KW (1973) Electro-focusing of grain proteins from wheat genotypes. *Ann NY Acad Sci* 209: 154–162.

Xiong H, Zhou C, Fu M, Guo H, Xie Y, Zhao L, Liu L (2022) Cloning and functional characterization of *Rht8*, a "Green Revolution" replacement gene in wheat. *Mol Plant* 15(3): 373–376.

Yamamori M, Fujita S, Hayakawa K, Matsuka J, Yusai T (2000) Genetic elimination of a starch granule protein, *SGP-1,* of wheat generates an altered starch with apparent high amylose. *Theor Appl Genet* 101: 21–29.

Yan L, Fu D, Li C, Blechl A, Tranquilli G, Bonafede M, et al. (2006) The wheat and barley vernalization gene *VRN3* is an orthologue of FT. *Proc Natl Acad Sci USA* 103(51): 19581–19586.

Yan L, Helguera M, Kato K, Fukuyama S, Sherman J, Dubcovsky J (2004) Allelic variation at the *VRN-1* promoter region in polyploid wheat. *Theoret Appl Genet* 109(8): 1677–1686.

Yan L, Loukoianov A, Tranquilli G, Helguera M, Fahima T, Dubcovsky J (2003) Positional cloning of the wheat vernalization gene *VRN1*. *Proc Natl Acad Sci USA* 100(10): 6263–6268.

Yoshioka M, Iehisa JCM, Ohno R, Kimura T, Enoki H, Nishimura S, Nasuda S, Takumi S (2017) Three dominant awnless genes in common wheat: Fine mapping, interaction and contribution to diversity in awn shape and length. *PLoS One* 12(4): e0176148

Youssefian S, Kirby EJM, Gale MD (1992) Pleiotropic effects of the GA-insensitive Rht dwarfing genes in wheat. 2. Effects on leaf, stem, ear and floret growth. *Field Crops Res* 28(3): 191–210.

Zanetti S, Winzeler M, Feuillet C, Keller B, Messmer M (2001) Genetic analysis of bread-making quality in wheat and spelt. *Plant Breed* 120: 13–19.

Zhai H, Feng Z, Du X, Song Y, Liu X, Qi Z, Ni Z (2018) A novel allele of *TaGW2-A1* is located in a finely mapped QTL that increases grain weight but decreases grain number in wheat (*Triticum aestivum* L). *Theor Appl Genet* 131(3): 539–553.

Zhang J, Li B, Yang Y, Mu P, Qian W, et al. (2016). A novel allele of *L-galactono-1,4-lactone dehydrogenase* is associated with enhanced drought tolerance through affecting stomatal aperture in common wheat. *Sci Rep* 6: 30177.

Zhang W, Zhang L, Qiao L, Wu J, Zhao G, Jing R, Jia J (2013) Cloning and haplotype analysis of *TaSTE*, which is associated with plant height in bread wheat (*Triticum aestivum* L). *Mol Breed* 31(1): 47–56.

Zhang X, Jia H, Li T, Wu J, Nagarajan R, et al. (2022) *TaCol-B5* modifies spike architecture and enhances grain yield in wheat. *Science* 376 (6589): 180–183.

Zhang Y, Li D, Zhang D, Zhao X, Cao X, et al (2018). Analysis of the functions of *TaGW2* homoeologs in wheat grain weight and protein content traits. *Plant J* 94(5):857–866.

Zhao LA, Zhang KP, Liu B, Deng ZY, Qu HL, et al (2010) A comparison of grain protein content QTLs and flour protein content QTLs across environments in cultivated wheat. *Euphytica* 174: 325–335.

Zulfiqar S, Ishfaq S, Ikram M, Nawaz MA, Rahman MU (2021) Characterization of gamma-rays-induced spring wheat mutants for morphological and quality traits through multivariate and gt bi-plot analysis. *Agronomy* 11(11): 2288.

4 Allele Mining in Wheat
Prospects and Potentials

Vijay Gahlaut, Hemant Sharma, Sachin Kumar, and Vandana Jaiswal*
*Corresponding author: zone4vijay@gmail.com

4.1 INTRODUCTION

Wheat (*Triticum aestivum* L.), an important cereal is a potential food source for the majority of the population (Yin et al., 2018). An important crop, bread wheat (*Triticum aestivum* L), arose from a two-step cross-hybridization (Pourkheirandish et al. 2020). As a result, bread wheat has three different genomes (A, B, and D genomes) with three different copies of each functional gene. Due to their highly conserved sequences and structures, genes tend to have similar functions among homeologous sub-genomes. However, this genomic redundancy can lead to plasticity, causing functional gene diversity and sub-functionalization among homeologous copies undergoing natural and artificial selection. Advances in plant breeding techniques continuously develop superior, high-yielding wheat varieties. In other words, identifying and accessing allelic variations that affect plant phenotypes are critical for the efficient use of germplasm resources, such as the development of plant cultivars (Sharma et al. 2014; Kumari et al. 2018). The germplasm collection is believed to be rich in undisclosed allelic variants due to the large number of accessions held collectively by gene banks. Crop improvement in plant breeding aimed at developing high-yielding superior cultivar responses to biotic and abiotic stress. In nature, the greatest genetic variation exists in the plant gene pool for various biotic and abiotic resistance genes (Lamaoui et al. 2018). Taking advantage of these genetic variations to identify and refine novel and superior alleles of various agronomically important genes from the available gene pool and use them to develop improved high-yielding cultivars is important. However, a significant portion of these favorable/superior alleles were left behind during evolution and domestication and were therefore not used. This untapped genetic variation present in crop wild relatives and landraces is profitably used to develop agronomically superior varieties.

Appropriate plant germplasms are therefore considered as they form the backbone of "smart breeding" programs that produce improved varieties by introducing useful alleles from potential wild relatives. Although high-fidelity and extensive information about new alleles, individual germplasm facilitates "allele mining" to improve both qualitative and quantitative trait expression within new and improved cultivars (Raddy et al. 2014). Allele mining is one of the most widely used approaches in crop improvement to analyze unique allelic variation. Literally, this involves analysis of sequence variation in both coding and non-coding regions of candidate genes for agronomically important traits (Kumar et al. 2010). The DNA sequence of the tested genotype is used to identify the type and location of the chromosomal variation by comparing it to naturally occurring polymorphisms in the reference genotype. Allele mining helps to understand the latest allelic evolution, identify current haplotypes, and detect allele-specific molecular markers for further use in molecular breeding.

DOI: 10.1201/9781003385004-4

4.2 TYPES OF ALLELE MINING

Allele mining is the approach to identify useful alleles of the genes across a wide range of accessions of the same species, related species, and different species with the aim of identifying allelic variations relevant to trait distribution. Mutation in germplasm is thought to be the evolutionary driving force underlying the allelic diversity present in all crops, including wheat. Mutations occurring in the genome as single nucleotide polymorphisms (SNPs), or insertions and deletions (InDels) are essential for the generation of new alleles or changes in existing alleles and allele combinations. The procedure of allele mining is shown in Figure 4.1.

4.2.1 True Allele Mining

Mutations in coding and/or regulatory regions may have a significant effect on plant phenotype. The mutation may alter the structure and function of the encoded protein or may change the regulation of gene expression. Allele mining is one of the most popular approaches used in the field of plant improvement to analyze inherent allelic variation. For true allele mining, it is necessary to analyze sequence variations in both coding and non-coding regions of the genome that are associated with agronomically important traits (Kumar et al. 2010). The DNA sequences of the tested genotypes are used to identify the type and location of the point mutation by comparing it to naturally occurring polymorphisms in the reference genotype (Figure 4.1).

4.2.2 Promoter Allele Mining

Over the years, it has been clearly demonstrated that allele mining is also important for identifying non-coding SNP variants in the context of gene expression, particularly promoter. The promoter elements are known to play an important role in gene expression because transcription binding sites are present in the promoter region and initiate transcription of the template DNA only after the RNA polymerase enzyme binds to the promoter region. Genic regions are supposed to be highly conserved; however, significant mutations can be observed in promoter sequence, which ultimately regulates gene expression. Allele mining in promoter sequences is depicted in Figure 4.2.

4.3 APPROACHES FOR ALLELE MINING

Important alleles associated with desirable phenotypes can be mined by two approaches, including (i) Eco-Tilling, (ii) sequencing (Figure 4.3). Sequencing approach may be further divided into two categories: (a) targeted and (b) whole genome. To identify different alleles in targeted genes or genomic regions, Sanger sequencing may be followed; however, for the determination of alleles from the whole genome, next-generation sequencing platforms are useful.

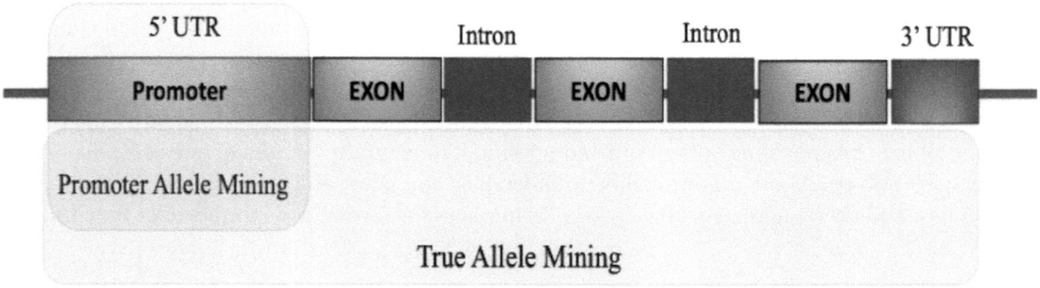

FIGURE 4.1 Schematic representation of the steps involves in allele mining.

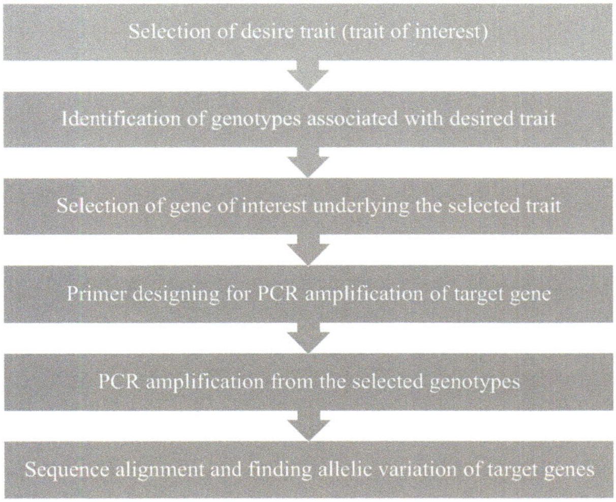

FIGURE 4.2 Schematic representation of the steps involves in allele mining.

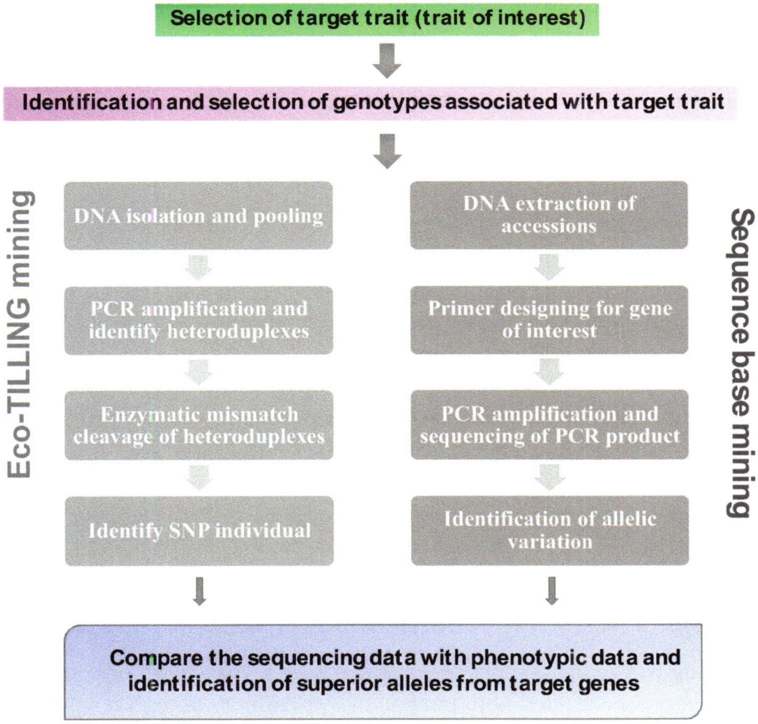

FIGURE 4.3 Steps and comparison of allele mining techniques.

4.3.1 Eco-TILLING

Eco-TILLING (Targeting Induced Original Lesions in Genomes), a modified tillage method, is used to identify single-nucleotide allelic variants (more specifically, convincing point mutations) in target genes (Comai et al. 2004; Barkley and Wang 2008). This approach identifies mutations in the gene of interest (Till et al. 2003). A large number of individual variations can be discovered quickly, and for each haplotype, the only need to sequence one existing variation once, making it cost-effective. Tillage uses chemical mutagens to introduce random mutations. Seeds are mutagenized with a chemical mutagen, ethyl methanesulfonate (EMS), to induce a single-nucleotide point mutation (G:C to A/T) (Nagy et al. 2006). M1 seeds self-pollinate to produce the M2 seed population. Screen for point mutations using M2 progeny from a single seed line. In screening, DNA is pooled in various ways to maximize the efficiency of mutation detection. Polymerase chain reaction (PCR) is performed using 5' prime end-specific primers targeting the locus of interest, and the PCR products are heated and cooled to form heteroduplexes. The CEL I nuclease enzyme is used to cleave the base mismatch, and the products representing the induced mutations are visualized using denaturing polyacrylamide gel electrophoresis. These PCR products are denatured and reannealed to allow mismatched or heteroduplex conformations representing natural SNPs and convincing SNPs. In the case of wheat, Eco-TILLING is successfully utilized for the identification of alleles in *VRN-A1* gene associated with vernalization (Liang et al. 2011) and *Pina/Pin b* genes associated with kernel hardness (Ma et al. 2017).

4.3.2 Sequencing-Based Allele Mining

With the advancements in sequencing techniques, allele mining through sequencing became cheaper and more precise. It provides higher resolution, i.e., at the DNA base level (Margulies et al. 2005).

4.3.2.1 Targeted Sequencing for Allele Mining in Target Genes

In targeted sequence-based allele mining, the gene/genomic region of interest is amplified, and sequencing of the amplicon takes place in different genotypes. Sequences of target gene/genomic region of different genotypes are aligned, and bioinformatic tools are utilized to identify different alleles, mainly in the form of SNPs or InDels.

4.3.2.2 Next-Generation Sequencing for Whole Genome Allele Mining

Next-generation sequencing allows the identification of different alleles from the whole genome. Although, the number of alleles at a particular locus depends on the number of accessions used for allele mining, utilization of a greater number of accessions will increase the chances of getting a higher number of alleles.

4.4 ASSOCIATION MAPPING: AN APPROACH TO MINE DESIRABLE ALLELES

Association mapping is a potential approach for the genetic dissection of complex traits. It identifies the genomic regions associated with phenotypic variation with high resolution. Along with the identification of genomic regions, it also allows the identification of desirable alleles associated with superior phenotypes. In wheat, several superior alleles have been reported using the genome-wide association study (GWAS) approach, which are associated with yield-related traits and Fusarium head blight resistance (Kollers et al. 2013; Li et al. 2020; Gahlaut et al. 2021). For instance, Li et al. (2020) identified 32 superior allele loci, each exhibiting a positive effect for two or three yield-related traits. Gahlaut et al. (2021) reported 30 alleles associated with four-grain weight-related traits (days to anthesis, grain filling duration, grain number per ear, and grain weight per ear). Kollers et al. (2013) detected 323 SSR alleles (associated with 794 MTAs) controlling Fusarium head blight resistance in a panel of 372 European wheat varieties.

4.5 CONSIDERATIONS FOR ALLELE MINING

Successful and efficient allele mining is based on several parameters, which need attention. Genetic and genomic resources may vary from crop to crop. In some crops including wheat, thousands of accessions are available. These include released varieties, modern cultivars, breeding lines, landraces and wild relatives, etc. Such kind of germplasm collection may lead to identification of a large number of alleles, including desirable alleles.

Similarly, in wheat, genomic resources are also well available. These genomic information eases the process of allele mining with higher precision. Overall, the following parameters are found critical for precise and successful allele mining.

1 *Germplasm selection*: The selection of germplasm is crucial for the identification of representative alleles of a plant species. Germplasm should be highly diverse and contain the maximum diversity. One may also construct a core collection representing whole diversity from a huge germplasm collection and use it for allele mining. Development of core collection reduces the workload of genotyping and is therefore cost-effective too. In the case of wheat, there are more than 850,000 accessions of wheat, and its wild relatives are available worldwide in different genebanks (Guzzon et al., 2022).
2 *Size and relativeness*: Chances of getting more alleles is directly proportional to the sample size and diversity among the individual of the studied population. For mining for a higher number of alleles, individuals within the population should be less related.
3 *Information of candidate genes*: Knowledge of genomic/gene sequences of the target regions facilitates the primer designing and precise amplification and sequencing of the target region.
4 *Phenotypic characterization*: Precise phenotyping enables identifying the genomic regions associated with the trait of interest along with the desirable marker/gene's allele for an improved phenotype. In wheat, phenomic platforms are evolving that help in high throughput and precise phenotyping (Yang et al. 2020); followed by the identification of desirable alleles for wheat improvement.

4.6 BIOINFORMATICS TOOLS REQUIRED FOR ALLELE MINING

A number of bioinformatics tools and databases have been developed to ease the different steps and processes related to allele mining (Table 4.1). These tools can be utilized for different analysis, such as processing of large and complex genome, alignment of sequences, SNP calling, prediction of regulatory elements and binding cites in promoter, identification of polymorphism, and their effect of gene and protein structure and function, etc.

4.7 STATUS OF ALLELE MINING IN WHEAT

In wheat, the number of alleles has been identified for different genes responsible for different traits (Table 4.2). For instance, over 100 *Pm* genes/alleles (*Pm1–Pm68, Pm8* is the allele of *Pm17, Pm18* = *Pm1c, Pm22* = *Pm1e, Pm23* = *Pm4c,* and *Pm31* = *Pm21*) have been identified in soft wheat and its relatives; which indicates that there are abundant genetic resources for controlling powdery mildew (Blanco et al. 2008; Perugini et al. 2008; McIntosh et al. 2020; He et al. 2021). However, most of these documented Pm genes/alleles cannot be used directly for wheat breeding due to the observed resistance to intergenic coupling. Resistance breeding requires constant efforts to enrich the reservoir of resistance genes in wheat. More than 37 powdery mildew resistance genes have been characterized in wheat.

TABLE 4.1
List and Details of Different Bioinformatics Tools Used in Allele Mining

S.No.	Tool	Function	URL
1	MEME	Multiple EM for Motif Elicitation. Analyzing DNA and protein sequence motifs for similarities among them.	meme.nbcr.net/meme/cgi-bin/meme.CGI (Bailey et al. 2006)
2	JASPAR	Database of Transcription Factor Binding Site (TFBS)	http://jaspar.genereg.net/ (Bryne et al. 2008)
3	AGRIS	Arabidopsis gene regulatory information server. A new information resource of Arabidopsis cis-promoter	http://arabidopsis.med.ohio-state.edu (Davuluri et al. 2003)
4	FastPCR	FastPCR software for PCR primers or probes design and *in silico* PCR, oligonucleotide assembly	http://en.bio-soft.net/pcr/FastPCR.html (Kalendar et al. 2009)
5	PlantCARE	A database of plant cis-acting regulatory elements	http://bioinformatics.psb.ugent.be/webtools/plantcare/html (Lescot et al. 2002)
6	Primer 3	Primer design. A computer program that suggests PCR primers for a variety of applications	http://frodo.wi.mit.edu/primer3/ (Rozen and Skaletsky, 1999)
7	Plantprom DB	Database of plant promoter sequences	http://mendel.cs.rhul.ac.uk/mendel.php?topic=plantprom (Shahmuradov et al. 2003)
8	Clustal Omega	Multiple alignment of DNA and protein sequences/Multiple sequence alignment programs	www.ebi.ac.uk/Tools/msa/clustalo/ (Sievers et al. 2011)
9	T-Coffee	Multiple alignment of DNA and protein sequences/ Multiple sequence alignment programs	www.ebi.ac.uk/Tools/msa/tcoffee/ (Notredame 2000)
10	MAFFT	Multiple alignment of DNA and protein sequences/Multiple sequence alignment programs	www.ebi.ac.uk/Tools/msa/mafft/ (Katoh et al. 2002)
11	MUSCLE	Multiple alignment of DNA and protein sequences/Multiple sequence alignment programs	www.ebi.ac.uk/Tools/msa/muscle/ (Edgar 2004)
12	DnaSP	DNA Sequence Polymorphism analysis, to identify SNPs and InDels and Phylogeny analysis	www.ub.edu/dnasp/ (Rozas et al. 2017)
13	MEGA	Integrated Software Molecular Evolutionary Genetics Analysis and Sequence Alignment	www.ncbi.nlm.nih.gov/pmc/articles/PMC2562624/ (Kumar 2008)
14	FrameWorker	A complex software tool that allows users to extract a common framework of elements from a set of DNA sequence	www.genomatix.de/online_help/help_gems/ (Cartharius et al. 2005)
15	DiAlign TF	DiAlign TF displays transcription factor (TF) binding site matches within a multiple alignment	www.genomatix.de/online_help/help_dialign/ (Cartharius et al. 2005)
16	Multalin	Multiple alignment of DNA and protein sequences/ Multiple sequence alignment programs	http://multalin.toulouse.inra.fr/multalin/ (Corpet 1988)
17	TRED (transcriptional regulatory element database)	Collection of mammalian regulatory elements	http://rulai.cshl.edu/cgi-bin/ TRED/tred.cgi?process=home (Jiang et al. 2007)
18	MAT inspector	To predict TFBS and promoter analysis	www. genomatix.de/products/MatInspector/(Cartharius et al. 2005)

TABLE 4.2
Status of Allele Mining in Wheat Crop

Gene/Allele	Trait	References
Pm3	Powdery mildew resistance	Yahiaoui et al. 2006
Wx-A1	Waxy gene and amylose biosynthesis	Saito and Nakamura 2005
Ap1 and PhyC	Vernalization response	Beales et al. 2005
SSII	Endosperm starch synthesis	Shimbata et al. 2005
Wx-B1	Waxy protein	Monari et al. 2005
PSY-1 and PSY-2	Grain yellow pigment content	Zhang and Dubcovsky 2008
Viviparous-1	Pre-harvest sprouting tolerance	Xia et al. 2008
TaALMT1	Aluminium resistant	Raman et al. 2008
TaPYL4	Grain yield	Wu et al. 2022
TaGRF-2D	Growth Regulating Factor	Kroupin et al. 2020

However, *Pm3* is the only wheat powdery mildew resistance gene cloned to date and is now known to occur in 15 functional alleles (*Pm3a* through *Pm3g*, *Pm3k* through *Pm3r*). The *Pm3* gene is a distinct member of a large cluster of nucleotide-binding site–leucine-rich repeat (NBS-LRR) genes on wheat chromosome 1A. The *Pm3* allele confers cultivar-specific resistance to *Graminis* f.sp. *tritici*, wheat powdery mildew fungus. Eight of the 15 *Pm3* alleles (*Pm3k* through *Pm3r*) were recently identified, with *Pm3k* being the only allele isolated from tetraploid wheat (Yahiaoui et al. 2004, 2006; Yahiaoui et al. 2009; Srichumpa et al. 2005; Bhullar et al. 2009). Separately, several other alleles responsible for various traits have been identified in wheat. Information on these alleles is presented in Table 4.2.

4.8 APPLICATION OF ALLELE MINING

"Mining" genes of interest from a variety of genetic resources enables efficient and accurate discovery of new and superior alleles. For example, the identification of new and distinct powdery mildew resistance alleles of the *Pm3* gene in wheat (Bhullar et al. 2010). Allele mining can be used to identify nucleotide variations in genomic regions of candidate genes associated with phenotypic variation in targeted traits. This allows us to assess the frequency, type, and occurrence of new haplotypes and the resulting phenotypic changes. Identification of sequence variants allows the development of relevant allele-specific markers for precise introgression of known superior alleles and/or new alleles into appropriate genetic backgrounds. In recent years, many case studies have been purported to demonstrate the existence of sequence variants in key genes, but with the advent of allele-specific molecular markers for marker-assisted selection (MAS), studies have undoubtedly demonstrated the utility of these variants. For example, in wheat, sequence analysis of the powdery mildew resistance (*Pm3*) locus identified different alleles in different lines (Srichumpa et al. 2005; Yahiaoui et al. 2006). For desirable alleles, functional markers may be developed for selection and wheat improvement (Tommasini et al. 2006).

4.9 CHALLENGES IN ALLELE MINING

1. A key challenge in allele mining is to select appropriate germplasms and determine the most cost-effective, efficient, and accurate phenotyping methods that can be used to characterize and evaluate the selected germplasms.
2. Handling the large complete genetic and genomic resources stored in gene banks requires multidisciplinary expertise. This is a major task in conventional breeding to develop allele-specific markers.

3. Delineation and identification of promoters in non-coding regions is a very difficult task in true allele mining studies. Delineating putative promoter regions, as opposed to defining coding regions for allele mining, is one of the major challenges in promoter mining analyses. Promoters and regulatory elements are distributed throughout the upstream region and their locations vary from gene to gene.
4. One of the key/major challenges in allele mining is minimizing the time and effort required while reducing costs. These challenges can be overcome by relying on inexpensive, high-throughput sequencing platforms to detect SNP variations.

4.10 CONCLUSION AND FUTURE PROSPECTS

Allele mining is a potential approach for the identification of naturally occurring allelic variations that control different traits, including biotic and abiotic traits, and thus have important roles and applications in crop improvement programs. It provides insight into the molecular basis of trait variation and identifies associated nucleotide sequence changes with superior alleles. Significant efforts are needed to develop tools and strategies to manage genetic and genomic resources. User-friendly and precise sampling strategies or tools need to be developed that may help in germplasm collection, followed by core development. The core collection should be representative of the entire genetic variability and thus should have maximum alleles of plant species. Further efforts are needed (qualitatively and quantitatively) to develop computational tools for efficiently identifying nuclear sets from large germplasm collections that provide accurate representations of all phenotypic classes. Sequencing-based allele mining is certainly expected to emerge as the method of choice for discovering natural allelic variations at loci of interest and discovering new effective alleles. It will therefore be the focus of all crop improvement activities. In addition, it has great potential applications in identifying allelic evolutionary rates, allelic similarities and dissimilarities of candidate genes, and allelic relationships with other family members. Allele mining may lead to the identification of novel alleles associated with desirable phenotypic characteristics, and these novel alleles may be introgressed into existing cultivars for the development of improved lines.

REFERENCES

Bailey TL, Williams N, Misleh C, Li WW (2006) MEME: Discovering and analyzing DNA and protein sequence motifs. *Nucleic Acids Res* 34:W369–73.

Barkley NA, Wang ML (2008) Application of TILLING and EcoTilling as reverse genetic approaches to elucidate the function of genes in plants and animals. *Curr Genom* 9:212–226.

Beales J, Laurie DA, Devos KM (2005) Allelic variation at the linked AP1 and PhyC loci in hexaploid wheat is associated but not perfectly correlated with vernalization response. *Theor Appl Genet* 110:1099–1107.

Bhullar NK, Street K, Mackay M, Yahiaoui N, Keller B (2009) Unlocking wheat genetic resources for the molecular identification of previously undescribed functional alleles at the Pm3 resistance locus. *Proc Natl Acad Sci USA* 106:9519–9524.

Bhullar NK, Zhang Z, Wicker T, Keller B (2010) Wheat gene bank accessions as a source of new alleles of the powdery mildew resistance gene Pm3: a large-scale allele mining project. *BMC Plant Biol* 10:1–3.

Blanco A, Gadaleta A, Cenci A, Carluccio AV, Abdelbacki AM, Simeone R (2008) Molecular mapping of the novel powdery mildew resistance gene Pm36 introgressed from Triticum turgidum var. dicoccoides in durum wheat. *Theor Appl Genet* 117(10):135–42.

Bryne JC, Valen E, Tang MHE, Marstrand T, Winther O, Piedade ID, et al. (2008) JASPAR, the open access database of transcription factor-binding profiles: New content and tools in the 2008 update. *Nucleic Acids Res* 36:D102–6.

Cartharius K, Frech K, Grote K, Klocke B, Haltmeier M, Klingenhoff A, et al. (2005) MatInspector and beyond: promoter analysis based on transcription factor binding sites. *Bioinformatics* 213:2933–42.

Comai L, Young K, Till BJ, Reynolds SH, Greene EA, Codomo CA (2004) Efficient discovery of DNA polymorphisms in natural populations by EcoTilling. *Plant J* 37:778–786.

Corpet F (1988) Multiple sequence alignment with hierarchical clustering. *Nucl Acid Res*, 25 16(22):10881–90.

Davuluri RV, Sun H, Palaniswamy SK, Matthews N, Molina C, Kurtz M, et al. (2003) AGRIS: Arabidopsis gene regulatory information server, an information resource of Arabidopsis cis-regulatory elements and transcription factors. *BMC Bioinformatic* 4:25.

Edgar RC (2004) MUSCLE a multiple sequence alignment method with reduced time and space complexity. *BMC Bioinformatics* 5(1):1–9.

Gahlaut V, Jaiswal V, Balyan HS, Joshi AK, Gupta PK (2022) Multi-Locus GWAS for Grain weight-related traits under rain-fed conditions in common wheat (*Triticum aestivum* L.). *Front Plant Sci* 21(12):758631. doi: 10.3389/fpls.2021.758631

Guzzon F, Gianella M, Giovannini P, Payne TS (2022) Conserving wheat genetic resources. In: Reynolds MP, Braun HJ (eds) *Wheat Improvement*. Springer, Cham. https://doi.org/10.1007/978-3-030-90673-3_17

He H, Liu R, Ma P, Du H, Zhang H, Wu Q, Yang L, Gong S, Liu T, Huo N, Gu YQ (2021). Characterization of Pm68, a new powdery mildew resistance gene on chromosome 2BS of Greek durum wheat TRI 1796. *Theor Appl Genet* 134(1):53–62.

Hutchison CA (2007) DNA sequencing: Bench to bedside and beyond. *Nucleic Acids Res* 35:6227–37.

Jiang C, Xuan Z, Zhao F, Zhang MQ (2007) TRED: A transcriptional regulatory element database, new entries and other development. *Nucleic Acid Res* 35:D137–40.

Kalendar R (2009) FastPCR software for PCR primer and probe design and repeat search. www. biocenter. helsinki.fi/bi/programs/fastpcr.htm2009

Katoh K, Misawa K, Kuma K-I, Takashi Miyata T (2002), MAFFT: A novel method for rapid multiple sequence alignment based on fast Fourier transform. *Nucleic Acids Res* 30(14):3059–3066.

Kollers S, Rodemann B, Ling J, Korzun V, Ebmeyer E, Argillier O, et al. (2013). Whole genome association mapping of Fusarium head blight resistance in European winter wheat (*Triticum aestivum* L.). *PLoS One* 8(2), e57500. doi: 10.1371/journal.pone.0057500

Kroupin PY, Chernook AG, Bazhenov MS, Karlov GI, Goncharov NP, Chikida NN, Divashuk MG (2020) Allele mining of TaGRF-2D gene 5'-UTR in *Triticum aestivum* and Aegilops tauschii genotypes. *PLoS One* 15(4):e0231704.

Kumar GR, Sakthivel K, Sundaram RM, Neeraja CN, Balachandran SM, Rani NS, Viraktamath BC and Madhav MS (2010). Allele mining in crops: Prospects and potentials. *Biotechnol Adv* 28:451–461.

Kumar S, Nei M, Dudley J, Tamura K (2008) MEGA: A biologist-centric software for evolutionary analysis of DNA and protein sequences. *Brief Bioinform* 9(4):299–306.

Kumari R, Kumar P, Sharma VK and Kumar H (2018). Allele mining for crop improvement. *Indian J Pure Appl Biosci* 6:1456–1465.

Lamaoui M, Jemo M, Datla R, Bekkaoui F (2018) Heat and drought stresses in crops and approaches for their mitigation. *Front Chem* 6:26.

Lescot M, Dehais P, thijs G, Marchal K, Moreau Y, peer YVD, et al. (2002) PlantCARE, a database of plant cis-acting regulatory elements and a portal to tools for in silico analysis of promoter sequences. *Nucleic Acids Res* 30(1):325–7. doi: 10.1093/nar/30.1.325

Li X, Xu X, Liu W, Li X, Yang X, Ru Z and Li L (2020) Dissection of superior alleles for yield-related traits and their distribution in important cultivars of wheat by association mapping. *Front Plant Sci* 11:175. doi: 10.3389/fpls.2020.00175

Liang C, Wang SQ, Hu YG (2011) Detection of SNPs in the *VRN-A1* gene of common wheat (*Triticum aestivum* L.) by a modified EcoTILLING method using agarose gel electrophoresis. *Aus J Crop Sci* 5:318–326.

Ma, X, Sajjad, M, Wang, J, Yang, W, Sun, J, Li, X, Zhang, A, Liu, D (2017) Diversity, distribution of Puroindoline genes and their effect on kernel hardness in a diverse panel of Chinese wheat germplasm. *BMC Plant Biol* 17:158.

Margulies M, Egholm M, Altman WE, Attiya S, Bader JS, Bemben LA, et al. (2005) Genome sequencing in microfabricated high-density picolitre reactors. *Nature* 2005:437.

McIntosh R, Dubcovsky J, Rogers JW, Xia X, Raupp W (2020). Catalogue of gene symbols for wheat. Available online at https://wheat.pw.usda.gov/GG3/WGC.

Monari AM, Simeone MC, Urbano M, Margiotta B, Lafiandra D (2005) Molecular characterization of new waxy mutants identified in bread and durum wheat. *Theor Appl Genet* 110:1481–9.

Nagy J, Sulyok D, Huzsvai L (2006) Effect of tillage on the yield of crop plants. *Cereal Res Commun* 34(1):255–8.

Notredame C, Higgins DG, Heringa J (2000) T-Coffee: A novel method for fast and accurate multiple sequence alignment. *J Mol Biol* 302(1):205–17.

Perugini LD, Murphy JP, Marshal D, Brown-Guedira G (2008) Pm37, a new broadly effective powdery mildew resistance gene from Triticum timopheevii. *Theor Appl Genet* 116:417–42.

Pourkheirandish M, Golicz AA, Bhalla PL, Singh MB. (2020) Global role of crop genomics in the face of climate change. *Front Plant Sci* 11:922.

Raddy AM, Gambhire VB and Raddy RT. (2014). Allele mining in crop improvement. *Int J Dev Res* 4:300–305.

Raman H, Ryan PR, Raman R, Stodart BJ, Zhang K, Martin P, Wood R, Sasaki T, Yamamoto Y, Mackay M, Hebb DM (2008) Analysis of TaALMT1 traces the transmission of aluminium resistance in cultivated common wheat (*Triticum aestivum* L.). *Theor Appl Genet* 116:343–54.

Rozas J, Ferrer-Matta A, Sanchez-Gracia A, Sánchez-DelBarrio JC, Guirao-Rico S, Librado P, Ramos-Onsins SE (2017) DnaSP 6: DNA sequence polymorphism analysis of large data sets. *Mol Biol Evol* 32(12):3299–302.

Rozen S, Skaletsky H (1999) Primer3 on the WWW for general users and for biologist programmers. *Bioinform Methods Protocols*, 365–86.

Saito M, Nakamura T (2005) Two point mutations identified in emmer wheat generate null Wx-A1 alleles. *Theor Appl Genet* 110(2):276–82.

Shahmuradov IA, Gammerman AJ, Hancock JM, Bramley PM, Solovyev VV (2003) PlantProm: A database of plant promoter sequences. *Nucleic Acids Res* 31(1):114–117. https://doi.org/10.1093/nar/gkg041

Sharma MK, Singh S and Senger RS (2014). Allele mining and its applications. *Biotech Today* 4:34–37.

Shimbata T, Nakamura T, Vrinten P, Saito M, Yonemaru J, Seto Y, Yasuda H (2005) Mutations in wheat starch synthase II genes and PCR-based selection of a SGP-1 null line. *Theor Appl Genet* 111(6):1072–9.

Sievers F, Wilm A, Dineen D, Gibson TJ, Karplus K, Li W, Lopez R, McWilliam H, Remmert M, Söding J, Thompson JD, Higgins DG (2011) Fast, scalable generation of high-quality protein multiple sequence alignments using Clustal Omega. *Mol Syst Biol* 7:539.

Srichumpa P, Brunner S, Keller B, Yahiaoui N (2005) Allelic series of four powdery mildew resistance genes at the Pm3 locus in hexaploid bread wheat. *Plant Physiol* 139(14):885–895.

Till BJ, Reynolds SH, Green EA, Codomo CA, Enns LC, Johnson JE, et al. (2003) Large-scale discovery of induced point mutations with high-throughput TILLING. *Genome Res* 13:524–30.

Tommasini L, Yahiaoui N, Srichumpa P, Keller B (2006) Development of functional markers specific for sever Pm3 resistance alleles and their validation in the bread wheat gene pool. *Theor Appl Genet* 114:165–75

Wu BB, Shi MM, Pourkheirandish M, Zhao Q, Ying Wang, Chen-Kang Yang, Qiao L, JZhao J-J, Yan S-X, Zheng X-W, Zheng J (2022) Allele mining of wheat ABA receptor at TaPYL4 suggests neo-functionalization among the wheat homoeologs. *J Integr Agric* 21(8):2183–2196.

Xia LQ, Ganal MW, Shewry PR, He ZH, Yang Y, Röder MS (2008) Exploiting the diversity of Viviparous-1 gene associated with pre-harvest sprouting tolerance in European wheat varieties. *Euphytica* 159:411–417.

Yahiaoui N, Brunner S, Keller B (2006) Rapid generation of new powdery mildew resistance genes after wheat domestication. *Plant J* 47(13):85–98.

Yahiaoui N, Kaur N, Keller B (2009) Independent evolution of functional Pm3 resistance genes in wild tetraploid wheat and domesticated bread wheat. *Plant J* 57:846–856.

Yahiaoui N, Srichumpa P, Dudler R, Keller B (2004) Genome analysis at different ploidy levels allows cloning of the powdery mildew resistance gene Pm3b from hexaploid wheat. *Plant J* 37:528–538, 12.

Yang, W, Feng, H, Zhang, X, Zhang, J, Doonan, JH, Batchelor, WD, Xiong, L, & Yan, J (2020). Crop phenomics and high-throughput phenotyping: Past decades, current challenges, and future perspectives. *Mol Plant* 13(2):187–214. https://doi.org/10.1016/j.molp.2020.01.008

Yin L, Dai X, He M (2018) Delayed sowing improves nitrogen utilization efficiency in winter wheat without impacting yield. *Field Crop Res* 15(221):90–7.

Zhang W, Dubcovsky J (2008) Association between allelic variation at the Phytoene synthase 1 gene and yellow pigment content in the wheat grain. *Theor Appl Genet* 116:635–45.

5 Alleles for an Optimum Phenology in Maize

*Venkata Rami Reddy Yannam, Maroua Doghri, and Bernardo Ordas**
*Corresponding author: bordas@mbg.csic.es

5.1 INTRODUCTION

Natural selection directs population mean flowering in wild plant species toward an optimum adapted to local conditions. Plants that flower early are smaller in size, have fewer resources to invest in reproductive organs and have less risk of mortality before seed maturation (Austen et al. 2017). For example, annual species with spring emergence can escape from summer droughts or premature autumn frosts by early flowering. The effect of flowering on adaptation has been proven in the model plant Arabidopsis (*Arabidopsis thaliana*). Thus, Ågren et al. (2017) grew Arabidopsis ecotypes from Italy and Sweden in Italy, and the local genotype had an advantage mainly due to a flowering time adapted to the region. Also, Taylor et al. (2019) planted flowering mutants of Arabidopsis in multiple environments and found that the mutations increased fitness in specific environments due to changes in flowering time. The importance of flowering time in adapting varieties to different environments is also evident in crops. One of the clearest proofs of it is the distribution of maize, a plant of tropical origin cultivated even in northern regions of North America, such as Alaska. This could have happened thanks to an intense selection for early flowering in temperate areas where the growing season is short. Artificial experiments provide direct examples of this adaptation. Two populations that grew in the Atlantic conditions of Northwestern Spain presented grains with elevated moisture at harvest, prone to deterioration during storage (Ordás et al. 2019). Those populations were subjected to mass selection for early flowering during 18 cycles, and a decrease of about 15 days in flowering time resulted in both populations. This was accompanied by the desired effect of reducing the moisture by about 3–4 % but with the undesired effect of reducing yield. We selected for grain yield in lowland environments (with longer seasons) a nearly low-yielding variety adapted to shorter seasons. The selection was very effective in increasing the yield, but the flowering was also delayed as an indirect effect of the selection. These results reveal that the tradeoff in natural wild populations among early and late flowering also exists in crops. The optimum flowering time for a variety depends on the specific environment where the variety is grown. The environments are dynamics and depends not only on the altitude and latitude but also on the weather and climate change. Regarding climate change, without adaptation, losses in production are expected for different crops, but with adaptation, the losses could be mitigated, especially in temperate environments (Challinor et al. 2014). In the specific case of Europe, (Moore and Lobell 2014) concluded that agricultural profit could increase under climate change if farmers adapt to the new environmental conditions, but could decrease with no adaptation. These authors also pointed out the high potential for the adaptation of maize. (Butler and Huybers 2013) come to the same conclusions analyzing the US maize cultivation: the potential loss of yield due to climate change can be averted with adaptation. An analysis of six field experiments of 121 genotypes and historical data from 59 sites over 36 years in Europe (Parent et al. 2018) concluded that maize grain would increase in the next

DOI: 10.1201/9781003385004-5

decades if crop cycle duration and sowing dates are optimized in each specific location. For that the farmers must select the genotypes most adequate to their specific conditions within the genetic variability in flowering time. The breeders need to make a large set of different genotypes available to farmers with different flowering times so they can select the most adapted to their conditions. Although the flowering time can be considered the most important trait for adapting a variety to a specific duration of the growing season, the onset and rate of senescence are also important in adapting varieties to the environment (Munaiz et al. 2020; Chibane et al. 2021). Thus, we found that the inbred line PHW79 stayed green for almost 2 months more than an isogenic line of B73 although both lines had a similar flowering time. Locating the genes responsible for flowering time and senescence and knowing their effects, interactions between alleles (dominance and epistasis) and between alleles and environments is critical for developing varieties with optimal phenological characteristics for each environment. Although these objectives were being achieved at a relatively slow rate due to the quantitative nature of the traits, huge advances have been made in these areas in the last decade with a large contribution to this progress due to new powerful genotyping and statistical methods. In this chapter, we summarize the advances in identifying and understanding the function of alleles controlling flowering and senescence timing in maize and the implications in the adaptation and evolution of the crop. Firstly, we explained what is known about the origin and expansion of maize from teosinte. Secondly, we summarized the knowledge about the genetics of maize flowering: identification and estimation of quantitative trait locus (QTL) effects and cloning and analysis of gene function. Thirdly, we explained in brief different aspects of the senescence in maize and summarized quantitative trait loci (QTLs) and differential expression experiments that have been carried out on this complex trait.

5.2 DOMESTICATION AND EXPANSION OF MAIZE

5.2.1 Teosintes Parviglumis and Mexicana

There are two close relatives of maize (*Zea mays* ssp. *mays*), the teosintes parviglumis (*Zea mays* ssp. *parviglumis*) and mexicana (*Zea mays* ssp. *mexicana*). Teosintes parviglumis and mexicana occupy complementary areas with low overlapping: parviglumis mainly in the coastal region of Western Mexico at low altitude and mexicana in interior regions of high altitude (Hufford et al. 2012). In greenhouse conditions at Ithaca (New York, US), teosinte flowered in a few months when planted in winter but in many months when planted in spring (Emerson 1924). Interestingly, independently of the planting date from March to June, flowering was in October, when days were short enough. That is, teosinte is photoperiod sensitive, and with long days, the plant delays the flowering or does not flower at all. When parviglumis grows under long-day conditions, only 7 days of short-day treatment is enough for conversion to a reproductive stage (Lazakis et al. 2011). The requirement for flowering can be short days or long nights. Under short-day conditions, the lack of response to the application of 1 hour of light at night (night break) suggests that the flowering is induced by the short days, not the long nights (Minow et al. 2018).

High-altitude adaptation includes lower oxygen availability, increased ultraviolet radiation, and cold temperatures (Barnes et al. 2022). As these authors explained, selection for cold temperature indirectly leads to selection for early flowering because the flowering time depends on the growing degree days (GDD). The genotypes grown at high altitude should be adapted to flowering with less GDD, which leads to early flowering in conditions of higher GDD as those usually found in lowlands. Pyhäjärvi et al. (2013) genotyped 11 populations of parviglumis and 10 of mexicana with the Ilumina MaizeSNP50 Beadchip, resulting in about 43,700 single nucleotide polymorphisms (SNPs) to analyze the genetic basis of local adaptation. These authors found that the divergence among subspecies is due to adaptation and that there is a limited gene flow among subspecies, mainly in populations in the border of the distributions where some overlap exists. Several environmental factors are involved in the differentiation of the subpopulations, for example, altitude, but the essential environmental factors for one population could not be for another. The complex genetic structure also hampered the identification of adaptation alleles, although some genes showed signals of local adaptation. Interestingly, inversions could have an essential role in the adaptation process. Aguirre-Liguori et al. (2019) completed the genetic analysis of parviglumis and mexicana with a comprehensive study of 9,780 DArtTseq SNPs for 47 populations and 33,450 MaizeSNP50 SNPs for 49 populations (there were 29 populations in common in the two sets). Some conclusions of these authors are consistent with Pyhäjärvi et al. (2013) as the existence of little gene flow among subspecies and the location in inversions of many of the significant SNPs related to adaptation. Aguirre-Liguori et al. (2019) also found that the environmental variables relevant to adaptation are related to temperature and phosphorus concentration in soil.

5.2.2 Early Domestication and Adaptation to the Highlands of Mexico

Maize was domesticated from parviglumis in the valley of Rio Balsas in Southern Mexico (Matsuoka et al. 2002; Doebley 2004). The early domestication started in lowland areas with seasonal wet and dry cycles (Hastorf 2009). Pre-domesticated maize migrated to other parts of Mexico and South America before all the domesticated traits were fixed (Kistler et al. 2018). The levels of diversity of maize populations decrease as they are located away from the center of domestication; in particular, the varieties from the Andes show a small effective population size indicating a solid founder event (Wang et al. 2017). Given that there is a gene flow between parviglumis and mexicana, gene flow is also expected among maize, a modified parviglumis, and mexicana in the varieties grown at high altitudes (Hufford et al. 2012). Hufford et al. (2013) assessed the pattern of mexicana introgression in maize by genotyping with the Ilumina MaizeSNP50 Beadchip nine sympatric populations pairs of mexicana and maize that are distributed over the area of mexicana in Mexico. These authors found substantial gene flow among maize and mexicana, although the introgression of mexicana into maize was more extensive than the opposite. Some introgressed regions overlap with QTLs for pigment intensity and micro hairs, which are traits associated with highland adaptation. One of the introgressed regions at chromosome 4 was mapped to an inversion of 13 Mb (Inv4m) that also differs among parviglumis and mexicana (Aguirre-Liguori et al. 2019). The authors conclude from their data analysis that the introgression from mexicana is limited to domestication loci involved in highland adaptation. Gonzalez-Segovia et al. (2019) used a different approach to study the introgressions of mexicana in maize. They sequenced three individuals from two varieties grown at high altitudes in Mexico (PalomeroToluqueño and Mushito de Michoacán). They crossed one individual of PalomeroToluqueño to a genotype derived from a lowland variety from Mexico (Reventador) to develop a mapping population and perform QTL analysis. To identify the introgressed regions, the authors compared the frequency of the mexicana allele in the highland maize and the *Tripsacum* allele in the lowland maize. They found several regions introgressed in maize from mexicana, many located in regions of low recombination, probably inversions consistent with Hufford et al. (2013). The authors identified 1,000 genes within the introgressed regions, but failed to associate the regions with agronomic or morphological traits. However, Romero Navarro et al. (2017), in a comprehensive genomic analysis of 4,471 landraces of America, found a significant association of Inv4m with flowering.

Inv4m was selected in mexicana because it confers adaptation to altitude through earlier flowering and was introgressed from mexicana into the varieties of maize in the highlands of Mexico. However, the effect of the inversion on phenotypic traits and adaptation could be much more complex than an effect on flowering. This inversion of long size contains many genes, which are expected to be involved in several processes. However, the extensive linkage disequilibrium within the inversion hinders the identification of functional genes responsible for phenotypic effects. Crow et al. (2020) developed BC5S1 families by backcrossing the highland populations of Palomero Toluqueño and Conico with B73. The families with the inversion of the highland populations in the B73 background were phenotyped in a growth chamber at low and high temperature for several traits at early developmental stages. In addition, samples were taken at different moments of early development and used for RNAseq analysis. The authors found diverse physiological effects of Inv4m with a substantial effect on photosynthesis and chloroplast development. This study also shows a considerable genotype × temperature interaction with the haplotype of highland populations showing more extreme differences among cold and warm conditions.

Another gene, *Phosphatidyl Choline 1* (*HPC1*), was also introgressed from mexicana and is involved in the highland adaptation of maize in Mexico. In a large sample of 2,700 maize landraces from Mexico, *HPC1* showed substantial changes in elevation-dependent allele frequency (Barnes et al. 2022). These authors corroborate the result in an independent experiment evaluating in highlands and lowlands, a BC1S5 population derived from a highland population from Mexico (Palomero Toluqueño) crossed to B73 as a recurrent parent. *HPC1* is a *phospholipase A1-Igamma1* classified as PC-hydrolyzing *PLA1* Class I that hydrolyzes phospholipids in the *sn-1* position and produces a lysophospholipid and a free fatty acid. In an analysis of 2,000 landraces F1s evaluated in 23 environments, Barnes et al. (2022) found that *HPC1* was associated with traits, such as flowering and yield. Like Inv4m, *HPC1* presented a significant genotype × environment interaction. Thus, for example, the Palomero Toluqueño allele derived from mexicana decreased and increased the flowering time for 1 day in highlands and lowlands, respectively. An analysis of optimal growth in prokaryotic species, followed by modeling by iterative threading assembly refinement, allowed Barnes et al. (2022) to firmly conclude that a mutation in the flap-lid domain is responsible for differences in *HPC1* activity. In the allele from mexicana – and highland populations derived – a replacement of isoleucine by valine at residue 211 was detected, which resulted in the loss of a methyl group that could affect the strength of hydrophobic interactions that, in turn, affect the substrate binding and, in general, the dynamics of the flap-lid domain. The replacement with valine in the mexicana allele is accompanied by a lower optimal growth temperature, which could confer better adaptation to high altitude.

5.2.3 Expansion Outside Mexico

Wang et al. (2017) resequenced individuals from regions outside Mexico (Southern US highlands, Central America highlands-Guatemala-, South America highlands-Andes-, and South America lowlands) to assess the impact of mexicana introgression outside Mexico. The analysis of these populations has shown that the populations of highlands of another part of Central America also have levels of introgression from Mexicana similar to the highlands of Mexico; however, the level of introgression in the highlands of Southern US is lower, and no evidence of introgression was found in the highlands of South America. Takuno et al. (2015) and Wang et al. (2021) addressed the question of convergence in the adaptation to different highland regions. Takuno et al. (2015) genotyped 94 accessions from Meso and South America with the MaizeSNP50 Beadchip and found that the convergence between the adaptative evolution of Meso and South America to highlands is minimal. The authors concluded from their data analyses that the mexicana introgression plays an essential role in adaptation to highlands in Mesoamerica but not in South America. Wang et al (2021) found that about 10% of the highland adaptative SNPs show a pattern of molecular parallelism, which indicates higher parallelism than the parallelism found by Takuno et al. (2015), but Wang et al. (2021) still, most of the genes selected for adaptation to altitude were different in Meso- and South America. Barnes et al. (2022) identified 23 glycerolipid genes that were selected across highland regions. One of them, *ZmLPCAT1*, encoded a Lyso-Phosphatidylcholine Acyl Transferase 1 that presented intense and repeated selection signals in all the highland populations, including the Andes, and could be an example of adaptative convergence.

Swarts et al. (2017) developed prediction equations for the flowering time that predicted landraces from Southern US with high accuracy. The authors applied the equations to a sample of cobs from

Southern US dated about 2,000 years BP and found that the flowering time of the ancient plants must have been similar to the flowering time of modern temperate maize. Besides, using SNPs with the highest differentiation index between maize from Mexico and Gaspé Flint from Canada (the earliest known population), the ancient samples grouped with temperate maize but separated from Mexican and South American populations. Therefore, it seems that early flowering or adaptation to temperate conditions in those varieties for Southern US was not provided by introgression of mexicana or Mexican highland populations, but by the selection *in situ* during 2,000 years.

The alleles in the ancient samples were already in parviglumis, indicating that the selection for early flowering used standing variation already present in the maize ancestor. However, the comparison of alleles among Southwestern US ancient samples and highland landraces from Mexico (Palomero de Jalisco) revealed a highland origin of the earliest Southwestern US maize (da Fonseca et al. 2015). Therefore, according to these authors and in contradiction to Swarts et al. (2017), the adaptation to the temperate conditions of the US was achieved not by a slow process of *in situ* selection but through a fast importation of Mexican highland varieties. The two processes, selection *in situ* and introgression from mexicana, probably have contributed to the adaptation of US temperate areas. The relative importance of each mechanism could vary depending on the specific region or variety. Irrespective of the relevance of mexicana in the adaptation to temperate areas, genomic regions derived from the introgression of mexicana are present in modern temperate inbred lines. For example, Yang et al. (2017) found that the genome of Mo17 and B73, two inbred lines of enormous importance in temperate maize breeding, has about 10% of introgression from mexicana. The authors remarked that the percentage could be higher with a higher density of markers. The identity by descent (IBD) of many regions of the genome of modern temperate lines was stronger with mexicana than with parviglumis, indicating that these regions were introgressed from mexicana (Hufford et al. 2013). These authors also found that introgression was higher in some germplasm groups than in others. Thus, the non-stiff-stalk and stiff-stalk groups showed 31% and 23% of the sites with greater IBS with mexicana, respectively. The allele of the *HPC1* gene that conferred adaptation to higher altitudes (introgressed from mexicana) was found, in addition to Mexican highland populations, in flint materials of Northern US and Canada (Barnes et al. 2022).

In temperate latitudes, North and South of parallel 30 in the Northern and Southern Hemisphere, respectively, there is an essential difference with the equatorial and tropical latitudes: long days and short nights during the maize growing season. Garner and Allard (1923) found that one Peruvian and one Mexican variety were sensitive to photoperiod, which is expected given that parviglumis is also photoperiod sensitive. This could have been a significant limitation for the expansion of maize to temperate areas of North and South America. However, some tropical maize varieties are not photoperiod sensitive, as shown by Gouesnard et al. (2002) in an analysis of 152 tropical populations grown under long- and short-day conditions. In this study, the effect of altitude was significant, and the populations of the higher altitude tended to have lower sensitivity to photoperiod, such as some varieties of highlands of Mexico, such as PalomeroToluqueño, or from the Andes as Morocho or Canguil.

The high-altitude varieties are adapted to flower with less GDD, which lead to earlier varieties under higher GDD. Still, this early flowering does not explain the loss of photoperiod sensitivity of highland varieties. If the earliest varieties introduced in Southern US are from the highland of Mexico, as proposed by da Fonseca et al. (2015), then their lack of sensitivity supposes an advantage for adaptation to temperate environments with long days. However, the precise relationship between photoperiod sensitivity and altitude is complex and has some exceptions. Thus, although most of the Caribbean accessions are photoperiod sensitive, some accessions from the Guadeloupean archipelago or British Virgin Island are insensitive to photoperiod. The loss of the photoperiod sensitivity should be advantageous for the adaptation to temperate areas, but this generalization could have many exceptions. Thus, many accessions of the Andes have a low yield in temperate environments, despite the lack of photoperiod sensitivity. In contrast, some Caribbean accessions have a high yield in those environments, regardless of their photoperiod sensitivity (Gouesnard et al. 2002).

5.3 QUANTITATIVE TRAIT LOCI (QTLS) FOR MAIZE FLOWERING

5.3.1 Biparental Linkage Mapping

Flowering time is a quantitative trait controlled by many genes. The availability of molecular markers allowed the study of the genetic architecture of the traits: the number and location of genes and the type and magnitude of the effects. The design most commonly used in the past for such studies was a biparental population derived from parental lines with contrasting values for the trait under study, in the case of flowering, an early and a late flowering line. The specific mapping population could be F2 F2:3, recombinant inbred lines (RILs), and different variants of these types. At the beginning of the genetic studies with molecular markers, it was thought that when a significant association between a marker and a trait, i.e., a quantitative trait locus (QTL), was found, the gene behind the association was located. However, the researchers were locating long regions of the genome associated with a trait, not specific genes, due to the lack of precision, which was due to the small size of the mapping populations and the extensive linkage disequilibrium between the markers. In maize, a typical precision of a QTL in a standard biparental design was about 20–30 cM in genetic distance or 20–30 Mb in physical distance. In regions of those sizes is expected to find hundreds, if not thousands, of genes.

Chardon et al. (2004) summarized the studies of flowering until that date and carried out a meta-analysis trying to combine the information of different studies. They analyzed 22 studies in which 313 quantitative trait loci (QTLs) were identified. The average number of QTLs per experiment was five, and most genome regions have at least one QTL for flowering. However, the distribution of the QTLs varied among regions, and thus some regions concentrated many QTLs, while others only one. The authors developed an overview statistic to identify the genomic regions behind the 313 QTLs that contribute to flowering variation. The application of the overview statistic reduced to 69 QTLs, the initial number of 313 found in previous studies. Six regions, at chromosomes 1, 8 (two regions very close named *Vgt1* and *Vgt2*), 9, and 10 (two regions close), presented a very high value of the overview statistic, indicating its relevance in controlling flowering variation. A meta-analysis further reduced the number of consensus QTLs to 62. The meta-analysis identified five consensus QTLs associated with several flowering traits, such as the number of leaves. Three of these QTLs for multiple traits (chromosome 1, 8, and 9) coincide with three of the six more significant regions detected with the overview statistic.

Salvi et al. (2009) updated the consensus map until 2009 and came to similar results and conclusions as Chardon et al. (2004). This study covers 29 studies with a total number of 441 QTLs and an average number of QTLs per experiment of 5, similar to the previous meta-analysis. The distribution of effects (proportion of phenotypic variance explained) was L-shaped, with most QTLs having minor effects. The regions with the highest frequency of QTLs were two at chromosome 8 (*Vgt1* and *Vgt2*) and one at chromosome 10. These three regions were already highlighted in the previous consensus analysis. The number of consensus regions estimated in this study (59) was very close to that of the former meta-analysis (62). Two interesting results of Salvi et al. (2009) are that there is not a clear relationship between population size and the number of detected QTLs, and the limited congruency among QTL studies, even with similar mapping populations. These results emphasize the limitations of the biparental designs with populations of relatively small size that were surpassed by multiparental designs with large populations.

5.3.2 Multiparental Linkage Mapping

A paper published by Buckler et al. (2009) represented a groundbreaking milestone in the study of quantitative traits, particularly in maize flowering time. The authors built a large population of 5,000 genotypes by crossing a reference parent (B73) with 25 diverse lines and developing 200 RILs from each cross (each of the 25 subgroups of RILs is considered a family). The mapping population was named nested mapping association (US-NAM). The joint analysis of combined information across families identified more than 50 QTLs, including minor QTLs that explained almost 90% of the total variance for the flowering-related traits. The QTL effects were minor and additive, without evidence of an essential role of epistasis. The magnitude of QTL × environment interaction was of secondary importance compared to the QTLs' main effects, that is, the QTLs' effects were consistent across environments. Also, the QTLs' effects were consistent across families, although the magnitude of the effects varied across families, indicating allelic series in many of the loci. The six hot spots identified by Chardon et al. (2004) were confirmed in this study.

5.3.3 Genome-Wide Association Studies (GWAS)

Another breakthrough that transformed the study of quantitative traits was the development of the genetic association analyses. These analyses allow a higher resolution than linkage mapping because the linkage disequilibrium is reduced due to historical recombination. Another advantage is that the analyses are done on panels of already developed genotypes, and there is no need for developing specific mapping populations. One disadvantage concerning linkage mapping is the need for more power to detect alleles at minor frequencies. In fact, the alleles at very low frequencies are usually discarded for analysis.

Romay et al. (2013) genotyped with genotyping by sequencing 2,815 maize inbred lines from the National Plant Germplasm of the US and produced 681,257 single nucleotide polymorphisms distributed along the whole genome. They analyzed with the powerful mapping source flowering time and found associations between SNPs and flowering with very high significance. The highest significance ($P=10^{-23}$) was for a SNP within a QTL repeatedly detected in previous studies at chromosome 10. This SNP is 2 kb away from *ZmCCT*, the gene responsible for the QTL, as we will explain later. This indicates that, as expected, the GWAS can detect associations with high precision. The second highest significance was for SNPs in chromosome 8, close to *Vgt1*, again an important QTL for flowering consistently found in many studies. These examples and others shown in the paper indicate the high resolution of GWAS. The number of significant SNPs was about 20. However, the threshold level was quite stringent, and with a more liberal threshold the number of significant associations would be probably like previous meta-analyses or the NAM. The authors concluded that although the number of markers was high, in the regions with high haplotype diversity, the linkage disequilibrium (LD) decay was so fast that more markers were needed. Therefore, although the analysis of temperate alleles was adequate, the analysis of tropical alleles was incomplete because the tropical germplasm is very diverse and is underrepresented in the panel. To overcome this limitation and to detect more variants, the diversity and size of the mapping populations should be increased as well as the density of the markers.

5.3.4 Joint Analysis of Different Experiments

Li et al. (2016) carried out a joint analysis of two NAMs populations (US-NAM and CN-NAM from China) and an association panel (Ames). In total, the population has 8,000 inbred lines genotyped with 1 million SNPs, which constitute an unprecedented source to analyze the genetic architecture of flowering time. Parallel linkage analysis was done in the two NAM populations, and 40 common QTLs were found in both the populations, 35 specifics from US-NAM, and 15 specifics from CN-NAM. In addition, an association genetic analysis was done with the three populations (US-NAM, CN-NAM, and Ames), and the results obtained were consistent with the QTLs' linkage analysis. The authors carried out cross-validation to check how well the associated SNPs from one population predict the phenotypes of other populations. US-NAM and Ames predicted each other well, which was expected given that the parents of the US-NAM were included in the Ames panel. In contrast, the associated SNPs of US-NAM and Ames did not predict well CN-NAM. Out of 919 candidates flowering genes selected from the literature, 220 candidates were at least 1 Mb close to a significant SNP. In those candidate genes, two regions were enriched in significant SNPs: the coding region and a region 5 kb away from the coding region. Therefore, the variants at promoters and other regulatory elements are also crucial in explaining the variation at flowering.

More recently, Swarts et al. (2021) complemented the previous study including in the analysis, in addition to US-NAM, CN-NAM, and Ames, two NAMs of European origin: EUNAM-Flint and EUNAM-Dent. This resulted in a whole population of about 10,000 genotypes. The number of markers also increased to 70 million, segregating across populations. Like the previous study, the best and worst prediction ability was for Ames and CN-NAM, respectively. The cross-population prediction ability was higher in this study than in the previous one; for example, the correlation between actual and predicted values when Ames predicted CN-NAM was high (r=0.78). Among the European NAMs, EUNAM-Flint had better prediction abilities than EUNAM-Dent. The lack of prediction ability of the CN-NAM is probably due to the presence of novel allelic variants at common loci that arose during the adaptation of maize of different origins, including tropical, to China environments. In the case of the EUNAM-Dent, the lack of prediction ability could be due to the presence in the population of many lines derived from recent current selection in Europe of germplasm derived from Minnesota 13 that is not present in other populations.

5.3.5 Mapping in Landraces

Mayer et al. (2017) genotyped 22–48 individuals from 35 European maize landraces (the total number of individuals was 952) with the 600k Affymetrix Axiom Maize Array, resulting in approximately 500,000 SNPs for the analyses. The authors found that the European maize landraces, although representing a small subset of the variability of the species in America, have high molecular variability, which highlights the relevance of the landraces as sources of variability. Another significant result of Mayer et al. (2017) is that LD decay distances were small (ranging from 56 to 73 kb) when the samples were obtained across landraces. This indicates that landraces are a powerful source for association mapping. The lack of studies in landraces may be due to the fact that they are not improved materials that limits its direct use in elite breeding and because the individuals of open-pollinated varieties are not immortal and cannot be replicated across environments. Hölker et al. (2019) proposed the development of large libraries of double haploids derived from landraces that capture their genetic diversity because it is easy to work with them in genetic studies or breeding programs. Alternatively, Romero Navarro et al. (2017) used a new design that the authors named F-one association mapping (FOAM) that consists in crossing individuals of different landraces to one or few parents, and the F1 are evaluated in multiple environments. According to the authors, two of the advantages of this design are the possibility to analyze thousands of alleles and the replication of alleles that increases the power and accuracy. The authors applied the method to the study of the allelic diversity of flowering time of a large sample of 4,471 American landraces genotyped with almost 1 million SNPs. The analysis of the FOAM design identified about 1,000 genes related to flowering. Many of them are involved in the adaptation to altitude and are included in large structural variants as inversions. One interesting result is that most of the genes that affect the timing of flowering are not from flowering pathways, but have an indirect effect on the trait.

5.4 GENES INVOLVED IN MAIZE FLOWERING

5.4.1 *Dlf1* and *Id1*

In Arabidopsis, the *FLOWERING LOCUS D* (*FD*) activates the *APETALA1* (*AP1*) gene at the apical meristem that initiates the flowering transition (Abe et al. 2005; Wigge et al. 2005). Muszynski et al. (2006) cloned and characterized delayed flowering1 (*dlf1*) in maize, which is a Leu zipper transcription factor orthologous to *FD* in Arabidopsis. The expression pattern of *dlf1* suggests that a minimum level of expression of this gene is needed in the apical meristem to initiate flowering. Plants in which *dlf1* is not functional have late flowering, as the gene's name indicates.

The *indeterminate1 gene (id1)* encodes a transcription factor with zinc finger motifs controlling the transition to flowering in maize (Colasanti et al. 1998). Mutants of *id1* delay flowering, but can also produce other effects, such as aberrant inflorescences. The expression of *id1* occurs in immature leaf tissues and is not affected by variations in dark and light nor by the leaves' photosynthesis capacity (Wong and Colasanti 2007). The analysis of differences in expression among wild and *id1* mutants with an oligonucleotide microarray chip allows the identification 55 genes that are potential targets of *id1* and probably connected to floral induction (Coneva et al. 2007). The authors also found that only a few genes show differences between floral-induced and uninduced parviglumis. Therefore, *id1* is developmentally induced, independent of external stimuli, such as photoperiod or the sink/source of the leaves. Wong and Colasanti (2007) concluded, in agreement with Coneva et al. (2007), that *id1* is part of an autonomous floral inductive pathway, different from the photoperiod pathway.

5.4.2 ZCN8 (QTL VGT2)

The *FLOWERING LOCUS T* (*FT*) gene in Arabidopsis is transcribed in the leaf phloem (Takada and Goto 2003), and the generated protein moves through the phloem to promote the transition from vegetative to reproductive phase in the meristem (Corbesier et al. 2007). Orthologous genes of *FT* in other species have a similar florigen function, for example, in tomato (Lifschitz et al. 2006), rice (Tamaki et al. 2007), etc. In Arabidopsis, the *FD* gene alone is not able to activate the floral transition; it is required that *FT* joins to *FD* in the meristem (Abe et al. 2005; Wigge et al. 2005).

The phosphatidylethanol amine-binding protein (*PEBP*) gene family that includes *FT* in Arabidopsis has been analyzed in maize by Danilevskaya et al. (2008). There are 24 genes in this family grouped into three main groups: *TFL1-like, MOTHER OF FT-like (MFT-like)*, and *FT-like*. *MFT-like* genes are expressed mainly in kernel tissues, while *TFL1-like* genes are expressed in different tissues, but not in leaves. The *FT*-like subfamily is complex and encompasses 15 genes, including *ZCN8, ZCN12*, and *ZNC26*, mainly expressed in leaf blades. The pattern of expression of *ZCN8* is compatible with being an activator of flowering as its level of expression increases before the floral transition and stays high after that moment. If *ZCN8* is homologous and works similarly to *FT*, it should interact with *dlfl*, which functions similarly to *FD* in Arabidopsis. Danilevskaya et al. (2008) confirmed that *ZCN8* protein interacts with *dlfl* protein in a yeast two-hybrid system in combination as bait and prey, indicating a strong interaction. The authors propose *ZNC8* as a florigen gene in maize, orthologous of *FT*.

Two studies published in 2011 in *Plant Cell* provided additional evidence that *ZNC8* is the primary florigen hormone in maize. Meng et al. (2011) examined the expression of seven *FT*-like genes. *ZCN7, ZCN8, ZCN12, ZCN14, ZCN18*, and *ZCN26* were chosen because they are expressed in leaves, and *ZCN15* because it is syntenic to the rice flowering *Hd3* QTL, which contains the Arabidopsis *FT* ortholog in rice. The expression was analyzed in two temperate genotypes (Gaspé Flint and B73) insensitive to photoperiod and two tropical genotypes (CML311 and CML436) sensitive to photoperiod, grown under short- and long-day conditions. In addition, there was a short-day inductive treatment, which consisted of long-day treatment, except for 7 days of short-daystreatment at stage V4-V5. Only *ZCN8, ZCN18*, and *ZCN26* were expressed in leaves before and after the floral transition. From those genes, only *ZCN8* responded to the floral inductive treatment; therefore, this gene seems, in agreement with Danilevskaya et al. (2008), the most consistent to act as florigen in maize. Meng et al. (2011) also found, by RNA *in situ* hybridization, that *ZCN8* is expressed in leaf phloem at or near the floral transition, as expected for a mobile signal that moves from leaves to meristems. In addition, those authors found in maize transgenic plants that ectopic expression of

ZCN8 in the shoot meristems produces early flowering. Silencing the gene by artificial microRNA produces late flowering, which also confirms the relevance of the gene on the timing of flowering. Finally, one characteristic of *ZCN8* in common with *FT* and *Hd3a*, the orthologous of *FT* in rice, is the circadian pattern of the level of expression in the photosensitive tropical lines. However, the circadian pattern is attenuated in the non-photosensitive temperate genotypes. Analyzing the expression of *id1* and *dlf1* mutants, Meng et al. (2011) placed *ZCN8* downstream of *id1* and upstream of *dlf1*. Finally, these authors hypothesized that during the adaptation to high altitude, mutations in the coding or promoter regions of *ZCN8* produced a disconnection from the circadian machinery, and, consequently, the diurnal regulation of *ZCN8* was reduced. Lazakis et al. (2011) analyzed the expression of B73 under long and short days and with a short-day inductive treatment of 7 days, as Meng et al. (2011) did, but Lazakis et al. (2011) also analyzed the expression of *ZCN8* in parviglumis. The expression of parviglumis is similar to tropical lines, much higher on short than on long days. One of the most interesting results of Lazakis et al. (2011) is that overexpression of *ZCN8* in Arabidopsis transgenic plants of Arabidopsis has the same effect as overexpression of *FT*, including early flowering and some developmental anomalies. *ZCN8* is even able to complement the *FT* mutation. This is clear evidence of *ZCN8/FT* being a plant universal florigen. These authors suggested that in the day-neutral varieties from temperate areas, *ZCN8* retained the function as an integrator of inductive signals.

Bouchet et al. (2013) analyzed a panel of 375 inbred lines that were genotyped with the maizeSNP50 Bead Cheap and evaluated under long-day conditions in Europe for flowering time. In addition to the whole genome association genetic analysis, loci with a significant signature of selection were detected. In most cases, the markers that were significant for the association were not significant for the selection test. As an exception, the marker PZE-108053909 had the highest significance of all markers in both analyses. This marker is located 5 kb away from *ZCN8*, and no other marker in the maizeSNP50 Beadchip is located closer to the gene. The authors claimed that their study was the most apparent evidence of the involvement of *ZNC8* in the natural variation of flowering time in maize.

In five mapping populations resulting from the cross of five accessions from parviglumis to the temperate inbred line W22 and one mapping population derived from mexicana × W22, a QTL of large effect for flowering was found: *qDTA8* (Guo et al. 2018). The supported intervals of *qDTA8* overlapped in a region of 495 kb, where 14 genes were annotated. One of those genes was *ZCN8*, the more evident candidate behind the QTL, given the relevance of *ZCN8* on flowering. Guo et al. (2018) sequenced the coding and promoter region of *ZCN8* in a panel of about 500 inbred lines that were evaluated for flowering time in a long-day environment. In the association analysis of those data, 13 sequence variants showed a significant association with flowering time on long days. The most significant SNP (SNP-1245) was in the promoter, 1.2 kb upstream of the coding region. For SNP-1245, the Adenine (A) allele genotypes were, on average, earlier than the genotypes with the Guanine (G) allele. In concordance with this result, the six teosinte parents of the mapping populations commented before had the late G allele, and the temperate line W22 had the early A allele. SNP-1245 was associated not only with flowering under long days but also with flowering under short days and photoperiod response. It was checked that the allele A increases the expression of *ZCN8* compared to the allele G in 20 random lines of the panel with the alleles A and G and two near-isogenic lines.

Guo et al. (2018) sequenced the promoter region around SNP-1245 in 35 parviglumis and found that 23.5% of the accessions carried the early A allele. However, the percentage was much higher in the tropical and temperate inbreds of the panel (82% and 96%, respectively). This result was confirmed in a large sample of about 1,000 accessions of maize landraces from America, in which the percentage of the early allele was about 90%. An analysis of diversity and a selection test showed that the allele A of the SNP-1245 was a target of selection. Based on the results, the authors concluded that the allele A of SNP-1245 derived from parviglumis was selected indirectly during early domestication.

Guo et al. (2018) also found that there were two clear linkage disequilibrium blocks embracing *ZNC8*: the first block includes SNP-1245 and most of the significant SNPs, and the second block includes another significant variant, indel-2339, which is a deletion of the three bases GAG. This variant is not in LD with SNP-1245 and is in strong LD with PZE-108053909, the most significant SNP detected by Bouchet et al. (2013). At difference of SNP-1245, indel-2339 has neither effect on short days nor in photoperiod response. The allele of indel-2339 with the deletion (indel-2339Del) increases the expression of *ZCN8* and produces an early flowering. Indel-2339Del is at a low frequency (5%) in the inbred lines of tropical origin, but at higher frequency in the temperate inbred lines (30%). In the landraces, Indel-2339Del is at high frequency in the Northern US but at low frequency in South America. Guo et al. (2018), to check the origin of the allele Indel-2339Del, sequenced 34 entries of parviglumis and 11 of mexicana and found that the allele Indel-2339Del only appears in mexicana. This indicates that the early flowering allele at indel-2339 could originate from introgression from mexicana during the adaptation of maize to highland environments in Mexico and was selected because it conferred advantages in those conditions. The fact that Indel-2339Del is more common in Northern US than in South America is consistent with a higher introgression of mexicana in the US than in South American varieties (Takuno et al. 2015).

5.4.3 *ZmCCT10*

Hung et al. (2012) evaluated the US NAM for flowering time in eight and three long- and short-day environments, respectively. With a joint linkage analysis, they detected 14 QTLs for photoperiod response and 13 colocalized with QTLs for flowering time, a trait in which a higher number of QTLs were detected. This indicates that, although the genetic control of photoperiod response is quantitative, it is simpler than flowering time. The most substantial effect for flowering time under long days and photoperiod response was in chromosome 10. However, the effect of this QTL was heterogeneous across NAM families. It was detected in three populations with a tropical parental line, but not in other populations of tropical origin. The QTL at chromosome 10 was confirmed in a backcross mapping population derived from teosinte crossed to a temperate maize line. The authors identified genes, in the B73 reference genome, with homology to flowering genes in Arabidopsis and to photoperiod regulatory genes in rice and sorghum. Although most genes were not located in the vicinity of the QTLs, homologs to rice photoperiod regulatory genes were located inside the QTL of chromosome 10. Three mapping populations with the parental lines of the NAM where the QTL was detected were developed to fine-map the QTL at chromosome 10. The position was refined by substitution mapping to a region of 443 kbp in one of the three populations. In this region, there were six genes, including *ZmCCT10*. The fine mapping was independently carried out in a maize-teosinte population. With this population, the QTL's position was narrowed to 202 kbp, with only one gene: *ZmCCT10*. The sequencing of the coding and promoter regions of *ZmCCT10* showed that the coding region is conserved, but the promoter region has significant variation. Two SNPs (SNP1411 and SNP1520) had alleles consistent with QTL results.

The region around these SNPs was sequenced in a diverse panel of 282 inbred lines. Again, these two SNPs had the maximum significance for the association with the photoperiod response. Five haplotypes were identified: four that were sensitive to photoperiod and maintained a high level of diversity and one not sensitive to photoperiod and with a reduced diversity. *ZmCCT10* is expressed at higher level in teosinte than in temperate maize when the plants are grown under long days. This suggests that *ZmCCT10* is acting in teosinte, repressing the expression of *ZCN8* on long days, which delays the flowering initiation, but in temperate lines, the effect of *ZmCCT10* on *ZCN8* is attenuated, which allows the plants to flower early on long days.

Yang et al. (2013) genotyped with 500,000 polymorphic sites and evaluated at different latitudes for different flowering-related traits, including photoperiod response, a panel of 368 inbred lines. The association genetic analysis of the panel identified 48 highly significant sites, of which 44 were in the promoter of *ZmCCT10*, in agreement with the results of Hung et al. (2012). Yang et al. (2013) identified a CACTA-like transposon element (CLT) 2543 bp away of the open reading frame. The authors identified seven haplotypes, two with the CLT insertion (Hap1 and Hap2) and the other five (Hap3-Hap7) without the CLT insertion. The haplotypes with the CLT insertion had early flowering and weak photoperiod sensitivity. In contrast, all haplotypes without CLT insertion had strong photoperiod sensitivity and late flowering, except Hap4. The phenotypic effect of the CLT insertion was validated by genetic transformation, transferring the promoter of *ZmCCT10* from a maize line without the CLT insertion to a maize line with the CLT insertion. The effect was consistent with previous results. The transgenic plants with a promoter without CLT had delayed flowering under long days compared to non-transgenic plants with only a promoter with CLT.

Yang et al. (2013) genotyped the promoter of 61 teosinte varieties in addition to the promoter of the 338 maize varieties and found both that the temperate maize had more CLT insertions than tropical maize and that within teosinte, the CLT insertion only appears in one entry of mexicana. Consistent with this result, in an independent association analysis of *ZmCCT10* in a panel of 513 inbred maize lines, about 50% of the tropical lines had the CLT insertion, but the number increased to 85% in the temperate lines (Huang et al. 2018). The fact that Yang et al. (2013) only found the CLT insertion in mexicanais in agreement with Hung et al. (2012), who proposed that the origin of the allele insensitive to photoperiod comes from introgression from mexicana.

At the gene expression level, the fragment of the promoter of *ZmCCT10* that does not contain CLT can drive GUS expression in tobacco leaves under long days, but when the fragment contains CLT cannot. Related to this last result, an inbred line without CLT had a low level of methylation in the promoter of *ZmCCT10*, while an inbred line with CLT had a high level of methylation. It can be concluded that *ZmCCT10* is a gene with a standard allele in parviglumis and tropical maize that represses *ZCN8* and, therefore, flowering in long days. However, an allele introgressed from mexicana into some maize varieties has a transposon insertion in the promotor that inhibits the expression of *ZmCCT10* and, therefore, this gene does not repress *ZCN8* and flowering in long days. This allele was selected during the adaptation of maize to temperate areas because it confers early flowering under long days, which is adaptively advantageous in those environments.

5.4.4 *ZmRap2.7 (QTL Vgt1)*

The QTL for flowering named Vegetative to generative transition 1 (*Vgt1*) was consistently found in different mapping populations and, in some cases, the estimated effect was large. This gene was one of the first cloned for a quantitative trait in maize by Salvi et al. (2007). An early allele from Gaspe Flint was introgressed in the inbred line N28, and a mapping population of near-isogenic lines was obtained crossing N28 and its near-isogenic line. This population was used to clone the gene by positional cloning. The gene responsible for *Vgt1* was *ZmRap2.7*, while the polymorphism responsible for flowering variation was located in the promoter region, about 70 kb upstream of the coding region. The significant variants of *ZCN8* and *ZmCCT10* are also upstream of the coding region, but much closer (less than 3 kb) to the coding region. The expression of *ZmRap2.7* was higher in N28 (the line with the late allele) and the near-isogenic lines with the allele from N28. The effect of increased and decreased *ZmRap2.7* expression on flowering was validated in transgenic plants using a *ZmRap2.7* cDNA and a RNAi based on *ZmRap2.7* cDNA, respectively, and a rice actin promoter. The results in transgenic plants indicated that *ZmRap2.7* is a negative regulator of flowering time probably acting as a repressor of *ZCN8*.

In addition, Salvi et al. (2007) genotyped the region in a panel of 95 inbred lines with 192 SNPs and indels and identified three sites highly associated with flowering: an inverted repeat transposable element (MITE), ATindel434, an indel of two bases and G/A/indel324 with three alleles (A, G and deletion). Ducrocq et al. (2008) conducted an association analysis in a region of 2.2 Mb around *Vgt1* on a broader panel of 375 inbred lines. Out of 269 polymorphisms, four were the most strongly associated with flowering: three already detected by Salvi et al. (2007) and one (CGindel587) with the highest significance (2×10^{-14}) not detected in the previous study. The level of *LD* was very high (r^2>0.7) within the *Vgt1* locus, making it challenging to identify the variant with an actual effect on the trait and not simply linked to the functional variant. Ducrocq et al. (2008) analyzed the MITE in a collection of landraces of America and Europe and found that European populations (except Southern Spanish and Italian), Northern Flints, and Northern Mexican populations generally have a high frequency of MITE insertion. In contrast, the Caribbean and South American (except Southern Chile) populations showed low frequency. Interestingly, within the Caribbean and South American populations, the highlands' populations have a higher frequency of MITE insertion than those of the lowlands. It is clear that *Vgt1* confers, through early flowering, adaptation to higher latitudes and altitudes. Guo et al. (2018), examined 15 cobs from 1,900 years ago found in Turkey Pen Shelter in Southern US and none of the archaeological samples had the MITE insertion at *Vgt1*, indicating that the selection for the MITE insertion occurred later than 1,900 years ago. However, SNP-1245, in the promoter of *ZCN8*, was selected earlier, during early domestication from parviglumis, and Indel-2339, also in the promoter of *ZCN8*, was selected later than SNP-1245, but earlier than *Vgt1* (during the adaptation of maize to Mexican highlands). Guo et al. (2018) concluded that the expansion of maize was a gradual process in which different loci were selected at different evolutionary times.

5.4.5 *ZmCCT9*

Huang et al. (2018) carried out comprehensive research to identify the *ZmCCT9* gene related to flowering time. First, a QTL for flowering was detected in chromosome 9 (*qDTA9*) in a large mapping population of 866 RILs derived from a cross teosinte × maize. Then, for fine mapping, a heterogeneous inbred family (HIF) was chosen that was heterozygous at the QTL, and a population of 5394 near-isogenic lines was generated. With this population, the location of the QTL could be reduced to a region of 2.4 kb in a non-coding section of the genome. This region of 2.4 kb was sequenced in a panel of 513 inbred maize lines and evaluated under long-day conditions. Out of 41 SNPs and indels, 13 had a significant association with flowering time. The strongest association was at a 360-bp Harbinger-like transposon (HLT) that was in strong linkage disequilibrium with all other significant variants. The maize parent of the teosinte × maize mapping population carries the HLT, while the teosinte parent does not. The comparison of two near isogenic lines (NILs) with different alleles at HLT showed that the NIL with the maize allele (carrying the HLT) flowers earlier than the NIL with the teosinte allele (not carrying the HLT) under long days, but under short days there were no differences among the NILs. Consistent with this result, when the panel of 513 maize lines was evaluated at different latitudes, the effect of the HTL was only detected at higher latitudes with longer days. The common parent of the US-NAM, B73, is a temperate line with the HLT, and the effect of *qDTA9* was consistently detected in the mapping populations of the US-NAM in which the other parent was a tropical line without the HLT.

HLT is located 57 kb upstream of a gene that Huang et al. (2018) named *ZmCCT9* because it encodes a CCT protein homologous to the rice *Ghd7* gene. The mutants with the function of *ZmCCT9* annulled using Targeting Induced Local Lesions IN Genomes/CRISPR-associated protein (CRISPR/Cas9) flowered earlier than the lines with the gene intact under long-day conditions, confirming the effect of the gene on flowering. The expression of *ZmCCT9* was analyzed in the US-NAM founders and the reference line, B73, and the results showed that the lines without HLT, mainly tropical lines, had higher expression of *ZmCCT9* under long-day conditions. The expression of *ZmCCT9* was analyzed in two NILs with different alleles at HLT, and the NIL with the allele of maize with HLT showed lower expression than the NIL with the allele of teosinte without HLT under long days. The expression of *ZmCCT9* showed a clear circadian pattern on long days in the NIL with the allele without HLT, but in the NIL with the allele with HLT the diurnal pattern is attenuated. The expression of *ZCN8* on the same NILs was the opposite of *ZmCCT9*. The results indicate that the standard allele of *ZmCCT9* without HLT in parviglumis and tropical maize represses *ZCN8* and therefore, the flowering on long days. On the other hand, the allele with the HLT insertion in temperate maize inhibits the expression of *ZmCCT9*, which cannot repress *ZCN8* under long days and therefore, the flowering is not inhibited.

HLT at *ZmCCT9*, similar to CLT at *ZmCCT10*, was not found in parviglumis and, therefore, should have arisen as a mutation after domestication. In the panel of 513 inbred lines, HLT at *ZmCCT9* is found in 10% of the tropical and 79% of the temperate lines. In contrast, CLT at *ZmCCT10* is in a much higher percentage in tropical lines. The analysis of about 1,000 maize landraces from America confirms that HLT at *ZmCCT9* occurs at low frequency at low latitudes but not *ZmCCT10*. Although both transposons arose after domestication, CLT at *ZmCCT10* arose earlier than HLT at *ZmCCT9*.

5.4.6 *ZmMADS69*

Liang et al. (2019) identified the gene *ZmMADS69* using methodology and material similar to previous research that we do not comment on in much detail. The initial QTL was detected in the 866 maize-teosinte BC2S3 mapping population. The teosinte parent was the CIMMYT accession 8,759 of parviglumis, and the maize parent was W22. The fine mapping of the QTL was accomplished by developing large mapping populations of near-isogenic lines derived from HIFs. The analysis of two NILs showed that an isogenic line with the allele coming from maize flowers 2 days earlier than the isogenic line with the teosinte allele, both in long and short days. That is, the QTL is insensitive to differences in day length. The fine mapping allowed the location of the QTL to be narrowed to a region of 192 kb where only one intact gene (*ZmMADS69*) is located, which encodes a MADS-box transcription factor.

The effect of the overexpression of the gene was studied in transgenic lines with the coding sequence of *ZmMADS69* and with a Ubiquitin promoter, and the downregulation of the gene was studied using a line from the Uniform Mu collection that has a Mu transposon insertion in *ZmMADS69* that down-regulated its expression. The lines with overexpression and down-regulation of *ZmMADS69* flowered earlier and later than not-modified plants. The authors developed knockout lines of *ZCN8* and *ZmRap2.7* and examined the expression of *ZmMADS69, ZCN8,* and *ZmRap2.7* in both the modified lines and in lines without modification. The results indicated that *ZmMADS69* represses *ZmRap2.7*. The wild line with the functional *ZmRap2.7* represses *ZCN8* and produces a delay in flowering, but if *ZmMADS69* represses *ZmRap2.7*, then *ZmRap2.7* does not repress *ZCN8*, and the flowering is early. So, the effect of *ZmMADS69* should be similar to the MITE or CGindel587 in the promoter of *ZmRap2.7* in temperate lines that represses the gene and relieves the expression of *ZCN8* in long days. As expected, the expression of *ZmMADS69* is higher in temperate lines than in tropical lines and could contribute to the adaptation of maize to temperate areas. This idea is reinforced by (Unterseer et al 2019) who found that *ZmMADS69* has a large effect on flowering in the analysis of a Dent-Flint introgression library.

5.5 GENERAL REVIEW OF LEAF SENESCENCE

Senescence in plants is the final phase of plant development, which starts with the death of a single cell, then the single leaf, and finally, the death of the complete plant (Noodén and Penney 2001). The leaf goes through three development stages in its lifecycle. The accumulated/uptaken plant nutrients are transported from roots to leaves in the first stage. Soon after, the leaf reaches its maximum photosynthetic capacity and turns into a vital plant organ; this second stage is known as the productive stage. In the end, the leaf starts dismantling the photosynthetic apparatus, and most of the stored nutrients in the leaf are translocated into fruits or ears in the case of maize. This last phase is known as the senescence stage (Hörtensteiner and Feller 2002; Buchanan-Wollaston et al. 2003). Senescence involves many processes: cytological and metabolism changes in the cell structure, the formation of anthocyanins in vacuoles, changes in gene expression, biochemical adjustments, etc. The senescence period involves the breakdown of macromolecules and the accumulation, redistribution, and remobilization of nutrients from senescence leaves to storage organs or reproductive organs (Barth et al. 2006). In addition to internal development, external factors, such as light limitation, low nitrogen supply, pathogen attacks, etc., can promote senescence too (Figure 5.2) (Smart 1994; Gregersen et al. 2013). The degradation of the photosynthetic apparatus during senescence results in a decrease in the chlorophyll conductance and the photosynthesis activity and a breakdown of proteins and RNA. This chlorophyll degradation changes the leaf color from green to yellow and finally to brown (Barth et al. 2006). There is a vertical variation profile of the leaf senescence in maize; two gradients of senescence are observed, one from the bottom to the center and the other from the top to the center. Therefore, the leaves in the center, mainly surrounding the ears, remain green for extended periods (Tollenaar and Daynard 1978; Wolfe et al. 1988). Husk leaves play a crucial part in the distribution of nitrogen during the grain-filling period by producing 42% of dry matter (Fujita et al. 1995; Oka et al. 1997). Husk leaves are efficient in translocating nutrients from leaves, mainly ear leaves, to ear, (Cantrell and Geadelmann 1981; Sawada et al. 1995). It has been proven that nitrogen absorption during the grain-filling stage is affected by husk leaves (Fujita et al. 1995), which could also have implications for the timing and rate of senescence.

5.5.1 Senescence and Abiotic Stress

The plant may undergo abiotic stress during its life cycle, which affects productivity and plant growth. During the aging of plants, the reactive oxygen species (ROS) detoxification compounds decrease, leading to an accumulation of ROS, which damages the chlorophyll and contributes to the initiation and progression of senescence (Liang et al. 2015). Therefore, the genes regulating the ROS detoxification agents contribute to the delay of senescence due to reducing chlorophyll degradation. ROS accumulation can happen earlier in cells during abiotic stresses. The plant response to abiotic stresses depends on factors like the stress intensity, the plant genotype, the development stage, and other factors that regulate senescence. High temperatures also lead to declining in photosynthetic pigments, antioxidant activity, and nitrate reductase activity. The increase of glutamate dehydrogenase and accumulation of H_2O_2 result in the acceleration of leaf senescence and also causes a disturbance in biochemical and physiological traits (Lea and Miflin 2003; de la Haba et al. 2014).

5.5.2 Metabolic and Biochemical Changes during Senescence

Chloroplast and chloroplast pigments are one of the earliest targets during senescence in all plant species (Biswal and Biswal 1988; Noodén et al. 1997). The level and amounts of photosynthetic molecules, such as rubisco, or glutamine synthetase II, decrease during senescence in maize, soybean, and Arabidopsis (Biswal and Biswal 1988; Chi et al. 2015). Vacuoles are also engaged in senescence; for example, senescence-associated vacuoles (SAVs) are present in mesophyll and guard cells and have intense proteolytic activity. The SAVs are present in chloroplast-containing cells and absent in the rest of the epidermis. An increase in their number during senescence results in an intensive breakdown of chloroplast proteins in soybean, Arabidopsis, and tobacco. The presence of vacuolar H+ pyrophosphatase in the vacuolar membrane makes them acidic and has extreme peptidase activity because of the cysteine protease (Martínez et al. 2008; Carrión et al. 2013; Carrión et al. 2014). During senescence, the chloroplasts' structure disrupts, leading to an unorganized dilation of the thylakoid membranes with the decline of grana stacking. These structural modifications of thylakoid membranes (along with the degradation of photosystem proteins, thylakoid galactolipids, and photosynthetic pigment) result in an imbalance in the electron transportation in photosynthesis (Krupinska and Humbeck 2004). The clustering of the thylakoid membrane and assembling of plastoglobuli occurs in the initial stage of senescence, so the proportion of plastoglobuli increases as the thylakoid membrane decreases in both C_3 and C_4 plants (Thomas et al. 2003). The proportions of stroma, thylakoid, and chlorophyll proteins can change due to the adjustment of the remaining parts of the leaf, which helps in photosynthetic activity in senescence and environmental stress (Hörtensteiner and Kräutler 2011). The ions and solutes from membranes are leaked due to the breakdown of lipid membranes caused by ROS. Malondialdehyde (MDH), as a result of lipid peroxidation, is liberated by polyunsaturated fatty acids (Dai et al. 2011) and can be used as a biomarker (Dai et al. 2011). MDH is negatively correlated with the yielding of the plant. During the vertical variation of leaf senescence, the MDH proportion in plants increases. So the delay in the initiation of leaf senescence will extend the leaf function and is essential to increase the yield of a plant (Song et al. 2013).

The chloroplast is the primarily affected organelle in senescence. Active chloroplasts in leaves transform into plastids (senescing chloroplasts) through large-scale structural modifications (unstack) of the thylakoid membrane and through an abundant accumulation of plastoglobuli. The developed plastid from chloroplast is known as a gerontoplast. Senescing chloroplasts lose their ability to perform biosynthetic processes (Thomas et al. 2003). During the initial stage of senescence, the chlorophyllase enzyme converts chlorophyll into chlorophyllide; due to this reaction, phytol is produced, and lipid globules of the senescing chloroplast are in ester form. The vacuolar enzymes are responsible for the breakdown of chlorophyll in leaves, which exposes the carotenoids, resulting in the change from green to yellow or orange. There is an accumulation of compounds, such as phenols, flavonoids, tannins, and many more in senescent cells that are produced by the phenylpropanoid pathway (Thakur et al. 2016).

5.5.3 Transportation of Organic Nitrogen and Sulphur during Senescence

Organic amides like glutamine and asparagine are liberated during the breakdown of proteins in leaf senescence. Organic amides, amino acids, and proteins play a vital role in the transportation of organic nitrogen and also sulphur, mainly in homoglutathione form (Buchanan-Wollaston 1997). The glutamine synthetase (GS) enzyme synthesizes glutamine expressed during natural senescence or senescence induced by dark (Nozawa et al. 1999). The products of the breakdown of proteins and nitrogen are remobilized to developing grains during senescence. In the senescence period endoproteolytic activity development increases the protein carbonylation and decreases the protein content (Havé et al. 2015).

5.5.4 Secondary Metabolites and Sugar Synthesis in the Senescence Stage

In the senescence stage, the plant faces sugar starvation due to the loss of photosynthetic activity. That activates the transformation of lipids (mainly released in the process of thylakoid membranes dismantling) into sugars by gluconeogenesis (Kim and Smith 1994). Sucrose takes part in key processes to regulate plant metabolism, accumulation of metabolites, and development in the leaf senescence stage. The breakdown of nucleic acids during senescence generates purines and pyrimidines that, in turn, are degraded to produce nitrogen-carbon compounds – which are easy to mobilize due to their low molecular weight. The concentration of secondary metabolites like flavonoids and phenylpropanoids differs based on response to light intensity and biotic and abiotic stress conditions (Agati and Tattini 2010; Watanabe et al. 2010; Samanta et al. 2011; Wu et al. 2012). Senescence leaves undergo many proactive mechanisms. For instance, the accumulation of anthocyanin occurs before the breakdown of chlorophyll. This is beneficial to protect the leaves that are sensitive to light-induced ROS because of chlorophyll absence and the decline in photosynthetic capacity (Nooden et al. 1996; Feild et al. 2001). Anthocyanins are formed by many nutritional deficiencies and high carbon stress in the cytoplasm and are transported to vacuoles (Bongue-Bartelsman and Phillips 1995; Shirley 1996; Park et al. 2007). The accumulation of anthocyanin occurs when there is an imbalance of carbon and nitrogen due to stress (Zheng et al. 2009). Anthocyanins are essential to safeguard the leaf from photo inhibitors during senescence, being a photoprotective mechanism.

5.5.5 The Senescence and Plant Hormones

Cytokinin (CK), auxins, and gibberellic acid are responsible for the delay in senescence, whereas other plant hormones like jasmonic acid, ethylene, and ABA (abscisic acid) promote senescence. Differential expressions of genes associated with several hormones are up-regulated or down-regulated in the senescence period (Thomas and Stoddart 1980). In monocotyledonous and dicotyledonous species, CK decreases during senescence, and senescence can be delayed by the external application of CK (Hwang et al. 2012). The enzyme cytokinin oxidase degrades cytokinin during leaf senescence. During this activity, the transcriptional genes associated with cytokinin biosynthesis and signaling are suppressed (Buchanan-Wollaston et al. 2005). The isopentenyl transferase (*ipt*) gene plays a vital role in the biosynthesis of cytokinin and increases the accumulation of cytokinin (Chang et al. 2003). Transgenic plants were transformed in the *ipt* gene for the overproduction of cytokinin. Plants regenerated with *ipt* show delay in senescence and a delay in chlorophyll and photosynthetic enzymes degradation. Kinetin is a type of cytokinin hormone that delays senescence, but new leaves are more reactive than old leaves towards kinetin (von Abrams and Pratt 1966). CK is not uniformly distributed all over the plant; in those parts with higher concentration, there is increased delay in senescence.

Active biological auxins, like indole acetic acid (IAA), decrease the PSII activity (Kelley and Estelle 2012; Gören-Sağlam et al. 2020). Genes induced by auxins like *SAUR36* are up-regulated in senescence leaves. Loss or gain of function of genes associated with auxins may postpone or bring forward senescence, respectively (Hou et al. 2013). Auxins and gibberellic acid (GA) have the same effect on senescence. On the other hand, ethylene is a potent initiator of senescence liberated at a high rate from quickly dividing and growing cells. As the senescence starts or when leaves become old, the production of ethylene increases, mainly in dark-induced senescence (Gepstein and Thimann 1981). ABA is also a hormone that contributes to leaf senescence, especially in water stress conditions (Zeevaart and Creelman 1988; Ren et al. 2007). It is important to remark that a high accumulation of cytokinin can decrease or terminate the senescence induced by ABA but not by gibberellic acid (Back and Richmond 1971). Jasmonic acid (JA) contributes to the increased expression of numerous enzymes involved in chlorophyll degradation (Reinbothe et al. 2009). In addition to its role in defense against pathogen attack, salicylic acid regulates transcriptional factors involved in the initiation of leaf senescence (Morris et al. 2000).

5.6 STAY-GREEN (SG)

There is a genetic variation in the timing and rate of senescence (Christopher et al. 2014; Pujol 2015). Stay-green (SG) genotype is a common denomination given to a heritable delay in senescence compared to a reference genotype (Figure 5.1) (Thomas and Howarth 2000). SG can increase carbon assimilation, resulting in higher biomass and grain yield. The duration of active photosynthesis in leaves is correlated to the production of plant residues in maize (leaves, stalks, and cobs) and therefore, has an impact on the utilization of residuals as feedstock for bioenergy (Sekhon et al. 2012). The stay-green trait also enhances genotype adaptation, and root health and helps to mitigate drought stress after flowering by maintaining water use efficiency (WUE) (Gregersen et al. 2013; Trachsel et al. 2016). It also enhances biotic stress resistance to stalk-rotting pathogens in sorghum and maize (Bertolini and Fonzo 1983; Thomson et al. 1987). The dry matter accumulation increased up to 63% in the new hybrid era of maize released after the twentieth century compared to earlier (1939–1999), partially due to the selection of the stay-green trait (Lee and Tollenaar 2007). However,

FIGURE 5.1 Degree of senescence at 45 days after flowering in late senescence genotype (left) and early senescence genotype (right).

the stay-green trait is also associated with an increment in grain and stover moisture (Caicedo et al 2016). Therefore, the SG characteristic could be detrimental in storing and transporting grain and agricultural residues for producing bioenergy.

5.6.1 Classifications of Stay Green

SG can be functional, in which the phenotype is green, the plants have a standard level of photosynthesis activity, or non-functional (cosmetic), which has low or no photosynthetic activity. Still, they stay-green due to the failure of chlorophyll to undergo its degradation pathway. The functional stay-green categorizes into type-A and type-B. In stay-green, type-A initiation of senescence is delayed while in type-B, the onset of leaf senescence is on time, but the rate of senescence is slow. Contrarily the non-functional stay-green, also known as cosmetic stay-green, is classified into type-C and type-D. Type-C has an average decline in photosynthesis rate during senescence, but due to alterations in the degradation path of chlorophyll, this maintains the color, although not the functionality.

In contrast, in type-D stay-green, the green color of the dead leaf is maintained because the leaf is killed by freezing or dying. In addition to the types as mentioned earlier of stay-green, type-E is just due to a high chlorophyll content at flowering (Thomas and Smart 1993; Thomas and Howarth 2000). In drought and standard conditions, type-A and type-B show a highly significant increase in yield. So, these two stay-green genotypes are essential for agronomical use (Renaud 2015).

5.7 ALLELE MINING FOR SENESCENCE-RELATED TRAITS IN MAIZE

As expected from such a complex process, several studies in cereal crops like rice, sorghum, and soybean showed that the senescence-related traits could be considered quantitative traits controlled by many locus/genes (Guiamet et al. 1990; Crasta et al. 1999; Tao et al. 2000; Jiang et al. 2004; Kumar et al. 2010). Polygenic inheritance was observed, for example, in segregation populations developed

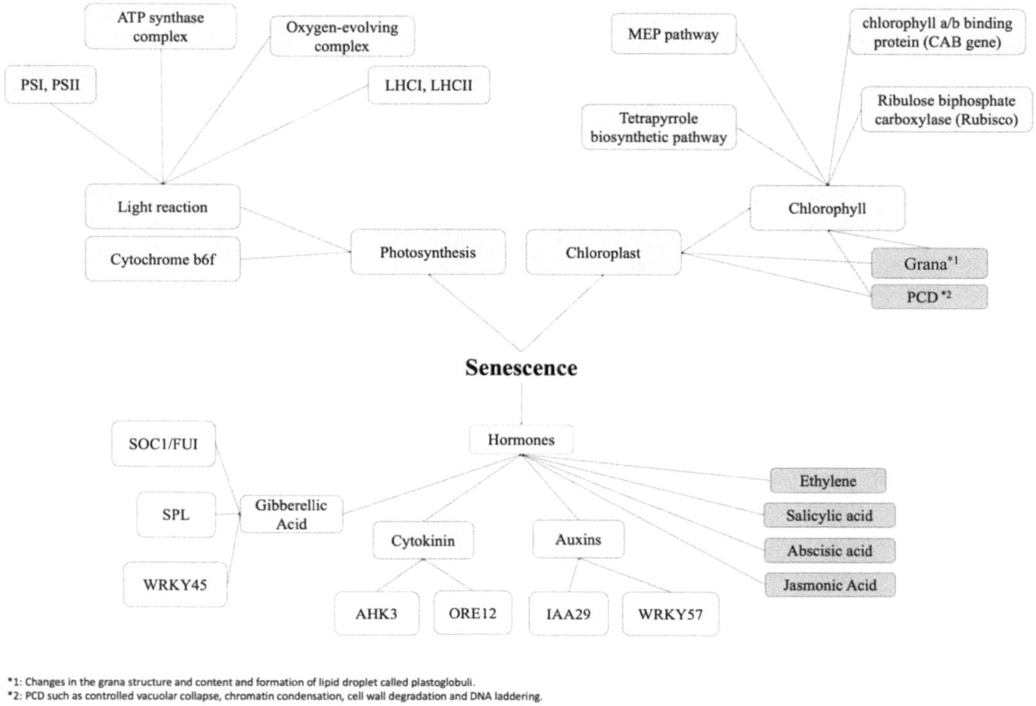

FIGURE 5.2 General overview of changes in senescent leaves; dotted boxes refer to a down-regulation, gray boxes refer to an up-regulation and white boxes indicate a non-clear trend.

by crossing late and standard or early senescence lines (Beavis et al. 1994; Yang et al. 2017). In contrast, some authors considered the inheritance of senescence-related traits not quantitative (Thomas 1987; Cha et al. 2002). The contradiction can be due to the study of alleles of significant effect that could separate the genotypes in two clear classes, stay- and no-stay-green, given the impression of a qualitative trait. The results of Yannam et al. (2022) are consistent with the quantitative nature of the senescence, although not as complex as flowering time. Identifying genetic determinants of the variation of leaf senescence will facilitate the optimization of senescence characteristics for grain, and biomass production, but also nitrogen and water use efficiency (Caicedo et al. 2016). The first and foremost dominant gene that delays leaf senescence in maize was discovered in inbred line LO87602 (Gentinetta et al. 1986). It was reported that the magnitude of additive effects was higher than the dominant effects by evaluating the genotypic variance of two elite stay-green tropical maize populations, proposing that selecting the late senescence genotypes has a significant impact on the increase in yield. The heritability of the traits can be influenced by the background of the population and by the environment (Guei and Wassom 1996).

Plant phenotyping is very important to quantify the effect of genotype and its interaction with the environment, particularly in traits like senescence traits. The high-throughput phenotyping (HTP) maximizes the probability of detecting favorable alleles for targeted traits. In the present situation, the availability of field-based phenotyping platforms, which could be user-friendly, multi-purpose, accurate, non-destructive, and quick, is critical to increment the quality and quantity of phenotypic data in general in plant breeding (Araus and Kefauver 2018) and specifically for senescence related traits.

Due to the complexity of the senescence traits and other biological phenomena involved in plant drying, it is challenging for breeders to develop an accurate and proper quantification of leaf senescence. The vertical pattern of senescence complicates the quantification of the leaf senescence. Almost two thirds of soluble proteins in a plant cell are dissolved or lost during senescence (Pyung et al. 2007). Quantifying the enzymes like RNAase and peroxidase activity involved in catabolic and hydrolytic activity is also a quantitative approach for measuring leaf senescence (Abeles et al. 1988; Taylor et al. 1993). One of leaf senescence symptoms is the degradation/rupture of plasma membranes, which is measured by ion leakage (Woo et al. 2001). This is one of the most promising approaches associated with cell death, although it is the last step of senescence. Leaf senescence is generated by a change in the expression of genes of many metabolic paths. The differential expression pattern of senescence-associated genes can be observed by RNA gene blot analysis, reverse transcription-polymerase chain reaction (RT-PCR), microarrays, or RNA sequencing (RNA-seq) (Woo et al. 2001).

The initial symptom of leaf senescence is the visual change of leaves' color due to chlorophyll degradation. Phenotyping can be done using the visual scales at the time the leaf changes color. One common, economical, and quick way of evaluating senescence is by the visual scores from one to five, where one corresponds to dried leaves and five to whole green leaves. The visual scales have a limitation when distinguishing the cosmetic and functional stay green. By making use of different sensors and platforms, the senescence can be quantified for several variables or parameters, such as chlorophyll content, canopy temperature, vegetation indices (VI), normalized difference vegetative indices (NDVI), green leaf area index, CO_2 assimilation rate and the maximum potential quantum efficiency of Photosystem II (Fv/Fm), etc.

5.7.1 QTL Analysis-Linkage and/or Association Mapping

QTLs were discovered using biparental populations through linkage mapping in inheritance studies for senescence-related traits measured by various methods in maize (Beavis et al. 1994; Guei and Wassom 1996; Câmara 2007; Zheng et al. 2009; Wang et al. 2012; Almeida et al. 2014; Belícuas et al. 2014; Kante et al. 2016; Trachsel et al. 2016; Yang et al. 2017; Zhang et al. 2019; Yannam et al. 2022).

QTLs related to senescence were found along the genome, but there are regions where consistently more QTLs were detected in different studies, particularly two regions at chromosomes 1 and 5. The variance explained by individual QTLs was, generally not high, ranging from 1–10%. It can be concluded from those studies that the genetic architecture of senescence in maize is quantitative, although not as complex as flowering. In the study of Zheng et al. (2009) the senescence was quantified by green leaf area at different stages during senescence in the F2:3 population derived from the Mo17 (late senescence) and Q319 (early senescence). The authors detected a high number of QTLs at 40 days after anthesis (DAA) compared to 20, 50, and 60 DAA. Câmara (2007) detected 53 minor effect QTLs by measuring the visual scale of plants (VSP) at 120 days after sowing. A biparental population derived from an SG line (PHG39) and an NSG line (EA1070) was used to detect QTLs by measuring senescence by visual scales of plants after 60 DAA (Kante et al. 2016). Beavis et al. (1994) identified three major loci with a phenotypic variance of 10–25% on chromosome 2,6,9 using biparental derived from B73 (standard senescence) and Mo17 (late senescence) genotypes. Belícuas et al. (2014) in tropical lines described 17 QTLs with an average genetic variance explained by the QTLs of 7.7%, but with a QTL that explained 38% of the genetic variance on chromosome 1. Almeida et al. (2014), quantifying the senescence using photosynthesis efficiency and chlorophyll rates, identified eight loci with phenotypic variance explained varying between 13% and 21%. Similar regions on chromosome 1 were detected by Kante et al. (2016) by quantifying chlorophyll levels using a SPAD meter. In a recent association study, we analyzed a large panel of 400 inbred lines of maize to evaluate senescence with visual scales of plant, husk leaf, and CO_2 assimilation rate. High correlations and consistent allelic affects were found between agronomical traits and senescence. Seven significant SNPs were found in the coding region of the core senescence genes and one in the promoter region of a gene. Among eight candidate genes three were up-regulated and five were down-regulated that is confirmed by an individual RNAseq analysis (Yannam et al. 2022).

5.7.2 Transcriptional Analysis of Senescence

The changes in gene expression are substantial during senescence (Guo et al. 2004; Buchanan-Wollaston et al. 2005; van der Graaff et al. 2006; Breeze et al. 2011). The differential expressed genes (DEGs) at senescence can be identified by comparing the level of expression of green leaves at flowering when the senescence has not started with the level of expression of leaves weeks after flowering, when the senescence has commenced. Studies of change in gene expression during senescence, under normal and stressful conditions, and dark and hormonal treatment suggest that the induction of senescence occurs in multiple pathways (Becker and Apel 1993; Oh et al. 1996; Chung et al. 1997; Gan and Amasino 1997; Park et al. 1998; Weaver et al. 1998). The pattern of expression and the signaling pathways are also different in the case of natural and induced leaf senescence (Weaver et al. 1998; Buchanan-Wollaston et al. 2005; van der Graaff et al. 2006).

Zhang et al. (2014) analyzed the transcriptional pattern during the maize natural leaf senescence and found 4,552 DEGs. A gene ontology analysis revealed that the DEGs are involved in multiple metabolic pathways. Photosynthesis was the only term enriched in the down-regulated cluster at early and late senescence. On the other hand, pathways related to transport (TP) were the only ones enriched in the upregulated genes at early and late senescence. Transporter genes involved in enzyme and amino acid transport were up-regulated during the early stage of senescence. In contrast, transport genes involved in sugar recycling were up-regulated in the late senescence stage. TP genes are essential in redistributing sugars/nutrients to plant tissue (Hörtensteiner and Feller 2002). Zhang et al. (2014) compared the DEGs in their study on natural maize senescence with the DEGs in Arabidopsis (Breeze et al. 2011) and with the DEGs in maize senescence induced by the prevention of pollination (Sekhon et al. 2012) and found a 25 % of coincidence in both cases. A substantial percentage of senescence genes are shared with a phylogenetically distant species, suggesting that a relatively high number of senescence genes are conserved across species. However, in the case of two senescence processes of the same species, *a priori*, we expected a higher percentage of shared genes, which indicates the complexity of the regulation of the senescence.

5.7.3 Combination of QTL – Linkage or Association Mapping and Transcriptional Analyses

The QTL experiments identify regions of the genome related to a trait's variation and quantify the effect's magnitude. Within those genomic regions, there are usually several genes, of which one has an actual effect on the variation of the trait and the rest do not, however, it is difficult to identify the gene with the actual effect due to the linkage disequilibrium in the QTL region. In contrast, differential expression analysis allows the identification of specific genes with a possible effect on a trait, but it is impossible to know if the gene affects the trait's variation, which is very relevant from an evolutionary or breeding point of view. In addition, the differential expression analysis does not allow the quantification of the effect of the gene on the trait. So, both methods give complementary information: the QTL experiments can quantify the effect on the trait and pinpoint several candidate genes in a genomic region, and the differential expression analysis can validate, among the candidate genes, the most likely associated with the trait. Initially, QTL and expression studies on senescence were carried out independently, but, recently, a few studies have begun to combine data from both types of experiments. Sekhon et al. (2019) analyzed a panel of 328 inbred lines from the two temperate heterotic stiff-stalk and non-stiff-stalk groups, genotyped with more than 400,000 SNPs, and evaluated in three temperate environments for the maximum quantum efficiency of PSII photochemistry (Fv/Fm) at 36 days after anthesis. Differences in Fv/Fm are expected to be due to differences in the onset or rate of senescence among the genotypes. The GWAS detected 62 SNPs significantly associated with senescence in at least one environment. The individual phenotypic variance explained by the SNPs ranged from 5 to 14 %. There were 411 candidate genes in a window of 200 kb around the significant SNPs. In addition, the authors compared the transcriptome of B73, a standard senescence line, and PHG35, an early senescence line, at four developmental stages and found 4,713 DEGs that were different in the senescence of the standard and the early senescence line. A coexpression network of the DEG genes identified a predominant network of 663 genes. Of those genes, 45 are transcription factors (TFs), of which eight are from the *NAC* family, highlighting the critical role of TFs, particularly the *NACs*, in regulating senescence. Eight of the 411 candidate genes detected in the GWAS were also detected in the coexpression network analysis. Therefore, they are strong candidates to be validated and studied in depth by reverse genetics. One of the conclusions of Sekhon et al. (2019) is that source-sink communication plays a vital role in senescence. Thus, the cell wall could be crucial in senescence, acting as a signaling molecules source or sink of excess sugars.

Caicedo et al. (2021) evaluated and genotyped with more than 200,000 SNPs a multiparental population of 672 recombinant inbred lines derived from eight parental lines of no stiff-stalk background to identify senescence-related QTLs. The population was evaluated in two environments for chlorophyll content and Fv/Fm at flowering and 2 months after flowering when the senescence had already started in most or all of the lines. Several SNPs were significantly associated with chlorophyll content or Fv/Fm 2 months after flowering. Still, they were grouped in a few regions, mainly in chromosome 1 (280 Mb, approximately) and particularly in chromosome 3 (201 Mb, approximately). In the region of chromosome 3, there were significant associations for chlorophyll content (201 Mb) and Fv/Fm (207 Mb), with a proportion of phenotypic variance explained of about 5 %. The significant region of chromosome 1 was also significant at flowering, suggesting that the differences in chlorophyll content at senescence could be due to the maintenance of differences already present at flowering. This is a type of senescence that was defined as type-E. On the other hand, there were no significant associations in the significant region of chromosome 3 at flowering but there was a significant association with Fv/Fm 2 months after flowering, reflecting an effect on photosynthesis, not just chlorophyll content. Therefore, it could be a QTL affecting senescence type-A (differences in onset) or B (differences in rates). However, it was not possible to distinguish between these two types of senescence due to the small number of samples over time. A candidate gene (*Zm00001d043586*) was located at about 200 kb of the significant SNPs of chromosome3 at 201 Mb that was orthologous to an Arabidopsis gene with an effect on senescence. This gene was validated in an RNAseq experiment with six inbred lines with different senescence timing, from early to late. The gene was up-regulated in the six inbred lines at senescence. The up-regulation started sooner in the earlier senescence lines, as expected of a gene with an effect on the variation in senescence. This suggests that the senescence controlled by this gene is of type-A.

5.7.4 Combination of QTL – Linkage or Association Mapping, Transcriptional Analyses, and Reverse Genetics

Although the combination of QTL and transcriptional analysis allows the identification of strong candidate genes, the final validation of those genes must be done with reverse genetics methods, increasing or decreasing the gene expression. In senescence, one of the few studies, to our knowledge, that reaches the final verification of the gene's function by reverse genetics was carried out by Zhang et al. (2019). The authors localized a QTL for senescence crossing two lines with different timing of senescence, then fine-mapped and cloned the gene developing a mapping population of 590 individuals starting from F6 plants in which the QTL region was heterozygous. This shortened the QTL region to 38 kb, where a single gene, named by the authors as nac7, was annotated. Transgenic plants with the expression down-regulated by RNAi delayed senescence and increased the yield, confirming the role of nac7 as a negative regulator of senescence. Differential expression analysis of plants with the normal functioning of nac7 and plants with the expression of gene down-regulated identified DEGs that are involved in proteolysis, particularly hydrolase activity and autophagy. Thus, autophagy could be a triggering factor of senescence or helpful in easing nutrient remobilization. Another function also controlled by nac7 is related to source-sink communications in agreement with Sekhon et al. (2019) about the relevance of those processes in the senescence. This study not only identified and analyzed a gene involved in senescence, but also showed its positive relationship with yield.

5.8 CONCLUSIONS AND FUTURE LINES OF RESEARCH

The mapping studies of flowering in maize have experienced a huge progress in terms of the representativeness of the germplasm, size of mapping populations, and genome coverage in the last 15 years. Linkage, association mapping, and/or positional cloning in large populations together to reverse genetics allowed some of the individual genes that controlled the quantitative trait to be studied in detail. *ZCN8*, orthologous of FT and connected to the circadian clock, was found to be a key integrator of flowering time in the varieties from tropical regions where maize was domesticated. Three critical genes involved in controlling flowering time, i.e. *ZmCCT9, ZmCCT10,* and *ZmRap2.7*, were repressors of *ZCN8*, and therefore of flowering time, in long days. In addition, one variant in the promoter of *ZCN8* also represses the gene over long days. Ancient grasses were short-day flowering with a photoperiod pathway that allowed them to flower only in the favorable period of the seasonal tropical climate to what they were adapted (Preston and Fjellheim 2020). The maize was domesticated precisely in Rio Balsas in Mexico, a typical seasonal wet and dry tropical environment (Hastorf 2009). Remarkably, three genes have a similar repressive function in *ZCN8*, and the knockout of the function of the three genes by transposons insertion has resulted in early flowering under long days.

Roux et al. (2006) analyzed the flowering genes in Arabidopsis restricted to those without pleiotropic adverse effects on fitness, which are the only ones expected to contribute to the evolutionary response to selection for early flowering. These authors' conclusions are consistent with the results found in maize. Thus, they concluded that the targets of selection for early flowering should be the promoters of floral pathway integrators, which is consistent with the results found in *ZCN8* in maize. In addition, they proposed floral repressors as the target of selection for early flowering, which is also the case with the repressors *ZmCCT9, ZmCCT10,* and *ZmRap2.7* in maize. The mutants that Roux et al. (2006) suggested that have the strongest effect on selection for early flowering are those which cause loss of function, as was shown in the repressor genes of maize. Finally, another interesting conclusion from these authors is that in Arabidopsis, late flowering is the ancestral type and early flowering types emerged independently several times. This is similar to what is thought to have happened in the evolution of maize.

The function of allelic variants of significant effects, such as those of *ZCN8* and its repressors in the flowering pathway, were analyzed in depth, and its biological fundament could be understood. This facilitates its use in breeding to optimize the flowering time of the genotypes to adapt them to specific environments. In fact, it was proved that those genes played an essential role in adapting maize to temperate areas. However, many other genes of small effect also influence flowering time, although they are not probably part of flowering pathways. The small size of its effect makes it difficult to locate them in the genome and quantify their effects. Despite this, it was proven that the proportion of alleles of small effect associated with early flowering was much higher (0.8 vs. 0.4) in temperate early germplasm, Northern and European Flints, than in late tropical germplasm (Bouchet et al. 2013). This shows the relevance of the many alleles of small effect distributed along the genome in the adaptation of maize to temperate areas. On the other hand, using individual alleles of minor effects in breeding is not feasible. It is more effective to use prediction equations that include many markers and consider the accumulated effects along the genome. The prediction ability of such equations was confirmed by Swarts et al. (2021), at least when the germplasm used for developing the prediction equations is related to the phenotypes that are going to be predicted.

The combination of an early Northern and a late Southern race successfully developed one of the most important races in maize temperate breeding: Corn Belt Dent. This success could be due to the combination of more production in the Southern race and adaptation in the Northern race, or just a combination of different favorable alleles that were fixed due to random drift in both races. In Europe, medium late genotypes have been combined with early genotypes and selected for early flowering to obtain early hybrids (Barrière et al. 2006). The study of the domestication and expansion of maize can give the breeder clues on where to look for new alleles of adaptation to temperate areas. Thus, the temperate maize of the US and Europe probably has several alleles of adaptation to temperate areas derived from varieties from the highlands of Mexico that, in turn, obtained those alleles from mexicana. However, the adaptation of maize to the Andes, southern South America, and China emerged independently from the highlands of Mexico. Therefore, in the varieties of those areas there could be new alleles, interesting for the US/European germplasm.

The circadian clock has an important role in the timing of flowering. The molecular components underlying the clock function are widely conserved across plant species (Hayes et al. 2010; Takata et al. 2010; McClung 2010). Maize core clock genes encompass *ZmCCA1*, *ZmLHY, ZmTOC1a, ZmTOC1b, ZmPRR73, ZmPRR37, ZmPRR59, GIGZ1a, GIGZ1b, ZmFKF1a*, and *ZmFKF1b*. All genes mentioned above showed circadian expression patterns with *ZmCCA1*, *ZmLHY, ZmPRR73,* and *ZmPRR59* peaking in the morning and *ZmTOC1a, ZmTOC1b GIGZ1a*, *GIGZ1b, ZmFKF1a,* and *ZmFKF1b* peaking in the evening (Hayes et al. 2010). A few maize clock genes are demonstrated to be implicated in flowering time. For instance, *ZmTOC1, ZmPRR73* and *ZmPRR95* overexpressing maize lines have late shoot apical meristem (SAM) development and differentiation into tassels, unlike the knockout lines of the same target genes, which promote tassel maturation (Fei et al. 2022).

Regarding future lines of research on maize flowering, Minow et al. (2018) suggested that the autonomous and the photoperiod pathway operate through distinct gene networks and that different component of the circadian clock influence both pathways. More investigation is needed to verify this hypothesis and, in general, the genetic mechanisms of the autonomous, the photoperiod pathway and the external and internal stimuli that affect the functioning of the pathways. Also, a more profound knowledge of the functional and adaptative significance of the pleiotropic effects of flowering genes on traits related to fitness and/or yield is needed to fully understand the adaptation of maize to different environments and to use efficiently the flowering variation on breeding (Auge et al. 2019). The study of the pleiotropic effects should be accompanied by the analysis of the effect of the environments and the genotype × environment interactions that could affect individual traits and the relationship between traits. An example of an allele whose effects varied with environments was given by Cui et al. (2021) in rice. Using CRISPR/Cas9, these authors generated a novel non-functional allele of the photoperiod sensitivity-13 (*Se13*) gene, which is aphytochrome-associated gene from the heading date pathway. This non-functional allele has a short period of growth in high latitudes but a long growth period in low latitudes. Because of this, the allele has a favorable pleiotropic effect on the yield in all environments.

Lower progress has been made in identifying and analyzing allelic variants responsible for senescence compared to flowering. In most mapping experiments, the precision was low, and only a general picture of the architecture of the trait can be obtained from those studies. The trait is controlled by many alleles of small effect distributed along the genome, but with hot spots on chromosomes 1 and 5. The trait is quantitative but controlled by fewer genes than flowering. On the other hand, the differential expression analyses have shown that senescence is very complex, and thousands of genes involved in many metabolic paths change the expression at senescence. Among all the processes, the photosynthesis genes were consistently down-regulated in all studies at senescence, and genes of catabolism and transport were up-regulated. A few strong candidate genes were identified with the combination of QTL and expression analyses. In addition to those genetic methodologies, reverse genetics was used to study in depth one of the few genes involved in senescence that has been fully validated in maize so far. This gene, named nac7, has an essential effect on yield and, therefore, can be used in breeding. The metabolic paths involved in senescence have been studied in Arabidopsis (Schippers 2015), but in maize, the knowledge about the metabolic paths related to senescence is reduced.

Regarding the future lines of exploration of maize senescence, more research, combining QTL, expression and reverse genetics analyses, is needed to identify genes involved in the trait and to analyze the functional variants. A better knowledge of the transcription factors that are involved at different moments of senescence is crucial for understanding the process. This includes, knowing which genes are modified by the transcription factors, how the transcription factors interact with each other and how the internal, for example, circadian clock, or external stimuli affect their functioning. Some tropical landraces with high sensitivity to photoperiod were able to produce grain in temperate environments, but others were not (Gouesnard et al. 2002). This could be due to differences in the onset or rate of the senescence among the varieties, so although some varieties flower late, they have rapid senescence and can produce mature grain very fast. This hypothesis, or, in general, the impact of senescence in the evolution of maize, also deserves further research.

As a final comment, the progress on the knowledge of the genetics of maize senescence and particularly, on the genetic of maize flowering has been impressive in the last two decades, but with the new techniques, like CRISPR/Cas9, that progress is expected to continue. The modification of the timing of flowering and/or senescence has been successful in adapting maize to many regions and is still successful in adapting maize to global weather change. The improved knowledge of flowering and senescence will allow a more efficient adaptation of maize to different regions and to global weather change.

REFERENCES

Abe M, Kobayashi Y, Yamamoto S, Daimon Y, Yamaguchi A, et al. (2005) FD, a bZIP protein mediating signals from the floral pathway integrator FT at the shoot apex. *Science*. 309(5737):1052–1056.

Abeles FB, Dunn LJ, Morgens P, Callahan A, Dinterman RE, et al. (1988) Induction of 33-kD and 60-kD peroxidases during ethylene-induced senescence of cucumber cotyledons. *Plant Physiol.* 87(3):609–615.

von Abrams GJ, Pratt HK. (1966) Interaction of naphthaleneacetic acid and kinetin in the senescence of detached leaves. *Plant Physiol.* 41(9):1525–1530.

Agati G, Tattini M (2010) Multiple functional roles of flavonoids in photoprotection. *New Phytol.* 186(4):786–793.

Ågren J, Oakley CG, Lundemo S, Schemske DW (2017) Adaptive divergence in flowering time among natural populations of *Arabidopsis thaliana*: Estimates of selection and QTL mapping. *Evolution (NY)*. 71(3):550–564. https://doi.org/10.1111/evo.13126

Aguirre-Liguori JA, Gaut BS, Jaramillo-Correa JP, Tenaillon MI, Montes-Hernández S, et al. (2019) Divergence with gene flow is driven by local adaptation to temperature and soil phosphorus concentration in teosinte subspecies (*Zea mays parviglumis* and *Zea mays mexicana*). *Mol Ecol.* 28(11):2814–2830.

Almeida GD, Nair S, Borém A, Cairns J, Trachsel S, et al. (2014) Molecular mapping across three populations reveals a QTL hotspot region on chromosome 3 for secondary traits associated with drought tolerance in tropical maize. *Mol Breed [Internet].* 34(2):701–715. https://doi.org/10.1007/s11032-014-0068-5

Araus JL, Kefauver SC. (2018) Breeding to adapt agriculture to climate change: Affordable phenotyping solutions. *Curr Opin Plant Biol.* 45:237–247.

Auge GA, Penfield S, Donohue K. (2019) Pleiotropy in developmental regulation by flowering-pathway genes: Is it an evolutionary constraint? *New Phytol.* 224(1):55–70.

Austen EJ, Rowe L, Stinchcombe JR, Forrest JRK. (2017) Explaining the apparent paradox of persistent selection for early flowering. *New Phytol.* 215(3):929–934. https://doi.org/10.1111/nph.14580

Back A, Richmond AE. (1971) Interrelations between gibberellic acid, cytokinins and abscisic acid in retarding leaf senescence. *Physiol Plant.* 24(1):76–79.

Barnes AC, Rodríguez-Zapata F, Juárez-Núñez KA, Gates DJ, Janzen GM, et al (2022) An adaptive teosinte mexicana introgression modulates phosphatidylcholine levels and is associated with maize flowering time. *Proc Natl Acad Sci USA*. 119(27):e2100036119.

Barrière Y, Alber D, Dolstra O, Catherine L, Motto M, et al. (2006) Past and prospects of forage maize breeding in Europe. II. History, germplasm evolution and correlative agronomic changes. *Maydica* 51:435–449.

Barth C, de Tullio M, Conklin PL. (2006) The role of ascorbic acid in the control of flowering time and the onset of senescence. *J Exp Bot.* 57(8):1657–1665.

Beavis WD, Smith OS, Grant D, Fincher R. (1994) Identification of quantitative trait loci using a small sample of topcrossed and F4 progeny from maize. *Crop Sci.* 34(4):882–896.

Becker W, Apel K. (1993) Differences in gene expression between natural and artificially induced leaf senescence. *Planta.* 189(1):74–79.

Belícuas PR, Aguiar AM, Bento DAV, Câmara TMM, de Souza Junior CL. (2014) Inheritance of the stay-green trait in tropical maize. *Euphytica [Internet]*. 198(2):163—173. http://europepmc.org/abstract/AGR/IND500900677

Bertolini M, Fonzo DI. (1983) *LO 87602, Linea Pura Di Mais" High Lysine*" Sviluppata in Italia.

Biswal UC, Biswal B. (1988) Ultrastructural modifications and biochemical changes during senescence of chloroplasts. *Int Rev Cytol.* 113: 271–321.

Bongue-Bartelsman M, Phillips DA. (1995) Nitrogen stress regulates gene expression of enzymes in the flavonoid biosynthetic pathway of tomato. *Plant Physiol Biochem (Paris)*. 33(5):539–546.

Bouchet S, Servin B, Bertin P, Madur D, Combes V, et al. (2013) Adaptation of maize to temperate climates: Mid-density genome-wide association genetics and diversity patterns reveal key genomic regions, with a major contribution of the Vgt2 (ZCN8) locus. *PLoS One* 8(8):e71377.

Breeze E, Harrison E, McHattie S, Hughes L, Hickman R, et al. (2011) High-resolution temporal profiling of transcripts during Arabidopsis leaf senescence reveals a distinct chronology of processes and regulation *Plant Cell.* 23(3):873–894.

Buchanan-Wollaston V. (1997) The molecular biology of leaf senescence. *J Exp Bot.* 48(2):181–199.

Buchanan-Wollaston V, Earl S, Harrison E, Mathas E, Navabpour S, et al. (2003) The molecular analysis of leaf senescence–a genomics approach. *Plant Biotechnol J.* 1(1):3–22.

Buchanan-Wollaston V, Page T, Harrison E, Breeze E, Lim PO, et al. (2005) Comparative transcriptome analysis reveals significant differences in gene expression and signalling pathways between developmental and dark/starvation-induced senescence in Arabidopsis. *Plant J.* 42(4):567–585.

Buckler ES, Holland JB, Bradbury PJ, Acharya CB, Brown PJ, et al. (2009) The genetic architecture of maize flowering time. *Science* (1979). 325(5941):714–718.

Butler EE, Huybers P. (2013) Adaptation of US maize to temperature variations. *Nat Clim Chang* 3(1):68–72.

Caicedo M, Barros J, Ordás B. (2016) Redefining agricultural residues as bioenergy feedstocks. *Materials* [Internet]. 9(8):635. https://doi.org/10.3390/ma9080635

Caicedo M, Munaiz ED, Malvar R, Jiménez-Galindo J, Ordás B. (2021) Precision mapping of a maize MAGIC population identified a candidate gene for the senescence-associated physiological traits. *Front Genet* 12(716821). https://doi.org/10.3389/fgene.2021.716821

Câmara TMM. (2007) Mapeamento de QTLs de caracteres relacionados à tolerância ao estresse hídrico em milho tropical. *Piracicaba: Universidade de São Paulo.* https://doi.org/10.11606/T.11.2007.tde-04042007-145739

Cantrell RG, Geadelmann JL. (1981) Contribution of husk leaves to maize grain Yield 1. *Crop Sci* 21(4):544–546.

Carrión CA, Costa ML, Martínez DE, Mohr C, Humbeck K, et al. (2013) In vivo inhibition of cysteine proteases provides evidence for the involvement of 'senescence-associated vacuoles' in chloroplast protein degradation during dark-induced senescence of tobacco leaves. *J Exp Bot.* 64(16):4967–4980.

Carrión CA, Martínez DE, Costa ML, Guiamet JJ. (2014) Senescence-associated vacuoles, a specific lytic compartment for degradation of chloroplast proteins? *Plants.* 3(4):498–512.

Cha K-W, Lee Y-J, Koh H-J, Lee B-M, Nam Y-W, et al. (2002) Isolation, characterization, and mapping of the stay green mutant in rice. *Theor Appl Genet.* 104(4):526–532.

Challinor AJ, Watson J, Lobell DB, Howden SM, Smith DR, et al. (2014) A meta-analysis of crop yield under climate change and adaptation. *Nat Clim Chang.* 4(4):287–291.

Chang H, Jones ML, Banowetz GM, Clark DG. (2003) Overproduction of cytokinins in petunia flowers transformed with PSAG12-IPT delays corolla senescence and decreases sensitivity to ethylene. *Plant Physiol.* 132(4):2174–2183.

Chardon F, Virlon B, Moreau L, Falque M, Joets J, et al. (2004).Genetic architecture of flowering time in maize as inferred from quantitative trait loci meta-analysis and synteny conservation with the rice genome. *Genetics.* 168(4):2169–2185. https://doi.org/10.1534/genetics.104.032375

Chi Y, Yang Y, Li G, Wang F, Fan B, et al. (2015) Identification and characterization of a novel group of legume-specific, Golgi apparatus-localized WRKY and Exo70 proteins from soybean. *J Exp Bot.* 66(11):3055–3070. https://doi.org/10.1093/jxb/erv104

Chibane N, Caicedo M, Martinez S, Marcet P, Revilla P, et al. (2021) Relationship between delayed leaf senescence (Stay-green) and agronomic and physiological characters in maize (*Zea mays* L.). *Agronomy.* 11(2):276.

Christopher J, Veyradier M, Borrell A, Harvey G, Fletcher S, et al. (2014) Phenotyping novel stay-green traits to capture genetic variation in senescence dynamics. *Funct Plant Biol.* 41:1035–1048. https://doi.org/10.1071/FP14052

Chung B-C, Lee SY, Oh SA, Rhew TH, Nam HG, et al. (1997) The promoter activity of sen 1, a senescence-associated gene of Arabidopsis, is repressed by sugars. *J Plant Physiol.* 151(3):339–345.

Colasanti J, Yuan Z, Sundaresan V. (1998) The indeterminate gene encodes a zinc finger protein and regulates a leaf-generated signal required for the transition to flowering in maize. *Cell.* 93(4):593–603.

Coneva V, Zhu T, Colasanti J. (2007) Expression differences between normal and indeterminate1 maize suggest downstream targets of ID1, a floral transition regulator in maize. *J Exp Bot.* 58(13):3679–3693.

Corbesier L, Vincent C, Jang S, Fornara F, Fan Q, et al. (2007) FT protein movement contributes to long-distance signaling in floral induction of Arabidopsis. *Science* (1979). 316(5827):1030–1033.

Crasta OR, Xu WW, Rosenow DT, Mullet J, Nguyen HT. (1999) Mapping of post-flowering drought resistance traits in grain sorghum: Association between QTLs influencing premature senescence and maturity. *Mol Gen Genet.* 262(3):579–588.

Crow T, Ta J, Nojoomi S, Aguilar-Rangel MR, Torres Rodríguez JV, et al (2020) Gene regulatory effects of a large chromosomal inversion in highland maize. *PLoS Genet.* 16(12):e1009213.

Cui Y, Xu Z, Xu Q. (2021) Elucidation of the relationship between yield and heading date using CRISPR/Cas9 system-induced mutation in the flowering pathway across a large latitudinal gradient. *Mol Breed.* 41(3):1–11.

Dai H-P, Zhang P-P, Lu C, Jia G-L, Song H, et al. (2011) Leaf senescence and reactive oxygen species metabolism of broomcorn millet (*'Panicum miliaceum'L.*) under drought condition. *Aust J Crop Sci.* 5(12):1655.

Danilevskaya ON, Meng X, Hou Z, Ananiev E v, Simmons CR. (2008) A genomic and expression compendium of the expanded PEBP gene family from maize. *Plant Physiol.* 146(1):250–264.

Doebley J. (2004) The genetics of maize evolution. *Annu Rev Genet.* 38(1):37–59.

Ducrocq S, Madur D, Veyrieras J-B, Camus-Kulandaivelu L, Kloiber-Maitz M, et al. (2008) Key impact of Vgt1 on flowering time adaptation in maize: Evidence from association mapping and ecogeographical information. *Genetics.* 178(4):2433–2437.

Emerson RA. (1924) Control of flowering in teosinte: short-day treatment brings early flowers. J Hered. 15:41–48.

Fei J, Jiang Q, Guo M, Lu J, Wang P, et al. (2022) Fine mapping and functional research of key genes for photoperiod sensitivity in maize. *Front Plant Sci.* 13. https://doi.org/10.3389/fpls.2022.890780

Feild TS, Lee DW, Holbrook NM. (2001) Why leaves turn red in autumn. The role of anthocyanins in senescing leaves of red-osier dogwood. *Plant Physiol.* 127(2):566–574.

da Fonseca RR, Smith BD, Wales N, Cappellini E, Skoglund P, et al. (2015) The origin and evolution of maize in the Southwestern United States. *Nat Plants.* 1(1):1–5.

Fujita K, Sato H, Sawada O, Sendo S. (1995) Husk leaves contribution to dry matter and grain production as well as n distribution in flint corn (*Zea mays* L.) genotypes differing in husk leaf area. *Soil Sci Plant Nutr.* 41(3):587–596. https://doi.org/10.1080/00380768.1995.10419620

Gan S, Amasino RM. (1997) Making sense of senescence (molecular genetic regulation and manipulation of leaf senescence). *Plant Physiol.* 113(2):313.

Garner WW, Allard HA. (1923) *Further studies in photoperiodism: The response of the plant to relative length of day and night.* [place unknown]: US government printing office.

Gentinetta E, Ceppl D, Lepori C, Perico G, Motto M, Salamini F. (1986) A major gene for delayed senescence in maize. Pattern of photosynthates accumulation and inheritance. *Plant Breed.* 97(3):193–203.
Gepstein S, Thimann K v. (1981) The role of ethylene in the senescence of oat leaves. *Plant Physiol.* 68(2):349–354.
Gonzalez-Segovia E, Pérez-Limon S, Cíntora-Martínez GC, Guerrero-Zavala A, Janzen GM, et al (2019) Characterization of introgression from the teosinte *Zea mays* ssp. *mexicana* to Mexican highland maize. *Peer J.* 7:e6815.
Gören-Sağlam N, Harrison E, Breeze E, Öz G, Buchanan-Wollaston V. (2020) Analysis of the impact of indole-3-acetic acid (IAA) on gene expression during leaf senescence in *Arabidopsis thaliana. Physiol Mol Biol Plants [Internet].* 26(4):733–745. https://doi.org/10.1007/s12298-019-00752-7
Gouesnard B, Rebourg C, Welcker C, Charcosset A. (2002) Analysis of photoperiod sensitivity within a collection of tropical maize populations. *Genet Resour Crop Evol.* 49(5):471–481.
van der Graaff E, Schwacke R, Schneider A, Desimone M, Flugge U-I, et al (2006) Transcription analysis of Arabidopsis membrane transporters and hormone pathways during developmental and induced leaf senescence. *Plant Physiol.* 141(2):776–792.
Gregersen PL, Culetic A, Boschian L, Krupinska K. (2013) Plant senescence and crop productivity. *Plant Mol Biol.* 82(6):603–622. https://doi.org/10.1007/s11103-013-0013-8
Guei RG, Wassom CE. (1996) Genetic analysis of tassel size and leaf senescence and their relationships with yield in two tropical lowland maize populations. *Afr Crop Sci J.* 4(3):275–281.
Guiamet JJ, Teeri JA, Nooden LD. (1990) Effects of nuclear and cytoplasmic genes altering chlorophyll loss on gas exchange during monocarpic senescence in soybean. *Plant Cell Physiol.* 31(8):1123–1130.
Guo L, Wang X, Zhao M, Huang C, Li C, et al. (2018) Stepwise cis-regulatory changes in ZCN8 contribute to maize flowering-time adaptation. *Curr Biol.* 28(18):3005–3015.
Guo Y, Cai Z, Gan S. (2004) Transcriptome of Arabidopsis leaf senescence. *Plant Cell Environ.* 27(5):521–549.
Hastorf CA. (2009) Rio Balsas most likely region for maize domestication. *Proc Natl Acad Sci USA.* 106(13):4957–4958. https://doi.org/10.1073/pnas.0900935106
Havé M, Leitao L, Bagard M, Castell J, Repellin A. (2015) Protein carbonylation during natural leaf senescence in winter wheat, as probed by fluorescein-5-thiosemicarbazide. *Plant Biol.* 17(5):973–979.
Hayes KR, Beatty M, Meng X, Simmons CR, Habben JE, et al. (2010) Maize global transcriptomics reveals pervasive leaf diurnal rhythms but rhythms in developing ears are largely limited to the core oscillator. *PLoS One.* 5(9):e12887. https://doi.org/10.1371/journal.pone.0012887
Hölker AC, Mayer M, Presterl T, Bolduan T, Bauer E, et al. (2019) European maize landraces made accessible for plant breeding and genome-based studies. *Theor Appl Genet.* 132(12):3333–3345.
Hörtensteiner S, Feller U. (2002) Nitrogen metabolism and remobilization during senescence. *J Exp Bot.* 53(370):927–937.
Hörtensteiner S, Kräutler B. (2011) Chlorophyll breakdown in higher plants. Biochim Biophys Acta (BBA) – *Bioenergetics [Internet].* 1807(8):977–988. https://doi.org/https://doi.org/10.1016/j.bbabio.2010.12.007
Hou K, Wu W, Gan S-S. (2013) SAUR36, a small auxin up RNA gene, is involved in the promotion of leaf senescence in Arabidopsis. *Plant Physiol.* 161(2):1002–1009.
Huang C, Sun H, Xu D, Chen Q, Liang Y, et al. (2018) *ZmCCT9* enhances maize adaptation to higher latitudes. *Proc Natl Acad Sci USA.* 115(2):E334–E341.
Hufford MB, Lubinksy P, Pyhäjärvi T, Devengenzo MT, Ellstrand NC, et al. (2013) The genomic signature of crop-wild introgression in maize. *PLoS Genet.* 9(5):e1003477. https://doi.org/10.1371/journal.pgen.1003477
Hufford MB, Martínez-Meyer E, Gaut BS, Eguiarte LE, Tenaillon MI. (2012) Inferences from the historical distribution of wild and domesticated maize provide ecological and evolutionary insight. *PLoS One* 7(11):e47659.
Hung H-Y, Shannon LM, Tian F, Bradbury PJ, Chen C, et al. (2012) ZmCCT and the genetic basis of day-length adaptation underlying the postdomestication spread of maize. *Proc Natl Acad Sci USA* 109(28):E1913–E1921.
Hwang I, Sheen J, Müller B. (2012) Cytokinin signaling networks. *Annu Rev Plant Biol.* 63:353–380.
Jiang GH, He YQ, Xu CG, Li XH, Zhang Q. (2004) The genetic basis of stay-green in rice analyzed in a population of doubled haploid lines derived from an indica by japonica cross. *Theor Appl Genet [Internet]* 108(4):688–698. https://doi.org/10.1007/s00122-003-1465-z

Kante M, Revilla P, de La Fuente M, Caicedo M, Ordás B. (2016) Stay-green QTLs in temperate elite maize. *Euphytica*. 207(2):463–473. https://doi.org/10.1007/s10681-015-1575-0

Kelley DR, Estelle M. (2012) Ubiquitin-mediated control of plant hormone signaling. *Plant Physiol.* 160(1):47–55.

Kim D-J, Smith SM. (1994) Expression of a single gene encoding microbody NAD-malate dehydrogenase during glyoxysome and peroxisome development in cucumber. *Plant Mol Biol.* 26(6):1833–1841.

Kistler L, Maezumi SY, Gregorio de Souza J, Przelomska NAS, et al. (2018) Multiproxy evidence highlights a complex evolutionary legacy of maize in South America. *Science* (1979). 362(6420):1309–1313. https://doi.org/10.1126/science.aav0207

Krupinska K, Humbeck K. (2004) Photosynthesis and chloroplast breakdown. In: Nooden LD (ed.) *Plant cell death processes*. Academic Press, San Diego, CA, pp 169–187.

Kumar U, Joshi AK, Kumari M, Paliwal R, Kumar S, et al (2010) Identification of QTLs for stay green trait in wheat (Triticum aestivum L.) in the 'Chirya 3' × 'Sonalika' population. *Euphytica [Internet]*. 174(3):437–445. https://doi.org/10.1007/s10681-010-0155-6

de la Haba P, de la Mata L, Molina E, Agüera E. (2014) High temperature promotes early senescence in primary leaves of sunflower (*Helianthus annuus* L.) plants. *Can J Plant Sci.* 94(4):659–669.

Lazakis CM, Coneva V, Colasanti J. (2011) ZCN8 encodes a potential orthologue of Arabidopsis FT florigen that integrates both endogenous and photoperiod flowering signals in maize. *J Exp Bot.* 62(14):4833–4842.

Lea PJ, Miflin BJ. (2003) Glutamate synthase and the synthesis of glutamate in plants. *Plant Physiol Biochem.* 41(6–7):555–564.

Lee EA, Tollenaar M. (2007) Physiological basis of successful breeding strategies for maize grain yield. *Crop Sci [Internet]*. 47:S-202–S-215. https://doi.org/10.2135/cropsci2007.04.0010IPBS

Li Yong-xiang, Li C, Bradbury PJ, Liu X, Lu F, et al. (2016) Identification of genetic variants associated with maize flowering time using an extremely large multi-genetic background population. *Plant J [Internet]*. 86(5):391–402. https://doi.org/10.1111/tpj.13174

Liang C, Zheng G, Li W, Wang Y, Hu B, et al. (2015) Melatonin delays leaf senescence and enhances salt stress tolerance in rice. *J Pineal Res.* 59(1):91–101.

Liang Y, Liu Q, Wang X, Huang C, Xu G, et al. (2019) Zm MADS 69 functions as a flowering activator through the ZmRap2. 7-ZCN 8 regulatory module and contributes to maize flowering time adaptation. *New Phytol.* 221(4):2335–2347.

Lifschitz E, Eviatar T, Rozman A, Shalit A, Goldshmidt A, Amsellem Z, et al. (2006) The tomato FT ortholog triggers systemic signals that regulate growth and flowering and substitute for diverse environmental stimuli. *Proc Natl Acad Sci USA* 103(16):6398–6403.

Martínez DE, Costa ML, Guiamet JJ. (2008) Senescence-associated degradation of chloroplast proteins inside and outside the organelle. *Plant Biol.* 10:15–22.

Matsuoka Y, Vigouroux Y, Goodman MM, Sanchez G J, Buckler E, et al. (2002) A single domestication for maize shown by multilocus microsatellite genotyping. *Proc Natl Acad Sci USA*. 99(9):6080–6084.

Mayer M, Unterseer S, Bauer E, de Leon N, Ordas B, et al. (2017) Is there an optimum level of diversity in utilization of genetic resources? *Theor Appl Genet.* 130(11):2283–2295.

McClung CR. (2010) A modern circadian clock in the common angiosperm ancestor of monocots and eudicots. *BMC Biol.* 8(1):55. https://doi.org/10.1186/1741-7007-8-55

Meng X, Muszynski MG, Danilevskaya ON. (2011) The FT-like ZCN8 gene functions as a floral activator and is involved in photoperiod sensitivity in maize. *Plant Cell.* 23(3):942–960.

Minow MAA, Ávila LM, Turner K, Ponzoni E, Mascheretti I, et al. (2018) Distinct gene networks modulate floral induction of autonomous maize and photoperiod-dependent teosinte. *J Exp Bot.* 69(12):2937–2952.

Moore FC, Lobell DB. (2014) Adaptation potential of European agriculture in response to climate change. *Nat Clim Chang.* 4(7):610–614.

Morris K, -Mackerness SA, Page T, John CF, Murphy AM, et al. (2000) Salicylic acid has a role in regulating gene expression during leaf senescence. *Plant J.* 23(5):677–685.

Munaiz ED, Martínez S, Kumar A, Caicedo M, Ordás B. (2020) The senescence (Stay-Green)—an important trait to exploit crop residuals for bioenergy. *Energies (Basel)*. 13(4). https://doi.org/10.3390/en13040790

Muszynski MG, Dam T, Li B, Shirbroun DM, Hou Z, et al. (2006) Delayed flowering1 encodes a basic leucine zipper protein that mediates floral inductive signals at the shoot apex in maize. *Plant Physiol.* 142(4):1523–1536.

Nooděn LD, Guiamét JJ, John I. (1997) Senescence mechanisms. *Physiol Plant.* 101(4):746–753. https://doi.org/10.1034/j.1399-3054.1997.1010410.x
Nooden LD, Hillsberg JW, Schneider MJ. (1996) Induction of leaf senescence in Arabidopsis thaliana by long days through a light-dosage effect. *Physiol Plant.* 96(3):491–495.
Nooděn LD, Penney JP. (2001) Correlative controls of senescence and plant death in Arabidopsis thaliana (Brassicaceae). *J Exp Bot [Internet]*. 52(364):2151–2159. https://doi.org/10.1093/jexbot/52.364.2151
Nozawa A, Ito M, Hayashi H, Watanabe A. (1999) Dark-induced expression of genes for asparagine synthetase and cytosolic glutamine synthetase in radish cotyledons is dependent on the growth stage. *Plant Cell Physiol [Internet]*. 40(9):942–948. https://doi.org/10.1093/oxfordjournals.pcp.a029626
Oh SA, Lee SY, Chung IK, Lee C-H, Nam HG. (1996) A senescence-associated gene of Arabidopsis thaliana is distinctively regulated during natural and artificially induced leaf senescence. *Plant Mol Biol* 30(4):739–754.
Oka T, Fujita K, Saneoka H, Sawada O, Sato H, et al. (1997) Physiological characteristics of husk leaves and its contribution to grain production in maize (*Zea mays* L.). In: Ando T, Fujita Kounosuke Mae T, Matsumoto H, Mori S, Sekiya J (eds.) Plant Nutrition for Sustainable Food Production and Environment: Proceedings of the XIII International Plant Nutrition Colloquium, 13–19 September 1997 Tokyo, Japan [Internet]. Springer, Dordrecht, Netherlands; pp 911–912. https://doi.org/10.1007/978-94-009-0047-9_295
Ordás B, Revilla P, Romay MC, Malvar RA, Butrón A, et al. (2019) Eighteen cycles of recurrent mass selection for early flowering in two maize synthetics. *Euphytica.* 215(3). https://doi.org/10.1007/s10681-019-2374-9
Parent B, Leclere M, Lacube S, Semenov MA, Welcker C, et al (2018) Maize yields over Europe may increase in spite of climate change, with an appropriate use of the genetic variability of flowering time. *Dordrecht* 115(42):10642–10647.
Park J-H, Oh SA, Kim YH, Woo HR, Nam HG. (1998) Differential expression of senescence-associated mRNAs during leaf senescence induced by different senescence-inducing factors in Arabidopsis. *Plant Mol Biol.* 37(3):445–454.
Park J-S, Choung M-G, Kim Jung-Bong, Hahn B-S, Kim Jong-Bum, et al. (2007) Genes up-regulated during red coloration in UV-B irradiated lettuce leaves. *Plant Cell Rep.* 26(4):507–516.
Preston JC, Fjellheim S. (2020) Understanding past, and predicting future, niche transitions based on grass flowering time variation1[OPEN]. *Plant Physiol [Internet]*. 183(3):822–839. https://doi.org/10.1104/pp.20.00100
Pujol B. (2015) Genes and quantitative genetic variation involved with senescence in cells, organs, and the whole plant. *Front Genet [Internet].* 6:57. https://doi.org/10.3389/fgene.2015.00057
Pyhäjärvi T, Hufford MB, Mezmouk S, Ross-Ibarra J. (2013) Complex patterns of local adaptation in teosinte. *Genome Biol Evol.* 5(9):1594–1609.
Pyung OL, Hyo JK, Hong GN. (2007) Leaf senescence. *Annu Rev Plant Biol.* 58:115–136. https://doi.org/10.1146/annurev.arplant.57.032905.105316
Reinbothe C, Springer A, Samol I, Reinbothe S. (2009) Plant oxylipins: role of jasmonic acid during programmed cell death, defence and leaf senescence. *FEBS J.* 276(17):4666–4681.
Ren H, Gao Z, Chen L, Wei K, Liu J, et al. (2007) Dynamic analysis of ABA accumulation in relation to the rate of ABA catabolism in maize tissues under water deficit. *J Exp Bot.* 58(2):211–219.
Renaud AL. (2015) *Genetic regulation of maize and sorghum under abiotic stress.* PhD Thesis, Purdue University.
Romay MC, Millard MJ, Glaubitz JC, Peiffer JA, Swarts KL, et al. (2013) Comprehensive genotyping of the USA national maize inbred seed bank. *Genome Biol [Internet]*. 14(6):R55. https://doi.org/10.1186/gb-2013-14-6-r55
Romero Navarro JA, Willcox M, Burgueño J, Romay C, Swarts K, et al. (2017) A study of allelic diversity underlying flowering-time adaptation in maize landraces. *Nat Genet.* 49(3):476–480.
Roux F, Touzet P, Cuguen J, le Corre V. (2006) How to be early flowering: an evolutionary perspective. *Trends Plant Sci.* 11(8):375–381. https://doi.org/10.1016/j.tplants.2006.06.006
Salvi S, Castelletti S, Tuberosa R. (2009) An updated consensus map for flowering time QTLs in maize. *Maydica.* 54(4):501.

Salvi S, Sponza G, Morgante M, Tomes D, Niu X, et al. (2007) Conserved noncoding genomic sequences associated with a flowering-time quantitative trait locus in maize. *Proc Natl Acad Sci USA*. 104(27):11376–11381.

Samanta A, Das G, Das SK. (2011) Roles of flavonoids in plants. *Carbon NY*. 100(6):12–35.

Sawada O, Ito J, Fujita K. (1995) Characteristics of photosynthesis and translocation of 13C-labelled photosynthate in husk leaves of sweet corn. *Crop Sci*. 35(2):480–485.

Schippers JHM. (2015) Transcriptional networks in leaf senescence. *Curr Opin Plant Biol*. 27:77–83.

Sekhon RS, Childs KL, Santoro N, Foster CE, Robin Buell C, et al. (2012) Transcriptional and metabolic analysis of senescence induced by preventing pollination in maize. *Plant Physiol*. 159(4):1730–1744. https://doi.org/10.1104/pp.112.199224

Sekhon RS, Saski C, Kumar R, Flinn BS, Luo F, et al. (2019) Integrated genome-scale analysis identifies novel genes and networks underlying senescence in Maize. *Plant Cell*. 31(9):1968–1989. https://doi.org/10.1105/tpc.18.00930

Shirley BW. (1996) Flavonoid biosynthesis:'new'functions for an 'old'pathway. *Trends Plant Sci*. 1(11):377–382.

Smart CM. (1994) Gene expression during leaf senescence. *New Phytol [Internet]*. 126(3):419–448. https://doi.org/10.1111/j.1469-8137.1994.tb04243.x

Song H, Gao X, Feng B, Zhang P, Gao J, et al. (2013) Leaf senescence and physiological characters in different adzuki bean (*Vigna angularis*) cultivars (lines). *Afr J Agric Res*. 8(29):4025–4032.

Swarts K, Bauer E, Glaubitz JC, Ho T, Johnson L, et al. (2021) Joint analysis of days to flowering reveals independent temperate adaptations in maize. *Heredity (Edinb)*. 126(6):929–941.

Swarts K, Gutaker RM, Benz B, Blake M, Bukowski R, et al. (2017) Genomic estimation of complex traits reveals ancient maize adaptation to temperate North America. *Science* (1979). 357(6350):512–515.

Takada S, Goto K. (2003) Terminal Flower2, an Arabidopsis homolog of Heterochromatin Protein1, counteracts the activation of Flowering Locus T by Constans in the vascular tissues of leaves to regulate flowering time. *Plant Cell*. 15(12):2856–2865.

Takata N, Saito S, Saito CT, Uemura M. (2010) Phylogenetic footprint of the plant clock system in angiosperms: evolutionary processes of Pseudo-Response Regulators. *BMC Evol Biol*. 10(1):126. https://doi.org/10.1186/1471-2148-10-126

Takuno S, Ralph P, Swarts K, Elshire RJ, Glaubitz JC, et al (2015) Independent molecular basis of convergent highland adaptation in maize. *Genetics*. 200(4):1297–1312.

Tamaki S, Matsuo S, Wong HL, Yokoi S, Shimamoto K. (2007) Hd3a protein is a mobile flowering signal in rice. *Science* (1979). 316(5827):1033–1036.

Tao YZ, Henzell RG, Jordan DR, Butler DG, Kelly AM, McIntyre CL. (2000) Identification of genomic regions associated with stay green in sorghum by testing RILs in multiple environments. *Theor Appl Genet*. 100(8):1225–1232.

Taylor CB, Bariola PA, DelCardayre SB, Raines RT, Green PJ. (1993) RNS2: a senescence-associated RNase of Arabidopsis that diverged from the S-RNases before speciation. *Proc Natl Acad Sci USA*. 90(11):5118–5122.

Taylor MA, Wilczek AM, Roe JL, Welch SM, Runcie DE, et al (2019) Large-effect flowering time mutations reveal conditionally adaptive paths through fitness landscapes in *Arabidopsis thaliana*. *Proc Natl Acad Sci USA*. 116(36):17890–17899. https://doi.org/10.1073/pnas.1902731116

Thakur N, Sharma V, Kishore K. (2016) Leaf senescence: an overview. *Indian J Plant Physiol*. 21(3):225–238. https://doi.org/10.1007/s40502-016-0234-3

Thomas H. (1987) Sid: a Mendelian locus controlling thylakoid membrane disassembly in senescing leaves of *Festuca pratensis*. *Theor Appl Genet*. 73(4):551–555.

Thomas H, Howarth CJ. (2000) Five ways to stay green. *J Exp Bot*. 51(suppl_1):329–337. https://doi.org/10.1093/jexbot/51.suppl_1.329

Thomas H, Ougham HJ, Wagstaff C, Stead AD. (2003) Defining senescence and death. *J Exp Bot*. 54(385):1127–1132.

Thomas H, Smart CM. (1993) Crops that stay green. *Ann Appl Biol*. 123(1):193–219. https://doi.org/10.1111/j.1744-7348.1993.tb04086.x

Thomas H, Stoddart JL. (1980) Leaf senescence. *Annu Rev Plant Physiol*. 31(1):83–111.

Thomson WW, Nothnagel EA, Huffaker RC (Eds.). (1987) Plant senescence: Its biochemistry and physiology. The American Society of Plant Physiologists Rockville, MD, pp. 43-53.

Tollenaar M, Daynard TB. (1978) Leaf senescence in short-season maize hybrids. *Can J Plant Sci*. 58(3):869–874.
Trachsel S, Sun D, Sanvicente FM, Zheng H, Atlin GN, et al. (2016) Identification of QTL for early vigor and stay-green conferring tolerance to drought in two connected advanced backcross populations in tropical maize (*Zea mays* L.). *PLoS One*. 11(3):1–22. https://doi.org/10.1371/journal.pone.0149636
Unterseer S, Pophaly SD, Peis R, WestermeierP, Mayer M, et al. (2016) A comprehensive study of the genomic differentiation between temperate Dent and Flint maize. *Genome Biol.* 17(1):137. https://doi.org/10.1186/s13059-016-1009-x
Wang AY, Li Y, Zhang CQ. (2012) QTL mapping for stay-green in maize (*Zea mays*). *Can J Plant Sci*. 92(2):249–256. https://doi.org/10.4141/CJPS2011-108
Wang L, Beissinger TM, Lorant A, Ross-Ibarra C, Ross-Ibarra J, et al. (2017) The interplay of demography and selection during maize domestication and expansion. *Genome Biol.* 18(1):1–13.
Wang L, Josephs EB, Lee KM, Roberts LM, Rellán-Álvarez R, et al. (2021) Molecular parallelism underlies convergent highland adaptation of maize landraces. *Mol Biol Evol.* 38(9):3567–3580.
Watanabe M, Hubberten H-M, Saito K, Hoefgen R. (2010) General regulatory patterns of plant mineral nutrient depletion as revealed by serat quadruple mutants disturbed in cysteine synthesis. *Mol Plant*. 3(2):438–466.
Weaver LM, Gan S, Quirino B, Amasino RM. (1998) A comparison of the expression patterns of several senescence-associated genes in response to stress and hormone treatment. *Plant Mol Biol.* 37(3):455–469.
Wigge PA, Kim MC, Jaeger KE, Busch W, Schmid M, et al. (2005) Integration of spatial and temporal information during floral induction in Arabidopsis. *Science* (1979). 309(5737):1056–1059.
Wolfe DW, Henderson DW, Hsiao TC, Alvino A. (1988) Interactive water and nitrogen effects on senescence of maize. II. photosynthetic decline and longevity of individual leaves. *Agron J.* 80(6):865–870. https://doi.org/10.2134/agronj1988.00021962008000060005x
Wong AYM, Colasanti J. (2007) Maize floral regulator protein Indeterminate1 is localized to developing leaves and is not altered by light or the sink/source transition. *J Exp Bot.* 58(3):403–414.
Woo HR, Chung KM, Park J-H, Oh SA, Ahn T, et al. (2001) ORE9, an F-box protein that regulates leaf senescence in Arabidopsis. *Plant Cell.* 13(8):1779–1790.
Wu A, Allu AD, Garapati P, Siddiqui H, Dortay H, et al. (2012) Jungbrunnen1, a reactive oxygen species–responsive NAC transcription factor, regulates longevity in Arabidopsis. *Plant Cell.* 24(2):482–506.
Yang N, Xu X-W, Wang R-R, Peng W-L, Cai L, et al. (2017) Contributions of Zea mays subspecies mexicana haplotypes to modern maize. *Nat Commun.* 8(1):1–10.
Yang Q, Li Z, Li W, Ku L, Wang C, Ye J, Li K, Yang N, Li Y, Zhong T. (2013) CACTA-like transposable element in ZmCCT attenuated photoperiod sensitivity and accelerated the postdomestication spread of maize. *Proc Natl Acad Sci USA*. 110(42):16969–16974.
Yang Z, Li X, Zhang N, Wang X, Zhang Y, et al. (2017) Mapping and validation of the quantitative trait loci for leaf stay-green-associated parameters in maize. *Plant Breed.* 136(2):188–196.
Yannam VRR, Caicedo M, Malvar RA, Ordás B. (2022) Genome-wide association analysis of senescence-related traits in maize. *Int J Mol Sci.* 23(24):15897. https://doi.org/10.3390/ijms232415897
Zeevaart JAD, Creelman RA. (1988) Metabolism and physiology of abscisic acid. *Annu Rev Plant Physiol Plant Mol Biol.* 39(1):439–473.
Zhang J, Fengler KA, van Hemert JL, Gupta R, Mongar N, et al. (2019) Identification and characterization of a novel stay-green QTL that increases yield in maize. *Plant Biotechnol J.* 17(12):2272–2285.
Zhang WY, Xu YC, Li WL, Yang L, Yue X, et al. (2014) Transcriptional analyses of natural leaf senescence in maize. *PLoS One*. 9(12):1–31. https://doi.org/10.1371/journal.pone.0115617
Zheng HJ, Wu AZ, Zheng CC, Wang YF, Cai R, et al. (2009) QTL mapping of maize (Zea mays) stay-green traits and their relationship to yield. *Plant Breed.* 128(1):54–62. https://doi.org/10.1111/j.1439-0523.2008.01529.x
Zheng Y, Tian L, Liu H, Pan Q, Zhan J, et al. (2009) Sugars induce anthocyanin accumulation and flavanone 3-hydroxylase expression in grape berries. *Plant Growth Regul.* 58(3):251–260.

6 Allelic Variation Corresponds to Yield Component Traits

Novel Genetic Variation for Maize (Zea mays L.) Crop Improvement

Ganapati Mukri, Samriti Sharma, Chandra Prabha, Shilpa, Jayant S. Bhat, RN Gadag, and Chandu singh*
*Corresponding author: ganapati4121@gmail.com

6.1 INTRODUCTION

Maize (*Zea mays* L.) is a staple food crop that fulfills the diverse demands of industrial requirements, food and feeds worldwide. The genetic enhancement of grain yield is the most important breeding objective in maize and which is composed of the number of ears per plant, ear weight and grain weight, etc. (Qin et al. 2016). Ear weight is determined by the kernel row number (KRN), kernel number per row (KNPR), and kernel weight (KW). These traits are quantitative in nature and are influenced by environments, genotypes, and environment-by-genotype interactions. Morphologically, maize has one dominant axis of growth, whereas its progenitor, teosinte is highly branched. The axillary branches in maize are short and feminized, whereas, the axillary branches of teosinte are long and end in a male inflorescence under normal growth conditions. There are several hypotheses regarding the transition of bushy teosinte to a single-stem maize plant. One of them catastrophic sexual transmutation theory (Iltis 1983). According to this, maize evolved through a sudden epigenetic sexual transmutation, involving condensation of primary branches, which brought their tassels into the zone of female expression, leading to strong apical dominance and a catastrophic shift in nutrient allocation. The theory emphasized that environmental factors, including cold growing seasons, were the likely triggers, potentially producing a rapid phenotypic transformation from teosinte- to maize-type branching and inflorescence sexuality without human involvement. But, subsequent studies showed that when lateral branches were shortened and female ears were translocated to the maize cob position at the ends of the branches, they were of normal teosinte type with hard cupulate fruit cases and did not possess characteristics, such as soft glumes, that would have been derived from tassels if the latter were involved (Doebley et al. 1995). Doebley and co-workers dissected the genetic architecture of branching by putting the teosinte allele for the same gene into corn, which resulted in corn with more branches. These data suggested that the branching gene plays a large role in the genetic control of this trait. With genetic studies, it is found that *teosinte branched 1 (tb1)* and *teosinte glume architecture 1 (tga1)* genes played a central role in transforming teosinte into modern-day maize. Early in teosinte ear development, there are two spikelet primordia on each rachis segment; however, one is aborted early. Thus, maize evolution involved the re-expression of the second spikelet primordium. Also, in teosinte, the cupulate fruit cases are borne in two ranks on opposite sides of the longitudinal axis of the ear. The cupules are borne in four (or more) ranks. Thus, maize evolution involved a switch from two to four ranks of cupules. In present-day maize, one ear of corn contains hundreds of kernels

DOI: 10.1201/9781003385004-6

arranged in up to 26 rows. The kernel row number (KRN) significantly contributed to grain yield which has high broad-sense heritability (H^2) (Dhillon et al. 1977). Understanding genetics and genetic enhancement of such yield attributes would help to understand the genetic and molecular basis of grain yield and aid in the development of high-yielding maize hybrids (Doebley 2004). So, understanding the genetic architecture of KRN not only increases our knowledge of genetic mechanisms underlying ear development but also facilitates the improvement of the maize grain yield (Yang et al. 2015; Chen et al. 2016).

6.2 IDENTIFICATION OF NOVEL GENES FOR KRN AND KERNEL-RELATED TRAITS

A number of quantitative traits loci have been identified that control quantitative variations in KRN (Brown et al. 2011; Liu et al. 2015). On the basis of analysis of diverse bi-parental genetic populations of maize, many quantitative trait loci (QTLs) were mapped to find out the genetic architecture of the KRN (Li et al. 2010; Lu et al. 2015; Yang et al. 2015; Calderon et al. 2016). Five QTLs related to KRN were detected in the $F_{2:3}$ population derived from crossing K10HEX206 and Nong513 inbred lines with four rows (Jiao et al. 2014). A total of 12 QTLs were identified for KRN by using 325 recombinant inbred lines (RILS) (in crosses between SICAU1212 and B73) from which three of them *viz.,qERN2-1*, *qERN8-1*, and *qERN8-2* were constantly detected across four environments (Yang et al. 2015). The 397 $F_{2:3}$ populations derived from Ye478 and Dan 340 inbreds were utilized in the identification of QTL, which explains 17.90% phenotypic variance on the seventh chromosome, across seven different environments (Lu et al. 2010). Major quantitative trait loci (QTL) that affect 51% of the phenotypic variance were mapped into a 203 kb interval in a set of recombinant chromosomes near-isogenic lines (W22 and teosinte inbred line) of *Zea mays* ssp. *parviglumis*. After combining with map-based cloning, it was observed that QTL mapping is a powerful approach for exploring the genes and genetic elements, which are responsible for the KRN (Bommert et al. 2013; Liu et al. 2015).

Nowadays, whole transcriptome RNA sequencing (RNA-seq) along with transcriptome analysis is used in various studies for the discovery of candidate genes and pathways responsible for target traits in several plant species. The *qHS3* is a QTL that controls genes related to starch content in maize and also their co-expression network in kernels, identified through transcriptome analysis of near-isogenic lines (NILs) (Xiao et al. 2016). In another study, genes involved in *Fusarium head blight* (FHB) resistance and Fhb1-mediated pathways in the wheat landrace "Wangshuibai" were identified through transcriptome analysis (Xiao et al. 2013). Therefore, the high sensitivity and accuracy of RNA-Seq help in the detection of important genes related to novel traits. Recently, the combination of association mapping along with linkage mapping has proven to be a reliable method for the identification of candidate loci viz., *qGW4.05, qKW7.02,* and *qKL1.07* related to yield traits in maize (Chen et al. 2016; Qin et al. 2016; Li et al. 2016). These three genes for kernel size and weight within three MQTL regions were identified successfully through MetaQTL analysis and regional association mapping (Chen et al. 2017).

6.2.1 Genetics of KRN and Kernel-Related Traits

The kernel row number (KRN) of an ear is a key trait related to the yield and domestication of maize. Exploring the KRN genetic mechanisms has great research significance and application potential. A total of three differentially expressed genes (DEGs), i.e., *Zm00001d015377*, *Zm00001d013277,* and *Zm00001d015310* were identified, which contained three SNPs associated with KRN were identified by regional association mapping. Transcription factors (TFs) also showed their role in KRN formation, i.e., *unbranched2 (ub2)* and *unbranched3 (ub3)* as SQUAMOSA promoter-binding protein (SBP) TFs affect KRN and tassel branch number by limiting cell differentiation rate of the lateral domains of meristems (An et al. 2019). In the last two decades, major emphasis has been on genome sequencing and DNA markers for rapid cloning of genes governing grain yield-related traits. To date, a total of 30 genes linked to grain yield components have been cloned in maize, and also many genes, which control maize kernel development have been functionally characterized. For example, *Emp10, Emp11, Emp12, Emp16, Dek2, Dek35, Dek37,* and *Zmsmk9* encode for P-type pentatricopeptide repeat (PPR) proteins are involved in alternative splicing of mitochondrial genes for regulating maize kernel development. Disruption of the function of these genes is responsible for damage to mitochondrial function and delays kernel development, which results in undersized kernels and a decrease in kernel yield (Cai et al. 2017; Ren et al. 2017;Chen et al. 2017; Dai et al. 2018; Sun et al. 2019; Xiu et al. 2016; Qi et al. 2016; Pan et al. 2019). The PLS-type pentatricopeptide repeat (PPR) genes, i.e., *Dek10, Dek36, Dek39, Emp4, Emp5, Emp7, Emp9, PPR2263,* and *Smk1*encode for proteins that function in the C-to-U editing of mitochondrial and chloroplasts transcripts (Sosso et al. 2012; Gutiérrez-Marcos et al. 2007; Liu et al. 2013; Sun et al. 2015; Wang et al. 2019). Loss of gene function is directly related to delay in embryo and endosperm development, which leads to reduced kernel yield (Doll et al. 2020).

The *ZmCEP1, ZmSWEET4c, ZmNPF7.9, ZmBAM1d, FEA2, qKW9, ids1/Ts6, KNR6,* and *UB3* were also involved in grain yield-related traits in different manners. The *ZmSWEET4c* is an ortholog of the *OsSWEET4* gene of rice, responsible for the synthesis of hexose transporter protein, which helps in the transport of sugar across the basal endosperm transfer cell layer (BETL) that results in an increase in seed filling. Loss of function of both genes, *ZmSWEET4c* and *OsSWEET4* is responsible for defective seed filling, which also results in a dramatic loss of endosperm (Baye et al. 2022). Recent studies revealed that the ZmNPF7.9/NRT1.5, a nitrate transporter is involved in maize kernel development (Guan et al. 2020). This *ZmNPF7.9* was highly expressed in the cells of the basal endosperm transfer cell layer (BETL). The loss of function of this gene was responsible for aberrant starch deposition along with faulty seed growth, which significantly results in the reduction of grain weight (Wei et al. 2021). The *ZmCEP1* is an ortholog of *OsCEP6.*, which encodes for a short C-terminal encoded peptide (CEP) that acts as a negative regulator for kernel weight in maize (Xu et al. 2021). Compared with the wild-type, overexpression of *ZmCEP1* significantly reduces kernel size, plant height, and 100-kernel weight, whereas knockout of this gene through clustered regularly interspaced short palindromic repeats/CRISPR-associated protein (CRISPR/Cas9) increases kernel yield-related traits (Sui et al. 2016). The *ZmCEP1* gene also showed its importance in nitrogen metabolism, sugar transport, nitrate, and auxin response pathways, which affect kernel development (Xu et al. 2021).

It was reported that CLV1/BAM family genes control the shoot meristems development (Nimchuk et al. 2015), fruit size, and floret number in plants (Somssich et al. 2016; Janocha et al. 2018). The *qHKW1*, a major QTL encode for CLAVATA1 (CLV1)/BARELY ANY MERISTEM (BAM)-related receptor kinase is responsible for 18.4% of phenotypic variation in maize. Sequence variation in the promoter region of *ZmBAM1d* revealed a positive selection signal with low DNA methylation along with an increase in gene expression of *HKW* during maize evolution (Yang et al. 2019). The presence of retromer complex subunit *ZmVPS29* and its overexpression provides slenderer kernel morphology with low kernel weight along with a higher kernel number per row (KNPR) and a higher grain yield per plant (YPP). The *ZmVPS29* also showed its importance in maize domestication (Chen et al. 2020). The *qKW9* is another major gene that encodes the DYW-PPR protein, which is essential for C-to-U editing of ndhB, a subunit of the chloroplast NADH dehydrogenase-like complex. A deletion of 13 bp in the coding region of *qKW9* results in a decrease in photosynthetic activity, which results in less maternal photosynthates availability during the grain-filling process and lowers kernel weight (Huang et al. 2020). The kernel yield in maize was also affected by the presence of transposable elements and long terminal repeat retrotransposons in the regulatory sequences of the transposable elements (Jia et al. 2020) (see Figure 6.1).

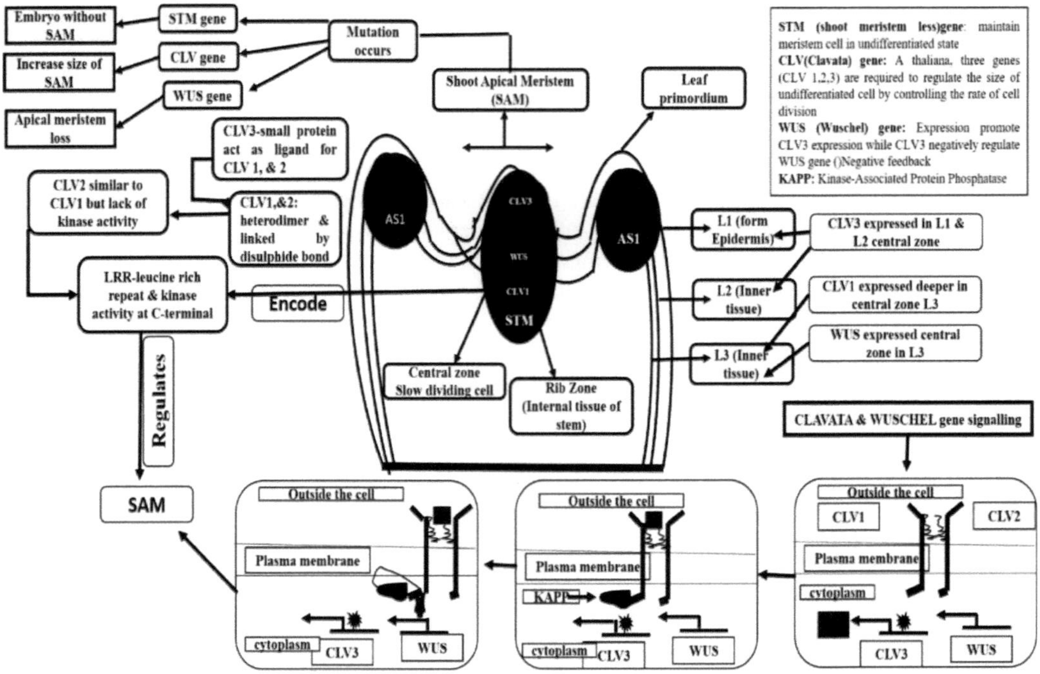

FIGURE 6.1 Genes involved in CLAVATA (CLV)-WUSCHEL (WUS) signaling pathway in maize.

6.3 UNDERSTANDING KRN VARIATION

Molecular marker technology is a continuously evolving strategy varying from hybridization-based techniques to sequencing-based high-throughput single-nucleotide polymorphisms (SNPs). High-throughput sequencing technologies proved to be potential strategies for sequence-based SNP genotyping, which are more effective for genetic mapping and genome analysis (Li et al. 2020). High-throughput analysis with reduced representation library (RRL) SLAF-seq high-resolution and efficient data generation for large-scale genotyping, was the key success to understanding the novel traits (Sun et al. 2013).

6.3.1 Allelic Variation in KRN and Kernel Related Genes

Maize is an excellent crop model to exploit the complementarity between quantitative and mutant allelic variation since it is phenotypically and genotypically diverse and possesses an excellent collection of well-characterized mutant lines (http://maizecoop.cropsci.uiuc.edu/). Naturally occurring allelic diversity in plants has been suggested to be an important genetic component for phenotypic variation (Doebley and Lukens 1998; Buckler and Thornsberry 2002). The allelic variation is attributed to differences in non-coding DNA sequences and does not involve epigenetic regulation or gene imprinting. This expression variation is suggested to play an important role in determining phenotypic diversity (Guo et al. 2004). The presence of alleles and haplotypes and their distribution in maize is found to vary among continents and countries (Kiss et al. 2014). Identification of allelic variation present in relevant candidate genes controlling key agronomic traits within the collection of genetic resources has a potential role in crop improvement (Kumari et al. 2018).

Multiple yield component traits such as the number of ears per plant, kernel weight (KW), kernel row number (KRN), and the number of kernels per row, are involved in determining the final yield of maize. Among them, KRN is the most important trait that, approaches uncovering the molecular causes of yield-related traits in maize (Trung et al. 2020). Initial numbers of spikelet-pair meristems (SPMs) on the female inflorescence meristem are responsible for determining kernel rows on the maize ear (Vollbrecht and Schmidt 2009). Different types of ear architecture of maize and the genes responsible have been dissected based on the strong loss-of-function mutation using the forward genetic approaches, which regulate the development of maize ears and modify the kernel row numbers. Identified genes, such as, the *ramose (ra1, ra2, ra3)*, *ramosa1 enhancer locus2 (rel2)*, *compact plants 2(ct2)*, *barren inflorescence 2(bif2)*, *unbranched2 (ub2)*, *unbranched3 (ub3)*, zea*floricaula 1 (zfl1)* and zea*floricaula 2 (zfl2)*, tasselsheath4 *(tsh4)*,*indeterminate spikelet1 (ids1)*, *thick tassel dwarf1 (td1)*, *corngrass1 (cg1)* and the *fasciated ear 2 (fea2)*, *fasciated ear 3(fea3)*and *fasciated ear 4 (fea4)*, etc. have a major role in determining KRN in maize (Chuck et al. 2007, 2014, Mei et al. 2021,Liu et al. 2019, Gallavotti et al. 2010). As most of these genes also regulate multiple other traits, the application potentials of natural alleles of these above genes in maize breeding need to be explored further.

A total of 20 genotypes with variability in KRN were characterized at ICAR-IARI, New Delhi, for possible allelic variation in *fea2, fea3, fea4, ids, td1* and *tsh4* through amplicon sequencing-based approaches. These genes combine together and harbor a total of 223 SNPs, which were identified through "the multiple sequence alignment" approach. Nucleotide sequence variation can potentially alter protein function if it occurs in the coding region and causes a qualitative difference. Alternatively, regulatory allelic variants can affect the level of gene expression and result in quantitative variants. Both types of allelic variation may produce functional changes and affect the phenotype of an organism (Mirth and Riddiford 2007).

To study the effect of these amino acid changes on the phenotypes, homology modeling was performed using the "swiss model" and identified the highest similarity with LRR receptor-like serine/threonine-protein kinase FLS2 in both genes. This protein is 72.43 kD in size and similar to receptor-like kinases (RLKs), which are involved in a diverse array of plant responses, including development, growth, hormone perception, and the response to pathogens. Expression studies for these genes indicated that, among the tested genes, *fea2* and *fea3* genes have high expression in high KRN genotypes as compared to low KRN genotypes. So, these allelic variations causing non-synonymous amino acid changes can be the region for the determination of KRN in studied genotypes (Trung et al. 2020).

The fasciation patterns are very important for genetic analysis as well as for improving crop yields, where abnormally flattened ears bear high kernel row numbers (Bommert et al. 2013; Je et al. 2016). Hence, ear fasciation is highly significantly correlated with KRN (Mei et al. 2021). Recently, TILLING (Targeting Induced Local Lesions in Genomes) approach was used to identify the alleles of *fea2 (fascinated ear 2)* (Trung et al. 2020).

A weak allele of *fea2*, which affect the inflorescence meristem size is the homozygous to *fea2*-1328 allele, having 2-5 more KRN. The *fea2*-1328 allele is found to have the strongest effect, with an increase of approximately three KRN, without reducing ear length (Trung et al. 2020). The *fea*3, encodes a leucine-rich-repeat receptor that functions in stem cell control and responds to a CLAVATA3/ESR-related (CLE) peptide expressed in organ primordia. Weak alleles of fascinated ear mutants could improve maize yield traits, such as kernel row number (Bommert et al. 2013), by increasing the meristem size and number of primordia while maintaining the structural integrity of the meristem. The B73 inbred line introgressed with mutant *fea3* showed wide SAMs, prominent inflorescence defects, including thick tassels and strongly fasciated ears, which shows fasciation and irregular rows, with normal female spikelet and floret architecture. Compared to wild *fea3* type plant that has narrow SAMs, male spikelet containing two florets, each floret with three anthers and female spikelet with a carpel, one degenerated floret, and three immature anthers, as well as regular kernel rows. The mutant *fea3* also encoded amino acid substitutions relative to the progenitor. Different hybrids with the weak *fea3-2* and *fea3-3* alleles have been found to enhance ear length, kernel row numbers, kernel numbers per ear, and ear weight (Je et al. 2016). Another important variant, the *fea4* is a semi-dwarf mutant with fasciated ears and tassels and greatly enlarged vegetative and inflorescence meristems. It provides a potential target for manipulation of inflorescence meristem size and kernel row number, as has been demonstrated for *fea2* (Bommert et al. 2013). Alleles *fea4-33* and *fea4*-ref (C-T) transition and *fea4-5171* (G-A) had early stop codons and fall after the bZIP domain which are associated with transcriptional activation and may mediate posttranslational regulation of bZIP proteins. There are orthologous to the Arabidopsis gene *PAN,* which increases in floral organ number, without a corresponding increase in the floral meristem (Running 1997, Meyerowitz 1997) and a mild increase in inflorescence meristem size (Paulter et al. 2015).

Similarly, *ub2* and *ub3* genes are involved in the function of specification of lateral primordia, such as leaves, tassel branches, and kernels, function together with *tsh4* to restrict the initiation of the lateral primordia, thus allowing cell renewal within the central domain of the meristem (Chuck et al. 2014). Elevation of *ub3* expression reduces KRN and ear diameter with enlargement in inflorescence meristem size, slightly reducing the tassel branch number (Liu et al 2015). It is the causative gene underlying a KRN and tassel branch number. Further, KRN4 is the major QTL, located on chromosome 4, and encodes an SPL14 domain, controlling the natural variation in KRN by acting as a distal regulator of *ub3* expression. The allelic variation, A/C SNP (Ser220Asn) in *ub3* is found to be significantly associated with the determination of KRN However, the favorable allele, 'A' of *ub3* is not detected in teosinte and has a low frequency in maize landraces, indicating that it might have emerged during the post-domestication improvement of modern maize (Chuck et al. 2014).

Bif2 is a maize co-ortholog of the Arabidopsis PINOID protein kinase (PID), which has a role in polar auxin transport in maize (Pressoir et al. 2009). The *Bif2* amplicons of Mo17 and B73 have been sequenced in the association panel of maize inbred lines and identified the allelic variation correlated with a reduced tassel branch. It shows lower expression in Mo17 and compared to that in B73, which is taller, late in flowering has longer tassels, and has leaves with more nodes. The B73 *bif2* allele is consistently expressed at a significantly higher level than the Mo17 allele.

Another critical gene *Td1* negatively modulates meristem size during maize inflorescence development and may also regulate ear-related traits and is significantly associated with kernel row number. This plays a significant role in the CLV-WUS pathway in the dominant negative regulation of inflorescence meristem (IM) size during inflorescence development. Three natural variations in the *Td1* locus were identified in diverse maize inbred lines, which are significantly associated with variation in KRN. Among them, SNP_2418 (T/C), detected in the kinase domain of *Td1,* is the non-synonymous SNP with an amino acid alteration from cysteine to arginine. Therefore, the KRN trait might be affected by the mutant kinase domain and SNP_2418 could be developed as functional markers related to KRN (Liu et al. 2019).

The function of *Tsh4* is primarily to repress lateral organ growth affecting phyllotaxy, axillary meristem initiation, and meristem determinacy within the floral phase. Down-regulation of *tsh4* by a combination of microRNAs and branching pathway genes allows the establishment of lateral meristems and the repression of leaf initiation. Wild-type *Tsh4 has* more tassel branches (13 tassel branches) compared to mutant *Tsh4,* which typically has one to five long tassel branches, and long bract leaves, clustered at the base of the tassel. Further, the phyllotaxy of the kernel rows was different in the tsh4 mutant, with near spiral phyllotaxy instead of straight-ordered rows of kernels in the wild-type. *Tsh4* gene is epistatic over ramosa genes *ra1, ra2*, and *ra3*. Ramose genes exhibit an enhanced number of tassel branches and impose SPM indeterminacy replacing spikelet pairs with branches containing extra spikelets (McSteen 2006). It operates with *ra2* to affect meristem determinacy, rather than stem growth of the pedicel (Chuck et al. 2010).

The ids1 gene is critical for the regulation of spikelet meristem determinacy in maize, including multiple pathways regulating spikelet pair meristem development, involving several key genes such as *fea3*, *fea4* and *ra3* (Wang et al. 2019). In the absence of the ids1 gene, the spikelet meristem becomes indeterminate and produces additional florets. Of the two alleles of *ids1* (ids1–*mum1*, ids1–*mum2*) the *ids1–mum1* female spikelets often contain more than one silk (Chuck et al. 1998) located at the ectopic position, which does not develop normally. They are often unfused as well as fail to elongate to their proper length in comparison to wild-type female spikelets, which contain only a single mature floret because of the abortion of the lower floret. The Ids1-*mum2* tassel spikelet shows four florets instead of two. Each floret has three stamens surrounded by palea and lemma similar to the wild-type. The spikelets of wild-type tassels showed lateral organs of upper and lower florets. The spikelets of the *ids1–mum2* tassels are larger than those of the wild-type because of the presence of extra florets (Chuck et al. 1998; Bensen et al. 1995; Meeley and Briggs 1995).

The maize*zfl* (zea*floricaula)* genes are involved in flower development that is associated with quantitative variation for KRN (Bomblies and Doebley 2006). The zfl1 and zfl2 are homologous of *FLO/LFY*-like genes upstream of the ABC genes conserved between maize and dicots. The *zfl1* maps near *umc44a* on chromosome 10, while *zfl2* maps near *umc6a* on chromosome 2. Alleles of *zfl1* (*zfl1-mum1* and *zfl1-mum2*) and *zfl2* (*zfl2-mum1, zfl2-mum2*, and *zfl2-mum4*) showed a strong association with aspects of flowering time, affecting lateral branch numbers and ear phyllotaxy respectively. Wild-type *zfl* genes have spikelet pairs showing two upper florets with glume, palea, stamen primordial, and a gynoecium ridge that surrounds the ovule. The *zfl1-mum1*; *zfl2-mum1* showed a quantitative decrease in inflorescence phyllotaxy, which was measured as the number of ranks of spikelet-pairs around the circumference of the ear or central tassel spike (Doebley 1992, Bomblies et al. 2003).

The *ZmCNR13* gene, encoding a protein of the *fw2.2-like* (*FWL- fruit weight-like*) family, is demonstrated to be involved in cell division, expansion, and differentiation. Four polymorphic sites – InDel413, SNP2286, SNP2305, and SNP2337 were captured within the *ZmCNR13* gene and were significantly associated with six ear traits (ear weight, ear glume weight, ear length, ear density, kernel row number, and kernel number per row). Of these, SNP2305 located in exon 2 jointly with SNP 2337 controlled four ear traits (EW, EGW ED, and KRN) and belongs to non-synonymous mutation, where the transition of T to A leads to the changes of amino acid serine to threonine. Inbred lines carrying SNP2305A possess higher EW, EGW, ED, and KRN than those carrying SNP2305T. So, the natural variations of ZmCNR13 might be involved in ear development and can be used in the genetic improvement of maize ear-related traits (Zuo et al. 2021).

6.4 ROLE OF KRN-RELATED TRAITS IN MAIZE IMPROVEMENT

6.4.1 QTL Mapping

The weight of the kernel is significantly correlated with the length of the kernel in maize (Li et al. 2009; Peng et al. 2011; Liu et al. 2014). Kernel size is also related to the quality of the maize (e.g., flour yield and protein content) (Gupta et al. 2006). Major progress has been made in the identification of major QTLs or genes for KRN related traits in maize. Currently, from the year 1994 to 2021, a total of 1920 QTLs associated with kernel weight and kernel size have been identified (Messmer et al. 2009; Li et al. 2013; Liu et al. 2016; Zhao et al. 2019; Han et al. 2020; Nie et al. 2019; Wang et al. 2020; Liu et al. 2020; Zhang et al. 2022; Wang et al. 2021). Out of these, 528 QTLs are associated with hundred kernel weight (HKW), followed by 386 QTLs for kernel width, 329 for kernel row number (KRN), 299 for kernel length, 250 for kernel thickness, and 118 QTLs for kernel number per row (KNPR). It was observed that all these QTLs were unequally distributed throughout the ten chromosomes of maize.

A total of 20 loci related to kernel traits were fine-mapped in maize (Table 6.1). The *qGW 1.05, gW4.05, qhkw5-3,* and *qKW9* were the four major QTLs of kernel weight, which were mapped on chromosomes 1, 4, 5, and 9, respectively. The *qGW4.05* were mapped in an interval of 279.6 kb between ND19 and ND16, which revealed 23.9% of phenotypic variation and explained the occurrence of two annotated genes. The *GRMZM2G039934* is a predicted candidate for *qGW4.05*,

TABLE 6.1
The QTLs Associated with Grain Yield-Related Traits

Sr. No.	Traits	QTL	Marker Interval	Chromosome	Distance	Number of Candidate Genes	References
1.	Kernel row number	*krn1*	SNP1-SNP2	1	6.6kb	1	Wang et al. 2019
2.	Kernel row number	*KRN4*	M6-M8	4	3kb	2	Liu et al. 2015
3.	Kernel length	*qKL1.07*	ML194-ML162	1	1.6Mb	1	Qin et al. 2016
4.	Kernel length	*qKL9*	C9/54-C9/58	9	942kb	24	Gong et al. 2021
5.	Kernel length	*qKL-2*	Mk3106-mk3114	9	1.95Mb	40	Wang et al.2020
6.	Kernel width	*qKW7b*	M115.8-M116.7	7	59kb	1	Tang et al. 2020
7.	Kernel row number	*KNE4*	Umc1086-M5	4	440kb	14	Zhan et al. 2018
8.	Grain weight	*qGW4.05*	ND16-ND19	4	279.6kb	2	Chen et al. 2016
9.	Grain weight	*qhkw5-3*	InYM20-InYM36	5	125.3kb	6	Li et al. 2019
10.	Grain weight	*qGW1.05*	Umc1601-umc1754	1	1.11 Mb	30	Zhou et al. 2017
11.	Grain weight	*qKW9*	M3484-M3506	9	20 kb	3	Huang et al. 2020

which mainly encodes for a putative leucine-rich repeat receptor-like protein kinase and *qGW4.05* showed the best pleiotropic effect on kernel length, weight, and width (Chen et al. 2016). Another major QTL *qhkw5-3* that was fine-mapped for kernel weight into a physical region of 125.3 kb between markers InYM36 and InYM20, which contained six annotated genes (Li et al. 2019). While *qGW1.05* was narrowed down to a 1.11 Mb target region flanking umcl601 and umcl754 simple sequence repeat (SSR) markers revealing 20.94% phenotypic variation in kernel weight and containing 30 predicted genes (Zhou et al. 2017), *gKW9* with 20 kb was recently localized between markers M3506 and M3484 with three annotated genes. Out of three, *Zm00001d048451* encodes for PLS-DYW type PRR protein, which is a causal gene for kernel weight (Huang et al. 2020).

Of the three major QTLs (*qKL1.07, qKL-2,* and *qKL9*) associated with the kernel length, *qKL1.07* on chromosome 1 harboring five genes were delimited into 1.6 Mb genomic fragments flanked between ML162 and ML194 markers. The *qKL1.07, GRMZM2G348452* (ZmCKX10) is responsible for the synthesis of cytokinin oxidase, which showed 11% of phenotypic variation in the kernel length (Qin et al. 2016). The major QTL, *qKL9*, present on chromosome 9, also associated with increased kernel length and HKW (hundred kernel weight) (Gong et al. 2021). Liu et al. (2014) reported that the *qKL9* was associated with kernel length in maize in various environments by using $F_{2:3}$ families. This was confirmed by mapping the BC_3F_1 and BC_2F_2 populations, which revealed a 16.09% phenotypic variance in kernel length. This locus was flanked by C9-58 and C9-54 markers, which delimited into a 942 kb region and have 24 annotated genes. Out of them, *Zm00001d046723* encodes for expansion-A20 protein and three insertion/deletion variations in the 5'UTR region of *Zm00001d046723* significantly reduce the expression levels of McqKL9-A among near-isogenic lines (NILs), which is also responsible for kernel length (Gong et al. 2021). Third QTL *qKL-2* was physically mapped recently on a 1.95 Mb intervalon chromosome 9 between mk3114 and mk3106 markers and contained 40 genes. Out of 40 genes, only one gene *GRMZM2G006080* was a putative candidate for *qKL-2* which codes for receptor-like protein kinase FERONIA (Wang et al. 2020). FERONIA itself is demonstrated as a signaling pathway that regulates the elongation of integument cells in a negative way and then controls seed size in *Arabidopsis thaliana* (Yu et al. 2014).

Two major QTLs for kernel width, i.e., *qKW9.2* and *qKW7* were fine-mapped on chromosomes 9 and 7, respectively. The *qKW7* was delimited into a 647 kb target region, which lies between 7F-5 and 7H-16 markers and contained four annotated genes harboring kinases. Among these kinases,

GRMZM2G114706 is responsible for the synthesis of ankyrin protein kinase, which regulates kernel width and kernel weight in maize in a positive way (Li et al. 2016). Further fine-mapping of major locus *qKW7* was fine-mapped using a series of backcross populations derived by crossing Huangzaosi and YE478. It was then divided into two tightly linked intervals *qKW7a* and *qKW7b* with opposite phenotypic effects. The *qKW7a* revealed a minor additive effect, which is mainly influenced by the environment. However, *qKW7b* had a high LOD value along with additive effects, which revealed large kernel width due to its shelter to the YE478 allele, which is the main locus for kernel width. The *Zm00001d020460* encodes a putative zinc finger homeodomain (ZF-HD) transcription factor, which controls the expression of genes associated with kernel development (Tang et al. 2020). The qKW-9.2 was associated with kernel width which was mapped into a 630 kb target region and flanked by MSR36 and FSR6 markers on chromosome 9 showing 5.23–11.26% of the phenotypic variance (Raihan et al. 2016).

6.4.2 Fine Mapping of QTLs

Of the four others major QTLs viz., *qKHR6, qKNPR6, KNE4,* and *qKN* associated with kernel number per row (KNPR). These were fine-mapped on chromosomes 6,4, and 10. *KNE4* was flanked between umc1667 and umc2135 markers and localized to a 440 kb genomic region. Only *Zm00001d052399* encodes for long-chain acyl-CoA synthetase, which revealed 26% of phenotypic variation in KNPR that is supposed to be a candidate gene (Zhan et al. 2018). The *qKNR6* was recently mapped and flanked between M8 and M6 markers into a 110 kb target region, which contained two candidate genes. Among these genes, *Zm00001d036602* (a causative gene for *qKNR6*) encodes for serine/threonine kinase, which regulates KNPR variation at various gene expression levels (Jia et al. 2020). The markers N6M19 and umc1257 were present on chromosome 6 narrowed into a 198 kb target region of *qKNPR6*, which contained six candidate genes. The *qKNPR6* revealed 56.3% of KNPR variation, which showed pleiotropic effects on kernel weight and ear length also (Liu et al. 2012).

Other major QTLs, *qkrnw4, KRN4, KRN1.4, krn1, qKRN5b,* and *qKRN8*, for KRN, were fine-mapped on chromosomes 4, 1, 5, and 8, respectively. The *KRN 1.4* is a major locus that explains 50.48% of phenotypic variance for KRN was localized into a 203 kb region containing seven predicted candidate genes. Among them, indeterminate spikelet 1 (*ids1*) is a probable candidate gene encoding for APETALA2 (AP2)-like transcription factor that regulates floral meristems determinacy, inflorescence branching, and spikelet meristems determinacy (Ren et al. 2013; Calderon et al. 2016). A major QTL, *qKRN5,* was reported previously on chromosome number 5 (Shen et al. 2019) and upon fine-mapping in advanced backcross population, it revealed that *qKRN5* was tightly linked with two loci; *qKRN5a* and *qKRN5b*. The *qKRN5b* was a major QTL with a high additive effect as that of *qKRN5a*, which was subsequently mapped into a target region of 147.2 kb by markers SC14631 and SC3603d1 and contained three putative candidate genes. Out of three, the *Zm00001d013603* encodes for exonuclease and/or endonuclease phosphatase family protein hydrolyzes phosphatidylinositol diphosphates and be characterized as the causal gene of *qKRN5b*, on the basis of its expression and sequence variation between two parental lines (Shen et al. 2019). The *krn1*, a major QTL on chromosome 1 present in 6.6 kb of the genomic region is flanked by SNP1 and SNP2 markers and *Zm00001d034629* is a single gene of krn1 encoding for the AP2 transcription factor that controls KRN in maize (Wang et al. 2019). Using fine-mapping approaches, it was understood that, *KRN4* is localized in a 3 kb region between M6 and M8 markers containing 1.2 kb of transposon insertion that may be enhancing the KRN and creating KRN variation in diverse maize inbred lines (Liu et al. 2015). The *qkrnw4* on chromosome 4 between the Ch4.200-1 and Ch4.K-2 target region at a 33 kb interval, contains two possible candidate genes. *Zm00001d052910* encoding for putative protein which could be assessed for KRN-based gene expression and bioinformatics analysis (Nie et al. 2019). Recently, a novel QTL, *qKRN8,* was fine-mapped on chromosome 8 with a 520 kb target region that localized between umc2571 and umc2593 markers and harbors six annotated genes. Among these, *Zm00001d010007* encodes for START domain-containing protein as a causal gene for qKRN8, giving differential expression pattern in immature ears of NILs qKRN8V54, and qKRN8Lian87 (Han et al. 2020).

6.5 GENOMIC SELECTION

Genomic selection (GS) is a specialized form of marker-assisted selection (MAS) in which information from genotypic data on marker alleles covering the entire genome forms the basis of selection and the performance of new breeding materials is predicted on the basis of genomic information (Meuwissen et al. 2001). It provides prospects to increase the genetic gain of complex traits per unit of time and cost.

The most significant reason for the efficient and accurate implementation of GS in crop species is the availability of genome-wide high-throughput, cost-effective, and versatile markers with less ascertainment bias, ideal for large population sizes as well as for both model and common crop species, with or without the reference genome sequence. By reducing the election cycle and increasing precision, strength, better assessment of performance, ensures higher gains per unit of time, GS offers a range of merits over the phenotypic selection and tends to produce more accurate and environmentally insensitive outcomes (Rutkoski et al. 2011; Desta and Ortiz 2014). Thus, GS enables the fast-tracking of genetic gain and the development of high-performing cultivars of crop species that can cope with many of the challenges including climate change and decreasing arable land (see Figure 6.2).

Genomic selection models have been shown to be beneficial for a wide range of complex quantitative traits, such as grain yield, quality, biotic and abiotic stresses (de los Campos et al. 2009; Crossa et al. 2010; Burgueño et al. 2012; GonzálezCamacho et al. 2012; Jannink et al. 2010). The key feature of this approach is the use of genome-wide high-throughput markers, which potentially explain all the genetic variance with at least one marker is in linkage disequilibrium (LD) with

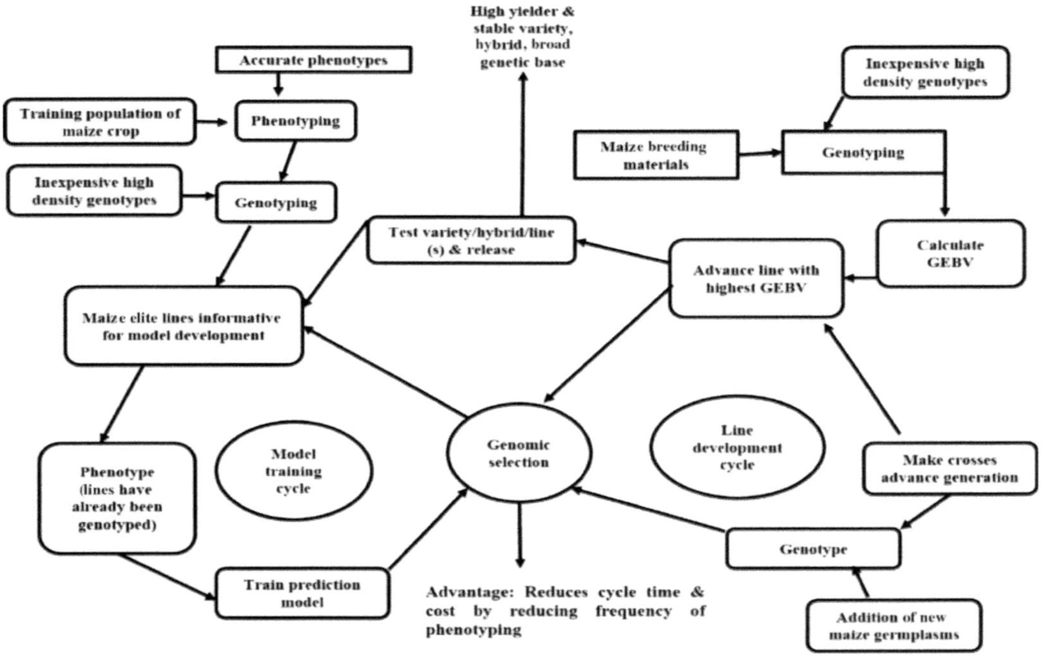

FIGURE 6.2 Schematic flowchart representing the components of genomic selection.

each QTL governing the trait of interest. The most apparent advantage of GS is that genotypic data collected from the seed or seedling can be used to determine the phenotypic output of mature individuals without the need for rigorous phenotypic assessment, thus raising the pace of varietal development in crop species (Bhat et al. 2016).

For instance, higher gains in grain yield under drought stress were achieved using genomic selection compared to conventional pedigree selection. The overall gain in average grain yield using GS was two to four folds higher than the previously reported gain through conventional phenotypic selection under drought stress in sub-Saharan Africa (Beyene et al. 2015). Further, a genetic gain of 100 tha^{-1} yr^{-1} by rapid cycling genomic selection (RCGS) was obtained, for four recombination cycles in a multi-parental tropical maize population (Zhang et al. 2017). Therefore, further studies in maize using tropical maize lines or other crops are needed to confirm the promising RCGS results obtained in multi-parental breeding populations (MPP), and to examine if the length of the selection cycle could be further reduced while simultaneously conserving the genetic diversity (Zhang et al. 2017b).

GS is a promising methodology that employs molecular markers to design novel and beneficial breeding programs and to generate new marker-based models for genetic evaluation. Also, it is a beneficial tool to enhance the prediction of phenotypes based on genomic, environmental, and management data. Genomic prediction works as a critical tool for large-scale maize breeding programs also around the world and has the potential to revolutionize smaller-scale public, private, government, and non-government breeding efforts in maize and other species (Washburn et al. 2020). The high-accuracy models that can predict and explain the performance of new genotypes in new environments are also needed, which would allow unprecedented gains in maize breeding efficiency, and the development of new maize cultivars that can survive and thrive in current and future climates.

6.6 GENOMIC RESOURCES TO ENHANCE GRAIN YIELD

The relationship between physical and biochemical features depends upon the combination of alleles of genes within an organism. Maize is a good example for deciphering this relationship because, maize is a staple crop worldwide and also a top production grain crop (http://faostat.fao.org/), has wider adaptability, and also serves as a model species. Presently, there are numerous quality reference genome assemblies of maize to discover the genotype-to-phenotype relationship (Lu et al. 2015; Hirsch et al. 2016; Jiao et al. 2017; Sun et al. 2018; Springer and Anderson 2018). In addition, by assessing the required information from these resources, one can link mutant phenotypes to genes or regions within the genome, i.e., QTLs (Schnable and Freeling 2011).

The Maize Genetics and Genomics database (Maize GDB, www.maizegdb.org) offers a wider range of valuable tools for different data types, including thousands of images in *Zea mays* ssp. *mays* (maize), and linked to mutant phenotypes. However, Cho et al. (2019) introduced and improvised the web-based software kit BioDIG (Biological Database of Images and Genomes) to expand the existing tools for discovering and visualizing these phenotypic relationships.

There is a database for genotype to phenotype relationships in maize (Table 6.2), which is user-friendly and extensible web-based resource, known as MaizeDIG (http://maizedig.maizegdb.org/). MaizeDIG has numerous advancements over the original BioDIG package, supporting multiple reference genome assemblies, and in order to accommodate custom tracks that display tagged images of mutant phenotypes in their genomic sense. It is also incorporated with genome browsers enabling custom tagging of images to highlight the phenotype. MaizeDIG is preloaded with 2,396 images, which are available for ten different maize reference genomes on genome browsers. Nearly, 90 images of classically specified maize genes have been manually annotated. This was done by an improved interface that helps users to create connections between images and genes and is accessible through the image search tool (https://githubcom/Maize-Genetics-and-Genomics-Database/maizedig) (Cho et al. 2019). These resources help in gathering information on genes, QTLs and

TABLE 6.2
Bioinformatics Databases for Crop Plants

Name	Website	Purpose
BLAST	www.ncbi.nml.nih.gov	Biological sequences information
PRIMER3,GENERUNNER, WEBCUTTER,Oligo Analysis and Plotting Tool	https://primer3.ut.ee/ www.generunner.net/ www.hsls.pitt.edu www.idtdna.com	Primer design and restriction mapping
GenBank	www.ncbi.nml.nih.gov	Nucleotide sequence
Pubmed	www.ncbi.nml.nih.gov	Literature references
PlantGDB	www.plantgdb.org/	Molecular sequence database
MaizeGDB	www.maizegdb.org	Maize genomics
GrainGenes	www.acedb.org	Database for *Triticeae* species
UK CropNet	www.ncbi.nml.nih.gov	Genome mapping in crop plants
NR	www.ncbi.nml/nih.gov	Protein sequences
SWISS-PORT	www.expasy.ch	Protein sequences
TAIR	www.arabidopsis.org	Arabidopsis database
InterPro	www.ebi.ac.uk	Protein domains
OMIM	www.ncbi.nml.nih.gov	Genetic diseases
Enzymes	www.chem.gmul.ac.uk	Enzymes
PDB	www.rcsb.org/pdb/	Protein sequences
KEGG	www.genome.ad.jp	Metabolic pathways

SNPs on desired trait and associations of SNPs with respective phenotypic traits through genome-wide association studies (Xiao et al. 2017).

6.7 CONCLUSION

Switching from vegetative development to reproductive development, in the transition process axillary meristems (AMs) turn into ear inflorescence meristems, which elongate and produce spikelet-pair meristems (SPMs). Subsequently every SPM form two spikelet meristems (SMs), which further produce floral meristems (FMs) and then kernels are produced after fertilization (Rankenberg et al. 2021). Initially, SPMs number on female inflorescence meristem reveals kernel rows number (KRN) on the maize ear, whereas meristematic activity on IMs determines the potential kernel number in each kernel row (KNPR). The initial SPMs numbers is related to inflorescence meristems size, which provides space for SPM development. The CLAVATA-WUSCHEL (CLV-WUS) feedback-signaling loop regulates IMsize by controlling stem cell proliferation and meristems activity. Various genes related to CLV-WUS, i.e., *thick tassel dwarf1* (td1), *COMPACT PLANT2* (*CT2*), and *fasciated ear2* (*fea2*) were isolated in maize. Grain weight is an important component of grain yield-related traits, which are mainly determined by genetic factors as well as environmental stress. The abiotic stress conditions, such as salt, drought, alkali, and barren were indirectly responsible for yield loss due to ionic stress, oxidative stress, and osmotic stress (Rankenberg et al. 2021). Drought conditions during the flowering stage in maize cause asynchronism in the anthesis and silking interval (ASI), which was responsible for serious yield losses. Genetic manipulation in *ZmEXPA4* of developing maize ears could significantly decrease ASI even under drought conditions (Liu et al. 2021).

Overall in this chapter, we have summarized the genetic basis of KRN-related traits in maize. The functionally characterized genes related to maize kernel yield *fea2, fea3, ub1,* etc., could be used for molecular markers development, which removes linked harmful/lethal mutations for crop production through a MAS. Furthermore, genome editing technology such as CRISPR, 5G-plant technology (a new kind of revolutionary technology), and a transgene (DNA)-free approach for the genetically modified organism (GMO) allows the gathering of various grain yield-related traits, such as kernel row number, kernel per row, ear length, etc., for developing ideal plant types designed by editing off-target sequences directly, accurately, and artificially in the future.

REFERENCES

An Y, Chen L, Li YX, Li C, Shi Y et al (2019) Candidate loci for the kernel row number in maize revealed by a combination of transcriptome analysis and regional association mapping. *BMC Plant Biol* 19(1):1–12.

Baye W, Xie Q, Xie P (2022) Genetic architecture of grain yield-related traits in sorghum and maize. *Int J Mol Sci* 23(5):2405.

Bensen RJ, Johal GS, Crane VC, Tossberg JT, Schnable PS et al (1995) Cloning and characterization of the maize *An1* gene. *Plant Cell* 7(1):75–84.

Beyene Y, Semagn K, Mugo S, Tarekegne A, Babu R et al (2015) Genetic gains in grain yieldthrough genomic selection in eight bi-parental maize populations under drought stress. *Crop Sci* 55(1):154–163.

Bhat JA, Ali S, Salgotra RK, Mir ZA, Dutta S et al (2016) Genomic selection in the era of next generation sequencing for complex traits in plant breeding. *Front Genet* 7:221.

Bomblies K, Doebley JF (2006) Pleiotropic effects of the duplicate maize Floricaula/Leafy genes *zfl1* and *zfl2* on traits under selection during maize domestication. *Genetics* 172(1):519–531.

Bomblies K, Wang RL, Ambrose BA, Schmidt RJ, Meeley RB et al (2003) Duplicate Floricaula/Leafy homologs *zfl1* and *zfl2* control inflorescence architecture and flower patterning in maize. *Development* 130(11):2385–2395.

Bommert P, Je BI, Goldshmidt A, Jackson D (2013) The maize Gα gene Compact Plant2 functions in Clavata signalling to control shoot meristem size. *Nature* 502(7472):555–558.

Bommert P, Nagasawa NS, Jackson D (2013) Quantitative variation in maize kernel row number is controlled by the Fasciated Ear2 locus. *Nat Genet* 45(3):334–337.

Brown PJ, Upadyayula N, Mahone GS, Tian F, Bradbury PJ et al (2011). Distinct genetic architectures for male and female inflorescence traits of maize. *PLoS Genetics* 7(11):e1002383.

Buckler IV ES, Thornsberry JM (2002) Plant molecular diversity and applications to genomics. *Curr Opin Plant Biol* 5(2):107–111.

Burgueño J, de los Campos G, Weigel K, Crossa J (2012) Genomic prediction of breeding values when modeling genotype × environment interaction using pedigree and dense molecular markers. *Crop Sci* 52(2):707–719.

Cai M, Li S, Sun F, Sun Q, Zhao H et al (2017) Emp10 encodes a mitochondrial PPR protein that affects the cis-splicing of nad2 intron 1 and seed development in maize. *Plant J* 91(1):132–144.

Calderon CI, Yandell BS, Doebley JF (2016) Fine mapping of a QTL associated with kernel row number on chromosome 1 of maize. *PLoS One* 11(3):e0150276.

Chen J, Zhang L, Liu S, Li Z, Huang R et al (2016) The genetic basis of natural variation in kernel size and related traits using a four-way cross population in maize. *PLoS One* 11(4):e0153428.

Chen L, An Y, Li YX, Li C, Shi Y et al (2017) Candidate loci for yield-related traits in maize revealed by a combination of metaQTL analysis and regional association mapping. *Front Plant Sci* 8:2190.

Chen L, Li YX, Li C, Shi Y, Song Y et al (2020) The retromer protein ZmVPS29 regulates maize kernel morphology likely through an auxin-dependent process(es). *Plant Biotechnol J* 18(4):1004–1014.

Chen L, Li YX, Li C, Wu X, Qin W et al (2016) Fine-mapping of qGW4. 05 a major QTL for kernel weight and size in maize. *BMC Plant Biol* 16(1):1–13.

Chen X, Feng F, Qi W, Xu L, Yao Det al (2017) Dek35 encodes a PPR protein that affects cis-splicing of mitochondrial nad4 intron 1 and seed development in maize. *Mol Plant* 10(3):427–441.

Cho KT, Portwood JL, Gardiner JM, Harper LC, Lawrence-Dill CJ et al (2019) MaizeDIG: Maize database of images and genomes. *Front Plant Sci* 10:1050.

Chuck G, Cigan AM, Saeteurn K, Hake S (2007) The heterochronic maize mutant Corngrass1 results from overexpression of a tandem microRNA. *Nat Genet* 39(4):544–549.

Chuck G, Meeley RB, Hake S (1998) The control of maize spikelet meristem fate by theAPETALA2-like gene indeterminate spikelet1. *Genes Dev* 12(8):1145–1154.

Chuck G, Whipple C, Jackson D, Hake S (2010) The maize SBP-box transcription factor encoded by tasselsheath4 regulates bract development and the establishment of meristem boundaries. *Development* 137(8):1243–1250.

Chuck GS, Brown PJ, Meeley R, Hake S (2014) Maize SBP-box transcription factors unbranched2 and unbranched3 affect yield traits by regulating the rate of lateral primordia initiation. *Proc Natl Acad Sci USA* 111(52):18775–18780.

Crossa J, de Los Campos G, Pérez P, Gianola D, Burgueño J et al (2010) Prediction of genetic values of quantitative traits in plant breeding using pedigree and molecular markers. *Genetics* 186:713–724.

Dai D, Luan S, Chen X, Wang Q, Feng Y et al (2018) Maize Dek37 encodes a P-type PPR protein that affects cis-splicing of mitochondrial nad2 intron 1 and seed development. *Genetics* 208(3):1069–1082.

de los Campos G, Naya H, Gianola D, Crossa J, Legarra A et al (2009). Predicting quantitative traits with regression models for dense molecular markers and pedigree. *Genetics* 182:375–385.

Desta ZA, Ortiz R (2014) Genomic selection: Genome-wide prediction in plant improvement. *Trends Plant Sci* 19:592–601.

Dhillon BS Singh J (1977) Estimation and inheritance of stability parameters of grain yield in maize. *J Agric Sci* 88:257–265.

Doebley J (1992) Mapping the genes that made maize. *Trends Genet* 8(9):302–307.

Doebley J (2004) The genetics of maize evolution. *Ann Rev Genet* 38(1):37–59.

Doebley J, Lukens L (1998) Transcriptional regulators and the evolution of plant form. *Plant Cell* 10(7):1075–1082.

Doebley J, Stec A, Gustus C (1995). Teosinte branched1 and the origin of maize: Evidence for epistasis and the evolution of dominance. *Genetics* 141:333e346.

Doll NM, Royek S, Fujita S, Okuda S, Chamot S et al (2020) A two-way molecular dialogue between embryo and endosperm is required for seed development. *Science* 367(6476):431–435.

Gallavotti A, Long JA, Stanfield S, Yang X, Jackson D et al (2010) The control of axillary meristem fate in the maize ramosa pathway. *Development* 137(17):2849–2856.

Gong D, Tan Z, Zhao H, Pan Z, Sun Q et al (2021) Fine mapping of a kernel length-related gene with potential value for maize breeding. *Theor Appl Genet* 134(4):1033–1045.

González-Camacho JM, de Los Campos G, Pérez P, Gianola D, Cairns JE et al (2012) Genome-enabled prediction of genetic values using radial basis function neural networks. *Theor Appl Genet* 125:759–771.

Guan HY, Dong YB, Lu SP, Liu TS, He CM et al (2020) Characterization and map-based cloning of miniature2-m1 a gene controlling kernel size in maize. *J Integr Agric* 19(8):1961–1973.

Guo M, Rupe MA, Zinselmeier C, Habben J, Bowen BA et al (2004) Allelic variation of gene expression in maize hybrids. *Plant Cell* 16(7):1707–1716.

Gupta PK, Rustgi S, Kumar N (2006) Genetic and molecular basis of grain size and grain number and its relevance to grain productivity in higher plants. *Genome* 49:565–571.

Gutiérrez-Marcos JF, Dal Prà M, Giulini A, Costa LM, Gavazzi G et al (2007) Empty pericarp4 encodes a mitochondrion-targeted pentatricopeptide repeat protein necessary for seed development and plant growth in maize. *Plant Cell* 19(1):196–210.

Han X, Qin Y, Sandrine AMN, Qiu F (2020) Fine mapping of qKRN8 a QTL for maize kernel row number and prediction of the candidate gene. *Theor Appl Genet* 133(11):3139–3150.

Hirsch CN, Hirsch CD, Brohammer AB, Bowman MJ, Soifer I et al (2016) Draft assembly of elite inbred line PH207 provides insights into genomic and transcriptome diversity in maize. *Plant Cell* 28(11):2700–2714.

Huang J, Lu G, Liu L, Raihan MS, Xu J et al (2020) The kernel size-related quantitative trait locus qKW9 encodes a pentatricopeptide repeat protein that affects photosynthesis and grain filling. *Plant Physiol* 183(4):1696–1709.

Iltis, Hugh H (1983) From teosinte to maize: the catastrophic sexual transmutation. *Science* 222(4626):886-894.

Jannink JL, Lorenz AJ, Iwata H (2010) Genomic selection in plant breeding: From theory to practice. *Brief Funct Genom* 9:166–177.

Janocha D, Lohmann JU (2018) From signals to stem cells and back again. *Curr Opin Plant Biol* 45:136–142.

JcSchnable JC, Freeling M (2011) Genes identified by visible mutant phenotypes show increased bias toward one of two subgenomes of maize. *PLoS One* 6, e17855.

Je BI, Gruel J, Lee YK, Bommert P, Arevalo ED et al (2016) Signaling from maize organ primordia via Fasciated Ear3 regulates stem cell proliferation and yield traits. *Nat Genet* 48(7):785–791.

Jia H, Li M, Li W, Liu L, Jian Y et al (2020) A serine/threonine protein kinase encoding gene kernel number per row6 regulates maize grain yield. *Nat Commun* 11(1):1–11.

Jiao FC, Li YX, Chen L, Liu ZZ, Shi YS et al (2014) Genetic dissection for kernel row number in the specific maize germplasm four-rowed waxy corn. *Sci Agric Sin* 47:1256–1264.

Jiao Y, Peluso P, Shi J, Liang T, Stitzer MC et al (2017) Improved maize reference genome with single-molecule technologies. *Nature* 546:524–527.

Kiss T, Balla K, Veisz O, Láng L, Bedő Z et al (2014) Allele frequencies in the VRN-A1 VRN-B1 and VRN-D1 vernalization response and PPD-B1 and PPD-D1 photoperiod sensitivity genes and their effects on heading in a diverse set of wheat cultivars (*Triticum aestivum* L.). *Mol Breed* 34(2):297–310.

Kumari R, Kumar P, Sharma VK, Kumar H (2018) Allele mining for crop improvement. *Int J Pure Appl Biosci* 6:1456–1465.

Li C, Li Y, Sun B, Peng B, Liu C et al (2013) Quantitative trait loci mapping for yield components and kernel-related traits in multiple connected RIL populations in maize. *Euphytica* 193(3):303–316.

Li G, Guo J, Luo W, Anjum T, Waheed A et al (2020) Development of high-density genetic map by specific-locus amplified fragment (SLAF) sequencing and identification of QTLS governing flowering and bolting time in Chinese kale. *Int J Agric Biol* 24(3):511–516.

Li M, Guo X, Zhang M, Wang X, Zhang G et al (2010) Mapping QTLs for grain yield and yield components under high and low phosphorus treatments in maize (*Zea mays* L.) *Plant Sci* 178(5):454–462.

Li W, Bai Q, Zhan W, Ma C, Wang S et al (2019) Fine mapping and candidate gene analysis of qhkw5-3 a major QTL for kernel weight in maize. *Theor Appl Genet* 132(9):2579–2589.

Li X, Li YX, Chen L, Wu X, Qin W et al (2016) Fine mapping of qKW7 a major QTL for kernel weight and kernel width in maize confirmed by the combined analytic approaches of linkage and association analysis. *Euphytica* 210(2):221–232.

Li YL, Li XH, Li JZ, Fu JF, Wang YZ et al (2009) Dent corn genetic background influences QTL detection for grain yield and yield components in high-oil maize. *Euphytica* 169(2):273–284.

Liu B, Zhang B, Yang Z, Liu Y, Yang S et al (2021). Manipulating ZmEXPA4 expression ameliorates the drought-induced prolonged anthesis and silking interval in maize. *Plant Cell* 33(6):2058–2071.

Liu C, Zhou Y, Zhang X, Zhang J, Zhou Z et al (2019) Natural variation in the THICK TASSEL DWARF1 (TD1) gene in the regulation of maize (*Zea mays* L.) ear-related traits. *Breed Sci* 69(2):323–331.

Liu L, Du Y, Huo D, Wang M, Shen X et al (2015). Genetic architecture of maize kernel row number and whole genome prediction. *Theor Appl Genet* 128(11):2243–2254.

Liu L, Du Y, Shen X, Li M, Sun W et al (2015) KRN4 controls quantitative variation in maize kernel row number. *PLoS Genetics* 11(11):e1005670.

Liu R, Jia H, Cao X, Huang J, Li F et al (2012) Fine mapping and candidate gene prediction of a pleiotropic quantitative trait locus for yield-related trait in *Zea mays*. *PLoS One* 7(11):e49836.

Liu Y, Wang L, Sun C, Zhang Z, Zheng Y et al (2014) Genetic analysis and major QTL detection for maize kernel size and weight in multi-environments. *Theor Appl Genet* 127(5):1019–1037.

Liu Y, Yi Q, Hou X, Hu Y, Li Y et al (2020) Identification of quantitative trait loci for kernel-related traits and the heterosis for these traits in maize (*Zea mays* L.). *Mol Genet Genom* 295(1):121–133.

Liu YJ, Xiu ZH, Meeley R, Tan BC (2013) Empty pericarp5 encodes a pentatricopeptide repeat protein that is required for mitochondrial RNA editing and seed development in maize. *Plant Cell* 25(3):868–883.

Liu Z, Cook J, Melia-Hancock S, Guill K, Bottoms C et al (2016) Expanding maize genetic resources with predomestication alleles: Maize–teosinte introgression populations. *Plant Genome* 9(1):2015–07.

Lu F, Romay MC, Glaubitz JC, Bradbury PJ, Elshire RJ et al (2015) High-resolution genetic mapping of maize pan-genome sequence anchors. *Nat Commun* 6:6914.

Lu M, Xie CX, Li XH, Hao ZF, Li MS et al (2010) Mapping of quantitative trait loci for kernel row number in maize across seven environments. *Mol Breed* 28(2):143–52.

McSteen P (2006) Branching out: The ramosa pathway and the evolution of grass inflorescence morphology. *Plant Cell* 18(3):518–522.

Meeley B, Briggs S (1995) Reverse genetics for maize. *Maize Genet Cooper Newsl* 69:67–67.

Mei X, Dong E, Liang Q, Bai Y, Nan J et al (2021) Identification of QTL for fasciated ear related traits in maize. *Crop Sci* 61(2):1184–1193.

Messmer R, Fracheboud Y, Bänziger M, Vargas M, Stamp P et al (2009) Drought stress and tropical maize: QTL-by-environment interactions and stability of QTLs across environments for yield components and secondary traits. *Theor Appl Genet* 119(5):913–930.

Meuwissen TH, Hayes BJ, Goddard M (2001) Prediction of total genetic value using genome-wide dense marker maps. *Genetics* 157(4):1819–1829.

Meyerowitz EM (1997) Genetic control of cell division patterns in developing plants. *Cell* 88(3):299–308.

Mirth CK, Riddiford LM (2007) Size assessment and growth control: How adult size is determined in insects. *Bioessays* 29(4):344–55. doi: 10.1002/bies.20552, PMID: 17373657

Nie N, Ding X, Chen L, Wu X, An Y et al (2019) Characterization and fine mapping of qkrnw4 a major QTL controlling kernel row number in maize. *Theor Appl Genet* 132(12):3321–3331.

Nimchuk ZL, Zhou Y, Tarr PT, Peterson BA, Meyerowitz EM (2015) Plant stem cell maintenance by transcriptional cross-regulation of related receptor kinases. *Development* 142(6):1043–1049.

Pan Z, Liu M, Xiao Z, Ren X, Zhao H et al (2019) ZmSMK9 a pentatricopeptide repeat protein is involved in the cis-splicing of nad5 kernel development and plant architecture in maize. *Plant Sci* 288:110205.

Pautler M, Eveland AL, LaRue T, Yang F, Weeks R et al (2015) Fasciated Ear4 encodes a bZIP transcription factor that regulates shoot meristem size in maize. *Plant Cell* 27(1):104–120.

Peng B, Li Y, Wang Y, Liu C, Liu Z et al (2011) QTL analysis for yield components and kernel-related traits in maize across multi-environments. *Theor Appl Genet* 122(7):1305–1320.

Pressoir G, Brown PJ, Zhu W, Upadyayula N, Rocheford T et al (2009) Natural variation in maize architecture is mediated by allelic differences at the PINOID co-ortholog barren inflorescence2. *Plant J* 58(4):618–628.

Qi W, J Zhu Q, Wu Q, Wang X, Li et al (2016) Maize reas1 mutant stimulates ribosome use efficiency and triggers distinct transcriptional and translational responses. *Plant Physiol* 170:971–988.

Qin W, Li YX, Wu X, Li X, Chen L et al (2016) Fine mapping of qKL1. 07 a major QTL for kernel length in maize. *Mol Breed* 36(1):1–9.

Qin X, Feng F, Li Y, Xu S, Siddique KH et al (2016) Maize yield improvements in China: Past trends and future directions. *Plant Breed* 135(2):166–176.

Raihan MS, Liu J, Huang J, Guo H, Pan Q et al (2016) Multi-environment QTL analysis of grain morphology traits and fine mapping of a kernel-width QTL in Zheng58× SK maize population. *Theor Appl Genet* 129(8):1465–1477.

Rankenberg T, Geldhof B, van Veen H, Holsteens K, Van de Poel B et al (2021) Age-dependent abiotic stress resilience in plants. *Trends Plant Sci* 26(7):692–705.

Ren D, Li Y, Zhao F, Sang X, Shi J, Wang N, ... He G (2013) Multi-Floret Spikelet1, which encodes an AP2/ERF protein, determines spikelet meristem fate and sterile lemma identity in rice. *Plant Physiol* 162(2):872–884.

Ren X, Pan Z, Zhao H, Zhao J, Cai M et al (2017) EMPTY PERICARP11 serves as a factor for splicing of mitochondrial nad1 intron and is required to ensure proper seed development in maize. *J Exp Bot* 68(16):4571–4581.

Running MP (1997) *Molecular genetics of floral patterning in Arabidopsis thaliana.* California Institute of Technology, Pasadena, California.

Rutkoski JE, Heffner EL, Sorrells ME (2011) Genomic selection for durable stem rust resistance in wheat. *Euphytica* 179:161–173.

Shen X, Zhao R, Liu L, Zhu C, Li M et al (2019) Identification of a candidate gene underlying qKRN5b for kernel row number in *Zea mays* L. *Theor Appl Genet* 132(12) 3439–3448.

Somssich M, Je BI, Simon R, Jackson D (2016) Clavata-Wuschel signaling in the shoot meristem. *Development* 143(18):3238–3248.

Sosso D, Mbelo S, Vernoud V, Gendrot G, Dedieu A et al (2012) PPR2263 a DYW-subgroup pentatricopeptide repeat protein is required for mitochondrial nad5 and cob transcript editing mitochondrion biogenesis and maize growth. *Plant Cell* 24(2):676–691.

Springer NM, Anderson SN (2018) The maize W22 genome provides a foundation for functional genomics and transposon biology. *Nat Genet* 50:1282–1288.

Sui Z, Wang T, Li H, Zhang M, Li Y et al (2016) Overexpression of peptide-encoding OsCEP6. 1 results in pleiotropic effects on growth in rice (*O. sativa*). *Front Plant Sci* 7:228.

Sun F, Wang X, Bonnard G, Shen Y, Xiu Z et al (2015) Empty pericarp7 encodes a mitochondrial E–subgroup pentatricopeptide repeat protein that is required for ccm FN editing mitochondrial function and seed development in maize. *Plant J* 84(2):283–295.

Sun F, Xiu Z, Jiang R, Liu Y, Zhang X et al (2019) The mitochondrial pentatricopeptide repeat protein EMP12 is involved in the splicing of three nad2 introns and seed development in maize. *J Exp Bot* 70(3):963–972.

Sun S, Zhou Y, Chen J, Shi J (2018) Extensive intraspecific gene order and gene structural variations between Mo17 and other maize genomes. *Nat Genet* 50 1289–1295.

Sun X, Liu D, Zhang X, Li W, Liu H et al (2013) SLAF-seq: An efficient method of large-scale de novo SNP discovery and genotyping using high-throughput sequencing. *PLoS One* 8(3):e58700.

Tang B, Li Y, Mu Z, Chen L, Guo H et al (2020) Fine mapping and candidate gene analysis of qKW7b a major QTL for kernel width in maize. *Mol Breed* 40(7):1–10.

Trung KH, Tran QH, Bui NH, Tran TT, Luu KQ et al (2020) A weak allele of Fasciated Ear 2 (FEA2) increases maize kernel row number (KRN) and yield in elite maize hybrids. *Agronomy* 10(11):1774.

Vollbrecht E, Schmidt RJ (2009) Development of the inflorescences. In: *Handbook of maize: Its biology*, pp 13–40. Springer, New York, NY.

Wang G, Zhao Y, Mao W, Ma X, Su C (2020) QTL analysis and fine mapping of a major QTL conferring kernel size in maize (*Zea mays*). *Front Genet* 11:603920.

Wang J, Lin Z, Zhang X, Liu H, Zhou L et al (2019) krn1 a major quantitative trait locus for kernel row number in maize. *New Phytol* 223(3):1634–1646.

Wang X, Liao C, Wang X, Yang R, Zhai L et al (2021) Construction of maize–teosinte introgression line population and identification of major quantitative trait loci. *Euphytica* 217(9):1–10.

Washburn JD, Burch MB, Franco JAV (2020) Predictive breeding for maize: Making use of molecular phenotypes machine learning and physiological crop models. *Crop Sci* 60(2):622–638.

Wei YM, Ren ZJ, Wang BH, Zhang L, Zhao YJ et al (2021) A nitrate transporter encoded by ZmNPF7. 9 is essential for maize seed development. *Plant Sci* 308:110901.

Xiao J, Jin X, Jia X, Wang H, Cao A et al (2013) Transcriptome-based discovery of pathways and genes related to resistance against Fusarium head blight in wheat landrace Wangshuibai. *BMC Genomics* 14(1):1–19.

Xiao Y, Liu H, Wu L, Warburton M, Yan J (2017) Genome-wide association studies in maize: praise and stargaze. *Mol Plant* 10 359–374.

Xiao Y, Thatcher S, Wang M, Wang T, Beatty M et al (2016) Transcriptome analysis of near-isogenic lines provides molecular insights into starch biosynthesis in maize kernel. *J Integr Plant Biol* 58(8):713–723.

Xiu Z, Sun F, Shen Y, Zhang X, Jiang R et al (2016) Empty Pericarp 16 is required for mitochondrial nad2 intron 4 cis-splicing complex I assembly and seed development in maize. *Plant J* 85(4):507–519.

Xu R, Li Y, Sui Z, Lan T, Song W, Zhang M, Zhang Y, Xing J (2021) A C-terminal encoded peptide ZmCEP1 is essential for kernel development in maize. *J Exp Bot* 72(15):5390–5406.

Yang C, Tang D, Zhang L, Liu J, Rong T (2015) Identification of QTL for ear row number and two-ranked versus many-ranked ear in maize across four environments. *Euphytica* 206(1):33–47.

Yang N, Liu J, Gao Q, Gui S, Chen L et al (2019) Genome assembly of a tropical maize inbred line provides insights into structural variation and crop improvement. *Nat Genet* 51(6):1052–1059.

Yu F, Li J, Huang Y, Liu L, Li D et al (2014) Feronia receptor kinase controls seed size in *Arabidopsis thaliana*. *Mol Plant* 7(5):920–922.

Zhan J, Wang F, Xing W, Liu J, Fan Z et al (2018) Fine mapping and candidate gene prediction of a major QTL for kernel number per ear in maize. *Mol Breed* 38(3):1–9.

Zhang M, Zheng H, Jin L, Xing L, Zou J, Zhang L, ... Wang L (2022) miR169o and ZmNF-YA13 act in concert to coordinate the expression of ZmYUC1 that determines seed size and weight in maize kernels. *New Phytol* 235(6):2270–2284.

Zhang X, Pérez-Rodríguez P, Burgueño J, Olsen M, Buckler E, Atlin G, ... Crossa J (2017) Rapid cycling genomic selection in a multiparental tropical maize population. *G3: Genes, Genomes, Genetics* 7(7):2315–2326.

Zhao Y, Su C (2019) Mapping quantitative trait loci for yield-related traits and predicting candidate genes for grain weight in maize. *Sci Rep* 9(1):1–10.

Zhou Q, Dong Y, Shi Q, Zhang L, Chen H et al (2017) Verification and fine mapping of qGW1. 05 a major QTL for grain weight in maize (*Zea mays* L.). *Mol Genet Genom* 292(4):871–881.

Zuo Z, Lu Y, Zhu M, Chen R, Zhang E et al (2021) Nucleotide diversity of the maize ZmCNR13 Gene and association with ear traits. *Front Genet* 12:2145.

7 Discover Genetic Diversity in Barley through Allele Mining

Davide Guerra, Luigi Cattivelli, and Alessandro Tondelli
Corresponding author: davide.guerra@crea.gov.it

7.1 INTRODUCTION

Following domestication from its immediate ancestor *Hordeum vulgare* spp. *spontaneum*, cultivated barley (*Hordeum vulgare* spp. *vulgare*) spread and adapted to a variety of ecogeographic conditions. Today the crop is grown on >50 million hectares around the world, from Nordic countries beyond the Arctic Circle to the limit of the deserts in the Middle East and to the highlands of Nepal and Ethiopia (Fischbeck 2003). This wide range of eco-geographical distribution is possible thanks to the overwhelming diversity of the barley gene pool. Such gene diversity is available in farmers' fields or in nearly 400,000 different barley accessions safely stored in more than 200 Genebanks around the world (www.genebanks.org/resources/crops/barley), which represent a reservoir of alleles with a potential for understanding the domestication and geographic expansion of the species, deciphering gene function and for current and future barley breeding (Mascher et al. 2018).

Landraces and wild barleys are of particular interest in resequencing-based studies because of their wide range of phenotypic responses deriving from natural or human selection (Camacho Villa et al. 2005). Other than adopting a traditional hit-and-miss approach, these genetic resources can be more efficiently scanned and valorized through a focused identification of germplasm strategy (FIGS, Mackay et al. 2005). Considering that the environment strongly influences natural selection, FIGS analyses passport data and agro-climatic information on the sample collection sites to select subsets of genetic materials that maximize allele diversity and have a higher probability of containing traits and genes of interest.

A strong prerequisite for allele mining studies is that the genes under investigation should have a clear and well-defined role, for example, as a candidate gene from a positional cloning or a genome-wide association study (GWAS) experiment. This way, detecting further allelic variants or testing known contrasting alleles in different genetic backgrounds can provide further validation of the proposed role or broaden the gene potential usefulness. Here we are going to review different allele mining approaches carried out in barley and how Genebank genomics (Mascher et al. 2019) will impact this powerful methodological approach.

7.2 THE PAST: ALLELE MINING THROUGH AMPLICON SANGER SEQUENCING

Allele mining in barley has a long and successful history even before the advent of high-throughput genomics. Studies targeting one or few genes were carried out by Sanger sequencing of small amplicons from hundreds or thousands of diverse accessions of worldwide origin with three main objectives, for which significant examples are listed.

DOI: 10.1201/9781003385004-7

(a) *Linking allelic diversity to barley evolution and domestication.* Early works sequencing and comparing multiple genes in relatively small sets of wild accessions and landraces from different origins (e.g. Kilian et al. 2006, Morrell and Clegg 2007) contributed to developing demographic models of barley domestication. Most importantly, cloning and allele mining of major genes directly involved in domestication traits allowed to reconstruct the genealogies of the selected mutations and to model the domestication process in a more targeted approach. For example, resequencing the *Vrs1* gene controlling row type supported the hypothesis of multiple and independent origins of six-rowed barley, with some six-rowed alleles clearly deriving from loss-of-function mutations in cultivated two-rowed or wild accessions that occurred in different regions and at different times (Komatsuda et al. 2007). More recently, sequencing the *Btr1* and *Btr2* genes responsible for grain dispersal in 505 *H. vulgare* spp. *spontaneum*, and 270 *H. vulgare* spp. *vulgare* accessions revealed temporally independent selections of non-brittle rachis during barley domestication in the southern and northern regions of the Levant (Pourkheirandish et al. 2015).
(b) *Linking allelic diversity to phenotypic diversity.* Allele mining usefulness was demonstrated in different works dealing with barley resistance to biotic and abiotic stress. Already in 2005, sequencing of all exons of the *HveIF4E* gene in a small collection of barley cultivars revealed nine non-synonymous single nucleotide polymorphisms (SNPs) that were associated with either the *rym4* or *rym5* multi-allelic recessive resistance to the barley yellow mosaic virus (Stein et al. 2005). Similarly, resequencing the *HvPDIL5-1* virus susceptibility factor in a large (*n*=1,732) worldwide barley collection revealed four distinct loss-of-function alleles in domesticated barley, but not in wild barley relatives. These recessive resistance alleles (named *rym11-a*, *-b*, *-c*, *-d*) originated probably from single mutation events of a most represented haplotype during barley migration to new environments characterized by a higher diversity of barley yellow mosaic virus strains (Yang et al. 2014). As for abiotic stress resistance, allelic variation at barley *CBF* genes was assessed in a diverse set of cultivars, landraces, and *H. Vulgare* spp. *spontaneum* accessions and correlated to their tolerance to freezing temperatures, thus revealing two nucleotide variants of *HvCbf14* as statistically associated with the trait (Fricano et al. 2009).
(c) *Linking allelic diversity to geographic/environmental/climatic diversity.* One of the first examples concerns the resequencing of a 3,508-bp region spanning the *Ppd-H1* gene controlling barley flowering under long-day conditions in a set of landraces from different origins (Jones et al. 2008). A large haplotype diversity was observed for this gene, with 121 haplotypes belonging to five haplotype groups. Most important, the geographic distribution of SNP48, considered the candidate causative variant, showed a latitudinal cline among European barley landraces, consistent with the higher frequency of long-day non-responsive barleys in the northern regions (Jones et al. 2008). A similar latitudinal cline was observed for the *HvCEN* gene involved in the control of barley flowering time (Comadran et al. 2012). Three major haplotypes were detected by resequencing a portion of the gene in a geographically referenced set of 1,143 cultivars, landraces, and wild barleys from worldwide origin, and haplotype frequency gradients were observed along south-north and west-east clines. Late-flowering haplotypes deriving from wild barley share a non-synonymous mutation that was selected during geographic range extension and is now fixed in northern European spring barley cultivars (Comadran et al. 2012). Resistance genes also contributed to barley adaptation to different environments. Regarding the *HveIF4E* gene previously reported, a total of 47 haplotypes were observed by sequencing of additional >1,000 barley landraces and cultivars from 84 countries, with most of these haplotypes specific to distinct geographic regions, suggesting a role of *HveIF4E* in barley local adaptation (Hofinger et al. 2011).

Despite their great value, these allele mining studies have targeted only one or a few genes or portions of genes at a time, in low-throughput experiments and with some obvious limitations in terms of resolution achieved. More recently, targeted enrichment and deep-sequencing approaches allowed the screening of more genes simultaneously, thus providing more informative datasets. For example, 666 loci involved in barley domestication and environmental adaptation were resequenced, together with 1,000 genome-wide neutral loci, in a panel of over 400 wild accessions and landraces from the Fertile Crescent, revealing more complex and reticulate models of domestication for the species (Pankin et al. 2018).

7.3 THE PRESENT: ALLELE MINING THROUGH WHOLE-GENOME SHOTGUN AND EXOME SEQUENCING

Mining the Genebank treasures is getting every day easier and easier thanks to the advancement in genomic science, and also for species characterized by large and complex genomes is nowadays possible to systematically search, identify and characterize new allelic variants for one or multiple genes.

The most ambitious effort has been undertaken at the Genebank of the Leibniz Institute of Plant Genetics and Crop Plant Research (IPK) in Gatersleben, Germany. Here genotyping-by-sequencing >22,000 barley accessions (almost the entire barley collection) highlighted the genetic composition and structure of the species at the global level and allowed high-resolution mapping of genotype-to-phenotype associations (Milner et al. 2019). In parallel, the BRIDGE data warehouse was developed to facilitate access to this digital barley resource. (https://bridge.ipk-gatersleben.de). Through BRIDGE it is possible to query the whole barley germplasm samples by passport and phenotypic data, inspect genetic diversity or GWAS results and explore >1 million SNP called against the latest version of the Morex reference genome (Konig et al. 2020). The Genomic Diversity Visualization tool integrates a SNP browser overlapped with >81,000 gene models (Figure 7.1). SNP matrices can be inspected and subset for different collections of germplasms, filtered for quality and allele frequencies, and exported in variant call files for further analyses. Based on gene models, SnpEff annotations predict the effects of genetic variants on genes and proteins (such as amino acid changes), providing quick and easy access to potentially relevant gene alleles. Custom collections

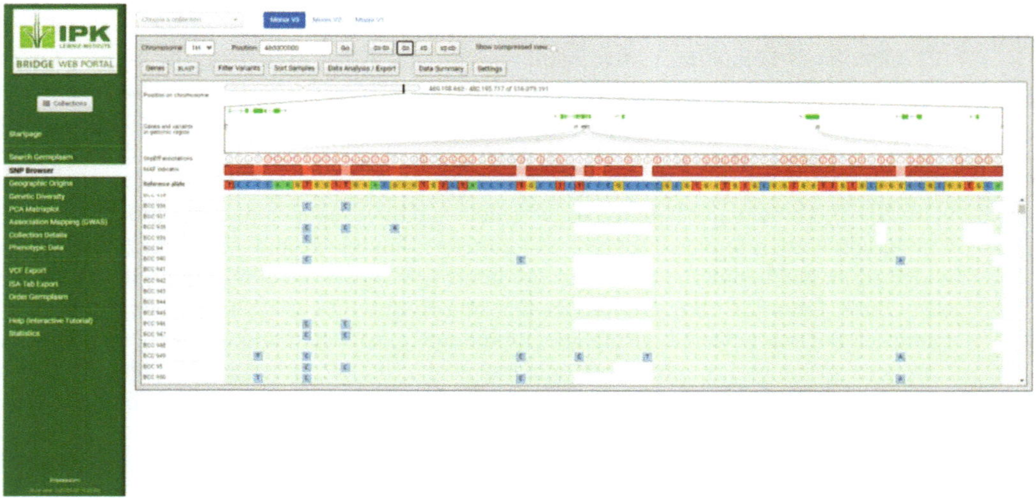

FIGURE 7.1 SNP browser from the BRIDGE web portal (https://bridge.ipk-gatersleben.de). SNPs can be mapped with respect to different versions of the Morex reference genome.

of germplasms harboring contrasting alleles can be built and finally ordered from the IPK Genebank for their validation and for utilization in barley breeding. As major drawbacks, genotyping-by-sequencing only offers a small representation of the genome, with genes being not covered for their entire length or not covered at all, and low depth of sequencing results in high rates of missing data in the final SNP matrices (Figure 7.1). Consequently, SNP haplotypes close to or surrounding a target candidate gene are most often considered instead of their allelic variants.

In crops with a very large genome, exome capture and sequencing represent a useful alternative, since it reduces genome complexity and sequencing costs by focusing only on the gene space. The exome sequencing approach in barley (Mascher et al. 2013) was first used to investigate the environmental adaptation displayed by a collection of 267 landrace and wild relatives (*H. vulgare* ssp. *spontaneum*) genotypes (Russell et al. 2016). The authors then focused their attention on allelic variation at 19 genes controlling flowering time and circadian clock and highlighted the existence of a geographical structuring for most of the genes. A later wider study used exome capture and sequencing to elucidate the genetic basis of adaptive traits in a barley collection of 403 accessions including cultivars, landraces, and wild relatives (Bustos-Korts et al. 2019). The large number of SNPs identified and their distribution over genes provides a resource for the high-resolution allele mining in barley. As an example, the authors analyzed the allelic diversity at the two major flowering time genes, *HvPPD-H1* (Turner et al. 2005) and *HvCEN* (Comadran et al. 2012), showing a clear geographical gradient in the distribution of their haplotypes.

With respect to whole-genome shotgun sequencing, these technologies coupled with more efficient algorithms are also valuable tools that can help to discover and characterize copy number variants (CNVs) and gene paralogs. With the ever-increasing number and quality of sequenced genomes, the idea has emerged that a fundamental contribution to phenotypic variability in plants is unaware of structural variations. Among structural variations, CNVs are of particular interest because of their role in shaping phenotypic variation in crop species (Saxena et al. 2014; Neik et al. 2017; Żmieńko et al. 2014; Schiessl et al. 2017). Additionally, the newly formed gene copies can evolve in parallel giving rise to genes that further increase the phenotypic and genetic variation of an individual (Fitch 1970; Hurles 2004).

The same resource from Bustos-Korts et al. (2019) has been recently exploited to detect CNVs either at genome-wide level (Bretani et al. 2020) or in a locus-specific manner (Guerra et al. 2022). Using exome sequencing and read count data, 16.605 deletions and duplications were detected when the exome capture sequences of a diverse panel of 397 accessions (cultivars, landraces, and wild relatives) were aligned to the Morex reference genome. The quest for segmental duplications in the reference sequence revealed many low-copy repeats, most of which overlap predicted coding sequences. Statistical analyses showed that the incidence of CNVs increases significantly in segmental duplication-rich regions, indicating that these sequence elements act as hot spots for the formation of CNVs (Bretani et al. 2020). A locus-specific analysis of CNV was applied to study the natural variation of *HvCBF2a* and *HvCBF4b* at the *Frost resistance-H2* locus (*Fr-H2*; Guerra et al. 2022). The authors have used whole-exome sequencing and depth of coverage analysis to study the extent of CNV for both genes. By counting the exome reads mapped at *Fr-H2* in each genotype of the collection, the authors identified a variation in the copy number of *HvCBF2a* and *HvCBF4b* and related it to the geographical origin.

The number of copies of a gene and even more the presence of paralogous genes, very often act as confounding elements in the assembly of genomic sequences and can be collapsed in a single locus, hampering the analysis of its allelic variation (Dou et al. 2012; Hohenlohe et al. 2012). In this case, the analysis of read depth is not enough to distinguish CNV from the presence of a gene paralog. Earlier studies proposed a method to identify gene paralogs from short reads relying on the combined analysis of read count and heterozygous SNPs (McKinney et al. 2017; Djedatin et al., 2017). Thereafter, a similar approach was used in barley to discriminate gene paralogs of NBS-LRR and *HvCBF2* genes from exome-wide sequencing data (Cantalapiedra et al., 2016; Guerra et al.,

2022). The rationale behind this approach relies on the unlikely chance of finding heterozygous SNPs clustered on a single or few loci. Instead, these heterozygous variants have originated from the piling up of closely related sequences (i.e. paralogous genes), which were captured by the exome baits, but for which the original locus is not present in the reference (Figure 7.2).

Overall, these findings highlight the importance of retaining high-confidence heterozygous SNPs, at least in allele mining studies, even when working in autogamous species, such as barley. Indeed, a common practice in diversity analysis is to filter out heterozygous SNPs, as they are often considered sequencing errors or the results of mixed DNA samples. The method described above is particularly useful if a pan-genome is not available; conversely, a high-quality pan-genome may be directly investigated for the presence of duplicated genes (as well as other structural variants) at candidate loci.

(a)

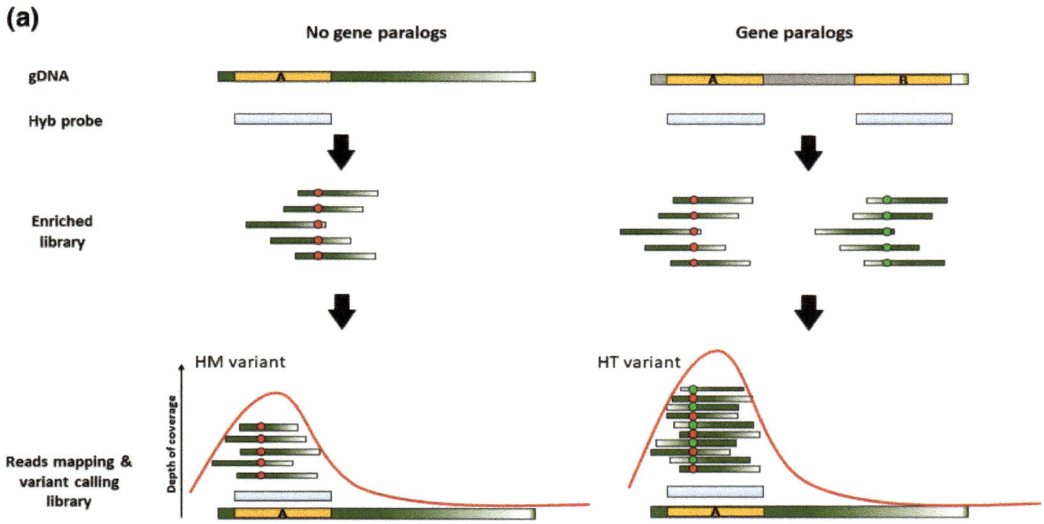

(b)

		16	21	29	43	63	76	632	640	661
	Morex 2A (M2A)	C	C	T	C	G	G	T	C	G
	Nure 2A (N2A)	C	C	T	C	G	G	T	C	A
	Nure 2B (N2B)	C	T	C	C	G	A	C	T	G
	Tremois 2A/B (T2AB)	C	C	T	C	G	G	C	T	G
HAP_1	M2A+N2B	C	Y	Y	C	G	R	Y	Y	G
	M2A+N2B+T2AB	C	Y	Y	C	G	R	Y	Y	G
HAP_2	N2A+N2B	C	Y	Y	C	G	R	Y	Y	R
	M2A+N2A+N2B	C	Y	Y	C	G	R	Y	Y	R
	N2A+N2B+T2AB	C	Y	Y	C	G	R	Y	Y	R
HAP_3	HAP3	C	Y	Y	C	Y	Y	Y	Y	G
HAP_4	HAP4	C	Y	Y	C	G	Y	T	Y	R
HAP_5	HAP5	Y	Y	T	C	G	G	Y	Y	G

FIGURE 7.2 (a) Schematic representation of heterozygous mapping for paralogs identification from targeted/exome resequencing data. Upon NGS sequencing of target-enriched libraries, heterozygous variants are caused by the piling up of closely related sequences (paralogous genes) captured by the exome baits but absent in the reference genome. Red and green circles are polymorphic positions between gene paralogs. HM and HT variants represent homozygous and heterozygous SNPs respectively. (b) Example of haplotypes of *HvCBF2* identified through the heterozygous mapping analysis in the WHEALBI barley collection (Bustos-Korts et al. 2019). Y and R refer to IUPAC ambiguity code to indicate heterozygous SNPs. Panel (b) is from Guerra et al. (2022). CC BY 4.0.

7.4 THE FUTURE: ALLELE MINING IN PANGENOMES

Increasing evidence is demonstrating that one or few reference genomes are not enough to cover the full genetic diversity of a species (Zhou et al. 2015; Hurgobin et al. 2018; Wang et al. 2018). Indeed, a detailed description and understanding of the alleles available for each locus and the structural variants requires the sequencing of many more accessions. Pan-genomics, defined as the collection of all genes and alleles of a given clade, is revolutionizing genomic research in crop plants since it allows a better estimation of gene content, its variation across a population, and how breeding shaped it over time (Liu et al. 2020). The release of new sequencing technologies characterized by improved capacity and reduced cost is fastening the release of an accurate pan-genome even in species with large genomes, such as barley (ca. 5.1 Gb, IBSC 2012).

A first pan-genome of barley was recently released based on sequencing 20 accessions selected within the IPK-Gatersleben Genebank to best represent the species diversity and including one *H. vulgare* ssp. *spontaneum* (Jayakodi et al. 2020). Such high-quality genome assemblies have allowed the detection of large structural variants, which are more difficult to detect by using methodologies described before. As an example, a 10 Mb inversion at the chromosome 2H centromeric region was observed in Western domesticated barleys, which also brings the late-flowering allele at the already mentioned *HvCEN* gene mapping close to this inversion. Such inversion might have an impact on *HvCEN* expression and explain the low genetic recombination frequency observed around this gene in elite germplasm. In addition, it was observed that each of the 20 barley assembled genomes includes on average 2.9 Mb of private sequence (Jayakodi et al. 2020).

A barley pangenome_v2 that includes high-quality sequencing of more accessions will become available in the next years (M. Mascher, N. Stein, personal communication). As already observed in other crops, such as rice (Shang et al. 2022), increasing the number of sequenced genomes represents a step forward to disclosing complex genetic variants. Similar to the *HvCEN* example reported above, structural variants affecting the expression of closely linked genes were associated with grain phenotypes (Shang et al. 2022). These variants most likely would not have been detected by analyzing only the coding portion of the genome. Moreover, lineage-specific haplotypes were more efficiently discovered using the super pan-genome, thus enabling to better describe how the genomes of different rice species have evolved.

To facilitate the visualization and the exploitation of such an extensive genomic resource, the Rice Super Pan-genome Information Resource Database (www.ricesuperpir.com) was built that hosts reference-free whole-genome alignment of all assemblies (Super Pan Browser, Figure 7.3) and a super pan-genome graph that can more effectively identify SNPs and structural variants.

In fact, the conventional linear representation of a genome sequence has shown weaknesses in the representation of large structural variants (Liu et al. 2020). In a pan-genome graph the iterative mapping and *de novo* assembly approaches are combined to present sequence variation across multiple genomes as different paths along a graph of sequence or variant nodes (Jayakodi et al. 2021). Pan-genome graphs are constructed through whole-genome alignment, unassembled read alignment, or *de novo* graph assembly. These methods enable to easily compare structural variants inside a population or among closely related species. Other than rice, a pan-genome graph has been constructed for soybean (Liu et al. 2020), leading to the identification of large DNA rearrangements and gene fusion events that have been linked to differences in gene expression and agronomic traits.

Currently, similar databases and graphs are not yet available for barley, and only the single genome assemblies can be browsed at the GrainGenes database (https://wheat.pw.usda.gov/GG3/genome_browser). However, CNVs are reported for each genome, while SNPs from the USDA the Triticeae Toolbox (T3)/Barley portal are mapped against the reference Morex_v3 genome assembly. Additional molecular variants can be found at EnsemblPlants (https://plants.ensembl.org/Hordeum_vulgare/Info/Index), which also provides resequencing alignments and variant effect predictor tools through dedicated software. Despite the low number of genomes available, these tools already represent a valuable resource for mining allelic diversity at candidate loci.

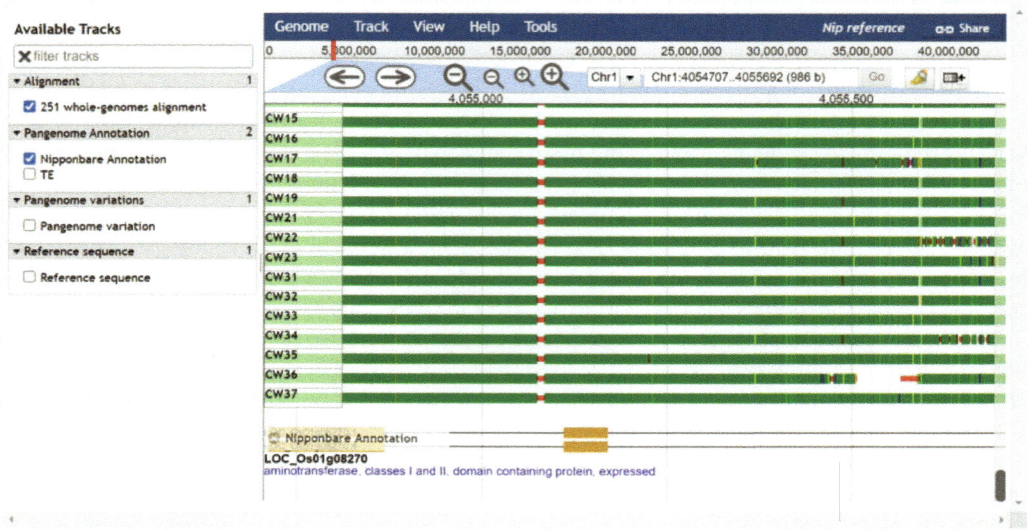

FIGURE 7.3 Super Pan Browser from the Rice Super Pan-genome Information Resource Database (RiceSuperPIRdb, www.ricesuperpir.com). Green bars represent consensus sequences from different accessions, and molecular variants (such as SNPs and InDels) are highlighted in colors.

7.5 CONCLUSIONS

Considering the progress in sequencing technologies, it is easy to predict that in the next years, more barley genomes will be fully sequenced and as soon as these sequences are compared and made accessible from a single website, allele mining will become a major approach to study genetic diversity.

Everybody is aware of the relevance of genetic diversity and many efforts have been carried out to conserve, characterize, and evaluate large sets of barley accessions. Moving into the "genome sequencing era", allele mining is now becoming a new paradigm for the characterization of genetic diversity. The traditional "forward approach" (from phenotype to alleles) in the analysis of genetic diversity will be coupled or replaced by a "reverse approach" (from genetic variants to phenotype) where the identification of all variants in the gene pool (SNPs, InDels, CNVs) represents the starting point for the selection of a set of accessions that will be subject to more deep and detailed phenotyping. An example is the work of Guerra et al. (2022), where extensive allele mining on about 400 diverse barley accessions was used to describe the genetic diversity at the *CBF* locus, the main factor controlling frost tolerance. This allowed the selection of about 20 accessions representing most of the genetic diversity available at the locus, which were then subjected to frost tolerance assay.

Allele mining is revolutionizing the analysis of the germplasm and in the future will be a key strategy for the exploitation of barley genetic diversity.

ACKNOWLEDGMENTS

The authors acknowledge the support from the projects PlantRed (MAECI, Accordo di Cooperazione nel campo della Ricerca e dello Sviluppo Industriale, Scientifico e Tecnologico tra Italia e Israele Bando scientifico 2021) and BARISTA (ERA-NET Cofund SusCrop Grant No. 771134, part of the Joint Programming Initiative on Agriculture, Food Security and Climate Change -FACCE-JPI, project ID: 77, MUR Italian Ministry of Research)

REFERENCES

Bretani G, Rossini L, Ferrandi C, Russel J, Waugh R, Kilian B, Bagnaresi P, Cattivelli L, Fricano A (2020) Segmental duplications are hot spots of copy number variants affecting barley gene content. *Plant J* 103:1073–1088.

Bustos-Korts D, Dawson IK, Russell J, Tondelli A, Guerra D, Ferrandi C, Strozzi F, Nicolazzi EL, Molnar-Lang M, Ozkan H, Megyeri M, Miko P, Çakır E, Yakışır E, Trabanco N, Delbono S, Kyriakidis S, Booth A, Cammarano D, Mascher M, Werner P, Cattivelli L, Rossini L, Stein N, Kilian B, Waugh R, van Eeuwijk FA (2019) Exome sequences and multi-environment field trials elucidate the genetic basis of adaptation in barley. *Plant J* 99:1172–1191.

Camacho Villa TC, Maxted N, Scholten M, Ford-Lloyd B (2005) Defining and identifying crop landraces. *Plant Genet Res* 3:373–384.

Cantalapiedra CP, Contreras-Moreira B, Silvar C, Perovic D, Ordon F, Gracia MP, Igartua E, Casas AM (2016) A cluster of nucleotide binding site-leucine-rich repeat genes resides in a barley powdery mildew resistance quantitative trait loci on 7HL. *Plant Genome* 9. doi: 10.3835/plantgenome2015.10.0101

Comadran J, Kilian B, Russell J, Ramsay L, Stein N, Ganal M, Shaw P, Bayer M, Thomas W, Marshall D, Hedley P, Tondelli A, Pecchioni N, Francia E, Korzun V, Walther A, Waugh R (2012) Natural variation in a homolog of Antirrhinum *CENTRORADIALIS* contributed to spring growth habit and environmental adaptation in cultivated barley. *Nat Genet* 44:1388–92.

Djedatin G, Monat C, Engelene S, François S (2017) Duplication Detector, a light weight tool for duplication detection using NGS data. *Curr Plant Biol* 9–10:23–28.

Dou J, Zhao X, Fu X, Jiao W, Wang N, Zhang L, Hu X, Wang S, Bao Z (2012) Reference-free SNP calling: Improved accuracy by preventing incorrect calls from repetitive genomic regions. *Biol Direct* 7:17.

Fischbeck G (2003) Diversification through breeding. In: von Bothmer R, van Hintum T, Knüpffer H, Sato K (eds) *Diversity in Barley Hordeum vulgare*, pp 29–52. Elsevier Science, Amsterdam.

Fitch WM (1970) Distinguishing homologous from analogous proteins. *SystZooll* 9:99.

Fricano A, Rizza F, Faccioli P, Pagani D, Pavan P, Stella A, Rossini L, Piffanelli P, Cattivelli L (2009) Genetic variants of *HvCbf14* are statistically associated with frost tolerance in a European germplasm collection of *Hordeum vulgare*. *Theor Appl Genet* 119:1335–48.

Guerra D, Morcia C, Badeck F, Rizza F, Delbono S, Francia E, Milc JA, Monostori I, Galiba G, Cattivelli L, Tondelli A (2022) Extensive allele mining discovers novel genetic diversity in the loci controlling frost tolerance in barley. *Theor Appl Genet* 135:553–569.

Hofinger BJ, Russell JR, Bass CG, Baldwin T, dos Reis M, Hedley PE, Li Y, Macaulay M, Waugh R, Hammond-Kosack KE, Kanyuka K (2011) An exceptionally high nucleotide and haplotype diversity and a signature of positive selection for the *eIF4E* resistance gene in barley are revealed by allele mining and phylogenetic analyses of natural populations. *Mol Ecol* 20:3653–68.

Hohenlohe PA, Catchen J, Cresko WA (2012) Population genomic analysis of model and non model organisms using sequenced RAD tags. *Methods Mol Biol* 888:235–260.

Hurgobin B, Golicz AA, Bayer PE, Chan C-KK, Tirnaz S, Dolatabadian A, Schiessl SV, Samans B, Montenegro JD, Parkin IAP, Pires JC, Chalhoub B, King GJ, Snowdon R, Batley J, Edwards D (2017) Homoeologous exchange is a major cause of gene presence/absence variation in the amphidiploid *Brassica napus*. *Plant Biotechnol J* 16:1265–1274.

Hurles M (2004) Gene duplication: The genomic trade in spare parts. *PLoSBiol* 2:e206.

IBSC-International Barley Sequencing Consortium (2012) A physical, genetic and functional sequence assembly of the barley genome. *Nature* 491:711–716.

Jayakodi M, Padmarasu S, Haberer G, Bonthala VS, Gundlach H, Monat C, Lux T, Kamal N, Lang D, Himmelbach A, Ens J, Zhang XQ, Angessa TT, Zhou G, Tan C, Hill C, Wang P, Schreiber M, Boston LB, Plott C, Jenkins J, Guo Y, Fiebig A, Budak H, Xu D, Zhang J, Wang C, Grimwood J, Schmutz J, Guo G, Zhang G, Mochida K, Hirayama T, Sato K, Chalmers KJ, Langridge P, Waugh R, Pozniak CJ, Scholz U, Mayer KFX, Spannagl M, Li C, Mascher M, Stein N (2020) The barley pan-genome reveals the hidden legacy of mutation breeding. *Nature* 588:284–289.

Jayakodi M, Schreiber M, Stein N, Mascher M (2021) Building pan-genome infrastructures for crop plants and their use in association genetics. *DNA Research* 28:1–9.

Jones H, Leigh FJ, Mackay I, Bower MA, Smith LMJ, Charles MP, Jones G, Jones MK, Brown TA, Powell W (2008) Population-based resequencing reveals that the flowering time adaptation of cultivated barley originated east of the Fertile Crescent. *Mol Biol Evol* 25:2211–2219.

Kilian B, Ozkan H, Kohl J, von Haeseler A, Barale F, Deusch O, Brandolini A, Yucel C, Martin W, Salamini F (2006) Haplotype structure at seven barley genes: Relevance to gene pool bottlenecks, phylogeny of ear type and site of barley domestication. *Mol Genet Genomics* 276:230–241.

Komatsuda T, Pourkheirandish M, He C, Azhaguvel P, Kanamori H, Perovic D, Stein N, Graner A, Wicker T, Tagiri A, Lundqvist U, Fujimura T, Matsuoka M, Matsumoto T, Yano M (2007) Six-rowed barley originated from a mutation in a homeodomain-leucine zipper I-class homeobox gene. *Proc Natl Acad Sci USA* 104:1424–1429.

König P, Beier S, Basterrechea M, Schüler D, Arend D, Mascher M, Stein N, Scholz U, Lange M (2020) BRIDGE – A visual analytics web tool for barley genebank genomics. *Front Plant Sci* 11:701.

Liu Y, Du H, Li P, Shen Y, Peng H, Liu S, Zhou G, Zhang H, Liu Z, Shi M, Huang X, Li Y, Zhang M, Wang Z, Zhu B, Han B, Liang C, Tian Z (2020) Pan-genome of wild and cultivated soybeans. *Cell* 182:162–176

Mackay M, von Bothmer R, Skovmand B (2005) Conservation and utilization of plant genetic resources: Future directions. *Czech J Genet Plant Breed* 41:335–344.

Mascher M, Richmond TA, Gerhardt DJ, Himmelbach A, Clissold L, Sampath D, Ayling S, Steuernagel B, Pfeifer M, D'Ascenzo M, Akhunov ED, Hedley PE, Gonzales AM, Morrell PL, Kilian B, Blattner FR, Scholz U, Mayer KF, Flavell AJ, Muehlbauer GJ, Waugh R, Jeddeloh JA, Stein N (2013) Barley whole exome capture: A tool for genomic research in the genus *Hordeum* and beyond. *Plant J* 76:494–505.

Mascher M, Sato K, Steffenson B (2018) Genomics approaches to mining barley germplasm collections. In: Stein N, Muehlbauer G (eds) *The barley genome. Compendium of plant genomes*, pp 155–169 Springer, Cham.

Mascher M, Schreiber M, Scholz U, Graner A, Reif JC, Stein N (2019) Genebank genomics bridges the gap between the conservation of crop diversity and plant breeding. *Nat Genet* 51:1076–1081.

McKinney GJ, Waples RK, Seeb LW, Seeb JE (2017) Paralogs are revealed by proportion of heterozygotes and deviations in read ratios in genotyping-by-sequencing data from natural populations. *Mol Ecol Resour* 17:656–669.

Milner SG, Jost M, Taketa S, Mazón ER, Himmelbach A, Oppermann M, Weise S, Knüpffer H, Basterrechea M, König P, Schüler D, Sharma R, Pasam RK, Rutten T, Guo G, Xu D, Zhang J, Herren G, Müller T, Krattinger SG, Keller B, Jiang Y, González MY, Zhao Y, Habekuß A, Färber S, Ordon F, Lange M, Börner A, Graner A, Reif JC, Scholz U, Mascher M, Stein N (2019) Genebank genomics highlights the diversity of a global barley collection. *Nat Genet* 51:319–326.

Morrell PL, Clegg MT (2007) Genetic evidence for a second domestication of barley (*Hordeum vulgare*) east of the Fertile Crescent. *Proc Natl Acad Sci USA* 104:3289–3294.

Neik TX, Barbetti MJ, Batley J (2017) Current status and challenges in identifying disease resistance genes in *Brassica napus*. *Front Plant Sci* 8:1788.

Pankin A, Altmüller J, Becker C, von Korff M (2018) Targeted resequencing reveals genomic signatures of barley domestication. *New Phytol* 218:1247–1259.

Pourkheirandish M, Hensel G, Kilian B, Senthil N, Chen G, Sameri M, Azhaguvel P, Sakuma S, Dhanagond S, Sharma R, Mascher M, Himmelbach A, Gottwald S, Nair SK, Tagiri A, Yukuhiro F, Nagamura Y, Kanamori H, Matsumoto T, Willcox G, Middleton CP, Wicker T, Walther A, Waugh R, Fincher GB, Stein N, Kumlehn J, Sato K, Komatsuda T (2015) Evolution of the grain dispersal system in barley. *Cell* 162:527–39.

Russell J, Mascher M, Dawson IK, Kyriakidis S, Calixto C, Freund F, Bayer M, Milne I, Marshall-Griffiths T, Heinen S, Hofstad A, Sharma R, Himmelbach A, Knauft M, van Zonneveld M, Brown JW, Schmid K, Kilian B, Muehlbauer GJ, Stein N, Waugh R (2016) Exome sequencing of geographically diverse barley landraces and wild relatives gives insights into environmental adaptation. *Nat Genet* 48:1024–30.

Saxena RK, Edwards D, Varshney RK (2014) Structural variations in plant genomes. *Brief Funct Genomics* 13:296–307.

Schiessl S, Huettel B, Kuehn D, Reinhardt R, Snowdon R (2017) Post-polyploidisation morphotype diversification associates with gene copy number variation. *Sci Rep* 7:41845.

Shang L, Li X, He H, Yuan Q, Song Y, Wei Z, Lin H, Hu M, Zhao F, Zhang C, Li Y, Gao H, Wang T, Liu X, Zhang H, Zhang Y, Cao S, Yu X, Zhang B, Zhang Y, Tan Y, Qin M, Ai C, Yang Y, Zhang B, Hu Z, Wang

H, Lv Y, Wang Y, Ma J, Wang Q, Lu H, Wu Z, Liu S, Sun Z, Zhang H, Guo L, Li Z, Zhou Y, Li J, Zhu Z, Xiong G, Ruan J, Qian Q (2022) A super pan-genomic landscape of rice. *Cell Res* 32:878–896.

Stein N, Perovic D, Kumlehn J, Pellio B, Stracke S, Streng S, Ordon F, Graner A (2005) The eukaryotic translation initiation factor 4E confers multiallelic recessive *Bymovirus* resistance in *Hordeum vulgare* L. *Plant J* 42:912–922.

Turner A, Beales J, Faure S, Dunford RP, Laurie DA (2005) The pseudo-response regulator *Ppd-H1* provides adaptation to photoperiod in barley. *Science* 310:1031–1034.

Wang W, Mauleon R, Hu Z, Chebotarov D, Tai S, Wu Z, Li M, Zheng T, Fuentes RR, Zhang F, Mansueto L, Copetti D, Sanciangco M, Palis KC, Xu J, Sun C, Fu B, Zhang H, Gao Y, Zhao X, Shen F, Cui X, Yu H, Li Z, Chen M, Detras J, Zhou Y, Zhang X, Zhao Y, Kudrna D, Wang C, Li R, Jia B, Lu J, He X, Dong Z, Xu J, Li Y, Wang M, Shi J, Li J, Zhang D, Lee S, Hu W, Poliakov A, Dubchak I, Ulat VJ, Borja FN, Mendoza JR, Ali J, Li J, Gao Q, Niu Y, Yue Z, Naredo MEB, Talag J, Wang X, Li J, Fang X, Yin Y, Glaszmann JC, Zhang J, Li J, Hamilton RS, Wing RA, Ruan J, Zhang G, Wei C, Alexandrov N, McNally KL, Li Z, Leung H (2018) Genomic variation in 3,010 diverse accessions of Asian cultivated rice. *Nature* 557:43–49.

Yang P, Lüpken T, Habekuss A, Hensel G, Steuernagel B, Kilian B, Ariyadasa R, Himmelbach A, Kumlehn J, Scholz U, Ordon F, Stein N (2014) Protein Disulfide Isomerase Like 5-1 is a susceptibility factor to plant viruses. *Proc Natl Acad Sci USA* 111:2104–2109.

Zhou Z, Jiang Y, Wang Z, Gou Z, Lyu J, Li W, Yu Y, Shu L, Zhao Y, Ma Y, Fang C, Shen Y, Liu T, Li C, Li Q, Wu M, Wang M, Wu Y, Dong Y, Wan W, Wang X, Ding Z, Gao Y, Xiang H, Zhu B, Lee SH, Wang W, Tian Z (2015) Resequencing 302 wild and cultivated accessions identifies genes related to domestication and improvement in soybean. *Nat Biotechnol* 33:408–414.

Zmienko A, Samelak A, Kozłowski P, Figlerowicz M (2014) Copy number polymorphism in plant genomes. *Theor Appl Genet* 127:1–18.

8 Avena Genetic and Genomics Resources for Allele Mining in Oats

Caterina Morcia, Valeria Terzi, Ilaria Carrara, Roberta Ghizzoni, Stefano Delbono, Fulvia Rizza, Franz-W. Badeck, Giorgio Tumino, and Franca Finocchiaro*
*Corresponding author: valeria.terzi@crea.gov.it

8.1 INTRODUCTION

Allele mining is an approach used to dissect the genetic variability naturally existing in relation to important agronomic and qualitative characteristics. This approach can be effectively used for the discovery of superior alleles through "mining" the gene of interest from diverse genetic resources. The allelic variation can be in fact found within genetic resource collections, including both cultivated accessions and wild relatives of a crop. Such an approach is therefore used to dissect naturally occurring allelic variations with the final aim to identify superior alleles and new haplotypes controlling key agronomic and qualitative traits. It should be emphasized that allele mining must address both the expressed and unexpressed regions of the gene, to cover most of the functional variations. The characterization of allelic variation can help evolutionary studies, whereas the identification of superior alleles can be exploited in breeding programs, after the development of allele-specific molecular markers for marker-assisted selection (MAS).

Three main roads, different from each other, have been identified as suitable for allele mining studies in crops, i.e., tilling techniques, sequencing, and association-mapping-based allele mining (Gokidi et al. 2017; Kumari et al. 2018) (Figure 8.1).

Moreover, bioinformatics and genomics applied to the study of germplasm, also called "genoplasmics" (Jia et al. 2017), play a leading role in modern allele mining. It is undeniable that the existence of genetic variation is the backbone of allele mining, therefore collections of germplasm constitute the main pillar of such studies. In this respect, oats are rich in materials.

8.2 OAT GENETIC RESOURCE COLLECTIONS

Cultivated oat (*Avena sativa* L.) is an allohexaploid (AACCDD, $2n = 6x = 42$) small grain cereal characterized by low carbon footprint, peculiar content in primary and secondary metabolites with positive impact on human health (Loskutov and Khlestkina 2021). It is an economically interesting plant used for the production of both forage and grains for animal husbandry. Oats were domesticated in Europe some 5,000 years ago from wild *Avena* species (Vavilov et al. 1992) and can therefore be considered as an indigenous crop to Europe. Oats are recognized as a low-input small grain cereal crop grown worldwide for human food consumption, animal feed and, more recently, also for non-food applications. Importantly, oats grow relatively well on marginal land or in situations unsuited to wheat, barley, or maize production. Average oat yields have increased over the last 50 years and northern Europe shows the highest average oat grain yields in the world

 DOI: 10.1201/9781003385004-8

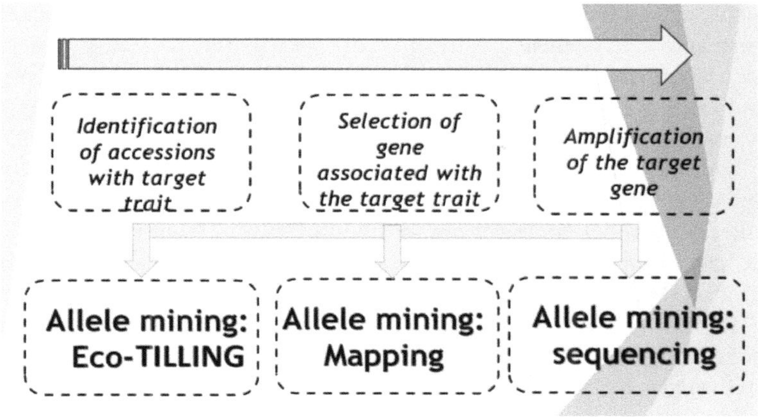

FIGURE 8.1 Steps in allele mining.

Nevertheless, the share of minor cereals (oats, rye, and triticale) in the total cereal area in Europe fell from 30% to 18% between 1961 and 2010 (http://faostat.fao.org/), while that of the major cereals increased. This trend towards increased monoculture has led to proposals to support crop diversification under the Common Agricultural Policy (CAP). In addition to contributing to diversification in many situations, oats are a low input cereal crop with higher nutrient use efficiency and lower pesticide requirements than maize or wheat and are competitive against weeds. Such a combination of traits makes oats highly adapted to sustainable cropping systems and of great relevance to agricultural policy development.

Even though they are cereals, oats are used in cereal-based crop rotations as a "break" crop, thereby reducing the build-up of diseases in the main crops. Much of Europe grows spring-sown oats, while winter oats (autumn-sown) are grown mostly in the United Kingdom (65% of the total oat acreage in UK) and southern Europe. Winter oats have high yields, deep roots, a long growth cycle, and early maturity but do not have sufficient winter hardiness suitable for cultivation elsewhere in Europe. In addition, in southern areas, winter sowing serves to often finish the crop cycle before harsh summer conditions occur. However, drought stress together with disease incidence are still the main causes of yield instability. Improved oat lines with higher yields and better end use quality are urgently needed to maintain the competitiveness with other cereals.

In recent years, there has been an increase in the demand for oats across Europe due to the unique properties of the grain. About 15% of the oat grain harvested in Europe is used for food in Europe although in some countries (e.g. the United Kingdom) up to 70% of oat grain is used for human consumption. Oat grains contribute to healthy diets due to the high levels of β-glucan, which is the major endosperm cell wall polysaccharide. Health claims for oat β-glucan have been approved by EFSA and the US FDA for their cholesterol-lowering effects. Several other health claims, specific to oats and related to reducing the impact of chronic diseases (such as type 2 diabetes, obesity, hypertension, immune-related diseases) are under study. Oats also have the highest oil (6–12%) and protein (12–20%) contents in de-hulled grains among cereals and unique types of antioxidants. Due to this grain composition, the use of oats has increased in recent years, especially in Europe and North America. The increasing knowledge of grain composition and the potential for various end-uses continues to lead to more oat-derived products.

All these features make oats the most promising "minor" small grain cereal crop with advantages over rye, triticale, or speciality types of wheat. The costs of producing oats are lower than for other cereals and its established and proposed health benefits will boost oat demand in Europe. Furthermore, there are many prospects for breeding new oat cultivars, tailoring crop husbandry, and developing supply chains for various uses to make the crop higher yielding, profitable, and sustainable in European agro ecosystems.

Wide collections of oat genetic resources are available worldwide in several private and public collections and Gene Banks, as the result of a decades-long effort by botanists and geneticists. Consequently, oat germplasm is included in the top 10 genetic resource collections held by Gene Banks, with more than 130.000 accessions available (Wambugu et al. 2018). According to De Carvalho et al. (2013), currently, 116 major Gene Banks conserve oats genetic resources. The major collections of oat landraces are located in Canada (Plant Gene Resources of Canada, PGRC, at the Saskatoon Research and Development Centre, Saskatoon, Saskatchewan), in the Russian Federation (Federal Research Center N. I. Vavilov All-Russian Institute of Plant Genetic Resources, VIR), USA (NSGC) and in Germany (Leibniz Institute of Plant Genetics and Crop Plant Research, IPK, Gatersleben). The collections include adapted cultivars, old landraces, marginally cultivated oats, such as *Avena strigosa*, *Avena abyssinica*, *Avena brevis*, and *Avena nuda*, crop wild relatives, and genetic stocks bearing defined traits. Different Gene Bank information systems have been developed (e.g. GRIN-CA/GRIN-Global-CA, GRIN-USA, EURISCO) that allow national and international clients and the public to inspect and access the Gene Bank holdings.

This heritage constitutes the starting point for the development of allele mining. The first step is the morpho-physiological and molecular characterization of a Gene Bank collection for diversity estimates. The subsequent step is the extraction of a core collection, i.e., a reduction of the number of accessions to about 10% of the total number of genotypes present in the Gene Bank. An efficient core collection is characterized by the least reduction of diversity estimates in comparison with the entire collection. At an additional level, a mini core set can be developed from core collection. Such mini core sets classically include about 10% of the core collection accessions and 1% of the accessions of the entire collection. The reduced size of the core collection and the mini core set is a relevant advantage to speed up and to reduce costs of allele mining.

A major condition for developing core collections consists in the availability of a great amount of genotypic and phenotypic data necessary to operate efficient selection of accessions to be included. This condition can be potentially met through the activation of collaborative studies.

The idea to identify sets of experimental germplasms to be used in collaborative trials was developed inside the Collaborative Oat Research Enterprise (CORE) effort, initiated in 2009 and ran until 2014. Under the CORE umbrella different coordinated projects, funded investigators, and collaborators were united by an over-arching goal of developing tools for genomics and molecular breeding in oat. The CORE activities also focused on the development of a sequence resource, a single nucleotide polymorphism (SNP) genotyping array, genotyping-by-sequencing methods, phenotyping data, a complete consensus linkage map, and information on population structure, linkage disequilibrium (LD), and adaptation in cultivated oat. To evaluate the impact of CORE activities on the scientific oat community, a survey was sent to scientists both involved or not in the CORE Project. The majority considered that CORE results were essential and/or had been used substantially in oat research, and had led to both essential and indirect derived benefits. The CORE activity was valorized through 33 key citations, and through a tabulation of 30 new research projects dependent on CORE results. The conclusion was that CORE has had a major impact in stimulating and facilitating oat research and in building an active oat research community (Tinker et al. 2016). In the frame of CORE, the study of Klos et al. (2016) addressed the assembly and analysis of a germplasm diversity panel constituted of about 600 accessions derived from public oat breeding programs in North America and beyond, as well as an additional set of lines with global or historical relevance. In this work, the authors evaluated genetic diversity, characterized population structure, to find subdivisions related to adaptation and to characterize genome-wide and chromosome-specific LD.

Several other works were focused on the development of specific germplasm panels.

Diederichsen (2008) reported that 10,105 accessions of hexaploid cultivated oat coming from 85 countries and conserved in the Plant Gene Resources of Canada (PGRC) collection were characterized at Saskatoon, Saskatchewan, Canada for eight environmentally stable morphological characters (panicle type, panicle erectness, panicle density, lemma color, dorsal awn of lemma, ligula, major infraspecific group), differentiated into 18-character states. It turned out that the 10,105 accessions represented 118 different morphological groups. The author suggested that "*to establish a core collection, the delimitation of morphological groups is useful, because the core collection should not only represent single character variation ranges, but also unique combinations of character states*".

An oat core collection has been proposed by Boczkowska et al. (2016) starting from 91 landraces preserved by the National Centre for Plant Genetic Resources in Radzików, i.e., the Polish Gene Bank. After morpho-physiological and molecular characterization (Dziubińska et al. 2022), the broadness of the gene pool turned out to be lower than previously expected but was still significantly higher than that of contemporary varieties. The authors reduced the number of accessions by 76.9% for the entire set selecting 21 accessions as representative of the diversity. Variability patterns of important agro-morphological traits are included in the landrace core collection, such as plant height, resistance to lodging, crown rust, and powdery mildew sensitivity.

With the advancement of genomic tools, relevant efforts have been made for the characterization of the genetic diversity in large and diverse collections of oats, as the first step for the development of core collections.

In the works of Tumino et al. (2016, 2017), in the frame of the European cooperative project AveQ, a wide collection of 655 oat accessions were selected, including current commercial cultivars as well as cultivated, marginally cultivated – and wild Gene Bank material – representing twelve Avena species, more than 100 years of breeding history, and more than 30 countries of origin. All the accessions were evaluated in 2-year field trials in seven European locations (Estonia, Sweden, Poland, France, Italy, Romania, and Bulgaria) for some morphological and agronomic traits – including yield and disease resistance. In each experiment 11 standard cultivars have been included. After this first step, a core collection of about 150 accessions was extracted from the entire collection and genotyped using the Illumina 6K Oat Array, and SNP data were analyzed as continuous intensity ratios, rather than as discrete genotypes with the ultimate aim to perform association mapping for cold tolerance and quality traits. Wide variability for cold tolerance was found, as exemplified in Figures 8.2 and 8.3. Low temperatures and light intensity affect the PSII excitation pressure, which plants are exposed to and, consequently, their tolerance to stress (Figure 8.2). A lab test has been done to explore the genetic variability for PSII stability and frost resistance in a set of 100 oat genotypes. The selected genotypes include standard varieties, winter and intermediate varieties and genotypes of different origin, representative of European areas of cultivation. One-week old plants were hardened at 3–1°C (day/night) 200 uE m-2 s-1 for 4 weeks and subjected to freezing treatments at –10°C and –12°C. The effect of freezing treatments on the variable to maximal fluorescence ratio (Fv/Fm) was measured in hardened plants immediately after freezing at different temperatures and during recovery. Starting from the lab and field data, high variability for frost resistance has been individuated in the oat genotypes characterized by different growth habit and origin (Figure 8.3).

Other studies focused on different collections, selected on the basis of geographical origin, breeding history, morpho-physiological characteristics. While far from being exhaustive, below we cite some examples.

Carlson et al. (2019) selected a diversity panel including 500 lines starting from more than 1000 accessions of different origins. After genotyping-by-sequencing (GBS) analysis an optimal and diverse subset of 500 lines were selected, as a core set of a worldwide collection of oat germplasm.

Canales et al. (2021) assembled a panel of 669 oat landraces from the Mediterranean rim and 40 cultivated red and white oat varieties. This oat collection was genotyped with 17,288 SNP loci to

FIGURE 8.2 Variability in oat accessions for frost resistance/susceptibility.

evaluate population structure and LD and to perform a genome-wide association study (GWAS) for heading date.

The population structure and genetic diversity were studied in a panel of 158 Chinese hulless oat lines by Yan et al. (2020).

A core collection of 50 Russian accessions was selected from the Russian germplasm collection held at the Vavilov Institute of Plant Genetic Resources by Bityutskii et al. (2020) and evaluated for high nutritive values of grain components, including iron (Fe), zinc (Zn), manganese (Mn), protein oil and fatty acids. Considerable genetic variation appeared to exist in almost all nutritional contents among genotypes grown in the same environment.

Accessions belonging to different Avena species, such as *A. sativa*, *A. subtilis*, *A. strigosa,* and *A. abyssinica* have been exploited in the search of pathogen resistance alleles. In particular, a major breeding target is the development of crown rust-resistant genotypes. Great effort has been made to understand the genetics underlying this trait, as recently reviewed by Park et al. (2022). Ninety-eight genes/alleles involved in *Puccinia coronata* resistance have been catalogued until now. Allelic relationships among them are largely unknown, but the recent availability of reference genomes open new perspectives in allele mining, diagnostic markers' development, and gene pyramiding aimed to increase rust resistance in oats.

8.3 TILLING AND ECO-TILLING

TILLING (Targeting Induced Local Lesions In Genomes) and Eco-TILLING (Ecotype Targeting Induced Local Lesions In Genomes) are well-known efficient tools for allele mining (Figure 8.4). They are non-transgenic reverse genetic approaches that allow the rapid screening for mutants in target genes in a plant mutagenized population (McCallum et al. 2000). This approach therefore helps in the direct identification of allelic variation/polymorphism resulting from induced mutations in a target gene (Till et al. 2003).

Pioneer studies in tetraploid and hexaploid wheats have demonstrated that it is possible to apply TILLING even to crop with complex genomes when used in combination with an efficient method

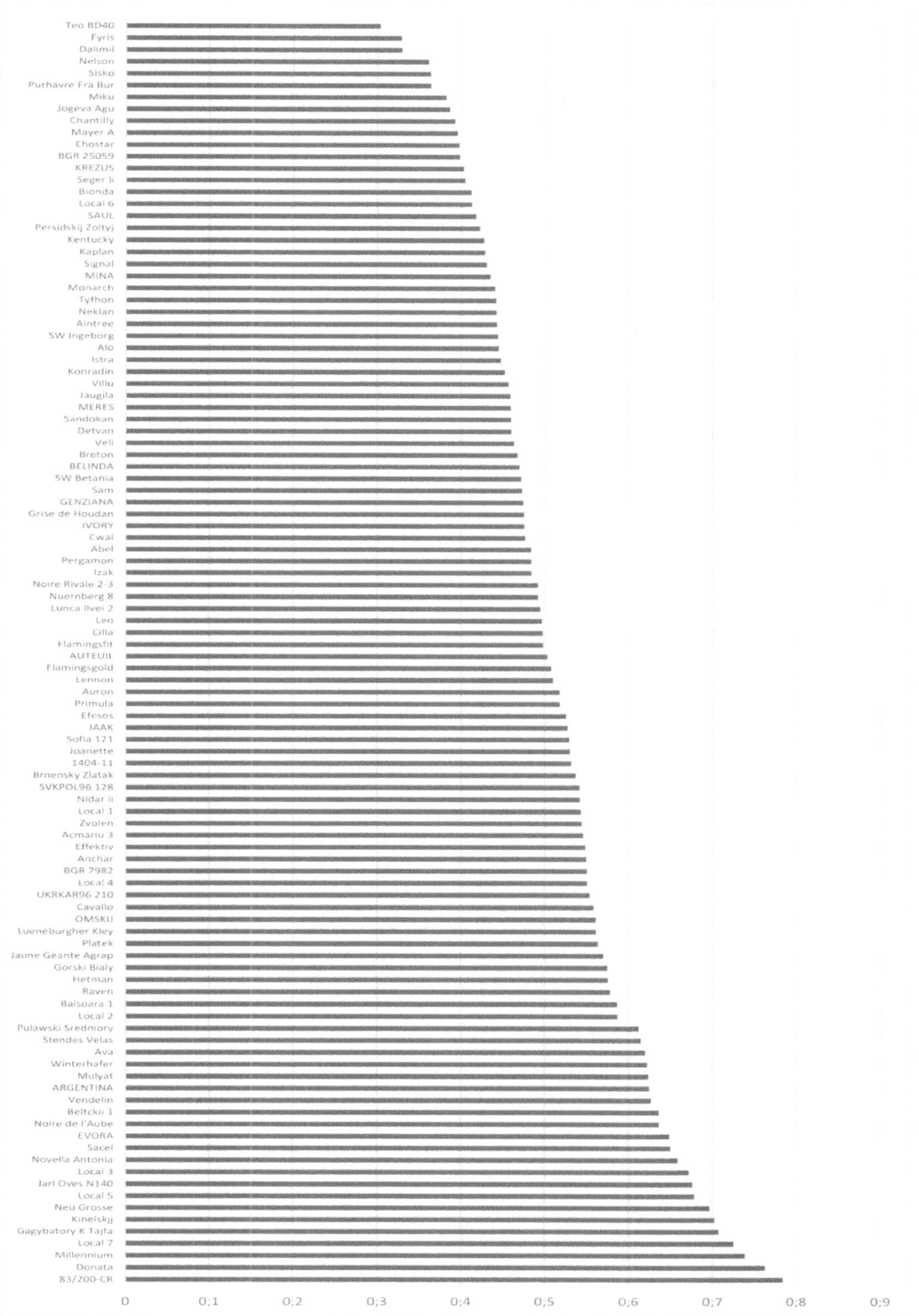

FIGURE 8.3 Variability in an oat core collection after freezing treatments. The frost resistance was measured as the variable to maximal fluorescence ratio (Fv/Fm) in hardened plants immediately after freezing at different temperatures and during recovery.

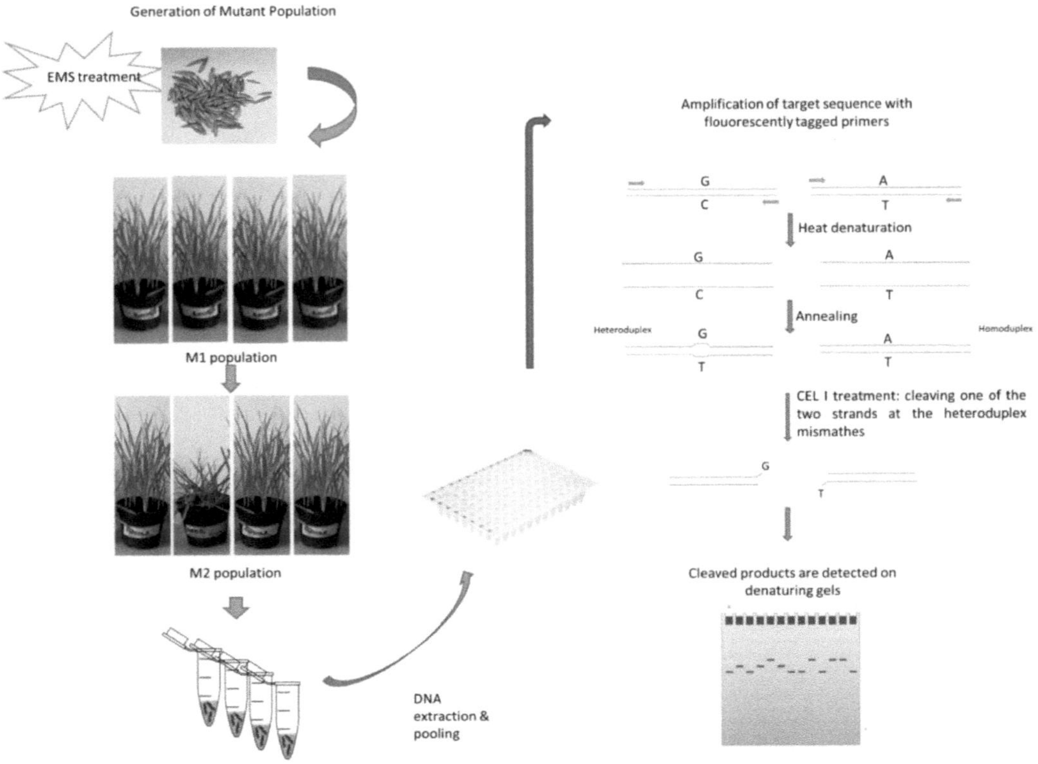

FIGURE 8.4 A typical TILLING scheme by CEL-1 enzyme.

to identify point mutations (Slade et al. 2005). The power of this approach was demonstrated by the identification of the number of waxy gene mutants in TILLING populations of both tetraploid and hexaploid wheats. By analogy, a TILLING-population in a hexaploid oat with a reasonable mutation frequency should therefore be obtained and used for molecular studies, including allele mining.

In oat, a useful TILLING population was developed by Chawade et al. (2010). By taking advantage of ethylmethane sulfonate (EMS)-induced mutagenesis, a mutagenized oat population was developed from seeds of the variety Belinda. The population was produced by random mutagenesis and therefore with a non-transgenic technique, the lines are not regulated as genetically modified organism (GMO) under the European parliament's directive 2001/18/EC. The TILLING population includes about 2800 mutagenized lines. Chawade et al. (2010) through molecular characterization of the population, showed that ca 1 million unique point mutations (SNPs) were introduced into each line and this means that about 3 billion mutations can be found in the whole population. This number of mutations is high enough to ensure that every single gene in the genome is mutated several times and therefore the mutation frequency of this TILLING population can provide observable phenotypic traits in the lines. In a successive step, the mutagenized lines were stabilized over several M-generations and the number of homozygous mutations was quantified as about 1.25 billion in the stabilized TILLING population.

Now, this TILLING population is managed by CropTailor, a plant biotechnology company, which aimed to develop oat materials with enhanced agronomic and health properties and methods through advanced technology (www.croptailor.com/about-us/). CropTailor offers to breeders the identification of specific mutations in practically any gene of interest using a proprietary Targeted Allele Mining pipeline, accompanied by the confirmed effect on the phenotype.

The Belinda TILLING population was used to explore phenotypic and allelic variations for some traits. An example is the study of Sikora et al. (2013).

These authors phenotyped 1700 lines of the mutagenized population for β-glucan levels in the seeds, using the quantitative Megazyme β-glucan assay. Variability for this character was found, with lines that were either higher or lower than the 4.9% seen in the Belinda variety. About 2% of all the screened lines had a very different β-glucan content in comparison with Belinda. The authors suggested that the low frequency found for high β-glucan phenotypes, despite the fact that in every line one induced mutation is present per 30 Kb, is coherent with the difficulties experienced by breeders to obtain breeding lines with high β-glucan levels.

Moreover, the stability of the high β-glucan trait in the mutated lines and the distribution of β-glucans in the different parts of the seed have been studied, together with the hypothesis that high levels of these secondary metabolites can affect the structure of the seed carbohydrates.

Finally, the DP3/DP4 ratios of all mutants as well as three market varieties were analyzed.

The population has been used successfully to generate new varieties with elevated β-glucan contents (Sikora et al. 2013), high and low lignin levels (Vivekanand et al. 2014), and high protein content (Sunilkumar et al. 2017).

A panel of 520 lines, randomly chosen from the Belinda derived TILLING population, was used to identify differences in seed lignin content, by Vivekanand et al. (2014). Seed lignin levels were found in the range from 20 to 63 g kg^{-1}, whereas Belinda, had seed lignin level around 41 g kg^{-1}. Structural analyses carried out with XRD, UV, and FT-IR spectroscopy showed that there were significant qualitative differences in seed lignin in the mutated lines and in Belinda, and a more digestible lignin in mutated lines was predicted.

A high protein quality with a favorable amino acid composition and a protein with a good bioavailability are also desirable characters, so if the oat protein content could be increased in combination with an improvement of other quality characters, oats would become an even more competitive source of vegetative proteins. Sunilkumar et al. (2017) screened 1050 lines from the oat mutagenized population developed and described by Chawade et al. (2010) for elevated protein levels. The authors identified several lines and further characterized both their protein content and quality as well as the content of other macromolecules like total dietary fiber (TDF), β-glucan and lipid content. Two hundred and thirty lines showed seed protein levels of 15% or higher; 15 lines reached protein levels ranging between 17 and 24%. In addition, the genetic stability of the high protein character was studied by preliminary field trials and back crossings to the original Belinda variety.

Oat is also characterized by the presence of another class of compounds, which have potential biological activities, the avenanthramides (AVNs). AVNs are phenolic alkaloids, formed by the covalent linking of hydroxycinnamic and anthranilic acids via an amide bond (de Bruijn et al. 2019; Schendel 2019). Using the same TILLING population developed by Chawade et al. (2010), Hernandez et al. (2021) selected 17 lines with elevated avenanthramide (AVN) content and their AVN structures, concentrations, and antioxidant potentials were determined by HPLC-MS2 and HPLC equipped with an on-line ABTS+ antioxidant detection system.

The authors showed qualitative and quantitative differences in the synthesis of AVNs in the different lines, with a total AVN concentration up to 227.5 μg/g oat seed flour, compared with 78.2 μg/g seed in the commercial variety, SW Belinda. These results indicate the potential of oat mutagenesis in combination with a high precision biochemical selection method for the generation of stable mutagenized lines with a high concentration of total and/or individual AVNs in the oat seed grain.

In conclusion, the authors demonstrated that TILLING populations are useful tools for the discovery of mutants in target traits and consequently for the identification of allelic variants. However, in oats, the exploitation of the TILLING population for allele mining was delayed until reference genomes were released. Such sequencing activities have been concluded only in the last 2 years.

8.4 REFERENCE OAT GENOMES

The availability of reference genomes for cultivated oats and wild relatives belonging to the *Avena* genus helps to leverage knowledge from other cereal genomes, improves understanding of basic oat biology and accelerates genomics-assisted breeding, re-analysis of quantitative trait studies and allele mining. Due to several reasons, including the high complexity of its genome, oat was sequenced later than other small grain cereals. However, genomic tools in oats were recently enriched by alternative sequencing works carried out by several groups.

A valuable example is the study of Bekele et al. (2018), which provides an in-depth evaluation of new software and approaches to genomic data analysis coupled with genotyping-by-sequencing (GBS) of a high number of accessions. A small and reproducible portion of the genome (<1%) was re-sequenced in the samples, the short sequences obtained were aligned and compared using the software "Haplotag". The software can identify single unique genetic markers from a duplicated genome even in absence of a fully-sequenced genome. Using such a strategy, the authors identified genomic regions that control heading date and the variants were used to predict the performance of oat lines.

In 2019, Maughan et al. released the fully annotated, chromosome-scale assemblies for *Avena atlantica* and *Avena eriantha*, which are the progenitor species of the A_s- and C_p-subgenomes, respectively. The diploid *Avena* species serve as important genetic resources for improving common oat's adaptive and food quality characteristics.

Very recently, a milestone in oat genomics has been reached with the sequencing of the hexaploid oat genome, cultivar Sang (Kamal et al. 2022). A fully annotated, high-quality reference genome of *A. sativa* and close relatives of its diploid (*Avena longiglumis*, AA, $2n = 14$) and tetraploid (*Avena insularis*, CCDD, $2n = 4x = 28$) progenitors have been produced and made publicly available. A mosaic structure of the oat genome has been found, as a result of large-scale genomic reorganizations in the polyploidization history of the plant. In the same study Kamal et al. (2022) analyzed gene families implicated in human health and nutrition. The results obtained suggested that the high content and quality of β-glucans in oat are linked to allelic variation and to transcription factor activities but not to copy number variation in the cellulose synthase superfamily genes as in other grasses. Oat storage proteins were found to be more similar with rice and dicotyledonous plants than with wheat and other gluten-rich cereals on the basis of their genomic organization, sequence characteristics, and expression patterns (Kamal et al. 2022), adding evidence that oat is safe in a gluten-free diet.

The PepsiCo released a v2 of OT3098 oat genome (*Avena sativa* – OT3098 v2, PepsiCo, https:// wheat.pw.usda.gov/jb?data=/ggds/oat-ot3098v2-pepsico" https://wheat.pw.usda.gov/jb/?data=%2Fggds%2Foat-ot3098v2-pepsico&loc=chr1A%3A216358827..324538239&tracks=&highlight=) and annotation work is currently running. The GFF, ORF nucleotide fasta, and ORF translated peptide fasta are available at the GrainGenes website (https://wheat.pw.usda.gov/jb/?data=%2Fggds%2Foat-ot3098v2-pepsico&loc=chr1A%3A216358827..324538239&tracks=&highlight=) Additionally, Dr. Nick Tinker has mapped relevant SNP markers to v2. No direct publication has been submitted for this genome assembly, and the authors opted for the Toronto Agreement, so researchers can freely use the data at the following link:

- Genome Browser: https://wheat.pw.usda.gov/jb?data=/ggds/oat-ot3098v2-pepsico
- BLAST: https://wheat.pw.usda.gov/blast/ (select "PepsiCo OT3098 Hexaploid Oat v2 pseudomolecules (2021)" under the "Oat Selections")
- Data Download: https://wheat.pw.usda.gov/GG3/content/pepsico-oat-oc3098-v2-files-2021

Peng et al. (2022) sequenced the genome of the *A. sativa* ssp. *nuda* Sanfensan, a hexaploid oat landrace that originated from the diversity center of hulless oat. Moreover, the reference genomes of its diploid and tetraploid ancestors have been released.

The next steps are the re-sequencing of a core collections, i.e., panels limited in number but representative of the genetic variability of the species, even if the current trend, at least in some crops such as rice, is to proceed with a re-sequencing of a wider number of accessions, included in the Gene Banks (Wambugu et al., 2018).

Among the first exploitations of the reference genomes is the work published by Khairullina et al. (2022) that focused on the characterization of UDP-glucosyltransferase (UGTs) genes involved in deoxynivalenol detoxification, a key resistance mechanism against Fusarium infection. Such genes have been identified in other small grain cereals, different from oats. Because the interactions between Fusaria and small grain cereals verisimilarly pre-dates the cereal speciation, the authors hypothesized that *UGT* orthologous genes have evolved in all cereals, including oats, to counteract the fungal virulence factor. Starting from the knowledge gained in other small grain cereals, *UGT* orthologous genes can therefore be individuated in oats. However, without a reference genome it is not trivial to individuate true orthologous *UGT* candidates among hundreds of highly homologous genes coming from three ancestral genomes in oat. The two high-resolution whole oat genome sequences made it possible to overcome the problem and to identify oat candidate genes involved in the conjugation of trichothecenes. The authors have identified the oat glucosyltransferase candidates, *AsUGT1* and *AsUGT2*, and verified and confirmed their trichothecene-detoxifying properties with transgenic experiments.

The next effort in oat genomics is the development of an oat Pan Genome, finalized to the discovery of complete set of genes, gene variants, and chromosome conformations in the species. To this aim, the PanOat project, coordinated by Martin Mascher, was activated by an international consortium, including more than 50 members. The consortium is sequencing and annotating about 30 oat genome reference sequences from diverse oat germplasms. The goals of PanOat include the characterization of genetic variants in core gene sets, including wild and cultivated accessions.

8.5 ANNOTATED REFERENCE GENOMES AND GENETIC RESOURCES EXPLOITATION

Re-sequencing in a representative or core collection is recognized as an effective strategy to evaluate the allelic richness at a given locus. This point has been clarified in several crops, in which the allelic isoforms of several genes associated with agronomic and qualitative traits have been mined.

Research on oats in this field is still in its infancy; however, the availability of reference genomes has recently empowered the allele mining approach.

The availability of an annotated reference genome amplifies the exploitation of genetic resources, including TILLING populations. As a proof-of-concept of this point Kamal et al. (2022) mapped the mutation responsible for the wax mutant *glossy.1*. The authors identified polymorphisms characteristic of the mutant and, after mapping to reference genomes, identified candidate genes hortologous to those cloned in other cereals and involved in wax synthesis. The authors concluded that the "*reference genome thus facilitated a major advance in understanding β-diketone biosynthesis in oat and can help breeders manipulate tissue-specific epicuticular wax composition in future oat cultivars adapted for hotter climates*".

8.6 GENETIC MAPPING AND GENOMICS INTEGRATION FOR ALLELE MINING

Several genes and quantitative trait loci (QTLs) have been identified in oats through genetic linkage mapping and genome-wide association (GWAS) approaches and several data deriving from such studies are publicly available in the literature and databases. Some efforts are now rising to realize the integration between the huge amount of genetic mapping data and the genomic reference sequences. Grain-Genes, a database for Triticeae and Avena, available at https://wheat.pw.usda.gov/GG3/ (accessed on 23 September 2022), is now steering the project to a more genome-centered approach. To this aim, GrainGenes is now curating tracks for GWAS results for the RefSeq browsers of the species.

Eighty-three oat QTLs, related to β-glucan content, were linked to the PepsiCo OT3098 v2 Hexaploid Oat (2021). In details, 56 QTLs from Fogarty et al. (2019) and 27 QTLs from Zimmer et al. (2020) were linked to such oat RefSeq, as the first step opening perspectives to allele mining.

8.7 CONCLUSIONS

Oats are known to fit into a healthy food diet and are also of great significance in circular agriculture and in sustainable crop production, however, it is a declining component of the more than 1400 million metric tons of mixed grains produced globally. The fragmentation of oats' production and utilization system has been indicated as a major limitation for this crop (Gilissen 2016). However, it deserves a role in food and feed production, especially since a transition is needed to sustainable agricultural practices and a healthier and more nutritional diet. Varietal innovation is a training factor in agriculture and allele mining is one of the pieces of the puzzle that supports breeding.

ACKNOWLEDGEMENTS

This study was funded by the Italian Ministry of Agriculture and Forestry in the DiBio–BIOPRIME project (Prot. 76381, MiPAAF PQAI I).

REFERENCES

Bekele WA, Wight CP, Chao S, Howarth CJ, Tinker NA (2018) Haplotype-based genotyping-by-sequencing in oat genome research. *Plant Biotechnol J* 16:1452–1463. https://doi.org/10.1111/pbi.12888

Bityutskii NP, Loskutov I, Yakkonen K, Konarev A, Shelenga T et al (2020) Screening of *Avena sativa* cultivars for iron, zinc, manganese, protein and oil content and fatty acid composition in whole grains. *Cereal Res Commun* 48:87–94. https://doi.org/10.1007/s42976-019-00002-2

Boczkowska M, Łapiński B, Kordulasińska I, Dostatny DF, Czembor JH (2016) Promoting the use of common oat genetic resources through diversity analysis and core collection construction. *PLoS One* 11(12):e0167855. https://doi.org/10.1371/journal.pone.0167855

Canales FJ, Montilla-Bascón G, Bekele WA, Howarth CJ, Langdon T, Rispail N et al (2021) Population genomics of Mediterranean oat (*A. sativa*) reveals high genetic diversity and three loci for heading date. *Theor Appl Genet* 134:2063–2077. https://doi.org/10.1007/s00122-021-03805-2

Carlson MO, Montilla-Bascon G, Hoekenga OA, Tinker NA, Poland J et al (2019) Multivariate genome-wide association analyses reveal the genetic basis of seed fatty acid composition in oat (*Avena sativa* L.). *G3-Genes Genom Genet* 9:2963–2975. https://doi.org/10.1534/g3.119.400228

Chawade A, Sikora P, Bräutigam M et al (2010) Development and characterization of an oat TILLING-population and identification of mutations in lignin and β-glucan biosynthesis genes. *BMC Plant Biol* 10:86. https://doi.org/10.1186/1471-2229-10-86

de Bruijn WJC, van Dinteren S, Gruppen H, Vincken JP (2019) Mass spectrometric characterization of avenanthramides and enhancing their production by germination of oat (*Avena sativa*). *Food Chem* 277:682–690. https://doi.org/10. 1016/j.foodchem.2018.11.013

de Carvalho MA, Bebeli PJ, Bettencourt E, Costa G, Dias S et al (2013) Cereal landraces genetic resources in worldwide GeneBanks. A review. *Agron Sustain Dev* 33:177–203. https://doi.org/10.1007/s13593-012-0090-0

Diederichsen A (2008) Assessments of genetic diversity within a world collection of cultivated hexaploid oat (*Avena sativa* L.) based on qualitative morphological characters. *Genet Resour Crop Evol* 55:419–440. https://doi.org/10.1007/s10722-007-9249-y

Dziubińska D, Bolc P, Kloc G, Podyma W, Boczkowska M (2022) A comprehensive dataset of *Avena sativa* L. landraces phenotypes and genotype. *Data Br* 41:107962. https://doi.org/10.1016/j.dib.2022.107962

Fogarty MC, Smith SM, Sheridan JL, et al. (2019). Identification of mixed linkage β-glucan quantitative trait loci and evaluation of CslF6 homoeologs in hexaploid oat. *Crop Sci* 60:914–933. https://doi.org/10.1002/csc2.20015

Gilissen L.(2016) *Trafoon TWS: From tradition to innovation in buckwheat, oats and gluten-free*. Warsaw, Poland, 29 June 2016.

Gokidi Y, Bhanu AN, Chandra K, Singh MN, Hemantaranjan A (2017) Allele mining–an approach to discover allelic variation in crops. *J Plant Sci Res* 33(2):167–180.

Hernandez-Hernandez O, Pereira-Caro G, Borges G, Crozier A, Olsson O (2021) Characterization and antioxidant activity of avenanthramides from selected oat lines developed by mutagenesis technique. *Food Chem* 343, 128408. https://doi.org/10.1016/j.foodchem.2022.132904

Jia J, Li H, Zhang X, Li Z, Qiu L (2017) Genomics-based plant germplasm research (GPGR). *Crop J* 5(2):166–174. https://doi.org/10.1016/j.cj.2016.10.006

Kamal N, TsardakasRenhuldt N, Bentzer J, Gundlach H, Haberer G et al. (2022) The mosaic oat genome gives insights into a uniquely healthy cereal crop. *Nature* 606:113–119. https://doi.org/10.1038/s41586-022-04732-y

Khairullina A, TsardakasRenhuldt N, Wiesenberger G, Bentzer J, Collinge DB et al (2022) Identification and functional characterisation of two Oat UDP-glucosyltransferases involved in deoxynivalenol detoxification. *Toxins* 14(7):446. https://doi.org/10.3390/toxins14070446

Klos EK, Huang YF, Bekele WA, Obert DE, Babiker E et al. (2016) Population genomics related to adaptation in elite oat germplasm. *Plant Genome* 9(2). https://doi.org/10.3835/plantgenome2015.10.0103

Kumari R, Kumar P, Sharma VK, Kumar H (2018) Allele mining for crop improvement. *Int J Pure Appl Biosci* 6:1456–1465. http://dx.doi.org/10.18782/2320-7051.6073

Loskutov IG, Khlestkina EK (2021) Wheat, barley, and oat breeding for health benefit components in grain. *Plant* 10(1):86. https://doi.org/10.3390/plants10010086

Maughan PJ, Lee R, Walstead R, Vickerstaff RJ, Fogarty MC et al (2019). Genomic insights from the first chromosome-scale assemblies of oat (*Avena* spp.) diploid species. *BMC Biol* 17:92. https://doi.org/10.1186/s12915-019-0712-y

McCallum CM, Comai L, Greene EA, Henikoff S (2000) Targeting Induced Local Lesions IN Genomes (TILLING) for plant functional genomics. *Plant Physiol* 123:439–442.

Park RF, Boshoff WHP, Cabral AL, Chong J, Martinelli JA, McMullen MS et al. (2022) Breeding oat for resistance to the crown rust pathogen *Puccinia coronata* f. sp. *avenae*: Achievements and prospects. *TheorAppl Genet* 135(11):3709–3734. doi: 10.1007/s00122-022-04121-z

Peng Y, Yan H, Guo L, Deng C, Wang C et al. (2022) Reference genome assemblies reveal the origin and evolution of allohexaploid oat. *Nat Genet* 54:1248–1258. https://doi.org/10.1038/s41588-022-01127-7

Schendel RR (2019) Phenol content in sprouted grains. In: Feng H, Nemzer B, DeVries J (eds) *Sprouted grains*, pp 247–315. Elsevier. https://doi.org/10.1016/ B978-0-12-811525-1.00010-5

Sikora P, Tosh SM, Brummer Y, Olsson O (2013) Identification of high β-glucan oat lines and localization and chemical characterization of their seed kernel β-glucans. *Food Chem* 137(1–4):83–91. https://doi.org/10.1016/j.foodchem.2012.10.007

Slade AJ, Fuerstenberg SI, Loeffler D, Steine MN, Facciotti D (2005) A reverse genetic, nontransgenic approach to wheat crop improvement by TILLING. *Nat Biotechnol* 23:75–81. https://doi.org/10.1038/nbt1043

Sunilkumar BA, Leonova S, Oste R, Olsson O (2017) Identification and characterization of high protein oat lines from a mutagenized oat population. *J Cereal Sci* 75: 100–107. http://dx.doi.org/10.1016/j.jcs.2017.03.003

Till BJ, Reynolds SH, Greene EA, Codomo CA, Enns LC, Johnson JE, Burtner C, Odden AR, Young K, Taylor NE, Henikoff JG, Comai L, Henikoff S (2003) Large-scale discovery of induced point mutations with high-throughput TILLING. *Genome Res* 13:524–530.

Tinker NA, Jackso EW, Snyder S, Tiede T, Azar C et al. (2016) Achievements and impact of the Collaborative Oat Research Enterprise (CORE). *Oat Newsl* 53:18.

Tumino G, Voorrips RE, Morcia C, Ghizzoni R, Germeier CU et al. (2017) Genome-wide association analysis for lodging tolerance and plant height in a diverse European hexaploid oat collection. *Euphytica* 213:163. https://doi.org/10.1007/s10681-017-1939-8

Tumino G, Voorrips RE, Rizza F, Badeck FW, Morcia C et al. (2016) Population structure and genome-wide association analysis for frost tolerance in oat using continuous SNP array signal intensity ratios. *Theor Appl Genet* 129:1711–1724. https://doi.org/10.1007/s00122-016-2734-y

Vavilov NI, Vavylov MI, Dorofeev VF (1992) *Origin and geography of cultivated plants.* Cambridge University Press.

Vivekanand V, Chawade A, Larsson M, Larsson A, Olsson O (2014) Identification and qualitative characterization of high and low lignin lines from an oat TILLING population. *Ind Crops Prod* 59:1–8. https://doi.org/10.1016/j.indcrop.2014.04.019

Wambugu PW, Ndjiondjop MN, Henry RJ (2018) Role of genomics in promoting the utilization of plant genetic resources in genebanks. *Brief Funct Genom* 17(3):198–206. https://doi.org/10.1093/bfgp/ely014

Yan H, Zhou P, Peng Y, Bekele WA, Ren C et al (2020) Genetic diversity and genome-wide association analysis in Chinese hulless oat germplasm. *Theor Appl Genet* 133:3365–3380. https://doi.org/10.1007/s00122-020-03674-1

Zimmer CM, McNish IG, Klos KE et al. (2020). Genome-wide association for β-glucan content, population structure, and linkage disequilibrium in elite oat germplasm adapted to subtropical environments. *Mol Breeding* 40:103. https://doi.org/10.1007/s11032-020-01182-0

9 Germplasm Resources for Allele Mining in Sorghum

Hugo E. Cuevas, Wilfred Vermerris, and Zhanguo Xin*
*Corresponding author: hugo.cuevas@usda.gov

9.1 INTRODUCTION

The domestication of plants is one of the most important achievements in human history that led to the establishment of agricultural systems (Purugganan and Fuller 2009). This evolutionary process driven by human selection is accomplished through the fixation of certain favorable alleles that provide benefits to agriculture but whose effects can be detrimental under natural conditions (Harlan 1992). During domestication a rapid divergence occurs between the crop and its wild ancestors, which is accompanied by a reduction in genetic diversity in the crop. Moreover, when the crop is adapted to other geographical regions (re-domestication), a second bottleneck occurs that drastically reduces the genetic diversity and increases the frequency of deleterious mutations (Gaut et al. 2018). This narrow genetic diversity in domesticated crops increases its vulnerability to abiotic and biotic factors, making it necessary to recuperate alleles lost during selection.

Sorghum [*Sorghum bicolor* (L.) Moench] is the fifth most important cereal in the world following wheat, maize, rice and barley. It is used as a source of food, feed, and fodder, and, more recently, is generating interest as a source of fermentable sugars that can be used for the production of renewable fuels and chemicals (Silva and Vermerris 2020; Umakanth et al. 2019). Sorghum was domesticated in the dry northeast region of Africa more than 5,000 years ago (Smith and Frederiksen 2000). From there it spread to western and southern regions of Africa, leading to the formation of five distinct races (Bicolor, Caudatum, Durra, Kafir, and Guinea) associated with distinct geographical regions. These re-domestication events resulted in a phenotypically highly diverse crop with multiple inflorescence types (Snowden 1936) with a broad genetic diversity for crop improvement. The adaptation of sorghum across environments involved the development of different maturity systems that increase its fitness in new habitats. Sorghum is a tropical short-day crop (<12 hours) that exhibits substantial photoperiod sensitivity, but during its migration to the temperate regions of southern Africa, mutants that flower independently of daylength (i.e., daylength-neutral) were selected (Morris et al. 2013a). This broadened adaptation permits sorghum production in more than 100 countries located in both tropical and temperate regions (FAOSTAT 2022). However, multiple research studies have demonstrated that plant breeding programs in tropical and temperate regions have a narrow genetic diversity that make the crop vulnerable against abiotic and biotic factors (Girma et al. 2020; Jordan et al. 2003; Murray et al. 2009). Moreover, sorghum cultivation for grain in temperate regions worldwide is limited to daylength-neutral varieties, while most of the crop genetic diversity resides in the tropical germplasm. Therefore, the integration and use of the genetic diversity of tropical germplasm in breeding programs in temperate regions has been one of the most important tasks of sorghum breeders.

DOI: 10.1201/9781003385004-9

Plant breeding programs are based on the effective use of genetic diversity to develop and improve crops (Swarup et al. 2021). *Ex situ* sorghum germplasm collections (i.e., gene banks) have been established in several countries as a resource for breeding programs. The largest worldwide sorghum collection is maintained by the United States Department of Agriculture-Agricultural Research Service (USDA-ARS) National Plant Germplasm Systems (NPGS) and includes >41,860 accessions from 114 countries. Another large sorghum collection, which includes 36,719 accessions, is located at the International Crops Research Institute for the Semi-arid Tropics (ICRISAT) in Hyderabad, India. These germplasm collections, in conjunction with other smaller germplasm collections located in China, Ethiopia, Australia, and Nigeria, capture important genetic diversity for crop improvement. Nonetheless, only a limited portion has been exploited in breeding programs. Certainly, the large size of these collections and the substantial photoperiod sensitivity of most of these accessions have limited their phenotypic characterization. Moreover, the low frequency of agriculturally beneficial alleles requires the use of multiple research strategies to uncover hidden genetic variation. The establishment of representative subsets (i.e., core collections) and the whole-genome characterization of accessions create germplasm and genomic resources for the scientific community that greatly enhance the screening efficiency of sorghum germplasm.

9.2 GERMPLASM RESOURCES FOR ALLELE MINING

9.2.1 Core Collections

A core collection is a representative subset of a large germplasm collection that captures most of the genetic diversity present in the larger germplasm collection (Brown 1989). Multiple methods have been developed to assemble core collections based on phenotypic and genotypic analyses and collection sites of the larger collection. The large size of both the ICRISAT and NPGS sorghum collection led to the establishment of two core collections (Dahlberg et al. 2004; Grenier et al. 2001).

The morpho-agronomic diversity of the ICRISAT collection was used to establish its core collection that includes 2,247 accessions (Grenier et al. 2001). The photoperiod response and race classification demarcated the accessions of this ICRISAT core into four main clusters. Nevertheless, 86.8% of the selected accessions fall in just two of those four clusters, and most accessions are photoperiod sensitive. The large size of this core, the small number of clusters, and the photoperiod sensitivity of the selected accessions have limited the strategic phenotypic evaluation. The subsequent evaluation and analysis of 11 qualitative and 10 quantitative traits divided the ICRISAT core collection into 21 clusters (Upadhyaya et al. 2009). A proportional sampling of ~10% of the accessions from each cluster led to the establishment of the ICRISAT mini core, a collection of 242 accessions. Certainly, the size of the ICRISAT mini core is suitable for extensive evaluation of economically important traits, particularly those with low heritability that require replicated trials. The genotyping-by-sequencing [GBS; (Elshire et al. 2011)] characterization of the ICRISAT mini core identified ~265,000 single nucleotide polymorphisms (SNPs) to facilitate the genomic dissection of economically important traits (Morris et al. 2013a).

The collection sites of the NPGS sorghum collection were used to assemble a core collection of 3,011 accessions (Dahlberg et al. 2004). However, the photoperiod-sensitivity of the accessions and large size of the NPGS core limited its immediate integration into sorghum breeding programs. A systematic phenotypic and GBS characterization of the NPGS core collection has been conducted using accessions from specific countries. To date, NPGS core sets from Ethiopia, Sudan, Yemen, West-Central Africa and sweet sorghum germplasm have been genotyped using GBS to examine their population structure and to establish strategies to preserve and utilize their genetic diversity in breeding programs (Cuevas and Prom 2020; Cuevas et al. 2017; Cuevas et al. 2023; Morales-Salva 2022). Recently, the entire NPGS collection from Niger (Maina et al. 2018), Nigeria (Olatoye et al. 2018), and Senegal (Faye et al. 2019) were genotyped using GBS to identify genomic regions associated with the agroclimatic zones of these countries. The population genetic analysis for each country demonstrates each one possesses plentiful genetic diversity for genome-wide association studies [GWAS; Botstein and Risch (2003)] and allele mining. Certainly, the genotypic characterization of these 3,401 tropical sorghum accessions generated an important genomic resource to uncover genetic variation and to establish an NPGS mini core collection for extensive evaluation of economically important traits.

9.2.2 Diversity Panels

The establishment of diversity panels for phenotyping and genetic analysis is the first step to broaden the genetic diversity of breeding programs (Zhu et al. 2008). A diversity panel is a group of cultivars or varieties that capture the largest genetic and phenotypic diversity for a particular application as the basis for GWAS. This is in contrast to the core collections, which are established to maximize the overall diversity of the germplasm collections (Nicolas et al. 2016). This distinction is particularly relevant for sorghum, because it is a multipurpose crop whose phenotypes (height, maturity, harvest index) vary considerably depending on the end use. Hence, diversity panels have been established to capture most of the diversity within each production group. The selection bias that was exerted during the assembly of the diversity panels led to different allele frequency distributions compared to the core collections. For example, the accessions in the sweet sorghum diversity panel (see Section 9.2.2.3) are on average taller than the accessions in the core collection or the sorghum association panel (see Section 9.2.2.1). A major benefit of the diversity panels is that once economically important alleles have been identified, they can be introgressed without the same degree of linkage drag of undesirable alleles present in the other sorghum production types, thereby leading to considerable time savings during the introgression process.

9.2.2.1 Sorghum Association Panel (SAP)

During a period of 25 years, the sorghum conversion program [SCP; Stephens et al. (1967)] exploited a backcross breeding scheme to introgress alleles conferring photoperiod insensitivity, early maturity and reduced height into approximately 1,100 sorghum accessions representing the breadth of the genetic diversity within NPGS. Certainly, these temperate-adapted lines have been the genetic base of the sorghum seed industry in the United States, Mexico, and southern Europe. The US sorghum association panel (SAP) was assembled using a selection of 377 accessions that represent the most important public and private breeding lines (Casa et al. 2008). This panel is mostly comprised of temperate-adapted lines developed through the SCP, but also contains historic breeding lines and some landraces. Population structure analyses revealed that the accessions comprising the SAP could be divided into four to nine clusters defined by sorghum races, but the Caudatum race is the most prevalent. The GBS characterization of the panel identified >250,000 SNPs that enabled the genetic dissection of several agronomically important traits (Morris et al. 2013a), some of which contributed to the domestication of sorghum. The subsequent whole-genome sequencing of the panel at an average coverage of 38× identified 43,983,694 variants including SNPs (~38 million), insertion-deletion (~5 million) and gene copy number variants (~170,000) (Boatwright et al. 2022a). The SAP has been used in multiple GWAS for plant height, drought tolerance, yield components, seed quality, resistance against stalk rot, anthracnose and grain mold, among others (Adeyanju et al. 2015; Boyles et al. 2016; Cuevas et al. 2018; 2019; Hayes et al. 2015; Morris et al. 2013a; 2013b; Rhodes et al. 2014). These studies have identified associations between genomic regions and particular traits, but typically only a limited portion of the observed phenotypic variation is explained by these regions. Therefore, a large portion of the genetic variation of the SAP is still undetected by GWAS, because the population structure and genetic relatedness among accessions are highly correlated to the observed phenotypic variation.

9.2.2.2 Bioenergy Association Panel (BAP)

Sweet and high-biomass sorghums serve as feedstocks for bioenergy and renewable chemicals. The SAP, however, does not capture the genetic diversity of these two types of sorghum. The importance and urgency to study traits relevant for bioenergy production led to the assembly of the Bioenergy Association Panel [BAP; (Brenton et al. 2016)]. This panel includes 152 sweet sorghum accessions and 238 high-biomass sorghums from the NPGS germplasm collection. The selection of these accessions was restricted to tall, photoperiod-sensitive, later maturing genotypes that displayed resistance to anthracnose in order to have the most relevant germplasm for bioenergy production. Indeed, the average flowering time, plant height, and vegetative biomass (dry weight) in the BAP is 2 to 3 times greater than observed for the SAP. The GBS characterization of the BAP identified 232,303 SNPs and highlighted genomic difference between the SAP and the BAP. Population structure analysis defined six clusters associated with sorghum race and geographical origin. The subsequent whole-genome sequencing of the BAP at an average coverage 15× identified ~13 million variants (Lozano et al. 2021). Certainly, the BAP is a critical germplasm and genomic resource for the study of bioenergy traits not found in grain sorghum types, but it is also valuable germplasm to study important agronomic traits relevant for all sorghum production types. For instance, a recent GWAS using both the SAP and BAP identified a gene encoding a WRKY transcription factor as a candidate gene for sugarcane aphid (*Melanaphis sacchari*) resistance (Poosapati et al. 2022).

9.2.2.3 Sweet Sorghum Diversity Panel (SWDP)

The NPGS is the primary source of genetic diversity for the development of new sweet sorghum varieties for biofuel production (Pederson and Spinks 2006). Approximately 2,180 accessions are classified as sweet sorghum based on historical and origin information, but upon closer examination of this collection it is evident that not all of these accessions are actual sweet sorghums. The evaluation of 1,801 accessions resulted in the identification of 737 accessions containing juice with soluble solids concentration values greater than 9.0 °Brix, determined with a refractometer to estimate the concentration of soluble sugars in the juice (Kawahigashi et al. 2013). A subset of 272 genetically and geographically diverse accessions were selected to establish a sweet sorghum diversity panel [SWDP; Cuevas et al. (2023)]. The panel contains germplasm from 19 countries and 135 accessions are advanced breeding materials, while only 21 accessions are shared with the BAP. The SWDP has been demonstrated to contain genes/alleles that are valuable for the production of commercially relevant sweet sorghums and it is a complementary germplasm resource to the BAP (Cuevas et al. 2023). The GBS characterization of the SWDP identified >132,000 SNPs suitable for GWAS. The genetic diversity of the SWDP is lower than what was observed for the SAP and BAP and it is comprised of four populations. Indeed, most of the elite germplasm is genetically highly related and distributed among three of the populations. Even though the genetic diversity of the SWDP is low, recent GWAS for anthracnose resistance and soluble sugar concentration have proven the panel useful for the genomic dissection of biofuel traits (Cuevas et al. 2023).

9.2.2.4 World Reference Panel

The world reference panels were assembled to represent the global sorghum diversity without the geographic bias or bias towards end use present in the other diversity panels (Morris et al. 2013a; Tao et al. 2020). The first sorghum world reference panel was assembled by using the 242 accessions of the ICRISAT mini core, the 377 accessions from the SAP, and another 342 accessions from ICRISAT germplasm core collection (Morris et al. 2013a). Later, a subset of 385 accessions from the SAP and the ICRISAT mini core were combined with 1,558 accessions from the larger ICRISAT collection to create a panel of 1,943 accessions that represent most of the worldwide geographic cultivation areas of the crop (Lasky et al. 2015). The GBS characterization of these accessions identified 404,627 SNPs of which 88% had a minor allele frequency <0.10. The vast genetic diversity of this large world reference panel has provided insights into the effect of the environment on the distribution of allelic variation and the evolution of economically important genes (Lasky et al. 2015). Recently, this world reference panel was refined to 767 accessions (377 from SAP and 390 from ICRISAT) that represent the five sorghum races and their intermediates (Muleta et al. 2022). Even though the genetic characterization of the world reference panel provides a formidable genomic resource for allele mining, its phenotyping is limited to tropical regions due to the large number of photoperiod-sensitive accessions from the ICRISAT germplasm collection.

A total of 1,176 temperate-adapted lines from the sorghum conversion program have been genetically characterized using different genotyping platforms (Hayes et al. 2015; Morris et al. 2013a; Tao et al. 2020; Thurber et al. 2013). The retrospective genomic analysis of the sorghum conversion program included the genotyping by sequencing of 579 temperate-adapted lines and their exotic parental lines (Thurber et al. 2013), and another 71 temperate-adapted lines that were included to perform a GWAS for dhurrin content (Hayes et al. 2015). The GBS characterization of these 650 temperate-adapted lines was performed using a double restriction enzyme digestion (*Pst*I/*Bfa*I), whereas the GBS characterization of the SAP was completed using one enzyme (*Ape*KI). Therefore, the 50,034 SNPs identified in these temperate-adapted lines could not be combined with those identified in the SAP. Recently, a diversity panel of 837 temperate-adapted lines were genotyped using the Diversity Arrays Technology [DArTseq; Jaccoud et al. (2001)] resulting in the identification of 111,089 SNPs (Tao et al. 2020). Nevertheless, this panel includes 225 lines from the SAP (Morris et al. 2013a) and 451 lines from the retrospective genomic analysis of the sorghum conversion program (Hayes et al. 2015; Thurber et al. 2013). Since several temperate-adapted lines have been genetically characterized using different methods, each subsequent GWAS involves different genomic regions.

9.2.3 Recombinant Populations

Even though GWAS have been proved to be useful for the genomic dissection of a variety of traits, they tend to explain a limited proportion of the observed phenotypic variation. Definitely, this type of analysis involves certain limitations, both in terms of statistical power (e.g., multiple testing problems, false discovery rates) and population genetics (e.g., population structure, familial relatedness, rare allele frequency). While the statistical limitations can be overcome with improved analysis (Benjamini and Hochberg 1995) and algorithms [e.g., TASSEL; Bradbury et al. (2007)], the limitations related to population genetics are difficult to resolve. Consequently, different approaches are needed, including the use of synthetic mapping populations that offer greater statistical power to detect small-effect loci with a lower false discovery rate (Yu et al. 2008). In sorghum, multiple mapping populations have been created to dissect economically important traits and to support GWAS based on diversity panels and core collections.

9.2.3.1 Nested Association Mapping

Nested association mapping (NAM) is based on a large mapping population that, due to its structure, integrates the advantages of linkage analysis and association mapping (Yu et al. 2008). NAM involves the development of a large set of related progenies derived from crossing a subset of highly genetically diverse lines with a common parental line. The subsequent genotyping of this large number of genetically related individuals and parental lines enables the dissection of complex quantitative traits without the limitations associated with GWAS, but the linkage disequilibrium (LD) in associated genomic regions is larger than typically observed in association mapping studies. Today, recombinant inbred lines (RILs) derived from F_2 populations or from the first backcross populations (BC) are the two most often used approaches for NAM. The NAM-RILs contain a larger number of recombination events and display transgressive segregation, whereas NAM-BC_1 populations produce lines most similar to the elite recurrent parent that can be included immediately into breeding programs.

9.2.3.1.1 Global NAM-RIL; RTx430

The global NAM population was designed using ten founder lines that represent the global genetic diversity present in the SAP (Bouchet et al. 2017; Perumal et al. 2021). These ten lines (Ajabsido, Macia, P898012, SC1103, SC1345, SC265, SC283, SC35, SC971, and Segaolane) were crossed with the elite inbred line RTx430 to generate 10 RIL populations with 199–232 individuals per population. This 2,214 RIL mapping population was genotyped using GBS (*Ape*KI) and showed the population harbors 57,411 recombination events with an average of two recombinations per Mb. Greenhouse and field screening of the 10 NAM founder lines has identified variation for heat stress, grain quality, and disease and pest resistance, among others (Cuevas et al. 2018; Grover et al. 2019; Impa et al. 2019; Sunoj et al. 2017). Certainly, the NAM-RTx430 is a powerful public germplasm and genomic resource for the genomic dissection of multiple traits.

9.2.3.1.2 Carbon-Partitioning NAM-RIL; Grassl

The carbon-partitioning NAM population was designed using 11 phenotypically and genetically diverse lines from the BAP that show a broad variance of their carbon-partitioning and bio-energy related traits (Boatwright et al. 2021; Kumar et al. 2022). These 11 lines (Chinese Amber, IS 2382, IS 13613, IS 13633, IS 11069, Mbonou, MA 38, AP79-714, Rio, Leoti, Pink Kafir) were crossed with Grassl, selected because it accumulates high levels of biomass and fermentable carbohydrates, and it is resistant to downy mildew (*Peronosclerospora sorghi*) and rust (*Puccinia purpurea*) (Kresovich et al. 1988). These crosses generated 11 RIL populations with 200–274 individuals per population for a total of 2,489 RILs. The genotypic characterization (GBS-*Pst*I/*Msp*I) of the complete NAM population identified 144,087 SNPs, with an average of one SNP per 5 kb and a genome-wide LD around ~100 kb. A subset of 1,201 RILs that represents the genetic diversity and most of the recombination events in the population was assembled to enhance the screening efficiency (Boatwright et al. 2022b). This NAM population provides a germplasm and genomic resource that is especially relevant for the dissection of bioenergy traits, but it could also be useful for dissecting agronomic traits shared by all sorghum types.

9.2.3.1.3 Chilling NAM-RIL; BTx623

Tolerance to chilling is an important trait that can prevent the death of seedlings in regions where spring temperatures can fluctuate. Chilling tolerance can extend the growing season, possibly enabling two successive crops. The chilling NAM population was developed to identify loci relevant for chilling tolerance (Marla et al. 2019). Three chilling-tolerant Chinese accessions [Niu Sheng Zui (PI 568016), Hong Ke Zi (PI 567946) and Kaoliang (PI 562744)] were crossed with the chilling-sensitive inbred line BTx623. The genetic diversity among these Chinese accessions is low and relates to durra sorghum from Asia and East Africa. However, this 771 RIL NAM population showed segregation for flowering time, plant height, disease susceptibility, and other agronomic traits. The genetic characterization (GBS-*Ape*KI) of this NAM population identified 43,320 SNPs and a genome-wide LD around ~500 kb. Even though the recurrent parent of this NAM population (BTx623) is different from the one used in the global NAM population [RTx430; Bouchet et al. (2017)], these two breeding lines showed genetic relatedness and both NAM populations could thus be integrated into a single analysis. Certainly, the chilling NAM population is a germplasm and genomic resource that provides additional genetic diversity for the identification of economically important alleles.

9.2.3.1.4 Australian NAM-BC_1; R931945-2-2

Fifty-six lines unadapted to Australian conditions were crossed with elite Australian cultivar R931945-2-2 (Jordan et al. 2011). The resulting F_1 plants were backcrossed to R931945-2-2 and segregating progenies were selected for agronomic traits (e.g., plant height, maturity, panicle shape, etc.). Selection and selfing was repeated four times to generate approximately 30–90 BC_1F_4 lines from each population. Therefore, each of these NAM-BC_1 populations contains a limited portion of the genetic diversity of the unadapted lines. Nevertheless, the integrated analysis using >1,300 BC_1F_4 lines from all populations provided enough statistical power to associate genomic regions to economically important traits (Mace et al. 2013a). Based on analysis of the whole-genome sequences of 44 of the NAM-BC_1 founder lines it was evident that much of the available genetic diversity is not used in breeding programs (Mace et al. 2013b). Remarkably, several of these founder lines are also present in the SAP and were also founder lines in the global NAM-RIL, providing a complementary germplasm and genomic resource that can benefit both breeding and research programs.

9.2.3.1.5 Crop Wild Relative NAM-BC_1; QL36 and Macia

As stated earlier, the genetic basis of commercial sorghum breeding programs is quite narrow. In light of novel uses of sorghum (e.g., renewable fuels and chemicals) and the need to adapt sorghum to the environmental conditions resulting from climate change (drought, heat, flooding, novel pathogens, and insect pests), it is important to broaden this genetic basis. Crop wild relatives (CWR) offer a valuable source of novel genetic variation for crop improvement (Hajjar and Hodgkin 2007). There are several challenges associated with this approach, including variation in flowering dynamics (timing; incompatibility) that can complicate cross-fertilization between the cultivated and wild species, low seed set and/or low seed number, spontaneous seed dispersal (shattering), and weedy characteristics (e.g., vegetative growth via rhizomes). As a consequence, the development of a CWR pre-breeding germplasm is necessary to identify desirable traits for use in breeding programs. Nine CWR of *Sorghum bicolor* belonging to (1) *S. bicolor* subsp. *verticilliflorum*, (2) *S. bicolor* subsp. *drummondii*, (3) *S. bicolor* subsp. *margaritiferum* were used to develop a novel panel intended to broaden the genetic diversity of breeding programs (Mace et al. 2021). These lines were developed by crossing the nine CWR with the adapted lines Macia or QL36, and the resulting F_1 progenies were either backcrossed to adapted lines or advanced to F_2 populations. During four generations of inbreeding (BC_1F_4 and $F_{2:5}$) progenies in both segregating populations were selected against shattering and photoperiod sensitivity. In total, 1,224 lines derived from 22 populations were developed, with most of the lines (>80%) derived from the backcross strategy. A subset of 1,155 lines from 15 populations were genotyped using DArTseq, resulting in the identification of 42,372 SNPs. Genomic analysis identified that the lines from the backcross and F_2 populations contained 25% and 48% of the CWR genome, respectively. Certainly, this CWR panel is an important germplasm resource that can broaden the genetic diversity of sorghum breeding programs and capture novel alleles that can be easily introgressed into elite sorghum germplasm.

9.2.3.1.6 Crop Wild Relative RIL Populations Derived from Crosses with S. propinquum and S. halepense

Recombinant inbred line (RIL) populations have also been developed from crossing cultivated *Sorghum bicolor* with the more distantly related wild sorghum species *S. propinquum* (Kunth) Hitchc., a perennial species native to South Asia, and *S. halepense* (L.) Pers., a noxious weed known as Johnsongrass that has spread to all six inhabited continents. The mapping populations derived from these wide crosses can assist in the dissection of domestication traits, including photoperiod sensitivity, plant height, seed yield, tillering, and rhizomatous growth. A better understanding of these traits, especially rhizomatousness, can aid attempts to develop perennial cultivated sorghum, which is of interest in areas where erosion control or nutrient management are a priority (Cox et al. 2018).

Chittenden et al. (1994) were the first to report an F_2 mapping population derived from a cross of inbred line BTx623 with an unnamed accession of *S. propinquum*. The main purpose of this population was to ensure there were enough polymorphisms to generate a map of the sorghum genome using the prevalent molecular markers at that time, restriction fragment length polymorphisms [RFLPs; Botstein et al. (1980)] and randomly amplified polymorphic DNA [RAPD; Williams et al. (1990)].

Using 370 F_2 individuals, Lin et al. (1995) mapped QTL related to plant height and flowering time. This population was subsequently advanced, and Washburn et al. (2013) used a population of 130 $F_{3:4}$ lines to map rhizomatousness and winter hardiness. The F_5 generation (161 RILs) was used to map flowering time at improved resolution (Kong et al. 2013) as well as rhizomatousness (Kong et al. 2015). An independent population of 191 $F_{3:5}$ RILs was developed by Govindarajulu et al. (2021) from an initial cross of an unnamed accession of *S. propinquum* with *S. bicolor* inbred line Tx7000. After whole-genome resequencing of the parent lines and RILs, a high-density SNP map was constructed to map 30 QTL for plant height, tiller number, stem diameter, flowering time, dry biomass, and aerial branch number. The QTL analysis was complemented by a transcriptome analysis that identified differentially expressed genes, some of which mapped to individual QTL. Variation in the expression of these genes may explain the observed phenotypic variation among the RILs, which forms the basis for follow-up studies, that can include introgression of specific alleles in cultivated sorghum. This population was also used by Hostetler et al. (2021) to map QTL for salinity tolerance, given that *S. bicolor* Tx7000 was highly tolerant to saline conditions, whereas *S. propinquum* was highly sensitive, based on the reduction in biomass following exposure to 75 mM NaCl for 12 weeks.

The development of mapping populations derived from crossing *S. bicolor* with *S. halepense* is more challenging, because the latter species is a tetraploid ($2n = 4x = 40$). Two strategies have been employed for making crosses between these species. One method is the generation of tetraploid *S. bicolor* with colchicine treatment. The tetraploid plant is then used as a female for crosses with *S. halepense* (Piper and Kulakow 1994). An alternative method is the use of cytoplasmic male-sterile *S. bicolor*, which generates a small proportion of unreduced female gametes, making these compatible with the pollen from tetraploid *S. halepense* (Hadley 1958). Kong et al. (2020) reported the development of two BC_1F_2 populations derived from backcrossing two F_1 hybrids (tetraploid *S. bicolor* BTx623 × *S. halepense* Gypsum 9E) to the tetraploid *S. bicolor* BTx623. These populations were used in a QTL study to map plant height and flowering time. The two populations showed substantial phenotypic differences, attributed to considerable genetic variation within *S. halepense*, possibly as a result of the genome duplication. The results suggest that *S. halepense* may, therefore, be of interest as a source of novel alleles for the improvement of cultivated sorghum (Kong et al. 2020).

9.2.3.2 Multi-Parent Advanced Generation Intercross

The crossing scheme design called multi-parent advanced generation intercross (MAGIC) is a powerful strategy that involves several rounds of recombination before inbreeding to promote LD decay and to increase mapping resolution (Mackay and Powell 2007). The first sorghum MAGIC population was developed by intercrossing 19 diverse founder lines that display broad phenotypic variability for economically important traits (Ongom and Ejeta 2018). These 19 founder lines were intercrossed with a gametic male sterile (GMS) population to produce 10 F_1 progenies per founder line (i.e., 190 F_1 progeny). The F_1 progenies were self-pollinated to produce F_2 populations segregating for GMS. These F_2 individuals were intermated randomly for nine generations by harvesting seeds from 1,000–1,500 male-sterile plants in each generation. In the final round of random mating 1,000 fertile plants were harvested and advanced for seven generations of self-pollination to produce inbred lines through the single-seed-descent method. The GBS characterization of a subset of 200 lines identified 79,728 SNPs, which were used to determine that the average LD of the population ranged from 40 to 220 kb. Moreover, the absence of population structure in this MAGIC population made it more suitable for GWAS, because the most powerful simple statistical models can be applied with a low false discovery rate. The high number of recombination events and allelic combinations present in the MAGIC population suggest the existence of multiple transgressive segregations for different economically important traits. Therefore, the phenotyping of the sorghum MAGIC population could result in the discovery of hidden genetic variation present in the 19 founder lines.

9.2.4 Mutagenized Populations

Targeting Induced Local Lesions IN Genomes (TILLING) is an important reverse genetics technique for allele mining through identification of a series of mutations in a gene of interest and associating these mutations with (consistent) phenotypic changes relative to the wild type (Gilchrist et al. 2006; McCallum et al. 2000; Till et al. 2003; Tsai et al. 2011; Wienholds et al. 2003; Winkler et al. 2005; Xin et al. 2008). Two sorghum TILLING populations have been developed by treating BTx623 seeds with the chemical mutagen ethylmethanesulfonate (EMS) followed by 4–5 generations of self-fertilization. The USDA-ARS in Lubbock, Texas, generated a mutant population of 1,600 lines (Xin et al. 2008). A second population of 568 lines was developed at Purdue University (West Lafayette, Indiana; Addo-Quaye et al., 2018).

The rapid development of NGS technologies with increased data output at a reduced cost enables whole-genome sequencing of large numbers of mutants to provide an online database for searching allelic series for genes of interest. Jiao et al. (2016) sequenced 256 mutant lines from the population generated by Xin et al. (2008) to an average genome coverage of 16×. This effort uncovered 1.8 million EMS-induced mutations (i.e., G→C or A→T transversions). The average mutation rate is 11.2/Mb, or about one SNP per 344 bp. The mutation frequency was consistent across all 10 chromosomes. About 236,000 (6.2 %) of the SNPs are located in 30,294 genes, which covers an estimated 92% of the genes in BTx623 genome (Paterson et al. 2009). A total of 111,850 non-synomynous SNPs are located in the exons of 25,605 (77.5%) genes, with an average of four mutations per gene. There were 8,043 potential knockout mutations resulting from the introduction of premature stop codons or changes to splice junctions. An additional 1,000 pedigreed mutant lines were sequenced and based on the primary analysis of the sequencing data, over 97% of the genes in sorghum genome were covered with on average >10 non-synonymous mutations (Zhanguo Xin, 2022, *personal communication*). More importantly, approximately 15,000 genes harbor at least one knockout mutation. Addo-Quaye et al. (2018) sequenced all 586 lines of their TILLING population to uncover the individual mutations. A total of 1,275,872 mutations were present in homozygous state and an additional 477,531 mutations were present in heterozygous state. In most cases mutations from the latter category will only result in observable phenotypes following an additional round of self-pollination to generate progenies that are homozygous for the mutant allele. The collection contains 56,514 homozygous missense mutations covering 23,227 genes. Among them, 4,035 are high-impact mutations (i.e., predicted to disrupt protein function) in 3,637 genes. These resources provide vital resources for allele mining and for validation of candidate genes identified by QTL mapping through linkage analysis and GWAS.

The mutant alleles can be searched on Gramene (www.gramene.org/) and sorghumbase (www.sorghumbase.org). Access to the mutants in the USDA-ARS collection can be provided through material transfer agreements. Small quantities (ten seeds) of the mutants in the population generated by Addo-Quaye et al. (2018) can be requested online from the USDA-ARS Germplasm Resource Information Network (GRIN; www.ars-grin.gov/) in Tifton, Georgia.

9.3 FUTURE PERSPECTIVES

Over the last 15 years great progress has been made in the development of germplasm and genomic resources for sorghum breeding programs (Figure 9.1). The sorghum reference genome (Paterson et al. 2009), the subsequent resequencing of additional sorghum genotypes (Cooper et al. 2019) and the availability of low-cost genotyping methods using NGS have enhanced the resources for allele mining. Indeed, these new resources provide unique opportunities to increase the genetic diversity of breeding trough phenomics and genomics research projects. To determine the actual benefits of these germplasm resources, it will be necessary to monitor the allelic diversity in breeding programs and ideally see overall genetic diversity increase over the next few years. The genomic dissection

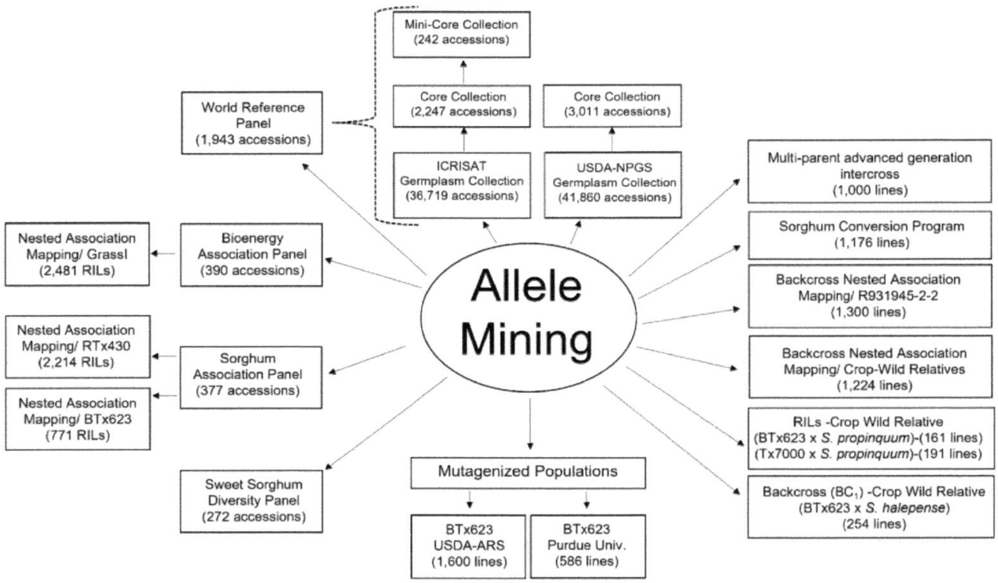

FIGURE 9.1 Germplasm resources for allele mining in sorghum.

of economically important traits and functional analysis of genes needs to be followed by the introgressions of alleles into new varieties.

The development of core collections for the different worldwide germplasm collections is the first step to organize the genetic diversity of the crop. The large size of most of the current germplasm collections imply the existence of redundant germplasm within and amongst collections. In this regard, the genetic characterization of multiple core collections using the same platform (i.e., GBS following restriction enzyme digest of the genomic DNA with *Ape*KI) could be used to compare the genetic diversity and redundancy amongst collections. The establishment of a unified core collection that represents the worldwide diversity of germplasm collections could become an important common resource for broadening the genetic base of breeding programs across the globe. For instance, the ICRISAT and USDA-ARS core collections could be subjected to a single analysis to determine their genetic relationships and to establish an initial unified core collection. The subsequent addition of other smaller germplasm collections could be used to expand or cover diversity gaps until a truly representative worldwide core collection is established. The development of such a common worldwide germplasm resource will require germplasm exchange agreements, which can be facilitated by the International Treaty on Plant Genetic Resources for Food and Agriculture spearheaded by the Food and Agriculture Organization (FAO). Understanding the genetic diversity across germplasm collections is vital to preserving the most important germplasm, and, therefore, worth pursuing.

The screening of new diversity panels should be the focus of research projects during the next few years. The SAP has been extensively phenotyped during the past 10 years for abiotic and biotic stresses (Adeyanju et al. 2015; Boyles et al. 2016; Cuevas et al. 2018; 2019; Hayes et al. 2015; Morris et al. 2013a; 2013b; Rhodes et al. 2014). Nevertheless, most of these GWAS studies have not continued to the allele introgression phase, which has made it difficult to assess their impact on breeding programs. The mode of inheritance and the linkage disequilibrium of these newly discovered loci with unfavorable alleles need to be addressed to determine their breeding values. The most recent diversity panels, such as the BAP and SWDP, provide additional germplasm for screening that could complement the SAP. Surely, diversity panels contain hidden genetic variation that needs to be uncovered using various methods. For instance, the whole-genome sequencing of the SAP (Boatwright et al. 2022a) identified vast genetic diversity that could be used to select and phenotype accessions harboring specific allelic variants. The development and use of precise phenotyping systems might also be also necessary to uncover alleles with smaller additive effects. The 1,039 accessions present among diversity panels could broaden the genetic diversity of breeding programs if multiple phenomics projects are utilized to uncover most of the identified genetic variation.

The effective use of the genetic diversity in plant breeding programs could be limited by a low number of recombination events in breeding populations (Taagen et al. 2020). The NAM RIL populations, backcross NAM, and the MAGIC populations comprise a total of 7,227 lines that could be used as parental germplasm for breeding programs. Even though most of the research efforts using these recombinant populations are directed towards uncovering the limitations of the GWAS, the existence of transgressive segregants provides unique opportunities for breeding programs. The phenotyping of these populations is imperative for identifying the most valuable germplasm for breeding programs. The phenotypic characterization of these lines across environments, the proper (preferably quantitative) documentation of the phenotypes, and the easy accessibility of the data are needed to integrate this germplasm in breeding programs. Likewise, the development of new mapping populations should be complementary to these populations to increase the statistical power and mapping resolution.

The genetic diversity of sorghum is now accessible for research and breeding programs. Genomic studies have provided an understanding of the population structure of the germplasm and shed light on how this diversity could be used for crop improvement. The large-scale phenotyping of the available germplasm is necessary for allele mining, particularly for traits with low heritability. In this regard, the sorghum research community might prioritize economically important traits and use common screening methods that ensure repeatability across research programs. Together, this germplasm should be exploited to safeguard the genetic diversity of cultivated sorghum and ensure that the challenges of the future can be met effectively and efficiently.

REFERENCES

Addo-Quaye C, Tuinstra M, Carraro N, Weil C, Dilkes BP (2018) Whole-genome sequence accuracy is improved by replication in a population of mutagenized sorghum. *G3: Genes, Genomes, Genet* 8:1079–1094.

Adeyanju A, Little C, Yu JM, Tesso T (2015) Genome-wide association study on resistance to stalk rot diseases in grain sorghum. *G3: Genes, Genome, Genet* 5:1165–1175.

Benjamini Y, Hochberg Y (1995) Controlling the false discovery rate – a practical and powerful approach to multiple testing. *J Roy Stat Soc B Met* 57:289–300.

Boatwright JL, Brenton ZW, Boyles RE, Sapkota S, Myers MT, Jordan KE, Dale SM, Shakoor N, Cooper EA, Morris GP (2021) Genetic characterization of a *Sorghum bicolor* multiparent mapping population emphasizing carbon-partitioning dynamics. *G3: Genes, Genomes, Genet* 11:jkab060.

Boatwright JL, Sapkota S, Jin H, Schnable JG, Brenton Z, Boyles RE, Kresovich S (2022a) Sorghum association panel whole-genome sequencing establishes cornerstone resource for dissecting genomic diversity. *Plant* 111:888–904.

Boatwright JL, Sapkota S, Myers M, Kumar N, Cox A, Jordan KE, Kresovich S (2022b) Dissecting the genetic architecture of carbon partitioning in sorghum using multiscale phenotypes. *Front Plant Sci* 13.

Botstein D, Risch N (2003) Discovering genotypes underlying human phenotypes: Past successes for mendelian disease, future approaches for complex disease. *Nat Genet* 33:228–237.

Botstein D, White RL, Skolnick M, Davis RW (1980) Construction of a genetic linkage map in man using restriction fragment length polymorphisms. *Am J Hum Genet* 32:314–331.

Bouchet S, Olatoye MO, Marla SR, Perumal R, Tesso T, Yu J, Tuinstra M, Morris GP (2017) Increased power to dissect adaptive traits in global sorghum diversity using a nested association mapping population. *Genetics* 206:573–585.

Boyles RE, Cooper EA, Myers MT, Brenton Z, Rauh BL, Morris GP, Kresovich S (2016) Genome-wide association studies of grain yield components in diverse sorghum germplasm. *Plant Genome* 9.

Bradbury PJ, Zhang Z, Kroon DE, Casstevens TM, Ramdoss Y, Buckler ES (2007) TASSEL: Software for association mapping of complex traits in diverse samples. *Bioinformatics* 23:2633–2635.

Brenton ZW, Cooper EA, Myers MT, Boyles RE, Shakoor N, Zielinski KJ, Rauh BL, Bridges WC, Morris GP, Kresovich S (2016) A genomic resource for the development, improvement, and exploitation of sorghum for bioenergy. *Genetics* 204:21–33.

Brown AHD (1989) Core collections: A practical approach to genetic resources management. *Genome* 31:818–824.

Casa AM, Pressoir G, Brown PJ, Mitchell SE, Rooney WL, Tuinstra MR, Franks CD, Kresovich S (2008) Community resources and strategies for association mapping in sorghum. *Crop Sci* 48:30–40.

Chittenden L, Schertz K, Lin Y, Wing RA, Paterson A (1994) A detailed RFLP map of *Sorghum bicolor* x *S. propinquum*, suitable for high-density mapping, suggests ancestral duplication of sorghum chromosomes or chromosomal segments. *Theor Appl Genet* 87:925–933.

Cooper EA, Brenton ZW, Flinn BS, Jenkins J, Shu S, Flowers D, Luo F, Wang Y, Xia P, Barry K, Duan C, Lipzen A, Yoshinaga Y, Schmutz J, Saski C, Vermeris W, Kresovich S (2019) A new reference genome for *Sorghum bicolor* reveals high levels of sequence similarity between sweet and grain genotypes: implications for the genetics of sugar metabolism. *BMC Genomics* 20:1–13.

Cox S, Nabukalu P, Paterson AH, Kong W, Nakasagga S (2018) Development of perennial grain sorghum. *Sustainability* 10:172.

Cuevas HE, Fermin-Perez, RA, Prom LK, Cooper EA, Bean S, Rooney WL (2019). Genome-wide association mapping of the grain mold resistance in the US sorghum association panel. *Plant Genome* 12(2):180070.

Cuevas HE, Knoll JE, Prom LK, Stutts LR, Vermerris W (2023). Genetic diversity, population structure and anthracnose resistance in a novel sweet sorghum diversity panel. *Front Plant Sci* 14:1249555.

Cuevas HE, Prom LK (2020) Evaluation of genetic diversity, agronomic traits, and anthracnose resistance in the NPGS Sudan sorghum core collection. *BMC Genomics* 21:88.

Cuevas HE, Prom LK, Cooper EA, Knoll JE, Ni X (2018) Genome-wide association mapping of anthracnose (*Colletotrichum sublineolum*) resistance in the US sorghum association panel. *Plant Genome* 11:170099

Cuevas HE, Rosa-Valentin G, Hayes CM, Rooney LW, Hoffmann L (2017) Genomic characterization of a core set of the USDA-NPGS Ethiopian sorghum germplasm collection: implications for germplasm conservation, evaluation, and utilization in crop improvement. *BMC Genomics* 18:108.

Dahlberg JA, Burke JJ, Rosenow DT (2004) Development of a sorghum core collection: Refinement and evaluation of a subset from Sudan. *Econ Bot* 58:556–567.

Elshire RJ, Glaubitz JC, Sun Q, Poland JA, Kawamoto K, Buckler ES, Mitchell SE (2011) A robust, simple genotyping-by-sequencing (GBS) approach for high diversity species. *Plos One* 6:e19379.

FAOSTAT (2022) Food and agriculture organization. 2022 edn. *FAOSTAT*, http://faostat.fao.org/site/339/default.aspx

Faye JM, Maina F, Hu Z, Fonceka D, Cisse N, Morris GP (2019) Genomic signatures of adaptation to Sahelian and Soudanian climates in sorghum landraces of Senegal. *Ecol Evolution* 9:6038–6051.

Gaut BS, Seymour DK, Liu QP, Zhou YF (2018) Demography and its effects on genomic variation in crop domestication. *Nat Plants* 4:512–520.

Gilchrist EJ, O'Neil NJ, Rose AM, Zetka MC, Haughn GW (2006) TILLING is an effective reverse genetics technique for Caenorhabditis elegans. *BMC Genomics* 7:1–16.

Girma G, Nida H, Tirfessa A, Lule D, Bejiga T, Seyoum A, Mekonen M, Nega A, Dessalegn K, Birhanu C, Bekele A, Gebreyohannes A, Ayana G, Tesso T, Ejeta G, Mengiste T (2020) A comprehensive phenotypic and genomic characterization of Ethiopian sorghum germplasm defines core collection and reveals rich genetic potential in adaptive traits. *Plant Genome* 13(3):e20055.

Govindarajulu R, Hostetler AN, Xiao Y, Chaluvadi SR, Mauro-Herrera M, Siddoway ML, Whipple C, Bennetzen JL, Devos KM, Doust AN (2021) Integration of high-density genetic mapping with transcriptome analysis uncovers numerous agronomic QTL and reveals candidate genes for the control of tillering in sorghum. *G3: Genes, Genomes, Genet* 11:jkab024.

Grenier C, Bramel-Cox PJ, Hamon P (2001) Core collection of sorghum: I. Stratification based on eco-geographical data. *Crop Sci* 41:234–240.

Grover S, Wojahn B, Varsani S, Sattler SE, Louis J (2019) Resistance to greenbugs in the sorghum nested association mapping population. *Arthropod-Plant Interact* 13:261–269.

Hadley HH (1958) Chromosome numbers, fertility and rhizome expression of hybrids between grain sorghum and Johnsongrass 1. *Agronomy J* 50:278–282.

Hajjar R, Hodgkin T (2007) The use of wild relatives in crop improvement: A survey of developments over the last 20 years. *Euphytica* 156:1–13.

Harlan JR (1992) *Crops and man*. American Society of Agronomy, Madison, WI, USA.

Hayes CM, Burow GB, Brown PJ, Thurber C, Xin ZG, Burke JJ (2015) Natural variation in synthesis and catabolism genes influences dhurrin content in sorghum. *Plant Genome* 8(2):plantgenome2014.09.0048.

Hostetler AN, Govindarajulu R, Hawkins JS (2021) QTL mapping in an interspecific sorghum population uncovers candidate regulators of salinity tolerance. *Plant Stress* 2:100024.

Impa S, Perumal R, Bean SR, Sunoj VJ, Jagadish SK (2019) Water deficit and heat stress induced alterations in grain physico-chemical characteristics and micronutrient composition in field grown grain sorghum. *J Cereal Sci* 86:124–131.

Jaccoud D, Peng K, Feinstein D, Kilian A (2001) Diversity arrays: A solid state technology for sequence information independent genotyping. *Nucleic Acids Res* 29:e25.

Jiao Y, Burke J, Chopra R, Burow G, Chen J, Wang B, Hayes C, Emendack Y, Ware D, Xin Z (2016) A sorghum mutant resource as an efficient platform for gene discovery in grasses. *Plant Cell* 28:1551–1562.

Jordan D, Mace E, Cruickshank A, Hunt C, Henzell R (2011) Exploring and exploiting genetic variation from unadapted sorghum germplasm in a breeding program. *Crop Sci* 51:1444–1457.

Jordan DR, Tao Y, Godwin ID, Henzell RG, Cooper M, McIntyre CL (2003) Prediction of hybrid performance in grain sorghum using RFLP markers. *Theor Appl Genet* 106:559–567.

Kawahigashi H, Kasuga S, Okuizumi H, Hiradate S, Yonemaru Ji (2013) Evaluation of Brix and sugar content in stem juice from sorghum varieties. *Grassl Sci* 59:11–19.

Kong W, Jin H, Franks CD, Kim C, Bandopadhyay R, Rana MK, Auckland SA, Goff VH, Rainville LK, Burow GB (2013) Genetic analysis of recombinant inbred lines for *Sorghum bicolor* × *Sorghum propinquum*. *G3: Genes, Genomes, Genet* 3:101–108.

Kong W, Kim C, Goff VH, Zhang D, Paterson AH (2015) Genetic analysis of rhizomatousness and its relationship with vegetative branching of recombinant inbred lines of *Sorghum bicolor* × *S. propinquum*. *Am J Bot* 102:718–724.

Kong W, Nabukalu P, Cox TS, Goff VH, Pierce GJ, Lemke C, Robertson JS, Compton R, Tang H, Paterson AH (2020) Transmission genetics of a *Sorghum bicolor* × *S. halepense* backcross populations. *Front Plant Sci* 11:467.

Kresovich S, Miller F, Monk R, Dominy R, Broadhead D (1988) Registration of 'Grassl' sweet sorghum. *Crop Sci* 28:194–195.

Kumar N, Brenton Z, Myers MT, Boyles RE, Sapkota S, Boatwright J, Cox W, Jordan KE, Kresovich S (2022) Registration of the sorghum carbon-partitioning nested association mapping (CP-NAM) population. *J Plant Regist* 16:656–663.

Lasky JR, Upadhyaya HD, Ramu P, Deshpande S, Hash CT, Bonnette J, Juenger TE, Hyma K, Acharya C, Mitchell SE, Buckler ES, Brenton Z, Kresovich S, Morris GP (2015) Genome-environment associations in sorghum landraces predict adaptive traits. *Sci Adv* 1:e1400218.

Lin Y-R, Schertz KF, Paterson AH (1995) Comparative analysis of QTLs affecting plant height and maturity across the Poaceae, in reference to an interspecific sorghum population. *Genetics* 141:391–411.

Lozano R, Gazave E, Dos Santos JP, Stetter MG, Valluru R, Bandillo N, Fernandes SB, Brown PJ, Shakoor N, Mockler TC (2021) Comparative evolutionary genetics of deleterious load in sorghum and maize. *Nat Plants* 7:17–24.

Mace E, Hunt C, Jordan D (2013a) Supermodels: Sorghum and maize provide mutual insight into the genetics of flowering time. *Theor Appl Genet* 126:1377–1395.

Mace ES, Cruickshank AW, Tao Y, Hunt CH, Jordan DR (2021) A global resource for exploring and exploiting genetic variation in sorghum crop wild relatives. *Crop Sci* 61:150–162.

Mace ES, Tai S, Gilding EK, Li Y, Prentis PJ, Bian L, Campbell BC, Hu W, Innes DJ, Han X (2013b) Whole-genome sequencing reveals untapped genetic potential in Africa's indigenous cereal crop sorghum. *Nat Commun* 4:1–9.

Mackay I, Powell W (2007) Methods for linkage disequilibrium mapping in crops. *Trends Plant Sci* 12:57–63.

Maina F, Bouchet S, Marla SR, Hu Z, Wang J, Mamadou A, Abdou M, Saïdou A-A, Morris GP (2018) Population genomics of sorghum (*Sorghum bicolor*) across diverse agroclimatic zones of Niger. *Genome* 61:223–232.

Marla SR, Burow G, Chopra R, Hayes C, Olatoye MO, Felderhoff T, Hu ZB, Raymundo R, Perumal R, Morris GP (2019) Genetic architecture of chilling tolerance in sorghum dissected with a nested association mapping population. *G3: Genes, Genom, Genet* 9:4045–4057.

McCallum CM, Comai L, Greene EA, Henikoff S (2000) Targeted screening for induced mutations. *Nat Biotechnol* 18:455–457.

Morales-Salva A (2022) *Genome-wide association analysis of anthracnose resistance in NPGS Yemen sorghum [Sorghum bicolor (L.) Moench] germplasm.* Department of Biology. University of Puerto Rico-Mayaguez Campus, Mayaguez.

Morris GP, Ramu P, Deshpande SP, Hash CT, Shah T, Upadhyaya HD, Riera-Lizarazu O, Brown PJ, Acharya CB, Mitchell SE, Harriman J, Glaubitz JC, Buckler ES, Kresovich S (2013a) Population genomic and genome-wide association studies of agroclimatic traits in sorghum. *Proc Natl Acad Sci USA* 110:453–458.

Morris GP, Rhodes DH, Brenton Z, Ramu P, Thayil VM, Deshpande S, Hash CT, Acharya C, Mitchell SE, Buckler ES, Yu JM, Kresovich S (2013b) Dissecting genome-wide association signals for loss-of-function phenotypes in sorghum flavonoid pigmentation traits. *G3: Genes, Genom, Genet* 3:2085–2094.

Muleta KT, Felderhoff T, Winans N, Walstead R, Charles JR, Armstrong JS, Mamidi S, Plott C, Vogel JP, Lemaux PG, Mockler TC, Grimwood J, Schmutz J, Pressoir G, Morris GP (2022) The recent evolutionary rescue of a staple crop depended on over half a century of global germplasm exchange. *Sci Adv* 8:eabj4633.

Murray SC, Rooney WL, Hamblin MT, Mitchell SE, Kresovich S (2009) Sweet sorghum genetic diversity and association mapping for Brix and height. *Plant Genome* 2:48–62.

Nicolas SD, Peros JP, Lacombe T, Launay A, Le Paslier MC, Berard A, Mangin B, Valiere S, Martins F, Le Cunff L, Laucou V, Bacilieri R, Dereeper A, Chatelet P, This P, Doligez A (2016) Genetic diversity, linkage disequilibrium and power of a large grapevine (*Vitis vinifera* L.) diversity panel newly designed for association studies. *BMC Plant Biol* 16:74.

Olatoye MO, Hu ZB, Maina F, Morris GP (2018) Genomic signatures of adaptation to a precipitation gradient in nigerian sorghum. *G3: Genes, Genom, Genet* 8:3269–3281.

Ongom PO, Ejeta G (2018) Mating design and genetic structure of a multi-parent advanced generation intercross (MAGIC) population of sorghum [*Sorghum bicolor* (L.) Moench]. *G3: Genes, Genom, Genet* 8:331–341.

Paterson AH, Bowers JE, Bruggmann R, Dubchak I, Grimwood J, Gundlach H, Haberer G, Hellsten U, Mitros T, Poliakov A, Schmutz J, Spannagl M, Tang HB, Wang XY, Wicker T, Bharti AK, Chapman J, Feltus FA, Gowik U, Grigoriev IV, Lyons E, Maher CA, Martis M, Narechania A, Otillar RP, Penning BW, Salamov AA, Wang Y, Zhang LF, Carpita NC, Freeling M, Gingle AR, Hash CT, Keller B, Klein P, Kresovich S, McCann MC, Ming R, Peterson DG, Mehboob-ur-Rahman, Ware D, Westhoff P, Mayer KFX, Messing J, Rokhsar DS (2009) The S*orghum bicolor* genome and the diversification of grasses. *Nature* 457:551–556.

Pederson GA, Spinks M (2006) *Utilization of old data to improve germplasm documentation: Sweet sorghum collection*, pp 1232b. ASA-CSSA-SSSA Annual Meeting, Indianapolis, IN.

Perumal R, Tesso TT, Morris GP, Jagadish SK, Little CR, Bean SR, Yu J, Prasad PV, Tuinstra MR (2021) Registration of the sorghum nested association mapping (NAM) population in RTx430 background. *J Plant Regist* 15:395–402.

Piper JK, Kulakow PA (1994) Seed yield and biomass allocation in *Sorghum bicolor* and F1 and backcross generations of *S. bicolor* × *S. halepense* hybrids. *Canadian J Bot* 72:468–474.

Poosapati S, Poretsky E, Dressano K, Ruiz M, Vazquez A, Sandoval E, Estrada-Cardenas A, Duggal S, Lim JH, Morris G, Szczepaniec A, Walse SS, Ni XZ, Schmelz EA, Huffaker A (2022) A sorghum genome-wide association study (GWAS) identifies a WRKY transcription factor as a candidate gene underlying sugarcane aphid (*Melanaphis sacchari*) resistance. *Planta* 255:37.

Purugganan MD, Fuller DQ (2009) The nature of selection during plant domestication. *Nature* 457:843–848.

Rhodes DH, Hoffmann L, Rooney WL, Ramu P, Morris GP, Kresovich S (2014) Genome-wide association study of grain polyphenol concentrations in global sorghum [*Sorghum bicolor (L.) Moench*] germplasm. *J Agric Food Chem* 62:10916–10927.

Silva TN, Vermerris W (2020) High-biomass sorghums as a feedstock for renewable fuels and chemicals. In Talwar HS and VA Tonapi (eds.), *Sorghum in the 21st century: Food–fodder–feed–fuel for a rapidly changing world*, pp: 723–754. Springer, Singapore.

Smith CW, Frederiksen RA (2000) *Sorghum; Origin, history, technology and production*. Wiley, New York.

Snowden JD (1936) *The cultivated races of sorghum*. Adlard and Son, London, England.

Stephens JC, Miller FR, Rosenow DT (1967) Conversion of alien sorghums to early combine genotypes. *Crop Sci* 7:396.

Sunoj VJ, Somayanda IM, Chiluwal A, Perumal R, Prasad PV, Jagadish SK (2017) Resilience of pollen and post-flowering response in diverse sorghum genotypes exposed to heat stress under field conditions. *Crop Sci* 57:1658–1669.

Swarup S, Cargill EJ, Crosby K, Flagel L, Kniskern J, Glenn KC (2021) Genetic diversity is indispensable for plant breeding to improve crops. *Crop Sci* 61:839–852.

Taagen E, Bogdanove AJ, Sorrells ME (2020) Counting on crossovers: Controlled recombination for plant breeding. *Trends Plant Sci* 25:455–465.

Tao YF, Zhao XR, Wang XM, Hathorn A, Hunt C, Cruickshank AW, van Oosterom EJ, Godwin ID, Mace ES, Jordan DR (2020) Large-scale GWAS in sorghum reveals common genetic control of grain size among cereals. *Plant Biotechnoly* 18:1093–1105.

Thurber CS, Ma JM, Higgins RH, Brown PJ (2013) Retrospective genomic analysis of sorghum adaptation to temperate-zone grain production. *Genome Biol* 14:R68.

Till BJ, Colbert T, Tompa R, Enns LC, Codomo CA, Johnson JE, Reynolds SH, Henikoff JG, Greene EA, Steine MN (2003) *High-throughput TILLING for functional genomics. Plant functional genomics*, pp 205–220. Springer.

Tsai H, Howell T, Nitcher R, Missirian V, Watson B, Ngo KJ, Lieberman M, Fass J, Uauy C, Tran RK (2011) Discovery of rare mutations in populations: TILLING by sequencing. *Plant Physiol* 156:1257–1268.

Umakanth A, Kumar AA, Vermerris W, Tonapi V (2019) Sweet sorghum for biofuel industry. In Aruna C, KBRS Visarada, BV Bhat, VA Tonapi (eds.), *Breeding sorghum for diverse end uses*, pp 255–270. Woodhead Publishing, Cambridge, UK.

Upadhyaya HD, Pundir RPS, Dwivedi SL, Gowda CLL, Reddy VG, Singh S (2009) Developing a mini core collection of sorghum for diversified utilization of germplasm. *Crop Sci* 49:1769–1780.

Washburn JD, Murray SC, Burson BL, Klein RR, Jessup RW (2013) Targeted mapping of quantitative trait locus regions for rhizomatousness in chromosome SBI-01 and analysis of overwintering in a *Sorghum bicolor* × *S. propinquum* population. *Mol Breeding* 31:153–162.

Wienholds E, Van Eeden F, Kosters M, Mudde J, Plasterk RH, Cuppen E (2003) Efficient target-selected mutagenesis in zebrafish. *Genome Res* 13:2700–2707.

Williams JGK, Kubelik AR, Livak KJ, Rafalski JA, Tingey SV (1990) DNA polymorphisms amplified by arbitrary primers are useful as genetic-markers. *Nucleic Acids Res* 18:6531–6535.

Winkler S, Schwabedissen A, Backasch D, Bökel C, Seidel C, Bönisch S, Fürthauer M, Kuhrs A, Cobreros L, Brand M (2005) Target-selected mutant screen by TILLING in *Drosophila*. *Genome Res* 15:718–723.

Xin Z, Li Wang M, Barkley NA, Burow G, Franks C, Pederson G, Burke J (2008) Applying genotyping (TILLING) and phenotyping analyses to elucidate gene function in a chemically induced sorghum mutant population. *BMC Plant Biol* 8:1–14.

Yu JM, Holland JB, McMullen MD, Buckler ES (2008) Genetic design and statistical power of nested association mapping in maize. *Genetics* 178:539–551.

Zhu CS, Gore M, Buckler ES, Yu JM (2008) Status and prospects of association mapping in plants. *Plant Genome* 1:5–20.

10 Progress in, Targets of, and Resources for Mining Alleles in Sorghum Candidate Genes

Richard E. Boyles, Davina H. Rhodes, Melinda K. Yerka, Chad M. Hayes, Zachary W. Brenton, Mary-Frances Behnke, Yves Emendack, and Jacobo Sanchez

10.1 INTRODUCTION

10.1.1 Sorghum's Global Relevance

Sorghum [*Sorghum bicolor* (L.) Moench] has a rich and diverse history of agricultural use across six of the world's seven continents (Antarctica excluded). Originating in sub-Saharan Africa (modern day Ethiopia and Sudan), sorghum was selected naturally, and later artificially, to withstand harsh environments, especially those that are water limited. Sorghum has been and remains a very important staple crop for some half billion people across developing countries of Africa and Asia. Elsewhere, the crop has myriad uses from gluten-free food products to animal feed to energy feedstock. Irrespective of end-use and production location, sorghum will be a key plant species of the future as the world experiences more challenging climatic conditions and more respect is given to sustainable farming practices. This certainty is a tribute to sorghum's versatility, perseverance, and genetic diversity.

10.1.2 Sorghum Diversity and Genomic Signatures

Part of sorghum's resilience as a crop can be attributed to its broad genetic diversity that was accelerated by the physical movement of sorghum (via seed) around the African continent and into Southeast Asia. This domestication and utility across contrasting regions, with respect to environment, led to the classification of five distinct botanical races: bicolor, caudatum, durra, guinea, and kafir, which were originally dictated by plant morphology (Harlan and de Wet 1972), with an emphasis on inflorescence architecture. The earliest perceived domesticate race was bicolor, which possessed an open panicle with long rachis and glumes that encapsulate the grain. This race contains the commonly named broomcorn type sorghum landraces such as "Black Spanish" (PI642998) and "Deer" (PI655995) (Martin and Washburn 1930). The guinea race was, from an inflorescence morphology perspective, not too dissimilar from bicolor; however, Guinea sorghums possess long but open glumes (as opposed to contracted like bicolor) and somewhat flattened seed shape (picture a pancake). Guinea sorghum thrived in West Africa because their open panicles (Wang et al. 2016) and extended glumes better protected the seed against fungal disease amid humid conditions. Durra would be the near opposite of guinea, with a compact inflorescence due to short branches (Harlan and Stemler 1976), minimal glume coverage of the grain, large grains in general, and an often-prominent recurved peduncle, which is the stalk just below the panicle. Most susceptible of the races to seedborne disease because of this architecture, durras tended to be grown in arid climates in

 DOI: 10.1201/9781003385004-10

Northern Africa and Southeast Asia. The caudatum race displayed an inflorescence that resembles sorghum varieties of hybrids of today: more circular and ample in shape that lends favorable to high seed number and more grain yield. Caudatum sorghums were persistent near the crop's origin but also spread into most of Africa, especially moving south (Harlan and Stemler 1976). Finally, kafir sorghums really dominated in South Africa, and these landraces exhibited a tall rachis and cylindrical inflorescence that was encouraged by short, consistent branch length across the panicle (picture a cattail inflorescence or maybe more relevant corndog of today's American cuisine). These botanical races best highlight sorghum's phenotypic diversity but shed light on its underlying genetic diversity; however, recent genetic evidence demonstrates some admixture among the five races (Brown et al. 2011; Sapkota et al. 2020a).

Sorghum has a relatively small but eclectic genome that is approximately 730 Mb in size (Paterson et al. 2009). Like maize, sorghum is diploid and possesses 10 ($2n = 20$) chromosomes (Ming et al. 1998). Unlike maize, sorghum has a perfect inflorescence and is predominantly self-pollinating (Schertz and Dalton 1980). Despite this mechanism, the sorghum species has established and maintained a wealth of genetic variation. This feat traces back to its earlier movement across wide geographic space and independent selection pressures, as it did not experience a large genetic bottleneck like soybean and peanut that underwent severe bottlenecks during domestication (Burow et al. 2001; Hyten et al. 2006). Therefore, sorghum theoretically possesses the natural genetic variation needed for large advances in crop improvement (Dillon et al. 2007), especially with regards to tolerance to abiotic and biotic stress (Howarth et al. 1996).

10.1.3 Sorghum Genomic Resources

The onus is on scientists to discover these genetic variants and exploit them for targeted trait enhancement. To aid this process, global sorghum repositories have individually maintained tens of thousands of accessions, many that have still yet to be rigorously mined for useful alleles. One limiting factor in characterizing sorghum accessions in temperate environments is photoperiod sensitivity. The Sorghum Conversion Program (Stephens et al. 1967; Klein et al. 2016) has successfully introgressed insensitivity and dwarfing alleles into more than a thousand exotic accessions, but *ex situ* repositories collectively house nearly 100,000 (USDA-NPGS: 42,000; ICRISAT: 39,000; National Genebank of China: 16,000) (Cuevas et al. 2017). Genomic sequencing is now cost-effective to genetically characterize untapped accessions and expedite the discovery of new alleles. This book chapter discusses previous accomplishments of identifying useful alleles in sorghum and forecasts how future discoveries can and should be expedited with the investment and utility of modern resources. After all, global crop improvement, agricultural productivity, and food security are all at stake. Fortunately, tremendous progress has been made in building genetic and genomic resources (Boyles et al. 2019) since the first *de novo* genome assembly of "BTx623" was completed in 2008 (Paterson et al. 2009). This includes multiple pangenomes in sorghum (see Section 10.3) that significantly expand our understanding of genome complexity and break the barrier of single or few reference genomes underrepresenting entire species diversity along with established mutant populations (Xin et al. 2008; Blomstedt et al. 2012) carrying induced mutations. Collectively, resources are no longer limiting and can sharpen resolution of genome investigation down to the genic level (Hao et al. 2021), which the plant science community should be welcoming and trying to exploit to the maximum potential.

10.1.4 Opportunity for Discovery

Association and linkage mapping, like with many other species, have been utilized *ad nauseam* to reveal a plethora of candidate genes in sorghum that warrant further investigation. With more than 660 quantitative trait loci (QTLs) reported insorghum (Mace et al. 2019) that encompass a much larger number of candidate genes, we appear to have surpassed the point of diminishing returns for trait mapping and science has already shifted toward genomic prediction using genome-wide markers, especially for complex, quantitative traits (Hunt et al. 2018; Sapkota et al. 2020a; 2020b). Sorghum scientists have built up this treasure trove of putative genes that will have little impact if not further examined for individual alleles or haplotypes that cause changes in gene function (or at the very least are tightly linked to causal mutations). While some studies have taken this extra step to validate genes and find causal alleles, most studies end at candidate gene discovery. This statement does not lessen the value of these previous findings but rather emboldens the scientific community to follow-up on the candidate genes that were previously identified. There are searchable databases in place, including for sorghum, that contain meta-information on QTLs and putative genes to streamline the process (Makita et al. 2015; Luo et al. 2016; Gladman et al. 2022). Gene coexpression network analysis has aided in identifying a collection of genes that are likely to be within the same molecular or biochemical pathway. Efforts should now be placed toward understanding gene function and identifying variants within genes that are responsible for altering gene expression. This chapter discusses candidate genes worthy of exploration and the current status of resources available for mining superior alleles in sorghum, including successful attempts of this method previously published and what progress can be expected moving forward given the resources now available.

10.2 EXAMPLES

Compared to other primary cereals such as maize, wheat, and rice, sorghum has fewer examples of successfully dissecting gene targets using classical forward genetics or other molecular approaches. That being said, several studies have elucidated gene and underlying allele function. This chapter highlights three important cases: *Dw2* for plant height, *Ma2* for plant maturity, and *RMES1* for sorghum aphid tolerance. These case studies provide a range of approaches, detail, and success that shed light on the usefulness and power of allele mining for traits that span monogenic to polygenic inheritance and cover both agronomic and biotic stress.

10.2.1 Dw2

Major effect QTLs for plant height in sorghum have been known since the 1950s (Quinby and Karper 1954), as there are four independent genes that act in the same Mendelian manner, with recessive alleles conferring dwarfness. Near-isogenic lines were developed thereafter to substantiate these four QTLs individually (Graham and Lessman 1966; Casady 1967). Of these four (labeled *Dw1-Dw4*), *Dw1* and *Dw3* have the greatest phenotypic effects (Morris et al. 2013; Higgins et al. 2014; Mural et al. 2021). *Dw2* typically demonstrates a moderate effect, and this effect gets swayed by the genetic background at other loci. Broomcorn sorghums were the only genotypes to show dominance at the last locus, *Dw4* (Hilley et al. 2017), and is therefore typically not found to play a meaningful role in plant height among adapted sorghum intended for grain and sweet sorghum use. Hilley et al. (2017) focused on *Dw2*, which is located on chromosome 6 and tightly linked to *Ma1*, a maturity gene that strongly affects photoperiod sensitivity and thus days to anthesis (i.e., flowering) in temperate, long-day environments. Fortunately for grain sorghum development, these two genes are coupled so that recessive allele of *Dw2* is linked to the recessive allele of *Ma1* (Lin et al. 1995). Hilley et al. (2017) fine-mapped the *Dw2* locus to a ~100 kb region at approximately 43.2 Mb using a recombinant inbred family. Sanger sequencing with an ABI 3130xl genetic analyzer was then used to sequence ten putative genes located in this 100 kb region. After mining individual bases across these genic sequences, Hilley et al. (2017) found only one variant between near-isogenic lines Dwarf Yellow Milo and Double Dwarf Yellow Milo that was positioned within Sobic.006G067700. This exonic insertion-deletion, based on predicted single nucleotide polymorphism (SNP) effects, caused a frameshift mutation that caused a premature stop codon, resulting in a severely truncated polypeptide (190 amino acids down from 809 in the wildtype). Sequencing of other genotypes with known *Dw2* action further confirmed this point mutation as being the causal allele (Hilley et al. 2017). Recently, functional validation efforts found that the *dw2* mutation in Double Dwarf Yellow Milo reduced cell division either prior to elongation at stem internodes or during elongation (Oliver et al. 2021).

10.2.2 Ma2

Maturity, commonly measured as the number of days to anthesis or flowering, is more polygenic than plant height but has seven well-known genes (*Ma1–Ma7*) with moderate effects. These genes have dominant action and interact at various degrees (Rooney and Aydin 1999), which combined regulate photoperiod sensitivity. Dominance at each locus will delay flowering in long days (Yang et al. 2014). A bit more recent than the dissection of *Dw2*, Casto et al. (2019) leveraged publicly available whole-genome sequence reads to mine for sequence disparities among *Ma2* and *ma2* genotypes. A total of 52 diverse genotypes with whole-genome sequencing (WGS) data were utilized for this analysis. Also, the putative *Ma2* locus (Sobic.002G302700; a putative SYMD domain-containing protein) was Sanger sequenced for confirmation of polymorphisms and indels. Further, an F_2 biparental population from a cross between near-isogenic lines 80M and 100M (primarily differing at *Ma2*) were sequenced using digital genotyping (Mitra et al. 2003; Morishige et al. 2013) to capture variant calls in progeny at sites not shared between the two isogenic parents. Because of the minimal recombination between 80M and 100M (included in the 52 genotypes with WGS data), allele mining of WGS reads between genotypes were able to identify a single base pair discrepancy. This polymorphism was a predicted non-synonymous point mutation in the third exon that caused a truncated polypeptide. Additional Sanger sequencing of a 500 bp region within *Ma2* as mentioned above corroborated WGS data on 11 of the 52 genotypes, and capillary sequencing did confirm results (Casto et al. 2019).

10.2.3 RMES1

The sorghum aphid (*Melanaphis sorghi*), previously known as the sugarcane aphid (*Melanaphis sacchari*) is a relatively new but major pest to sorghum after a biotype of the species switched its primary host from sugarcane to sorghum in 2013 (Nibouche et al. 2018). Infestations across the United States in the first years decimated the crop and caused substantial yield loss. Host plant genetics that confer tolerance or demonstrate partial resistance has been successfully identified and incorporated into adapted hybrids since pest outbreaks (Armstrong et al. 2015), but the sorghum aphid continues to be the predominant pest of sorghum in the Americas. The major locus, defined as Resistance to *Melanaphis sacchari 1* (*RMES1*), was identified by Wang et al. (2013) using biparental mapping with the Chinese resistant parent Henong 16. This QTL interval spanned 126 kb between microsatellite markers with greatest significance (i.e., LOD score), which encompassed five putative genes. Muleta et al. (2022) recently utilized whole-genome resequence (WGRS) data and *de novo* sequencing of genotypes with known resistance to sorghum aphid. Of the reported seven genes evaluated across that region on chromosome 6, three contained no intragenic polymorphisms. Four remaining loci had a high number of functional sequence variants, including ten non-synonymous mutations in two genes (Sobic.006G017200 and Sobic.006G017500) that both annotated as nucleotide binding leucine rich repeats (Muleta et al. 2022). However, WGS data was not powerful enough in this scenario to dissect the functional allele or even the causal gene underlying *RMES1* with high certainty. This example does present the limitation of utilizing existing WGS (resequencing and *de novo*) data for allele mining in regions of high recombination with many polymorphisms. The ability to parse out the causal SNP or indel can be a challenging and thus demand additional sequencing (either another sequencing method for existing lines or additional lines for higher power). Zhang et al. (2021)

deployed transcriptomics along with WGRS data to identify a large set of differentially expressed genes in the *RMES1* under aphid pressure. Among these genes were Sobic.006G017500 that was consistent with results from Muleta et al. (2022). Sequence data for the Chinese lines suggested copy number variation across genotypes that collectively led to significantly different expression levels of this gene cassette that included Sobic.006G017500 (Zhang et al. 2021). Nevertheless, both Zhang et al. (2021) and Muleta et al. (2022) found potential sequence variants for their respective genetic material for predicting the resistance or susceptible phenotype with high specificity. In the latter study, a kompetitive allele-specific PCR (KASP) marker was developed and validated across seven different USA breeding programs (Muleta et al. 2022).

10.3 THE PANGENOME ERA

10.3.1 Core- and Pangenomes

As highlighted in the introduction, the first published *de novo* assembled genome in sorghum was "BTx623" (Paterson et al. 2009), a seed parent commonly used for hybrid production in the USA. Completed in 2008, the "BTx623" genome became a reliable reference for sequence alignment of future genotyping projects, especially large genotyping-by-sequencing (GBS) efforts (Morris et al. 2013; Brenton et al. 2016; Hu et al. 2019). "BTx623" was later reassembled for improved sequence accuracy and downstream annotation (McCormick et al. 2018). A second reference genome was completed for the sweet sorghum cultivar "Rio" using Pacific Biosciences long-read single nucleotide sequencing (Cooper et al. 2019). Sequence comparison between "BTx623" and "Rio" found strikingly similar genome size (99.6%), sequence arrangement, and gene number (~1% difference; BTx623: 34,129; Rio: 35,467); however, certain observations followed trends now identified for other crop species, such as presence/absence variation for putative disease-related genes (e.g. nucleotide binding leucine rich repeats) and deletions in homologues important for sugar accumulation. This trend has the scientific community in large agreement that a single genome reference is inadequate and can restrict progress in genomics and genetic mapping (Sherman and Salzberg 2020; Tao et al. 2021). Missing indels and copy number variation can significantly inhibit the ability to map important causal variants in the absence of tight linkage with a polymorphism that is present in GBS datasets. Thus, discovery of structural changes between genomes warranted the development of a larger pangenome to capture more genetic variation and facilitate alignment of future next-generation sequencing reads.

Rapid investment of whole-genome sequencing across sorghum populations has resulted in an accumulation of publicly available pangenomes that better capture genome-wide sequence variation. There have been five pangenomes reported to date (Table 10.1), with more efforts undoubtedly still underway. The first published pangenome was specific to a Haitian sorghum breeding program (Jensen et al. 2020), which sequenced 24 founder lines representing the gene pool. Average sequencing coverage was 8×, adequate for building a Practical Haplotype Graph using GBS for the remaining individuals in the breeding population (Jensen et al 2020). Wang et al. (2021) *de novo* assembled two additional reference sequences for the inbred lines "RTx436" and "Tx2783" and combined preexisting assemblies of three other inbreds ("BTx623", "Rio", and "RTx430"). Further, this study built the pangenome using "Tx2783" as the principal reference, utilized the additional four assemblies to capture indels, and then took public whole-genome resequencing data of 62 sorghum accessions to report a core sorghum genome (and set of genes) along with a list of dispensable genes that were genotype-specific. The pangenome resource that used the largest number of *de novo* assemblies (>150× coverage from Illumina platform) that is currently available consisted of 12 geographically and racially diverse *Sorghum bicolor* accessions along with one *S. propinquum* genotype (Tao et al. 2021), along with incorporating the three public reference genomes mentioned above. Long reads for these 13 genotypes were also generated using PacBio sequencing (25–88× coverage) to capture more of the repetitive genome. Pangenome construction using these 16 total genomes was

TABLE 10.1
Published Pangenomes to Date Capture Sufficient Sequence Coverage and Extensive Genetic Diversity

					Genes	
Resource	**Genotypes**	**Reference**	**Coverage**	**Variants**	**Core**	**Dispensable**
Jensen et al. (2020)	24	BTx623 v3.0	8x	n/a	n/a	n/a
Wang et al. (2021)	2 (+3)#	Tx2783	80x	n/a	19,117	16,238
Tao et al. (2021)	13 (+3)#	BTx623 v3.0	174x	n/a	15,867	28,212
Ruperao et al. (2021)	176	BTx623 v3.0	10x	2,026,095	16,821	18,898
Boatwright et al. (2022)	400	BTx623 v3.1.1	38x	43,983,694	n/a	n/a

Additions refer to the study incorporating previously published *de novo* sequences of other genotypes.

built onto the "BTx623" reference to discover more than 200 Mb of new sequence, with a pangenome size of 955 Mb (Tao et al. 2021). Across genotypes, 62% of genome sequence represented the core genome (present in 15 or more genotypes). Ruperao et al. (2021) since sequenced 176 accessions at >10× coverage and mapped to the most recent "BTX623" sequence (McCormick et al. 2018). Of putative genes captured, 47% were core genes (compared to 62% in Tao et al. 2021), with more than 18,000 genes being dispensable, or variable among sequenced accessions. This study further mapped existing WGS data from previous studies (Guo et al. 2019; Valluru et al. 2019; Jensen et al. 2020) to the pangenome assembly to capture SNPs and facilitate genetic relatedness analysis and genome-wide mapping. Boatwright et al. (2022) generated >25× coverage for 400 accessions that comprise the Sorghum Association Panel (Casa et al. 2008). This resource generated over 43 million variants (SNPs, indels, and copy number variants), which drastically increased the marker saturation of the Sorghum Association Panel over existing GBS data. Specifically, 50% of variants detected via deep-coverage WGS were positioned in 5 kb genome bins that had not a single SNP from previous GBS efforts.

10.3.2 Utility in Allele Mining

While next generation sequencing methods such as GBS and skim sequencing, coupled with downstream tools like imputation, enabled the discovery of many candidate genes by facilitating linkage mapping and genome-wide association studies (GWAS), many causal variants, or even non-causal but tightly linked variants that could serve as reliable genetic markers, remain elusive. Pangenomes add sequence coverage and stability to more easily and reliably mine these publicly available resources for superior and/or causal alleles (Zanini et al. 2022). As evident in the gene examples highlighted in section 10.2, WGS does not always provide the clarity needed (see Section 10.2.3) but can be effective in given genomic regions and, at minimum, be used in tandem with functional validation (especially using gene editing to insert site-specific point mutations or delete single base pairs) or other reverse genetics methods to determine gene influence on a particular phenotype and identify causal alleles. This two-step approach can be more of the norm when studying crop species such as sorghum that possess dense variants positioned in gene-rich, euchromatic regions. In situations, large biparental or multiparent populations can reach the resolution for fine-mapping to narrow down the candidate region harboring the gene/allele(s) of interest, but this again is not guaranteed since fine-mapping is reliant on local recombination events within the QTL. What can be stated with confidence is that the public availability of multiple pangenomes gives scientists an extremely low-cost, straightforward first step to searching for functional variants across a QTL region or within a candidate gene. What is also known in plants is that structural variants such as copy number variation (i.e., number of gene duplications in the genome) do play a significant role in traits underlying adaptation (Tao et al. 2019). Zheng et al. (2011) and later Tao et al. (2021) confirmed extensive CNV in sorghum, and recent studies have found CNV to play a role in sugar accumulation (Brenton et al. 2020), disease resistance (Mewa et al. 2022), and grain quality (Sattler et al. 2009). Given that in sorghum there have been more than 1,000 individual candidate genes reported in the literature, there will undoubtedly be opportunities to exploit pangenome resources for (1) gene identification or confirmation (of prior mapping efforts) that will foster scientific knowledge of gene function and fill-in genetic pathways and (2) causal structural variant and allele identification that can translate into reliable genetic markers for crop improvement.

10.4 TARGET TRAITS AND CANDIDATE GENES

10.4.1 Biotic Stress Traits

10.4.1.1 Anthracnose Resistance

Anthracnose in sorghum is caused by *Colletotrichum sublineola*, which causes initial leaf blight (Khanal et al. 2022) but can infect the pith of the stem as well as the grain (Tesso et al. 2011). This fungal disease is a major threat to production globally, especially in humid regions (Burrell et al. 2015) such as West Africa (Thomas et al. 1996) and the southeastern USA (Stutts and Vermerris 2020). With rising temperatures, it is plausible that mean severity of anthracnose damage to sorghum may increase in future years (Gaffoor et al. 2021).

Many sources of anthracnose resistance have been discovered, primarily from field testing. Being a disease that causes visually distinct lesions on the leaves, observational ratings are effective. Some of the widely-used sources of resistance have also been crossed to a susceptible genotype to develop recombinant inbred line populations, and underlying QTLs have been identified. For a comprehensive list of anthracnose resistance QTLs, please refer to Stutts and Vermerris (2020) in addition to the following more recent studies: Cuevas et al. (2021), Mengistu et al. (2021), and Khanal et al. (2022). Marker development for certain QTLs has occurred but with little testing, validation, or utilization to date (Abreha et al. 2021). Lee et al. (2022) used a multistep approach to effectively identify and dissect a resistance QTL on chromosome 7. Whole genome resequencing data of resistant and susceptible bulk RILs from the biparental population of SC283 (resistant) × TAM428 (susceptible) were analyzed for sequence variants within the region, specifically targeting the named *ANTHRACNOSE RESISTANCE GENE1* (*ARG1*; Sobic.007G085400) and *CARRIER OF ANTHRACNOSE RESISTANCE GENE* (*CARG*; Sobic.007G085350). Putative functional variants within both loci were identified, and variants consistently co-segregated among the resistant bulk, SC283 (resistant parent), and other anthracnose resistant lines (Lee et al. 2022). Transcript profiling followed by virus-induced gene silencing (VIGS) in a separate study led to evidence of two loci (designated *Cs1* and *Cs2*) on chromosome 9 that, upon further inspection, were NB-LRR encoding genes that underwent duplication events as both *Cs1* and *Cs2* had duplicated genes nearby (Biruma et al. 2012). These gene clusters on chromosomes 7 and 9 warrant further investigation to elucidate superior resistance alleles within and across loci.

10.4.1.2 *Fusarium* Resistance

Not unlike anthracnose, *Fusarium spp.* can infect multiple sorghum tissues. The two predominant concerns are *Fusarium* stalk rot (Tesso et al. 2011; Adeyanju et al. 2015) and *Fusarium* grain mold (Ackerman et al. 2021; Wenndt et al. 2023). Host plant resistance to *Fusarium* grain mold infection has been studied in depth (Klein et al. 2001; Rooney et al. 2002; Cuevas et al. 2019; Nida et al. 2019; Prom et al. 2020) while stalk rot has received less attention, despite its ability to cause severe economic loss (Tesso et al. 2004). With regards to stalk rot, there has been a sharp increase in prevalence of disease in recent years, indicating that more frequent and extreme oscillation of weather patterns between hot/dry and cool/wet will indeed be more conducive to significant stalk rot outbreaks. *Fusarium* stalk rot resistance in sorghum is genetically controlled by incomplete dominance action of multiple genes (Tesso et al. 2004), but F_1 hybrids tend to have greater yield loss than their parents (Seetharama et al. 1991), which is suggested to be a function of higher sink strength in hybrids that remobilize non-structural carbohydrates from the stem to the grain (Tesso et al. 2004). Conversely, *Fusarium* grain mold resistance can be contributed to both additive, dominance, and, to a lesser extent, epistatic effects (Rodriguez-Herrera et al. 2000). The greatest mode of resistance to grain mold is the production of tannins within the testa layer (Menkir et al. 1996), but tannins have a significant antinutritional effect when used as an animal feed (Armstrong et al. 1974; Pan et al. 2019). Thus, finding resistance mechanisms that do not alter nutritional quality are a challenge, yet essential. Resistance mechanisms and genes have been difficult to detect, which can be at least partly attributed to the fact that grain mold in sorghum is typically a result of a broader disease complex composed of a plethora of fungal species (Ackerman et al. 2021). As such, effective sorghum resistance itself must be complex and quantitative (Cuevas et al. 2019), which results in numerous, minor effect QTLs that are often difficult to detect in mapping populations.

Outside of *Tan1* and *Tan2*, epistatic genes that regulate tannin production in the grain (Wu et al. 2012), few genes have been validated that impact sorghum *Fusarium* resistance. Similar to anthracnose, numerous studies have reported significant *Fusarium* resistance QTLs/MTAs (Klein et al. 2001; Upadhyaya et al. 2013; Adeyanju et al. 2015; Cuevas et al. 2019; Nida et al. 2019, 2021; Prom et al. 2020), with a number of these identifying positional candidate genes that affect grain composition and quality. Notably, Nida et al. (2021) found SNPs associated with grain mold resistance near the annotated genes *KAFIRIN* (S1_51860558) and *LATE EMBRYOGENESIS ABUNDANT 3* (*LEA3*; S3_15689447) offer targets for molecular characterization and allele mining. Kafirin proteins affect protein storage and kernel texture that can increase grain mold resistance but negatively impact grain quality attributes such as protein digestibility (Jambunathan et al. 1992). Other genetic variants that alter production of flavonoids and aspects of pigmentation in the pericarp (see Nida et al. 2019 and Section 10.4.3 for details about *YELLOW SEED* genes) and other grain tissues deserve exploration to find an optimal concentration of defense compounds without compromising grain quality and nutritional value.

10.4.1.3 Aphid Resistance

Sorghum is vulnerable to aphid feeding and has a long history of certain species threatening production. For example, reports of greenbug (*Schizaphis graminum*) infesting sorghum date back to the mid-19th century (Rondani 1852). Aphid resistance is included here as there are known resistance sources and mechanisms in sorghum for most species that target sorghum as a primary host.

With respect to the greenbug, much effort was done to incorporate resistance or tolerance into sorghum to eliminate or endure feeding of this pest, respectively. Crop losses from greenbug feeding were significant in the USA during 1968 as a result of a shift in biotype, and regional outbreaks of greenbug biotype Censued in succeeding years. A future prominent biotype shift from C→E led to major crop losses (Porter et al. 1982), and later shifts have led to additional greenbug infestation, albeit to a lesser degree (Harvey et al. 1991; Harvey et al. 1997). Biparental populations have been used to successfully map resistance QTLs for various biotypes, including a major effect locus on chromosome 9 that accounted up to 80% of phenotypic variation to biotype I feeding (Wu and Huang 2008; Punnuri et al. 2013). Little resistance was identified in sorghum against the yellow sugarcane aphid (*Sipha flava*), an sap-sucking aphid similar to *Melanaphis sacchari*, despite a single study screening 3,600 USDA accessions (Starks and Mirkes 1979), which is concerning. The most prominent resistance sources were later identified by Webster (1990) from a large Ethiopian collection, but the genetic underpinnings remain unknown.

Since mitigating severe greenbug and yellow sugarcane aphid (Breen and Teetes 1990) outbreaks, the sorghum aphid has become the predominant pest threat to sorghum (Rodríguez-del-Bosque and

Terán 2015; Bowling et al. 2016). Fortunately, cross-resistance from greenbug sources was found (Armstrong et al. 2015), and these sources or descendent lines were rapidly used to introgress into adapted, high-yielding sorghums. Refer to Section 10.2.3 in this chapter for details about sorghum aphid and how the *RMES1* locus is providing broad-spectrum tolerance in many modern hybrids. Outside of *RMES1*, there seem to be other gene(s) providing levels of tolerance based on germplasm source studies (Paudyal et al. 2019) that deserve exploration for identifying genetic control(s) for durable resistance or in the case that aphids become virulent to *RMES1*.

10.4.1.4 *Striga* Resistance

Striga spp., commonly known as witchweeds, are hemiparasitic plants that attach themselves to the roots of sorghum any many other staple crops, causing extensive damage and crop loss. Working underground, control of *Striga* using mechanical or chemical measures is largely ineffective. Sorghum exudes, through its roots, strigalactones that, as the name suggests, stimulates germination of *Striga* seeds (Netzly et al. 1988; Siame et al. 1993; Butler 1994). Thus, much research has been performed to measure differences in production of root exudates and manipulate exudation concentration and/or composition (Hess et al. 1991; Vogler et al. 1996; Ramaiah et al. 1990; Erickson et al. 2001; Mohemed et al. 2016, 2018). While various levels of resistance were reported decades ago (Saunders 1933; Williams 1959), much more work was needed (Rich et al. 2004) and remains needed to decrease sorghum production losses in sub-Saharan Africa, where sorghum is a staple for hundreds of millions of people (Ejeta 2007).

Sorgoleone, a major allelopathic chemical that is exuded through sorghum roots, has been a major target for manipulation to retard weed germination and growth (Einhellig and Souza 1992; Nimbal et al. 1996). Yang et al. (2004) first identified and isolated a single gene, *SOR1*, located on chromosome 5 (Shehzad and Okuno 2020). *SOR1* was correlated with sorgoleone production across *Sorghum* species and is likely a key gene in the pathway of sorgoleone synthesis (Yang et al. 2004). Haussmann et al. (2004) also identified significant QTLs, including *lgs*, from biparental mapping that explained reasonably large phenotypic variance of *Striga* resistance. Recently, *qtl10RL* was mapped on chromosome 10 that was linked to allelopathic effect that inhibited lettuce root length (Shehzad and Okuno 2020). These regions should be investigated for causal genes and mined for genetic variants that manipulate root exudate concentration and composition.

10.4.2 Abiotic Stress Traits

10.4.2.1 Stay-Green Drought Tolerance

The USA Sorghum belt extends from South Dakota to south Texas. As such, the production environments across this large swathe of the USA vary considerably with respect to soil moisture availability and seasonal temperature rhythms. In particular, USA sorghum production occurs primarily on dryland acres.

Sorghum is highly regarded for its ability to produce grain in marginal lands and under rainfed conditions. Of importance, the stay-green trait in sorghum has been associated with grain yield components (Boyles et al. 2016) under moisture limiting conditions at the post-flowering stage of development and related to chlorophyll concentration (Xu et al. 2000). As such, this trait has been advocated as a breeding target for the improvement of major cereal crops grown in increasingly variable environments due to climate perturbations (Kamal et al. 2019). Indeed, this trait has also proven useful in maintaining production under abiotic stress in other cereals like wheat and rice (Shin et al. 2020; Cook et al. 2021a). The ability of crop plants to maintain source strength (i.e., stay green) during grain formation and filling, especially during terminal drought stress, allows for source-sink dynamics to continue towards deposition of grain constituents (Sade et al. 2018).

The stay-green trait in crop plants has been well described with five mechanistic routes to staying green involving timing of senescence initiation and rate of progression (Thomas and Howarth 2000)

These include delayed initiation, normal initiation but diminished rate, normal initiation and rate coupled with chlorophyll degradation inhibition, rapid tissue mortality, and increased greenness in combination with normal initiation and rate of senescence. Indeed, the stay-green trait in sorghum has been shown to be exhibited phenotypically across genotypes in varying degrees based on timing and rate of senescence (Thomas and Howarth 2000).

In sorghum, the two most commonly used genetic sources of stay green by breeders are B35 and KS19, landraces from Ethiopia and northern Nigeria, respectively (Mahalakshmi and Bidinger 2002). Early inheritance studies of the trait in B35 showed large environmental effects on the expression of stay green (Walulu et al. 1994). Molecular analysis of an inbred population using the B35 background (BTx642) and a senescent type (Tx7000) led to the discovery of four major stay-green QTLs on chromosomes 2 (*Stg3*), 3 (*Stg1 and Stg2*), and 5 (*Stg4*) (Xu et al. 2000). The genetics controlling the phenotypic expression of the stay-green trait are complex and the effectiveness of each of the four QTLs in driving expression also varies in importance. A study by Harris et al. showed that onset of post-flowering drought induced senescence varied among near isogenic lines of all four *Stg* (Harris et al. 2007). In addition, the implications of individual *Stg* QTLs to crop development and yield were studied by Borrell et al. (2014) also using near-isogenic lines for each of the *Stg* QTLs. These authors found that, relative to the recurrent senescent parent (Tx7000), growth of the *Stg* near-isogenic lines was characterized by diminished canopy size, lower water extraction prior to anthesis, increased moisture extraction during grain filling, and increased grain number and yield. The effectiveness of the stay-green phenotype from B35 in improving grain yield under moisture deficit during grain fill had been previously established in other genetic backgrounds (Kassahun et al. 2010). As such, breeding programs in Australia and the USA have largely used the B35 background for stay-green trait introgression. However, other sources of stay green, from Ethiopian backgrounds, with similar retention of green leaf area, relative to B35, but increased panicle weight under post-flowering water deficit stress have been identified, agronomically characterized, and publicly released (Emendack et al. 2017; Burke et al. 2018).

While the stay-green trait has been shown to be beneficial for lowering the grain yield penalty incurred from water deficits occurring post-anthesis, consideration should be made as to how this trait impacts deposition of grain constituents important for human and animal nutrition. A recent study of 23 grain sorghum inbred lines with varying levels of stay green found that stay-green lines contained higher concentrations of kafirin proteins relative to senescent lines, irrespective of water-deficit treatment (irrigated, pre-flowering water deficit, post-flowering water deficit) (Emendack et al. 2018a). These proteins are known to be nutritionally poor due to their deficiency of several essential amino acids, including lysine (Virupaksha and Sastry 1968; Wall and Blessin 1970). In addition to poor nutritional quality, sorghum grains with high kafirin contents have been shown to have reduced starch digestibility and, thus, low metabolizable energy (Salinas et al. 2006). Nonetheless, variation does exist in sorghum for high nutrient content under water deficit and stay-green status (Queiroz et al. 2015) such that with targeted breeding strategies, varieties could be developed to overcome the drag in nutritional quality that can accompany the stay-green trait.

10.4.2.2 Dhurrin

As a cyanogenic glucoside, dhurrin [(S)-*p*-hydroxymandelonitrile-b-D-glucopyranoside] in sorghum leaves and stems has been associated with several important biological functions including: defense against herbivory, resistance to charcoal rot, osmoprotectant, and nitrogen storage and turnover (Burke et al. 2015; Møller 2010). A recent GWAS analysis of the 700 lines in the Sorghum Conversion Panel determined that existing natural variation in the concentration of dhurrin in sorghum leaves was driven by variation in alleles of biosynthetic and catabolic genes (Hayes et al. 2015). Additionally, a dhurrin QTL (*Dhu1*) on chromosome 1 (SBI01) was discovered that colocalized with dhurrin biosynthetic genes (*CYP79A1, CYP71E1, UGT85B1*) and was also aligned with a stay-green QTL (*Stg5*) that had gone undiscovered in previous published analysis of the BTx642/Tx7000 population used to map *Stg1-4* QTLs (Xu et al. 2000). The discovery of *Stg5* and its alignment with *Dhu1* confirmed previous results by Burke et al.(2015) that stay-green genotypes generally contained higher concentrations of leaf dhurrin compared to their senescent counterparts and that this character was associated with the level of pre- and post-flowering drought tolerance (Burke et al. 2013). The importance of elevated levels of dhurrin in stay-green lines was exemplified among a diverse genotypic panel with known leaf dhurrin levels in a study of seedling growth under nitrogen starvation (Emendack et al. 2017). In this study, stay-green seedlings exhibited higher growth metrics (shoot fresh weight, shoot dry weight, seedling length) compared to seedlings of senescent types. The authors postulated that the large differences in growth between stay-green (high dhurrin) and senescent (low dhurrin) genotypes could be catabolism of dhurrin as a nitrogen source for growth under N starvation. The direct applicability of the high dhurrin trait would be beneficial to sorghum producers in third-world countries where fertility is limited or non-existent.

Similar to the stay-green trait, trade-offs also exist for the high dhurrin trait. Because dhurrin contains a cyanide moiety, degradation of dhurrin by b-glucosidase (dhurrinase) leads to the production of hydrogen cyanide (HCN), which has been the culprit of livestock poisoning (Hunt and Taylor 1976). As such, the use of sorghum for forage has been considered risky.

The development of sorghum acyanogenic mutants, *totally cyanide deficient class 1* (*tcd1*) and *adult cyanide deficient 1* (*acdc1*), has provided a potential opportunity for the use of sorghum as a forage crop (Blomstedt et al. 2012); however, the *acdc1* mutant was shown to accumulate dhurrin normally at the early seedling stage. Additionally, a small drag in seedling growth was seen in the acyanogenic *tdc1,* which could be a consequence of the lack of N from dhurrin catabolism. Conversely, a recent study evaluating the progression of dhurrin concentration reported the presence of several natural germplasms within the Sorghum Conversion Panel that remained near-acyanogenic across the entire developmental time scale and under water deficit (Emendack et al. 2018b). These naturally occurring variants with low to near-zero dhurrin content provide the best sources of acyanogenicity in sorghum towards the development of high value forage without the risk of prussic acid (HCN) poisoning.

10.4.3 Grain Pigmentation Traits

Sorghum grain pigments are of considerable interest due to their influence on both seed quality and nutrition. The primary pigments found in sorghum grain are polyphenols and carotenoids, each of which encompass a variety of individual compounds. In sorghum, grain polyphenols have been a major focus of study for their potential role in the treatment or prevention of chronic diseases (Awika and Rooney 2004; Abugri et al. 2013; Rhodes and Kresovich 2016; Girard and Awika 2018; Smolensky et al. 2018). In contrast, sorghum grain carotenoids have received less attention, but have been studied primarily for their potential impact on vitamin A deficiencies (Salas Fernandez et al. 2009; Cardoso et al. 2015; Cruet-Burgos et al. 2020). Both the flavonoid and the carotenoid biosynthesis pathways have been extensively studied and are well understood. This allows for a comparative genomics approach to leverage available genomic resources in maize and Arabidopsis to narrow down gene targets.

10.4.3.1 Polyphenols

Two of the major polyphenol compounds in sorghum are proanthocyanidins (condensed tannins) and 3-deoxyanthocyanidins. Condensed tannins are brown pigments constitutively produced in the pericarp (outer seed coat) and the testa (inner layer of tissue between the pericarp and the endosperm) of the grain, whereas 3-deoxyanthocyanidins are red phytoalexins – produced in response to biotic stressors – found in the pericarp (Nicholson et al. 1987; Lo et al. 1999). The 3-deoxyanthocyanidins are particularly high in black-grain sorghum (Rooney et al. 2013a; 2013b) and have been shown to be upregulated as protection in response to ultraviolet B radiation exposure

(Fedenia et al. 2020). They are highly stable during processing, so have potential to be used as a natural food colorant (Geera et al. 2012; Yang et al. 2014; Herrman et al. 2020). The sorghum *yellow seed1* gene (Sobic.001G397900) is a MYB transcription factor required for the biosynthesis of 3-deoxyanthocyanidins and is an ortholog to the maize *P1* gene (Boddu et al. 2005; Ibraheem et al. 2010). Association studies for variation in sorghum 3-deoxyanthocyanidins have identified significant MTAs near known flavonoid-pathway genes, however, the associations are weak and have not been validated (Rhodes et al. 2014; Elango et al. 2022). It may be that, as a phytoalexin, there is too much environmental variation to successfully map the genetic controls underlying its variation.

The bitter flavor imparted by condensed tannins is usually undesirable to both humans and animals, so non-tannin sorghums are preferentially grown in many regions of the world. Sorghum is also used in beer-making around the world, and non-tannin sorghum is preferred for brewing because it is thought that the condensed tannins interfere with the brewing process (Beta et al. 2000; Taylor et al. 2006; Holmes et al. 2017). Additionally, condensed tannins bind to proteins, carbohydrates, and minerals, reducing the nutritional value of the grain. However, tannin-containing sorghum varieties are preferred in some regions for their ability to reduce bird predation (Xie et al. 2019), for their perceived health effects (Awika and Rooney 2004; Girard and Awika 2018), and also for the flavor they impart in liquor brewing (Zhou et al. 2020).

Classical genetics studies have uncovered the genetic controls of presence/absence of condensed tannins in the testa (dominant alleles at the *B1* and *B2* loci), and in the pericarp [dominant allele at the spreader (*S*) locus, as well as the *B1* and *B2* loci] and are described extensively elsewhere (Rooney and Smith 2000). More recently, quantitative studies have identified genes underlying variation of condensed tannins. The gene underlying the *B2* locus has been identified as *Tannin1* (Sobic.004G280800), which encodes a WD40 protein and is homologous to *TTG1* In Arabidopsis and *PAC1* in maize (Wu et al. 2012). The gene underlying *B1* has been identified as *Tannin2* (Sobic.002G076600), which encodes a bHLH protein and is homologous to *TT8* in Arabidopsis, *IN1* in Maize, and *Rc* in rice (Wu et al. 2019). *Tannin1* and *Tannin2* loss of function alleles have also been identified: *tan1-a*, *tan1-b*, *tan1-c*, *tan2-a*, *tan2-b*, *tan2-c*, and *tan2-d* (Wu et al. 2012; 2019), and competitive allele-specific PCR (KASP) markers have been developed (Terry Felderhoff, personal communication).

10.4.3.2 Carotenoids

Carotenoids are red, orange, or yellow pigments produced in all photosynthetic organisms (Cazzonelli and Pogson 2010). In photosynthetic tissue, carotenoids are involved in light harvesting and photoprotection, whereas in non-photosynthetic tissues such as seeds, carotenoids act as antioxidants to reduce membrane deterioration, and are also precursors to abscisic acid, a plant hormone that regulates seed dormancy and germination (Esteban et al. 2015). In humans, some carotenoid compounds, such as lutein and zeaxanthin, are thought to protect against age-related macular degeneration (Wu et al. 2015); whereas other compounds – primarily β-carotene, β-cryptoxanthin, and α-carotene – are precursors to vitamin A, an essential nutrient that is severely deficient in many communities around the world.

The carotenoid biosynthesis genes have been largely characterized and are highly conserved across species. In photosynthetic tissues, carotenoids are tightly regulated; however, in non-photosynthetic tissues there are less constraints, so there is greater diversity of carotenoid compounds and concentrations. Although the biosynthesis genes are well-known, regulators of the biosynthesis genes are not well described and may be more species-specific (Stanley and Yuan 2019). In sorghum, association studies have identified loci underlying carotenoid variation, some of which are near or within *a priori* biosynthesis gene candidates (Salas Fernandez et al. 2008; Cruet-Burgos et al. 2020). Perhaps the most promising is MTAs identified within the *zeaxanthin epoxidase* gene (*ZEP*; Sobic.006G097500). ZEP catalyzes the conversion of zeaxanthin to antheraxanthin and violaxanthin, a key step in the biosynthesis of abscisic acid (Finkelstein 2013), and has been implicated in carotenoid variation in many species (Wolters et al. 2010; Owens et al. 2014; Jourdan et al. 2015; Gonzalez-Jorge et al. 2016; Cruet-Burgos et al. 2020). Several other MTAs have been identified near *a priori* gene candidates in sorghum, but none of the genes have been validated. KASP markers have been developed for the *ZEP* gene, as well as for five other loci associated with carotenoid variation (unpublished).

10.4.4 Grain Quality Traits

10.4.4.1 End-Use Quality

Commercial breeders have long prioritized breeding for yield over quality due to the lack of sufficiently high-throughput methods to assess quality. Recent technological advances in the throughput of grain phenotyping have made breeding for quality much more feasible; hence, there is tremendous emerging interest in studying quality traits, especially those related to human and animal nutrition and their relationship with yield and economics. End-use quality traits, in particular, are those that influence the ways in which end-users interact with the crop. For example, the end use of grain for silage could be ruminant consumption or dairy cow [*Bos taurus* L.] nutrition, and the end use of grain for bread could include loaf volume, flavor, or texture.

End-use traits of grain sorghum specific to human consumption improve food or beverage characteristics and can be marketed as a specialty crop. Most sorghum consumed as a food is used as a whole grain, including the seed coat, embryo, and endosperm. Direct consumption of whole grains can occur when kernels are popped like popcorn, puffed as an addition to cereals or snack bars, or boiled and added to salads like wheat bulgur. Consumption of grain sorghum as a food ingredient may include decortication to remove the outer layers of the grain (including most tannins) and enrich flours for starch content. Sorghum as a food ingredient is typically ground and used with other starches for gluten-free baking. Consumption of grain sorghum as a beverage is predominantly through malting and brewing to make beer or distilling to make spirits.

10.4.4.2 Grain Characteristics and Alleles Known to Impact Nutrition and End-Use Quality

The major chemical components of whole sorghum grain include starch, protein, and oil. The chemical compositions of these macromolecules, and the ways in which they are physically arranged within the embryo and endosperm, impact kernel structure (vitreosity and the ratio of hard to soft endosperm), human and animal nutrition, and amenability to various food and beverage applications.

Advances in mapping population design, genomics-assisted breeding, and the use of mutant (e.g., TILLING) populations have vastly improved our understanding of the genetic control of sorghum grain chemical composition in recent years (Boyles et al. 2019). The grain Sorghum Association Panel (SAP) (Casa et al. 2008) and numerous derived breeding or recombinant inbred lines have been used in GWAS to identify potential SNPs associated with grain quality related traits, including starch, protein, and oil contents (Sukumaran et al. 2012; Boyles et al. 2016; Rhodes et al. 2017). GWAS combined with WGS of 196 sorghum accessions identified significant SNPs associated with levels of tannins, starch, and amino acids, respectively (Kimani et al. 2020). Boyles et al. (2017) mapped 25 QTLs related to amylose, starch, crude protein, crude oil, and gross energy by linkage mapping.

So far, only a few grain quality genes have been functionally validated, mostly those involved in protein and starch pathways. A single-point mutation in the α-kafirin peptide leads to the formation of irregular protein bodies, which promotes protein digestibility and indirectly increases lysine content (Wu et al. 2013). Clarifying the genetic architecture of high-protein digestibility was challenging because there are 20 α-kafirin gene family members in the sorghum genome. Gene editing was used to create lesions in all 20 genes, and the resulting plants had high grain lysine content and improved protein digestibility (Li et al. 2018).

Functionally validated alleles impacting starch traits of sorghum occur at low copy numbers. Lesions in genes participating in starch synthesis, including *STARCH BRANCHING ENZYME IIb* (*SBEIIb;Sobic.004G163700*) and *STARCH SYNTHASE IIa* (*SSIIa;Sobic.010G093400*), affect the alkali spreading value, starch composition, various thermal properties, and flour viscosity (Griebel et al. 2019). Naturally-occurring deletions in sorghum *PULLULANASE(Sb06g001540)*, a starch metabolism gene, improve starch digestibility and nutrition with no impact on plant yield (Gilding et al. 2013, Li et al. 2019). Several functional mutations have been identified in the *GRANULE BOUND STARCH SYNTHASE* (*GBSS*; often referred to as *Waxy, Wx*) gene that controls amylose synthesis. The wx^a and wx^b alleles (Sattler et al. 2009) confer grain with very low amylose content that may or may not be associated with yield drag in inbred lines or hybrids (Jampala et al. 2012; Yerka et al. 2015a; Yerka et al. 2015b; Yerka et al. 2016; Kent and Rooney 2021). Two additional *waxy* alleles have since been similarly validated (Kawahigashi et al. 2013; Lu et al. 2013). No studies have assembled biosynthetic pathways of sorghum grain starch, protein, or oil contents, so it is likely that many more alleles will be functionally validated for their impact on human and animal nutrition, and end-use quality, in coming years.

10.4.4.3 Future Considerations for Improved Breeding for Grain End-Use Quality

Despite the great strides made in recent years to identify alleles impacting sorghum grain end-use qualities, much more remains to be done. This is especially the case in the western half of USA, which is experiencing accelerated rates of climate change and rapid declines in agricultural water availability in the form of precipitation, snowmelt, or aquifers. Many growers throughout the country will face water shortages in coming years and crops like grain sorghum that can mature on 15–20 inches of water per growing season will have a competitive advantage over crops like alfalfa and maize that require 30–40 inches. Switching to sorghum will be incentivized when there is an option for selling grain having high-value traits for human consumption. However, significant effort is needed to clarify the ways in which production environments impact end-use quality and harvestable yields of the specific grain characteristics being sought (i.e., baking quality, malt flavor, etc.). Evidence from barley [*Hordeum vulgare* L.], wheat [*Triticum aestivum* L.], and rice [*Oryza sativa* L.] breeding programs indicates that adapting specific end-use qualities to specific production environments is feasible, but it requires large regional trials, appropriate mapping populations, heterotic group development, and cropping system designs that optimize plant population densities and plant nutrition.

Methods for multi-environment trials and cropping system designs are well-established in sorghum, but no mapping populations or heterotic group evaluations specific to grain end-use qualities have been published. These are key gaps in the existing literature that must be filled. Future research should focus on mapping population designs that are best able to capture both trait expression and local adaptation to clarify the genotype-by-environment-by-management (GxExM) interaction influencing observed relationships between production environments and quality. Similarly, heterotic group design may need to be re-evaluated for different end-use traits, as current heterotic groups in most sorghum breeding programs have focused on grain yield, but not harvestable yield of specific grain constituents (protein, lysine, digestible protein, starch, digestible starch, fat, etc.).

The most common mapping population structures currently being used by plant breeders to study quality traits include diversity panels (like the SAP) and biparental, nested association mapping (NAM), and MAGIC populations (Figure 10.1). In grain sorghum, several mapping populations have been developed, including three NAMs (Boatwright et al. 2021), several BC-NAMs, the Sorghum Association Panel (SAP), and a structured MAGIC (Boyles et al. 2019; Kumar et al. 2022). However, unlike the other staple cereal crops like maize, wheat [*Triticum aestivum* L.], rice [*Oryza sativa* L.], and barley [*Hordeum vulgare* L.], no grain quality mapping population has been published. No single population can fill every function in a pre-breeding program, but recent research indicates that

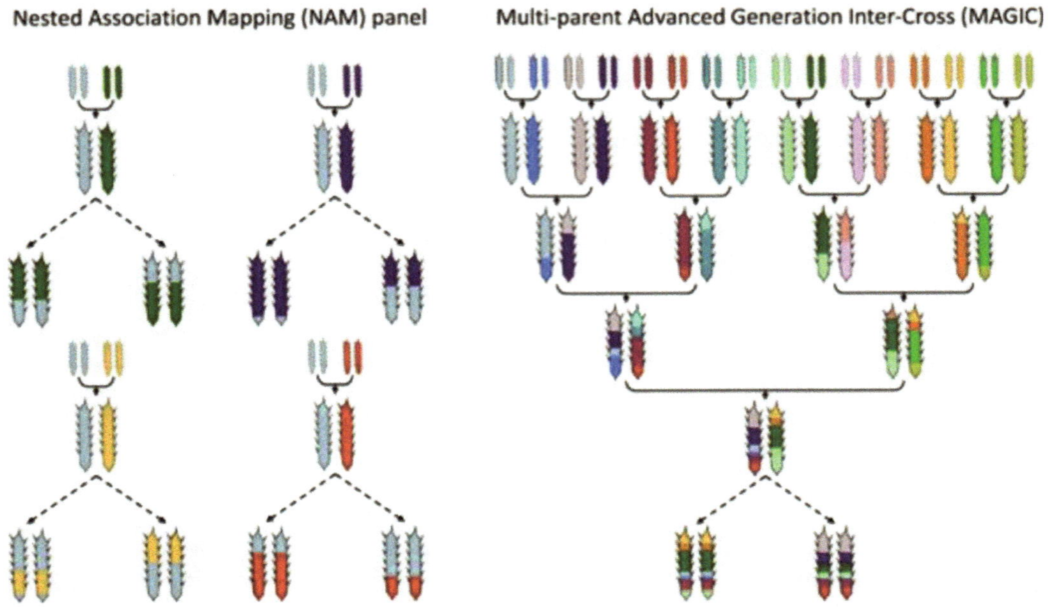

FIGURE 10.1 Modified schematic from Scott et al. (2020), the NAM design consists of a series of biparental crosses against a common founder from which RILs are generated through self-pollination and single-seed descent (dashed arrows). In the MAGIC design, a series of equally balanced crosses are made between founders before RILs are developed, so that each RIL from an eight-parent MAGIC population (for example) possesses ~12.5% of each parental genome. This enables many more combinations of traits to be evaluated in a common genetic background than is possible in other mapping population designs, which is essential for studying physiological tradeoffs among traits and trait adaptation to production environments (Bülow et al. 2019).

gene discovery for molecular marker development is enhanced when different population structures are used because they provide increasing degrees of complexity (biparental<NAM<MAGIC) for robust statistical analyses (Cook et al. 2021b; Handakumbura et al. 2019; Kremling et al. 2019; Guadagno et al. 2020).

10.4.4.4 Considerations for Ensuring High Grain Quality under Climate Change

As climate modeling and ecophysiological crop modeling become more sophisticated, it will be worthwhile for researchers to focus on assessing the physiological limits of sorghum to produce and maintain high harvestable yields of grain chemical constituents impacting end-use qualities. Here again is an excellent opportunity to use precision stress phenotyping of grain quality mapping populations to identify alleles impacting trait stability. Once the environmental limits are identified that result in decreased harvestable yields of a given constituent, USA regions that are least likely to exceed those limits can be targeted for long-term breeding.

10.4.5 Bioenergy, Bioindustrial, and Carbon Dioxide Removal (CDR)

In addition to its importance in providing calories for both people and animals, *Sorghum bicolor* is also widely touted as a preferred crop feedstock for bioenergy or bioindustrial processes. Sorghum's ability to fix carbon efficiently per unit area per unit input per unit time on relatively marginal soils makes the crop an attractive target for processes that utilize biomass. Moreover, the ability for sorghum to either produce high levels of simplified non-structural sugars or high quantities of cellulosic biomass allows its incorporation into a diverse array of bioindustrial applications. As world economies move to de-carbonize various sectors of the economy, the agricultural sector is not only targeted because of its increasing annual emissions, but also because agriculture has the unique ability to remove anthropogenic sources of CO_2 and sequester carbon in soils. Sorghum is touted as a key crop to mitigate climate change for many of the same reasons it is also preferred for bioenergy conversion.

Much like cereals, improvement of biomass crops can largely be lumped into two categories: yield or quality. Since (1) biomass is not traded as a commodity and (2) the purchaser of the biomass is also the end-user, improvement in biomass quality – however defined – is easily commercialized. This differs considerably from cereals that are traded as commodities and oftentimes the purchaser is not the end-user; this dichotomy creates difficulties in extracting value from nutritional improvement in cereals and thus has limited widespread adoption of nutritionally-enhanced cereals. For biomass crops, however, a large portion of research and development has focused on increasing the quality of biomass for specific industrial processes.

Despite limited commercialization of biomass sorghum, the scientific community has succeeded in utilizing genetic or genomic tools to drastically improve the quantity and quality of biomass produced largely because the industry leaned heavily on the more mature and established forage sorghum breeding programs. The two most widely commercialized traits in biomass sorghum improvement are the utilization of *Ma5/Ma6* maturity system and the brown midrib (*bmr*) mutants; these two traits, serendipitously, represent both yield and quality target traits, respectively.

10.4.5.1 The *Ma5/Ma6* Maturity System

The easiest way to increase biomass yield is to increase the length of vegetative growth and delay the development of reproductive tissues (i.e., grain). The *Ma5/Ma6* maturity system does just that: it allows for temperately adapted dwarfed photoperiod insensitive material to produce a photoperiod sensitive hybrid (Mullet et al. 2012). This drastically increases the ease of seed production, which aids in the commercialization of biomass hybrids. This also allows sorghum to take advantage of heterosis, or hybrid vigor, and allows for the creation of higher yielding and more resilient hybrids. The *Ma5/Ma6* system works by leveraging homozygous recessive alleles at complimentary early flowering time locations (Rooney and Aydin 1999; Yang et al. 2014). The two parental lines with complementary recessive alleles create a hybrid with complementary dominant alleles that allows for seed production of a photoperiod-sensitive biomass hybrid from two temperately adapted photoperiod insensitive grain hybrids. This system greatly enhances the yield of biomass hybrids while also significantly reducing the overall cost for hybrid seed production.

Ma6, also referred to as *Ghd7*, was previously mapped to chromosome 6 (Sobic.006G057866; Sb06g000570) by Murphy et al. (2014). The *Ma6* locus regulates photoperiod sensitivity and delays reproductive initiation by inhibiting floral activator *SbEhd1* (Murphy et al. 2014). The dominant allele is present in tropically adapted material and is utilized in forage and bioenergy sorghum to delay flowering and accumulate additional biomass (Rooney and Aydin 1999). *Ma5*(Sobic.001G087100) is located on chromosome 1 that likely encodes phytochrome C (Yang et al. 2014). However, molecular markers and allele mining are widespread throughout sorghum bioenergy and forage breeding initiatives (Kawahigashi et al. 2022).

10.4.5.2 Brown Midrib (*Bmr*) Genes

The brown midrib (BMR) phenotype is characterized by the brown coloration of the midribs on leaves. In addition to the visual phenotype, bmr lines have a recessive mutation that inhibits lignin formation increasing the nutritional value for ruminants and certain conversion processes (Bucholtz et al. 1980; Cherney et al. 1991; Dien et al. 2009). The creation of bmr lines relied heavily on mutagenesis techniques (Porter et al. 1978). Upon the discovery and characterization of these mutants, naturally occurring bmr traits were later identified (Vogler et al. 1994; Saballos et al. 2008); however, the vast majority of commercial bmr varieties or hybrids were derived from chemical mutagenesis (Bout and Vermerris 2003).

The *bmr* designation includes several mutations that not only led to significant commercialization but also elucidated the genetic mechanisms of lignin synthesis in the Poaceae grass family. Despite the wide range of genetic mechanisms, two recessive *bmr* mutations (*bmr6* and *bmr12*) have dominated early scientific analysis and industry adoption (Sattler et al. 2010). The wildtype allele restores proper lignin formation. Non-functional *bmr6* interrupts the cinnamyl alcohol dehydrogenase (CAD) activity (Sattler et al. 2009; Li and Huang 2017) while the recessive *bmr12* allele inhibits caffeic O-methyl transferase activity (Eudes et al. 2017). These two enzymes are important in monolignol biosynthesis, and a mutation in either of these genes disrupts lignin formation. Further, *bmr6* significantly impacts three lignin types (*p-hydroxyphenyl* lignin, guaiacyl lignin and syringyl lignin); whereas, *bmr12* alleles only slightly reduce *p-hydroxyphenyl* and guaiacyl lignins while greatly reducing syringyl lignin. Although the mutations create a visual phenotype that can easily be selected upon, molecular markers have been developed to aid in the selection of F_2 or F_3 plants as well as create stacked *bmr6 bmr12* mutations (Pedersen et al. 2008). In addition to its commercial viability, the characterization and development of *bmr* lines have greatly enhanced the knowledge of lignin biosynthesis in plant species.

10.4.5.3 Opportunities for CO_2 Removal (CDR)

Overall, the impact of bioenergy crops has not met forecasted expectations so far in the early 21st century. In 2005, USA Congress passed the Energy Policy Act of 2005, which created the Renewable Fuel Standard (RFS). In 2007, the RFS was revised by the Energy Independence and Security Act of 2007. The 2007 law revised the RFS and dictated that by 2022, the RFS would utilize 16 billion gallons of cellulosic biofuels throughout the USA. On 3 June 2022, the EPA revised their final volume requirements of cellulosic biofuel to just 630 million gallons, a mere 4% of the total Congress envisioned (US EPA 2022). Needless to say, bioenergy research, specifically research and development for cellulosic biofuels, has not had the grand societal impact that one might have expected when listening to experts in the early 2000s. Nevertheless, governments around the world continue to invest large sums of monies into bioindustrial feedstocks and are now increasingly repurposing this research to respond to the new carbon economy and the potential to use crops to aid in atmospheric CDR. Hopefully, leveraging crops for active CDR will be more successful than the adoption of cellulosic bioenergy or bioindustrial feedstocks.

Bioenergy crops have many analogous principles to the emerging field of crops for CDR (Conant et al. 2011). Both ideotypes need efficient fixation of carbon into biomass per unit time per unit input per unit area. Crops for CDR can also be primary crops – those whose main economic benefit is the sequestration of carbon – or dual-purpose crops where seed or grain is harvested and the stover is removed for supplemental income. Bioenergy and CDR crops will probably be tailored for marginal lands to limit their impact on the food supply system. With their highly analogous systems, the past 20 years of bioenergy research will greatly aid in the development of these new crop paradigms.

Despite the similarities, there are considerable differences between the bioenergy and CDR cropping systems. The difference is "yield", or what is defined as yield. For bioenergy, yield is defined as harvestable carbon; in CDR, this will probably be a percentage of fixed versus sequestered carbon, which is in large part driven by the mass of below-ground biomass. Root traits will be heavily prioritized in CDR systems. Given the lack of selection of root traits and difficulty in phenotyping root mass, it is entirely possible that specific alleles can be identified that increase root mass or depth and that genetic markers for these traits would be widely adopted for crop improvement. Bioenergy breeding has focused on biomass compositional traits because they impacted conversion rates. CDR traits will also focus on composition, but in the other direction. Bioenergy crop improvement focused on increased microbial digestion whereas CDR crops will be less amenable to microbial degradation to increase the permanence of carbon fixed by the plants. Instead of low lignin plants crops for CDR might be high lignin, which increases the longevity of soil-accumulated biomass. Regardless, allele mining for CDR-related traits is poised to become a growing field of interest – and its practical application spans both the academic and commercial sectors.

10.5 CHAPTER SUMMARY

Mining for allelic variation in diverse germplasm, specifically those in genebanks, was first described in the literature by Warburton and Hoisington (2001) and further explained by Kresovich et al. (2002). At the turn of the century, these scientists understood the value of genomics in the discovery of useful genotypes and genes to leverage for global crop improvement. Twenty years later, we have discovered many genetic regions, and candidate genes therein, associated with meaningful traits for plant health, quality, and resiliency, including a plethora of such in sorghum (Mace and Jordan 2010; Boyles et al. 2019; Mace et al. 2019). This chapter suggests that the obvious next step for plant scientists is to increase resolution to identify individual alleles within these genes that alter their expression or alternative function (e.g., regulatory). What are the grounds for this argument? Fully sequenced genomes and increased feasibility of precision gene editing will enable molecular breeders to soon manipulate individual genes and their alleles at an appreciable scale and affordable cost. This technology will seemingly work very well in a diploid sorghum crop. But what will serve as targets for this technology? Obviously, site-specific gene editing is only effective when guide RNA knows where to position nucleases. Thus, an emphasis in the near-term should be centered on elucidating specific genetic variants that could be high-confidence targets for editing. This work in sorghum should utilize established pangenomes that capture genome-wide polymorphisms and structural variants, further building onto *de novo* assembled reference genomes.

What is being proposed is not groundbreaking since molecular characterization and functional validation of genes have indeed found causal alleles for crop trait variation. Success stories in sorghum conveyed here included *Dw2* (Hilley et al. 2017) and *Ma2* (Casto et al. 2019). What will be new is the utility of more elaborate and detailed genome sequences that capture much greater coverage of genome-wide structural and non-structural (i.e., point) mutations. This chapter detailed what are to be considered prime targets (i.e., candidate genes) for discovering valuable allelism in sorghum that contributes to significant changes in phenotype, and how it is expressed across environment and genetic background. These underlying genetic targets span agronomic, abiotic and biotic stress, grain quality, and sustainability traits that warrant improvement for sorghum to be a productive cereal grain crop, resilient in extreme environments, and valuable for end-users. Targeting the identification and dissection of major effect loci provides a step-change means to improvement that would complement ongoing breeding activities focusing on marginal (though consistent) gains year-over-year.

Discoveries of superior alleles that drive phenotypic variance remain limited in sorghum despite the number of candidate variants and positional genes found to be significantly associated with meaningful traits. We argue here that this limitation is a function of inadequate investment to "mine the gold" for these alleles (Kresovich et al. 2002) in sorghum, and until recently, a lack of genomic resources publicly available to do so, not a limitation of deficient genetic diversity or its impermeable complexity. Therefore, we expect research in this area to provide fruitful breakthroughs throughout the 21st century to accelerate sorghum crop improvement at a pace far greater than the historical average of genetic gains for this underserved and underutilized crop.

REFERENCES

Abreha KB, Ortiz R, Carlsson AS, Geleta M (2021) Understanding the sorghum–*Colletotrichum sublineola* interactions for enhanced host resistance. *Front Plant Sci* 12(641969).

Abugri DA, Tiimob BJ, Apalangya VA, Pritchett G, McElhenney WH (2013) Bioactive and nutritive compounds in *Sorghum bicolor* (Guinea corn) red leaves and their health implication. *Food Chem* 138:718–723.

Ackerman A, Wenndt A, Boyles R (2021) The sorghum grain mold disease complex: Pathogens, host responses and the bioactive metabolites at play. *Front Plant Sci* 12: 893.

Adeyanju A, Little C, Yu J, Tesso T (2015) Genome-Wide association study on resistance to Stalk Rot diseases in grain Sorghum. *Genes Genomes Genet* 5(6): 1165–1175.

Armstrong SJ, Rooney WL, Peterson GC, Villenueva RT, Brewer MJ, Sekula-Ortiz D (2015) Sugarcane aphid (Hemiptera: Aphididae): Host range and sorghum resistance including cross-resistance from greenbug sources. *J Econ Entomol* 108:576–582.

Armstrong WD, Featherston WR, Rogler JC (1974) Effects of bird resistant sorghum grain and various commercial tannins on chick performance. *Poult Sci* 53:2137–2142.

Awika JM, Rooney LW (2004) Sorghum phytochemicals and their potential impact on human health. *Phytochemistry* 65:1199–1221.

Beta T, Rooney LW, Marovatsanga LT, Taylor JRN (200) Effect of chemical treatments on polyphenols and malt quality in sorghum. *J Cereal Sci* 31:295–302.

Biruma M, Martin T, Fridborg I, Okori P, Dixelius C (2012) Two loci in sorghum with NB-LRR encoding genes confer resistance to Colletotrichum sublineolum. *Theor Appl Genet* 124:1005–1015.

Blomstedt CK, Gleadow RM, O'Donnell N, Naur P, Jensen K, Laursen T, et al. (2012) A combined biochemical screen and TILLING approach identifies mutations in Sorghum bicolor L. Moench resulting in acyanogenic forage production. *Plant Biotechnol J* 10:54–66.

Boatwright JL, Brenton ZW, Boyles RE, Sapkota S, Myers MT, Jordan KE, et al. (2021) Genetic characterization of a Sorghum bicolor multiparent mapping population emphasizing carbon-partitioning dynamics. *Genes Genomes Genet* 11(4):jkab060.

Boatwright JL, Sapkota S, Jin H, Schnable JC, Brenton Z, Boyles R, et al. (2022) Sorghum Association Panel whole-genome sequencing establishes cornerstone resource for dissecting genomic diversity. *Plant J* 111(3):888–904.

Boddu J, Svabek C, Ibraheem F, Jones AD, Chopra S (2005) Characterization of a deletion allele of a sorghum Myb gene *yellow seed1* showing loss of 3-deoxyflavonoids. *Plant Sci* 169:542–552.

Borrell AK, van Oosterom EJ, Mullet JE, George-Jaeggli B, Jordan DR, Klein PE, et al. (2014) Stay-green alleles individually enhance grain yield in sorghum under drought by modifying canopy development and water uptake patterns. *New Phytol* 203:817–830.

Bout S, Vermerris W (2003) A candidate-gene approach to clone the sorghum Brown midrib gene encoding caffeic acid O-methyltransferase. *Mol Genet Genomics* 269:205–214.

Bowling RD, Brewer MJ, Kerns DL, Gordy J, Seiter N, Elliott NE, et al. (2016) Sugarcane aphid (Hemiptera: Aphididae): A new pest on sorghum in North America. *J Integr Pest Manag* 7(1):12.

Boyles RE, Brenton ZW, Kresovich S (2019) Genetic and genomic resources of sorghum to connect genotype with phenotype in contrasting environments. *Plant J* 97:19–39.

Boyles RE, Cooper EA, Myers MT, Brenton Z, Rauh BL, Morris GP, et al. (2016) Genome-wide association studies of grain yield components in diverse sorghum germplasm. *Plant Genome* 9(2).

Boyles RE, Pfeiffer BK, Cooper EA, Rauh BL, Zielinski KJ, Myers MT, et al. (2017) Genetic dissection of sorghum grain quality traits using diverse and segregating populations. *Theor Appl Genet* 130:697–716.

Breen JP, Teetes GL (1990) Economic injury levels for yellow sugarcane aphid (Homoptera: Aphididae) on seedling sorghum. *J Econ Entomol* 83(3):1008–1014.

Brenton ZW, Cooper EA, Myers MT, Boyles RE, Shakoor N, Zielinski KJ, et al (2016) A genomic resource for the development, improvement, and exploitation of sorghum for bioenergy. *Genetics* 204:21–33.

Brenton ZW, Juengst BT, Cooper EA, Myers MT, Jordan KE, Dale SM, et al. (2020) Species-specific duplication event associated with elevated levels of nonstructural carbohydrates in Sorghum bicolor. *Genes Genomes Genet* 10: 1511–1520.

Brown PJ, Myles S, Kresovich S. (2011) Genetic support for phenotype-based racial classification in sorghum. *Crop Sci* 51:224–230.

Bucholtz DL, Cantrell RP, Axtell JD, Lechtenberg VL (1980) Lignin biochemistry of normal and brown midrib mutant sorghum. *J Agric Food Chem* 28:1239–1241.

Bülow L, Nachtigall M, Frese L (2019) A MAGIC population as an approach to the conservation and development of genetic diversity of winter barley for breeding purposes by on-farm management. *J Cult Plant* 71(11):286–298.

Burke JJ, Chen J, Burow G, Mechref Y, Rosenow D, Payton P, et al. (2013) Leaf dhurrin content is a quantitative measure of the level of pre- and postflowering drought tolerance in sorghum. *Crop Sci* 53:1056–1065.

Burke JJ, Emendack Y, Hayes C, Xin Z, Burow G (2018) Registration of four postflowering drought-tolerant grain sorghum lines with early-season cold tolerance. *J Plant Regist* 12:386–390.

Burke JJ, Payton P, Chen J, Xin Z, Burow G, Hayes C (2015) Metabolic responses of two contrasting sorghums to water-deficit stress. *Crop Sci* 55:344–353.

Burow MD, Simpson CE, Starr JL, Paterson AH (2001) Transmission genetics of chromatin from a synthetic amphidiploid to cultivated peanut (*Arachis hypogaea* L.). broadening the gene pool of a monophyletic polyploid species. *Genetics* 159:823–837.

Burrell AM, Sharma A, Patil NY, Collins SD, Anderson WF, Rooney WL, et al. (2015) Sequencing of an anthracnose-resistant sorghum genotype and mapping of a major QTL reveal strong candidate genes for anthracnose resistance. *Crop Sci* 55:790–799.

Butler LG (1994) Chemical communication between the parasitic weed *Striga* and its crop host. In Inderjit, Dakshini KMM, Einhillig FA (eds). *Allelopathy: organisms, processes, and applications*, pp 158–168. American Chemical Society, Washington, DC, USA.

Cardoso L de M, Pinheiro SS, da Silva LL, de Menezes CB, de Carvalho CWP, Tardin FD, et al (2015) Tocochromanols and carotenoids in sorghum (*Sorghum bicolor* L.): diversity and stability to the heat treatment. *Food Chem* 172:900–908.

Casa AM, Pressoir G, Brown PJ, Mitchell SE, Rooney WL, Tuinstra MR, et al. (2008) Community resources and strategies for association mapping in sorghum. *Crop Sci* 48:30–40.

Casady AJ. (1967) Effect of a single height gene (*Dw3*) of *Sorghum vulgare* Pers. on certain culm and leaf blade characteristics. *Crop Sci* 7(6):595–598.

Casto AL, Mattison AJ, Olson SN, Thakran M, Rooney WL, Mullet JE (2019) *Maturity2*, a novel regulator of flowering time in *Sorghum bicolor*, increases expression of SbPRR37 and SbCO in long days delaying flowering. *PLoS One* 14:e0212154.

Cazzonelli CI, Pogson BJ (2010) Source to sink: Regulation of carotenoid biosynthesis in plants. *Trends Plant Sci* 15:266–274.

Cherney JH, Cherney DJR, Akin DE, Axtell JD (1991) Potential of brown-midrib, low-lignin mutants for improving forage quality. In: Sparks DL (ed) *Advances in Agronomy*, pp 157–198. Academic Press, Cambridge, MA, USA.

Conant RT (2011) Sequestration through forestry and agriculture. *Wiley Interdiscip Rev Clim Change* 2:238–254.

Cook JP, Acharya RK, Martin JM, Blake NK, Khan IJ, Heo H-Y, et al (2021a) Genetic analysis of stay-green, yield, and agronomic traits in spring wheat. *Crop Sci* 61:383–395.

Cook JP, McMullen MD, Holland JB, Tian F, Bradbury P, Ross-Ibarra J, et al. (2021b) Genetic architecture of maize kernel composition in the nested association mapping and inbred association panels. *Plant Physiol* 158:824–834.

Cooper EA, Brenton ZW, Flinn BS, Jenkins J, Shu S, Flowers D, et al. (2019) A new reference genome for Sorghum bicolor reveals high levels of sequence similarity between sweet and grain genotypes: Implications for the genetics of sugar metabolism. *BMC Genom* 20:420.

Cruet-Burgos C, Cox S, Ioerger BP, Perumal R, Hu Z, Thomas J. Herald, et al. (2020) Advancing provitamin A biofortification in sorghum: Genome-wide association studies of grain carotenoids in global germplasm. *Plant Genome* 13:e20013.

Cuevas HE, Cruet-Burgos CM, Prom LK, Knoll JE, Stutts LR, Vermerris W (2021) The inheritance of anthracnose (*Colletotrichum sublineola*) resistance in sorghum differential lines QL3 and IS18760. *Sci Rep* 11:20525.

Cuevas HE, Fermin-Pérez RA, Prom LK, Cooper EA, Bean S, Rooney WL (2019) Genome-wide association mapping of grain mold resistance in the US sorghum association panel. *Plant Genome* 12:180070.

Cuevas HE, Rosa-Valentin G, Hayes CM, Rooney WL, Hoffmann L (2017) Genomic characterization of a core set of the USDA-NPGS Ethiopian sorghum germplasm collection: Implications for germplasm conservation, evaluation, and utilization in crop improvement. *BMC Genomics* 18:108.

Dien BS, Sarath G, Pedersen JF, Sattler SE, Chen H, Funnell-Harris DL, et al. (2009) Improved sugar conversion and ethanol yield for forage sorghum (*Sorghum bicolor* L. Moench) lines with reduced lgnin contents. *Bioenergy Res* 2:153–164.

Dillon SL, Shapter FM, Henry RJ, Cordeiro G, Izquierdo L, Lee LS (2007) Domestication to crop improvement: Genetic resources for sorghum and *Saccharum* (Andropogoneae). *Ann Bot* 100:975–989.

Einhellig FA, Souza IF (1992) Phytotoxicity of sorgoleone found in grain Sorghum root exudates. *J Chem Ecol* 18:1–11.

Ejeta G (2007) Breeding for Striga resistance in sorghum: Exploitation of an intricate host–parasite biology. *Crop Sci* 47(S3):S–216–S–227.

Elango D, Wang X, Bhatnagar RS, Tan Q, Gaffoor I, Hu Z, et al. (2022) Association genetics of early season cold and late season frost tolerance in *Sorghum bicolor*. *Crop Sci* 62(5):1844–1865.

Emendack Y, Burke J, Bean S, Wilson J, Hayes C, Laza H (2018a) Composition, functional components, and physical characteristics of grain from staygreen and senescent sorghum lines grown under variable water availability. *Cereal Chem* 95:634–645.

Emendack Y, Burke J, Laza H, Sanchez J, Hayes C (2018b) Abiotic stress effects on sorghum leaf dhurrin and soluble sugar contents throughout plant development. *Crop Sci* 58:1706–1716.

Emendack Y, Hayes CM, Chopra R, Sanchez J, Burow G, Xin Z, et al. (2017) Early seedling growth characteristics relate to the staygreen trait and dhurrin levels in sorghum. *Crop Sci* 57:404–415.

Erickson J, Schott D, Reverri T, Muhsin W, Ruttledge T (2001) GC-MS analysis of hydrophobic root exudates of sorghum and implications on the parasitic plant *Striga asiatica*. *J Agric Food Chem* 49:5537–5542.

Esteban R, Moran JF, Becerril JM, García-Plazaola JI (2015) Versatility of carotenoids: An integrated view on diversity, evolution, functional roles and environmental interactions. *Environ Exp Bot* 119:63–75.

Eudes A, Dutta T, Deng K, Jacquet N, Sinha A, Benites VT, et al. (2017) SbCOMT (*Bmr12*) is involved in the biosynthesis of tricin-lignin in sorghum. *PLoS One* 12:e0178160.

Fedenia L, Klein RR, Dykes L, Rooney WL, Klein PE (2020) Phenotypic, phytochemical, and transcriptomic analysis of black sorghum (*Sorghum bicolor* L.) pericarp in response to light quality. *J Agric Food Chem* 68:9917–9929.

Finkelstein R (2013) Abscisic Acid synthesis and response. *Arabidopsis Book* 11:e0166.

Gaffoor I, Sandoya GV, Xavier KV, Nuckles EM, Pinnamaneni SR, Vaillancourt LJ, et al. (2021) Performance of novel sorghum germplasm in Pennsylvania and their response to anthracnose. *Crop Sci* 61:2612–2627.

Geera B, Ojwang LO, Awika JM (2012) New highly stable dimeric 3-deoxyanthocyanidin pigments from sorghum bicolor leaf sheath. *J Food Sci* 77(5):C566–C572.

Gilding EK, Frere CH, Cruickshank A, Rada AK, Prentis PJ, Mudge AM, Mace ES, Jordan DR, Godwin ID (2013) Allelic variation at a single gene increases food value in a drought-tolerant staple cereal. *Nat Commun* 4(1):1483.

Girard AL, Awika JM (2018) Sorghum polyphenols and other bioactive components as functional and health promoting food ingredients. *J Cereal Sci* 84:112–124.

Gladman N, Olson A, Wei S, Chougule K, Lu Z, Tello-Ruiz M, et al. (2022) SorghumBase: A web-based portal for sorghum genetic information and community advancement. *Planta* 255:35.

Gonzalez-Jorge S, Mehrshahi P, Magallanes-Lundback M, Lipka AE, Angelovici R, Gore MA, et al. (2016) ZEAXANTHIN EPOXIDASE activity potentiates carotenoid degradation in maturing seed. *Plant Physiol* 171:1837–1851.

Graham D, Lessman KJ (1966) Effect of height on yield and yield components of two isogenic lines of *Sorghum vulgare* Pers. *Crop Sci* 6(4):372–374.

Griebel S, Webb MM, Campanella OH, Craig BA, Weil CF, Tuinstra MR (2019) The alkali spreading phenotype in Sorghum bicolor and its relationship to starch gelatinization. J Cereal Sci 86:41–47.

Guadagno CR, Millar D, Lai R, Mackay DS, Pleban JR, McClung CR, et al. (2020) Use of transcriptomic data to inform biophysical models via Bayesian networks. *Ecol Modell* 429:109086.

Guo H, Jiao Y, Tan X, Wang X, Huang X, Jin H, et al. (2019) Gene duplication and genetic innovation in cereal genomes. *Genome Res* 29:261–269.

Handakumbura PP, Stanfill B, Rivas-Ubach A, Fortin D, Vogel JP, Jansson C (2019) Metabotyping as a stopover in genome-to-phenome mapping. *Sci Rep* 9:1858.

Hao H, Li Z, Leng C, Lu C, Luo H, Liu Y, et al. (2021) Sorghum breeding in the genomic era: Opportunities and challenges. *Theor Appl Genet* 134:1899–1924.

Harlan JR, de Wet JMJ (1972) A simplified classification of cultivated sorghum. *Crop Sci* 12:172–176.

Harlan JR, Stemler A (1976) The races of sorghum in Africa. In: Harlan JR (ed) *Origins of African plant domestication*, pp 465–478. De Gruyter, Berlin, Germany.

Harris K, Subudhi PK, Borrell A, Jordan D, Rosenow D, Nguyen H, et al. (2007) Sorghum stay-green QTL individually reduce post-flowering drought-induced leaf senescence. *J Exp Bot* 58:327–338.

Harvey TL, Kofoid KD, Martin TJ, Sloderbeck PE (1991) A new greenbug virulent to E-biotype resistant sorghum. *Crop Sci* 31:1689–1691.

Harvey TL, Wilde GE, Kofoid KD. (1997) Designation of a new greenbug, biotype K, injurious to resistant sorghum. *Crop Sci* 37(3):989–991.

Haussmann BIG, Hess DE, Omanya GO, Folkertsma RT, Reddy BVS, Kayentao M, et al. (2004) Genomic regions influencing resistance to the parasitic weed *Striga hermonthica* in two recombinant inbred populations of sorghum. *Theor Appl Genet* 109:1005–1016.

Hayes CM, Burow GB, Brown PJ, Thurber C, Xin Z, Burke JJ (2015) Natural Variation in synthesis and catabolism genes influences Dhurrin content in Sorghum. *Plant Genome* 8(2).

Herrman DA, Brantsen JF, Ravisankar S, Lee K-M, Awika JM (2020) Stability of 3-deoxyanthocyanin pigment structure relative to anthocyanins from grains under microwave assisted extraction. *Food Chem* 333:127494.

Hess DE, Ejeta G, Butler LG (1991) Research into germination of *Striga* seed by sorghum root exudates. In: Ransom JK, Musselman LJ, Worsham AD, Parker C (eds) *Proceedings of the 5th International symposium of parasitic weeds*, pp 217–222. Nairobi, Kenya, 24–30 June 1991.

Higgins RH, Thurber CS, Assaranurak I, Brown PJ (2014) Multiparental mapping of plant height and flowering time QTL in partially isogenic sorghum families. *Genes Genomes Genet* 4:1593–1602.

Hilley JL, Weers BD, Truong SK, McCormick RF, Mattison AJ, McKinley BA, et al. (2017) Sorghum *Dw2* encodes a protein kinase regulator of stem internode length. *Sci Rep* 7:4616.

Holmes CP, Casey J, Cook DJ (2017) Mashing with unmalted sorghum using a novel low temperature enzyme system: Impacts of sorghum grain composition and microstructure. *Food Chem* 221:324–334.

Howarth CJ, Rattunde EW, Bidinger FR, Harris D (1996) Seedling survival of abiotic stress: Sorghum and pearl millet. In: *Proceedings of the International conference on genetic enhancement of Sorghum and pearl millet*, pp 379–399. Lubbock, TX, USA, 22–27 September 1996.

Hu Z, Olatoye MO, Marla S, Morris GP (2019) An integrated genotyping-by-sequencing polymorphism map for over 10,000 sorghum genotypes. *Plant Genome* 12:180044.

Hunt BJ, Taylor AO (1976) Hydrogen cyanide production by field-grown sorghums. *NZ J Exp Agric* 4:191–194.

Hunt CH, van Eeuwijk FA, Mace ES, Hayes BJ, Jordan DR (2018) Development of genomic prediction in sorghum. *Crop Sci* 58:690–700.

Hyten DL, Song Q, Zhu Y, Choi I-Y, Nelson RL, Costa JM, et al. (2006) Impacts of genetic bottlenecks on soybean genome diversity. *Proc Natl Acad Sci USA* 103:16666–16671.

Ibraheem F, Gaffoor I, Chopra S (2010) Flavonoid phytoalexin-dependent resistance to anthracnose leaf blight requires a functional *yellow seed1* in *Sorghum bicolor*. *Genetics* 184:915–926.

Jambunathan R, Kherdekar MS, Stenhouse JW (1992) Sorghum grain hardness and its relationship to mold susceptibility and mold resistance. *J Agric Food Chem* 40:1403–1408.

Jampala B, Rooney WL, Peterson GC, Bean S, Hays DB (2012) Estimating the relative effects of the endosperm traits of waxy and high protein digestibility on yield in grain sorghum. *Field Crops Res* 139:57–62.

Jensen SE, Charles JR, Muleta K, Bradbury PJ, Casstevens T, Deshpande SP, et al. (2020) A sorghum practical haplotype graph facilitates genome-wide imputation and cost-effective genomic prediction. *Plant Genome* 13:e20009.

Jourdan M, Gagné S, Dubois-Laurent C, Maghraoui M, Huet S, Suel A, et al. (2015) Carotenoid content and root color of cultivated carrot: A candidate-gene association study using an original broad unstructured population. *PLoS One* 10:e0116674.

Kamal NM, Alnor Gorafi YS, Abdelrahman M, Abdellatef E, Tsujimoto H (2019) Stay-green trait: A prospective approach for yield potential, and drought and heat stress adaptation in globally important cereals. *Int J Mol Sci* 20(23):5837.

Kassahun B, Bidinger FR, Hash CT, Kuruvinashetti MS (2010) Stay-green expression in early generation sorghum [*Sorghum bicolor* (L.) Moench] QTL introgression lines. *Euphytica* 172:351–362.

Kawahigashi H, Oshima M, Nishikawa T, Okuizumi H, Kasuga S, Yonemaru JI (2013) A novel waxy allele in sorghum landraces in East Asia. *Plant Breed* 132(3):305–10.

Kawahigashi H, Yonemaru J-I, Kiyosawa A, Mizuno H, Kasuga S (2022) DNA marker analysis of flowering time, semi-dwarf, fertility restorer, and brown midrib genes in sorghum (*Sorghum bicolor* [L.] Moench). *Grassl Sci* 68:101–113.

Kent MA, Rooney WL (2021) Registration of Tx3483 through Tx3488 sorghum germplasm with waxy endosperm. *J Plant Regist* 15(3):606-11.

Khanal A, Adhikari P, Kaiser C, Lipka AE, Jamann TM, Mideros SX (2022) Genetic mapping of sorghum resistance to an Illinois isolate of *Colletotrichum sublineola*. *Plant Genome* 15(3):e20243.

Kimani W, Zhang LM, Wu XY, Hao HQ, Jing HC (2020) Genome-wide association study reveals that different pathways contribute to grain quality variation in sorghum (Sorghum bicolor). *BMC Genom* 21:1-9.

Klein RR, Miller FR, Bean S, Klein PE (2016) Registration of 40 converted germplasm sources from the reinstated sorghum conversion program. *J Plant Regist* 10:57–61.

Klein RR, Rodriguez-Herrera R, Schlueter JA, Klein PE, Yu ZH, Rooney WL (2001) Identification of genomic regions that affect grain-mould incidence and other traits of agronomic importance in sorghum. *Theor Appl Genet* 102:307–319.

Kremling KAG, Diepenbrock CH, Gore MA, Buckler ES, Bandillo NB (2019) Transcriptome-wide association supplements genome-wide association in *Zea mays*. *Genes Genomes Genet* 9:3023–3033.

Kresovich, S, Luongo, AJ and Schloss, SJ (2002) Mining the gold: Finding allelic variants for improved crop conservation and use. In: Engels JMM, Rao VR, Brown AHD, Jackson MT (eds) *Managing plant genetic diversity*, pp 379–386. CABI, Wallingford, England.

Kumar N, Brenton Z, Myers MT, Boyles RE, Sapkota S, Boatwright JL, et al. (2022) Registration of the sorghum carbon-partitioning nested association mapping (CP-NAM) population. *J Plant Regist* 16(3):656–663.

Lee S, Fu F, Liao C-J, Mewa DB, Adeyanju A, Ejeta G, et al. (2022) Broad-spectrum fungal resistance in sorghum is conferred through the complex regulation of an immune receptor gene embedded in a natural antisense transcript. *Plant Cell* 34:1641–1665.

Li A, Jia S, Yobi A, Ge Z, Sato SJ, Zhang C, Angelovici R, Clemente TE, Holding DR (2018) Editing of an alpha-kafirin gene family increases, digestibility and protein quality in sorghum. Plant Physiol 177(4):1425-1438.

Li H, Huang Y (2017) Expression of brown-midrib in a spontaneous sorghum mutant is linked to a 5'-UTR deletion in lignin biosynthesis gene SbCAD2. *Sci Rep* 7:11664.

Lin YR, Schertz KF, Paterson AH (1995) Comparative analysis of QTLs affecting plant height and maturity across the Poaceae, in reference to an interspecific sorghum population. *Genetics* 141:391–411.

Lo SCC, de Verdier K, Nicholson RL (1999) Accumulation of 3-deoxyanthocyanidin phytoalexins and resistance to Colletotrichum sublineolum in sorghum. *Physiol Mol Plant Pathol* 55:263–273.

Lu Y, Zhao G, Li Y, Fan J, Ding G, Zhao J, Ni X, Xu Y, Wang W (2013) Identification of two novel waxy alleles and development of their molecular markers in sorghum. Genome 56(5):283-8.

Luo H, Zhao W, Wang Y, Xia Y, Wu X, Zhang L, et al. (2016) SorGSD: A sorghum genome SNP database. *Biotechnol Biofuels* 9:6.

Mace ES, Jordan DR (2010) Location of major effect genes in sorghum (*Sorghum bicolor* (L.) Moench). *Theor Appl Genet* 121:1339–1356.

Mace E, Innes D, Hunt C, Wang X, Tao Y, Baxter J, et al. (2019) The Sorghum QTL Atlas: A powerful tool for trait dissection, comparative genomics and crop improvement. *Theor Appl Genet* 132:751–766.

Mahalakshmi V, Bidinger FR (2002) Evaluation of stay-green sorghum germplasm lines at ICRISAT. *Crop Sci* 42:965–974.

Makita Y, Shimada S, Kawashima M, Kondou-Kuriyama T, Toyoda T, Matsui M (2015) MOROKOSHI: Transcriptome database in *Sorghum bicolor*. *Plant Cell Physiol* 56:e6.

Martin JH, Washburn RS (1930) *Broomcorn growing and handling*. U.S. Department of agriculture farmer's bulletin No. 1631, pp 1–37. Washington, DC, USA.

McCormick RF, Truong SK, Sreedasyam A, Jenkins J, Shu S, Sims D, et al. (2018) The *Sorghum bicolor* reference genome: Improved assembly, gene annotations, a transcriptome atlas, and signatures of genome organization. *Plant J* 93:338–354.

Mengistu G, Shimelis H, Assefa E, Lule D (2021) Genome-wide association analysis of anthracnose resistance in sorghum [*Sorghum bicolor* (L.) Moench]. *PLoS One* 16:e0261461.

Menkir A, Ejeta G, Butler L, Melakeberhan A (1996) Physical and chemical kernel properties associated with resistance to grain mold in sorghum. *Cereal Chem* 73(5):616–627.

Mewa DB, Lee S, Liao CJ, Souza AM, Adeyanju A, Helm M, et al. (2022) *ANTHRACNOSE RESISTANCE GENE2* confers fungal resistance in sorghum. *Plant J* 113(2):308–326.

Ming R, Liu SC, Lin YR, da Silva J, Wilson W, Braga D, et al. (1998) Detailed alignment of saccharum and sorghum chromosomes: Comparative organization of closely related diploid and polyploid genomes. *Genetics* 150:1663–1682.

Mitra RD, Butty VL, Shendure J, Williams BR, Housman DE, Church GM (2003) Digital genotyping and haplotyping with polymerase colonies. *Proc Natl Acad Sci USA* 100:5926–5931.

Mohemed N, Charnikhova T, Bakker EJ, van Ast A, Babiker AG, Bouwmeester HJ (2016) Evaluation of field resistance to *Striga hermonthica* (Del.) Benth. in *Sorghum bicolor* (L.) Moench: The relationship with strigolactones. *Pest Manag Sci* 72:2082–2090.

Mohemed N, Charnikhova T, Fradin EF, Rienstra J, Babiker AGT, Bouwmeester HJ (2018) Genetic variation in *Sorghum bicolor* strigolactones and their role in resistance against *Striga hermonthica. J Exp Bot* 69:2415–2430.

Møller BL (2010) Dynamic metabolons. *Science* 330(6009):1328–1329.

Morishige DT, Klein PE, Hilley JL, Sahraeian SME, Sharma A, Mullet JE (2013) Digital genotyping of sorghum – a diverse plant species with a large repeat-rich genome. *BMC Genomics* 14:448.

Morris GP, Ramu P, Deshpande SP, Hash CT, Shah T, Upadhyaya HD, et al. (2013) Population genomic and genome-wide association studies of agroclimatic traits in sorghum. *Proc Natl Acad Sci USA* 110:453–458.

Muleta KT, Felderhoff T, Winans N, Walstead R, Charles JR, Armstrong JS, et al. (2022) The recent evolutionary rescue of a staple crop depended on over half a century of global germplasm exchange. *Sci Adv* 8:eabj4633.

Mullet JE, Rooney WL, Klein PE, Morishige D, Murphy R, Brady JA (2012) Discovery and utilization of sorghum genes (Ma5/Ma6). United States. www.osti.gov/servlets/purl/1080323

Mural RV, Grzybowski M, Miao C, Damke A, Sapkota S, Boyles RE, et al. (2021) Meta-analysis identifies pleiotropic loci controlling phenotypic trade-offs in sorghum. *Genetics* 218(3):iyab087.

Murphy RL, Morishige DT, Brady JA, Rooney WL, Yang S, Klein PE, et al. (2014) *Ghd7* (*Ma*$_6$) represses sorghum flowering in long days: *Ghd7* alleles enhance biomass accumulation and grain production. *Plant Genome* 7(2).

Netzly DH, Riopel JL, Ejeta G, Butler LG (1988) Germination stimulants of witchweed (*Striga asiatica*) from hydrophobic root exudate of sorghum (*Sorghum bicolor*). *Weed Sci* 36:441–446.

Nibouche S, Costet L, Holt JR, Jacobson A, Pekarcik A, Sadeyen J, et al. (2018) Invasion of sorghum in the Americas by a new sugarcane aphid (*Melanaphis sacchari*) superclone. *PLoS One* 13:e0196124.

Nicholson RL, Kollipara SS, Vincent JR, Lyons PC, Cadena-Gomez G (1987) Phytoalexin synthesis by the sorghum mesocotyl in response to infection by pathogenic and nonpathogenic fungi. *Proc Natl Acad Sci USA* 84:5520–5524.

Nida H, Girma G, Mekonen M, Lee S, Seyoum A, Dessalegn K, et al. (2019) Identification of sorghum grain mold resistance loci through genome wide association mapping. *J Cereal Sci* 85:295–304.

Nida H, Girma G, Mekonen M, Tirfessa A, Seyoum A, Bejiga T, et al. (2021) Genome-wide association analysis reveals seed protein loci as determinants of variations in grain mold resistance in sorghum. *Theor Appl Genet* 134:1167–1184.

Nimbal CI, Pedersen JF, Yerkes CN, Weston LA, Weller SC (1996) Phytotoxicity and distribution of sorgoleone in grain sorghum germplasm. *J Agric Food Chem* 44:1343–1347.

Oliver J, Fan M, McKinley B, Zemelis-Durfee S, Brandizzi F, Wilkerson C, et al. (2021) The AGCVIII kinase *Dw2* modulates cell proliferation, endomembrane trafficking, and MLG/xylan cell wall localization in elongating stem internodes of *Sorghum bicolor. Plant J* 105:1053–1071.

Owens BF, Lipka AE, Magallanes-Lundback M, Tiede T, Diepenbrock CH, Kandianis CB, et al. (2014) A foundation for provitamin A biofortification of maize: Genome-wide association and genomic prediction models of carotenoid levels. *Genetics* 198:1699–1716.

Pan L, Ma X, Hu J, Liu L, Yuan M, Liu L, et al. (2019) Low-tannin white sorghum contains more digestible and metabolisable energy than high-tannin red sorghum if fed to growing pigs. *Anim Produc Sci* 59:524–530.

Paterson AH, Bowers JE, Bruggmann R, Dubchak I, Grimwood J, Gundlach H, et al. (2009) The *Sorghum bicolor* genome and the diversification of grasses. *Nature* 457:551–556.

Paudyal S, Armstrong JS, Giles KL, Payton ME, Opit GP, Limaje A (2019) Categories of resistance to sugarcane aphid (Hemiptera: Aphididae) among sorghum genotypes. *J Econ Entomol* 112:1932–1940.

Pedersen JF, Toy JJ, Funnell DL, Sattler SE, Oliver AL, Grant RA (2008) Registration of BN611, AN612, BN612, and RN613 sorghum genetic stocks with stacked bmr-6 and bmr-12 genes. *J Plant Regist* 2:258–262.

Porter KB, Peterson GL, Vise O (1982) A new greenbug biotype. *Crop Sci* 22:847–850.

Porter KS, Axtell JD, Lechtenberg VL, Colenbrander VF (1978) Phenotype, fiber composition, and in vitro dry matter disappearance of chemically induced brown midrib (*bmr*) mutants of sorghum. *Crop Sci* 18:205–208.

Prom LK, Cuevas HE, Ahn E, Isakeit T, Rooney WL, Magill C (2020) Genome-wide association study of grain mold resistance in sorghum association panel as affected by inoculation with *Alternaria alternata* alone and *Alternaria alternata*, *Fusarium thapsinum*, and *Curvularia lunata* combined. *Eur J Plant Pathol* 157:783–798.

Punnuri S, Huang Y, Steets J, Wu Y (2013) Developing new markers and QTL mapping for greenbug resistance in sorghum [Sorghum bicolor (L.) Moench]. *Euphytica* 191(2):191–203.

Queiroz VAV, da Silva CS, de Menezes CB, Schaffert RE, Guimarães FFM, Guimarães LJM, et al. (2015) Nutritional composition of sorghum [Sorghum bicolor (L.) Moench] genotypes cultivated without and with water stress. *J Cereal Sci* 65:103–111.

Quinby JR, Karper RE (1954) Inheritance of height in sorghum. *Agron J* 46:211–216

Ramaiah KV, Chidley VL, House LR (1990) Inheritance of Striga seed-germination stimulant in sorghum. *Euphytica* 45:33–38.

Rhodes DH, Hoffmann Jr L, Rooney WL, Ramu P, Morris GP, Kresovich S (2014) Genome-wide association study of grain polyphenol concentrations in global sorghum [Sorghum bicolor (L.) Moench] germplasm. *J Agric Food Chem* 62(45):10916–10927.

Rhodes DH, Hoffmann L Jr, Rooney WL, Herald TJ, Bean S, Boyles R, et al. (2017) Genetic architecture of kernel composition in global sorghum germplasm. *BMC Genom* 18:15.

Rhodes DH, Kresovich S (2016) Sorghum [*Sorghum bicolor* (L.) Moench] genotypes with contrasting polyphenol compositions differentially modulate inflammatory cytokines in mouse macrophages. *J Chem* 2016(9640869), pp.10. https://doi.org/10.1155/2016/9640869

Rich PJ, Grenier C, Ejeta G (2004) Striga resistance in the wild relatives of sorghum. *Crop Sci* 44:2221–2229.

Rodríguez-del-Bosque LA, Terán AP (2015) *Melanaphis sacchari* (Hemiptera: Aphididae): A new sorghum insect pest in Mexico. *Southwest Entomol* 40:433–434.

Rooney WL, Aydin S (1999) Genetic control of a photoperiod-sensitive response in *Sorghum bicolor* (L.) Moench. *Crop Sci* 39:397–400.

Rooney WL, Collins SD, Klein RR, Mehta PJ, Fedriksen RA, Rodriguez-Herrera R (2002) Breeding sorghum for resistance to anthracnose, grain mold, downy mildew and head smuts. In: Leslie JF (ed) *Sorghum and millets diseases*, pp 273–280. Iowa State Press, Ames, IA, USA.

Rooney WL, Portillo O, Hayes C (2013a) Registration of ATx3363 and BTx3363 black sorghum germplasms. *J Plant Regist* 7:342–346.

Rooney WL, Rooney LW, Awika J, Dykes L (2013b) Registration of Tx3362 sorghum germplasm. *J Plant Regist* 7:104–107.

Rooney WL, Smith CW (2000) Genetics and cytogenetics. In: Smith CW, Fredriksen RA (eds) *Sorghum: Origin, history, technology, and production*, pp 261–306. John Wiley & Sons, New York, NY, USA.

Ruperao P, Thirunavukkarasu N, Gandham P, Selvanayagam S, Govindaraj M, Nebie B, et al. (2021) Sorghum pan-genome explores the functional utility for genomic-assisted breeding to accelerate the genetic gain. *Front Plant Sci* 12:666342.

Saballos A, Vermerris W, Rivera L, Ejeta G (2008) Allelic association, chemical characterization and saccharification properties of brown midrib mutants of sorghum (*Sorghum bicolor* (L.) Moench). *Bioenergy Res* 1:193–204.

Sade N, Del Mar Rubio-Wilhelmi M, Umnajkitikorn K, Blumwald E (2018) Stress-induced senescence and plant tolerance to abiotic stress. *J Exp Bot* 69:845–853.

Salas-Fernandez MG, Hamblin MT, Li L, Rooney WL, Tuinstra MR, Kresovich S (2008) Quantitative trait loci analysis of endosperm color and carotenoid content in sorghum grain. *Crop Sci* 48:1732–1743.

Salas-Fernandez MG, Kapran I, Souley S, Abdou M, Maiga IH, Acharya CB, et al. (2009) Collection and characterization of yellow endosperm sorghums from West Africa for biofortification. *Genet Resour Crop Evol* 56:991–1000.

Salinas I, Pró A, Salinas Y, Sosa E, Becerril CM, Cuca M, et al. (2006) Compositional variation amongst sorghum hybrids: Effect of kafirin concentration on metabolizable energy. *J Cereal Sci* 44:342–346.

Sapkota S, Boyles R, Cooper E, Brenton Z, Myers M, Kresovich S (2020a) Impact of sorghum racial structure and diversity on genomic prediction of grain yield components. *Crop Sci* 60:132–148.

Sapkota S, Boatwright JL, Jordan K, Boyles R, Kresovich S (2020b) Multi-trait regressor stacking increased genomic prediction accuracy of sorghum grain composition. *Agronomy* 10(9):1221.

Sattler SE, Funnell-Harris DL, Pedersen JF (2010) Brown midrib mutations and their importance to the utilization of maize, sorghum, and pearl millet lignocellulosic tissues. *Plant Sci* 178:229–238.

Sattler SE, Saathoff AJ, Haas EJ, Palmer NA, Funnell-Harris DL, Sarath G, et al. (2009) A nonsense mutation in a cinnamyl alcohol dehydrogenase gene is responsible for the Sorghum *brown midrib6* phenotype. *Plant Physiol* 150:584–595.

Saunders (1933) *Studies in phanerogamic parasitism*, with particular reference to *Striga lutea* Lour. South African Department of Agriculture Science Bulletin 128:1–56. https://cir.nii.ac.jp/crid/1571698600466712704#citations_container

Schertz KF, Dalton LG (1980) Sorghum. In: Fehr W, Hadley H (eds) *Hybridization of crop plants*, pp 577–588. American Society of Agronomy, Crop Science Society of America, Madison, WI, USA.

Scott MF, Ladejobi O, Amer S, Bentley AR, Biernaskie J, Boden SA, et al. (2020) Multi-parent populations in crops: A toolbox integrating genomics and genetic mapping with breeding. *Heredity* 125:396–416.

Seetharama N, Sachan RC, Huda AKS, Gill KS, Rao KN, Bidinger FR, et al. (1991) Effect of pattern and severity of moisture-deficit stress on stalk-rot incidence in sorghum II: Effect of source/sink relationships. *Field Crops Res* 26: 355–374.

Shehzad T, Okuno K (2020) Genetic analysis of QTLs controlling allelopathic characteristics in sorghum. *PLoS One* 15:e0235896.

Sherman RM, Salzberg SL (2020) Pan-genomics in the human genome era. *Nat Rev Genet* 1:243–254.

Shin D, Lee S, Kim TH, Lee JH, Park J, Lee J, et al. (2020) Natural variations at the stay-green gene promoter control lifespan and yield in rice cultivars. *Nat Commun* 11:2819.

Siame BA, Weerasuriya Y, Wood K, Ejeta G, Butler LG (1993) Isolation of strigol, a germination stimulant for *Striga asiatica*, from host plants. *J Agric Food Chem* 41:1486–1491.

Smolensky D, Rhodes D, McVey DS, Fawver Z, Perumal R, Herald T, et al. (2018) High-polyphenol sorghum bran extract inhibits cancer cell growth through ROS induction, cell cycle arrest, and apoptosis. *J Med Food* 21:990–998.

Stanley L, Yuan Y-W (2019) Transcriptional regulation of carotenoid biosynthesis in plants: So many regulators, so little consensus. *Front Plant Sci* 10:1017.

Starks, Mirkes (1979) Yellow sugarcane aphid: Plant resistance in cereal crops. *J Econ Entomol* 72(4):486–488.

Stephens JC, Miller FR, Rosenow DT (1967) Conversion of alien sorghums to early combine genotypes. *Crop Sci* 7:396–396.

Stutts LR, Vermerris W (2020) Elucidating anthracnose resistance mechanisms in sorghum — A review. *Phytopathology* 110:1863–1876.

Sukumaran S, Xiang W, Bean SR, Pedersen JF, Kresovich S, Tuinstra MR, Tesso TT, Hamblin MT, Yu J (2012) Association mapping for grain quality in a diverse sorghum dollection. *Plant Genome* 5(3):126.

Tao Y, Luo H, Xu J, Cruickshank A, Zhao X, Teng F, et al. (2021) Extensive variation within the pan-genome of cultivated and wild sorghum. *Nat Plants* 7:766–773.

Tao Y, Zhao X, Mace E, Henry R, Jordan D (2019) Exploring and exploiting pan-genomics for crop improvement. *Mol Plant* 12:156–169.

Taylor JRN, Schober TJ, Bean SR (2006) Novel food and non-food uses for sorghum and millets. *J Cereal Sci* 44: 252–271.

Tesso TT, Claflin LE, Tuinstra MR (2004) Estimation of combining ability for resistance to Fusarium stalk rot in grain sorghum. *Crop Sci* 44:1195–1199.

Tesso T, Perumal R, Little CR, Adeyanju A, Radwan GL, Prom LK, Magill CW (2011) Sorghum pathology and biotechnology-a fungal disease perspective: Part II. Anthracnose, stalk rot, and downy mildew. *Eur J Plant Sci Biotechnol* 6(1):31–44.

Thomas H, Howarth CJ (2000) Five ways to stay green. *J Exp Bot* 2000 51 :329–337.

Thomas MD, Sissoko I, Sacko M (1996) Development of leaf anthracnose and its effect on yield and grain weight of sorghum in West Africa. *Plant Dis* 80(2):151–153.

Upadhyaya HD, Wang YH, Sharma R, Sharma S (2013) SNP markers linked to leaf rust and grain mold resistance in sorghum. *Mol Breed* 32:451–462.

US EPA (2022) Renewable fuel standard program: RFS annual rules. 86(242):72436–72501, 21 December 2021. www.federalregister.gov/documents/2022/07/01/2022-12376/renewable-fuel-standard-rfs-program-rfs-annual-rules

Valluru R, Gazave EE, Fernandes SB, Ferguson JN, Lozano R, Hirannaiah P, et al. (2019) Deleterious mutation burden and its association with complex traits in sorghum (*Sorghum bicolor*). *Genetics* 211:1075–1087

Virupaksha TK, Sastry LVS (1968) Protein content and amino acid composition of some varieties of grain sorghum. *J Agric Food Chem* 16:199–203.

Vogler RK, Ejeta G, Butler LG (1996) Inheritance of low production of Striga germination stimulant in sorghum. *Crop Sci* 36:1185–1191.

Vogler RK, Ejeta G, Johnson K, Axtell JD (1994) *Characterization of a new brown midrib sorghum line* Agronomy Society of America, Madison, WI.

Wall JS, Blessin CW (1970) Composition of sorghum plant and grain. In: Wall JS, Ross WM (eds) *Sorghum production and utilization*, pp 118–166. Avi Publishing, Westport, CT,USA.

Walulu RS, Rosenow DT, Wester DB, Nguyen HT (1994) Inheritance of the stay green trait in sorghum. *Crop Sci* 34:970–972.

Wang B, Jiao Y, Chougule K, Olson A, Huang J, Llaca V, et al. (2021) *Pan-genome analysis in sorghum highlights the extent of genomic variation and sugarcane aphid resistance genes. bioRxiv*. https://doi.org/10.1101/2021.01.03.424980

Wang F, Zhao S, Han Y, Shao Y, Dong Z, Gao Y, et al. (2013) Efficient and fine mapping of *RMES1* conferring resistance to sorghum aphid *Melanaphis sacchari*. *Mol Breed* 31:777–784.

Wang YH, Upadhyaya HD, Dweikat I (2016) Sorghum. In: Singh M, Upadhyaya HD (eds) *Genetic and genomic resources for grain cereals improvement*, pp 227–251. Academic Press, San Diego, CA, USA.

Warburton M, Hoisington D (2001) Applications of molecular marker techniques to the use of international germplasm collections. In: Henry RJ (ed) Plant Genotyping–The DNA fingerprinting of plants, pp 83–93. CABI, Wallingford, England.

Webster JA (1990) Yellow sugarcane aphid (Homoptera: Aphididae): Detection and mechanisms of resistance among Ethiopian sorghum lines. *J Econ Entomol* 83(3):1053–1057.

Wenndt AJ, Boyles RE, Ackerman AJ, Sapkota S, Repka AC, Nelson RJ (2023) Host determinants of fungal species composition and symptom manifestation in the sorghum grain mold disease complex. *Plant Dis* 107(2):315–325..

Williams CN (1959) Resistance of sorghum to witchweed. *Nature* 184:1511–1512.

Wolters AMA, Uitdewilligen JG, Kloosterman BA, Hutten RCB, Visser RGF, van Eck HJ (2010) Identification of alleles of carotenoid pathway genes important for zeaxanthin accumulation in potato tubers. *Plant Mol Biol* 73:659–671.

Wu J, Cho E, Willett WC, Sastry SM, Schaumberg DA (2015) Intakes of lutein, zeaxanthin, and other carotenoids and age-related macular degeneration during 2 decades of prospective follow-up. *JAMA Ophthalmol* 133:1415–1424.

Wu Y, Huang Y (2008) Molecular mapping of QTLs for resistance to the greenbug *Schizaphis graminum* (Rondani) in *Sorghum bicolor* (Moench). *Theor Appl Genet* 117:117–124.

Wu Y, Guo T, Mu Q, Wang J, Li X, Wu Y, et al. (2019) Allelochemicals targeted to balance competing selections in African agroecosystems. *Nat Plants* 5:1229–1236.

Wu Y, Li X, Xiang W, Zhu C, Lin Z, Wu Y, et al. (2012) Presence of tannins in sorghum grains is conditioned by different natural alleles of *Tannin1*. *Proc Natl Acad Sci USA* 109:10281–10286.

Wu Y, Yuan L, Guo X, Holding DR, Messing J (2013) Mutation in the seed storage protein kafirin creates a high-value food trait in sorghum. *Nat Commun* 4(1):2217.

Xie P, Shi J, Tang S, Chen C, Khan A, Zhang F, et al. (2019) Control of bird feeding behavior by *Tannin1* through modulating the biosynthesis of polyphenols and fatty acid-derived volatiles in sorghum. *Mol Plant* 12:1315–1324.

Xin Z, Wang ML, Barkley NA, Burow G, Franks C, Pederson G, et al. (2008) Applying genotyping (TILLING) and phenotyping analyses to elucidate gene function in a chemically induced sorghum mutant population. *BMC Plant Biol* 8:103.

Xu W, Rosenow DT, Nguyen HT (2000) Stay green trait in grain sorghum: relationship between visual rating and leaf chlorophyll concentration. *Plant Breed* 119:365–367.

Yang L, Dykes L, Awika JM (2014) Thermal stability of 3-deoxyanthocyanidin pigments. *Food Chem* 160:246–254.

Yang S, Murphy RL, Morishige DT, Klein RR (2014) Sorghum phytochrome B inhibits flowering in long days by activating expression of *SbPRR37*and *SbGHD7*, repressors of *SbEHD1*, *SbCN8* and *SbCN12*. *PLoS One* 9(8):e105352.

Yang X, Scheffler BE, Weston LA (2004) *SOR1*, a gene associated with bioherbicide production in sorghum root hairs. *J Exp Bot* 55:2251–2259.

Yerka MK, Toy JJ, Funnell-Harris DL, Sattler SE, Pedersen JF (2015a) Registration of A/BN641 and RN642 waxy grain sorghum genetic stocks. *J Plant Regist* 9:258–261.

Yerka MK, Toy JJ, Funnell-Harris DL, Sattler SE, Pedersen JF (2015b) Registration of N619 to N640 grain sorghum lines with waxy or wild-type endosperm. *J Plant Regist* 9:249–253.

Yerka MK, Toy JJ, Funnell-Harris DL, Sattler SE, Pedersen JF (2016) Evaluation of interallelic waxy, heterowaxy, and wild-type grain sorghum hybrids. *Crop Sci* 56:113–121.

Zanini SF, Bayer PE, Wells R, Snowdon RJ, Batley J, Varshney RK, et al. (2022) Pangenomics in crop improvement-from coding structural variations to finding regulatory variants with pangenome graphs. *Plant Genome* 15:e20177.

Zhang J, Li W, Lv P, Yang X, Xu W, Ni X, et al. (2021) Whole-genome resequencing and transcriptome analysis provide insights on aphid-resistant quantitative trait loci/genes in *Sorghum bicolor*. *Plant Breed* 140:618–629.

Zheng LY, Guo XS, He B, Sun LJ, Peng Y, Dong SS, et al. (2011) Genome-wide patterns of genetic variation in sweet and grain sorghum (*Sorghum bicolor*). *Genome Biol* 12:R114.

Zhou L, Wang C, Gao X, Ding Y, Cheng B, Zhang G, et al. (2020) Genome-wide variations analysis of sorghum cultivar Hongyingzi for brewing Moutai liquor. *Hereditas* 157:19.

11 Allele Mining for Genome Designing in Pearl Millet

C. Tara Satyavathi, Supriya Ambawat, Charu Lata, Radha Shivhare, Chandan Kapoor, S. Mukesh Sankar, R. Gobu, Soumya SL, Chandra Nayaka Siddaiah, and Rakesh K. Srivastava*
*Corresponding author: csatyavathi@gmail.com

11.1 INTRODUCTION

Development of high yielding and superior agricultural crop varieties is made feasible by accretion of advantageous alleles from the extensive plant genetic resources that exist in the world. Allele mining is a technique for analyzing naturally occurring allelic variations or any acceptable alleles of a candidate gene influencing important agronomic properties that may be exploited to improve crops (Kumari et al. 2018). It is helpful in developing functional molecular markers for marker-assisted selection (MAS), gene prediction, expression analysis, evolutionary research, finding superior alleles, identifying new haplotypes and similarity analysis between and within species (Sarika and Amruta 2014). Allele mining is important for enhanced yield and improved quality, biotic and abiotic stress resistance, better nutrient efficiency etc. In order to account for the majority of functional changes relevant to the genes, true allele mining also examines the regulatory and non-coding areas of the candidate gene(s) in addition to the sequence differences in their coding regions.

Pearl millet [*Pennisetum glaucum* (L.) R. Br.] with chromosome number ($2n = 2x = 14$) is an important dry land cereal crop and ranks fifth in the world followed by rice, wheat, maize, and sorghum. It is mainly cultivated over 30 million ha globally with major crop area in Africa (>18 million ha) and Asia (>10 million ha). It is an important staple food for human consumption, as feed and fodder in livestock sector and in different industries. It is a climate-resilient crop and hence extremely beneficial to mitigate adverse conditions of changing climatic scenario and is able to also facilitate food and income security among farmers of arid regions. In India, it is the fourth most extensively cultivated food crop after rice, wheat and maize. During 2020–2021, pearl millet was cultivated on 7.57 million ha with a productivity of 1436 kg/ha and a production estimate of 10.86 million tons (Directorate of Economics and Statistics, Ministry of Agriculture and Farmers Welfare, Govt. of India). It grows rapidly with least inputs, possesses higher photosynthetic efficiency, balanced and good nutritional profile, tolerance to various abiotic and biotic stresses and can even survive under adverse agro-climatic conditions. It provides gluten-free grains, vitamins, higher and superior quality protein, antioxidants, and crucial micronutrients like zinc and iron with a higher nutritional value and a balanced essential amino acid profile in comparison to wheat, rice, maize and sorghum (Govindaraj et al. 2019; Satyavathi et al. 2021a,d). It is useful for protection against certain types of cancer and cardiovascular diseases.

Pearl millet possesses high out crossing rate of around 90% with huge diversity at both phenotypic and genotypic levels (Satyavathi et al. 2013; Singh et al. 2013). It has enormous gene potential, which can be explored to enhance its productivity. Crop wild relatives (CWRs) are a very useful resource of valuable genes and alleles for resistance against different biotic and abiotic stresses and

 DOI: 10.1201/9781003385004-11

various nutritional and agronomic traits can prove useful for enriching the gene pool (Kilian et al. 2011; Sharma et al. 2020b). Diverse alleles of important CWR features can significantly contribute to boost up genetic gain in pearl millet. CWRs of pearl millet are useful for enhancing richness of alleles among cultivated germplasm by means of evaluation of variety at haplotype, gene or allele level using high-end genomic resources and next-generation genotyping platforms (Sharma et al. 2020b). The available germplasm can be used to enhance the adaptability of pearl millet by targeted pre-breeding activities by multidisciplinary efforts from breeders, physiologists and pathologists. Various trait-specific genes and alleles can be utilized in different breeding programs to speed up breeding cycle and assimilation of traits. Genomic technologies must be used for specific gene introgressions and trait mapping for developing different biotic/abiotic resistant cultivars.

The genome resequencing project of pearl millet has proven very useful as it provided a database of >29 million genome-wide single nucleotide polymorphisms (SNPs) (Varshney et al. 2017). Additionally, 31 wild *Pennisetum* accessions collected in Sahel from Senegal and Sudan and were also resequenced under this project. The resequencing provided significant information about genetic diversity among wild *Pennisetum* species and relative fitness and adaptation of correlated genes at the time of domestication of pearl millet. Diversity of wild germplasm and variety of alleles and genes available in pearl millet CWRs may now be determined using the existing extensive genomic resources. Effectiveness of simple sequence repeats (SSRs) for research on genetic diversity was reported by Mariac et al. (2006). They found 46 wild accessions with more alleles and more diverse allelic composition than 421 cultivated lines from Niger. Beneficial genes and alleles can be selected for trait introgression and can be characterized at molecular level before their introgression in the crop plants. Various approaches like SNPs, next-generation sequencing, genotyping-by-sequencing (GBS) and whole-genome sequencing can be used for trait mapping in suitable genetic backgrounds. These techniques along with molecular markers are useful to reduce negative linkage drag usually reported among wide cross derivatives via genome-wide background selection. Exploitation of present markers will be useful in enriching the lost genetic diversity of pearl millet during domestication process. Molecular markers are able to support gene banks to determine inter- and intra-accession variability amid various regeneration cycles and characterize them at molecular level. In this chapter, we discuss the concepts, strategies and approaches, tools and techniques and application potential of allele mining in genome elucidation and improvement along with different challenges and emphasize the requirement of other refined "mining" approaches to accelerate the course of action of allele discovery and deployment in studies of gene expression and gene prediction along with molecular breeding.

11.2 PEARL MILLET GENOMIC RESOURCES

Pennisetum genus is comprised of around 140 extremely different species (Brunken et al. 1977). This genus is a diverse collection of species with ploidy levels varying from diploid to octoploid, annual, biennial or perennial life cycles, apomictic and sexual reproductive strategies. The species in the genus have four diverse basic chromosome numbers (n = 5, 7, 8, and 9) as well as sexual and apomictic reproductive strategies (Martel et al. 1997). The genus is classified into five sections based on morphological differences: Eupennisetum Stapf, Heterostachya Stapf & C. E. Hubb., Penicillaria (Willd) Benth & Hook f (= *Pennisetum*), Gymnothrix (P. Beauv.) Benth & Hook f (Stapf and Hubbard 1934). There are 55 taxa in the genus *Pennisetum*, of which two are intra-specific taxa and 53 are wild species.

Enormous numbers of accessions available at gene banks indicate that genetic resource collections are considered to possess a wealth of unidentified allelic variants. In 52 gene banks across 38 nations, approximately 4,900 accessions of 56 wild *Pennisetum* species are preserved for the genetic advancement of cultivated pearl millet (Sharma et al. 2017). Approximately 91% of the collected wild germplasm is preserved among ten gene banks in eight different nations. The two major pearl millet germplasm banks – International Crops Research Institute for Semi-Arid Tropics (ICRISAT) and the United States Department of Agriculture's National Plant Germplasm System (USDA-NPGS) – have around 26,000 accessions, which include historical accessions, wild relatives, landraces and breeding lines. The most accessions (963) are available at Plant Genetic Resources Conservation Unit at the Southern Regional Plant Introduction Station, University of Georgia, followed by Laboratoire des Ressources Génétiques et Amélioration des Plantes Tropicales, ORSTOM, France (798 wild accessions) and RS Paroda genebank at ICRISAT, Patancheru, India (794 wild accessions). With 24,373 accessions, RS Paroda genebank has the leading pearl millet germplasm collection in the world including landraces, breeding lines, extinct varieties, cultivars and CWRs from 51 different nations (genebank.icrisat.org). Additionally, Canadian Genetic Resources Program, Saskatoon has 3821 accessions, the Institut de Recherche Pour le Dévelopment (IRD), France, has 3968 accessions and National Bureau of Plant Genetic Resources (NBPGR), India, has 8291 collections (Singh and Nara 2022). At GRIN, 1283 active pearl millet accessions are accessible (Serba et al. 2017). More than 75% of them came from Burkina, Faso, India, Nigeria, and Zimbabwe. The plant genetic resources are endowed with abundant natural genetic variations, which might be used in pearl millet improvement programs (Serba et al. 2020; Sharma et al. 2020a, b).

Crop germplasm banks are critical for preserving genetic diversity but a sizeable proportion of beneficial/superior alleles present among these huge germplasm were not explored. Hence, it is believed that genetic resource collections contain a treasure of unexplored allelic variations, which requires unlocking. In this respect, allele mining becomes the means to identify unique, superior and favorable polymorphisms from the gene bank or natural population as well as their variants within related species that are involved in expression of a particular trait. Allele mining has made it easy to use available gene pools in global germplasm collections specifically consisting of wild allied species and local land races for almost all major crops. It facilitates the functional discovery of the gene(s) responsible for a specific trait as well as the extraction of numerous beneficial alleles available from the germplasm collected over periods and over geographical locations. Thus, shortlisting massive gene/seed bank into less redundant and more diverse DNA banks, which harbor unique germplasm containing novel genes and allele combinations is required. Nowadays, various bioinformatics databases/tools are available, which can assist in identification of superior alleles or their combinations for crop improvement. Many core subsets were developed in pearl millet since last two decades with the help of molecular and phenotypic data in order to indicate structure and range of genetic variability available among large collections (Upadhyaya 2011; Srivatava et al., 2020 a,b; Govindaraj et al., 2020). The ability to accurately characterize germplasm at the allele level for their duplicability and economic value remains still a bottleneck (Varshney et al. 2021). In this situation, resequencing in conjunction with an upsurge in reference sequences can make it easier to collect molecular data on huge collections of germplasm, lead to the discovery of novel alleles utilizing functional genomics (Varshney et al. 2017).

Microsatellite markers depicted that only 74% genetic diversity of wild relatives is present among domesticated groups indicating that domestication resulted in a reduction in genetic diversity (Oumar et al. 2008). Mariac et al. (2006) also found that cultivated pearl millet accessions had significantly fewer alleles and less gene diversity than wild accessions. These findings reveal that wild populations can be a valuable source for capturing novel allelic variations linked to climate resilient traits, which can be used to broaden the genetic base of cultivated pearl millet. These findings also make it clear that major loss of diversity among cultivated accessions has been caused due to domestication and selection processes. Diverse and novel sources of variation are required to introduce high frequency of valuable alleles and genes in order to continuously improve crop growth. Wild *Pennisetum* species are very important source of useful and novel alleles of various abiotic and biotic stresses, cytoplasmic male sterility (CMS) quality-related traits and forage yield. CWRs also have huge potential and can be used to discover new alleles from gene pools for novel variations (Suszkiw 2005; Tarwacka et al. 2013; Nazeer et al. 2014; Chandel et al. 2015; Sharma et al. 2020 a,b).

11.3 SEARCH FOR NEW ALLELES IN PEARL MILLET

Pre-breeding has emerged as a crucial link between conservation of plant genetic resources (PGRs) and their utilization in crop improvement programs. Pre-breeding aims at broadening the genetic base of crop species through detection of desired or beneficial traits in non-adapted germplasm of the crop and then transferring these traits to better adapted or cultivated species for further utilization in crop improvement programs. Landraces and wild relatives of the crop are widely used for searching available diversity for the trait. Enhancing the genetic base in cultivated crops is achieved by making available desired traits in usable form from the Crop wild relatives (CWRs) by making the crossing easy with the cultivated species without much genetic constraints. CWRs are storehouse of new alleles providing resistance against various biotic and abiotic stresses, which are generally encountered during crop production (Hajjar and Hodgkin 2007). These are the source population for broadening the genetic base of cultivated species. One of the main reasons for reduced fitness of modern varieties is their narrow genetic base due to "domestication bottleneck" having reduced genetic variation as compared to their wild progenitors (Ross-Ibarra et al. 2007; van Heerwarden et al. 2011). Introgressing alleles from wild gene pool have been used as a pre-breeding approach for improvement of cultivated species for limiting traits (Prohens et al. 2017). Pre-breeding was effectively utilized for transferring desired traits from gene pool diversity, which was previously a tedious task due to genetic incompatibilities and distinct geographic locations (Dwivedi et al. 2008; Ogbonnaya et al. 2013). Pre-breeding has been a necessary footstep towards linking desired traits in CWRs with the on-going variety development programs through making available the traits to be in a usable form without any genetic intricacies (Sharma et al. 2013).

Introgression lines (ILs) and chromosome substitution lines (CSLs) having genome fragments from wild relatives have been used most commonly following backcrossing the hybrid with cultivated recurrent parent (Prohens et al. 2017). A set of chromosome segment substitution lines (CSSLs) of 863B (drought-tolerant landrace) were developed in ICMB 841 (elite maintainer line) background to map complex traits (Kumari et al. 2014). Major marker loci linked to GCA and SCA of traits were established using 85 CSSLs for stover and grain yield traits under moisture deficit and irrigated conditions (Basava et al. 2019). Unadapted germplasm has been a source for new alleles/ genes for which advanced backcross-QTL strategy has been utilized to discover and transfer desired quantitative trait loci (QTLs) into cultivated varieties (Tanksley and Nelson 1996). Analysis of QTL is delayed until BC_2 or BC_3 generation and negative selection is employed to reduce frequency of unwanted donor alleles.

Pearl millet is blessed with considerable genetic diversity. Strong genetic structure of wild pearl millet accessions as compared to weak structure in cultivated pearl millet put forward strong unique and untapped diversity for breeding from wild populations of pearl millet (Varshney et al. 2017). These reports revealed that wild populations of pearl millet may probably be a potent source for new alleles related to traits conferring climate resilience. Moreover, due to more use of elite × elite germplasm for gaining higher yields, gene frequency of defensive traits is reduced (Cobb et al. 2019). Among them drought, heat and blast disease have been of concern causing significant production losses. Climate change has already shown its ill effects causing in-congenial temperature and other climatic variables causing yield reduction and frequent episodes of pest and disease incidence with more severity. Pearl millet is already a marginal crop under tremendous heat and drought stress, climate change further enhances the stress levels causing severe economic losses to the crop. With more emphasis on hybrids and strictly homozygous lines, narrowing of genetic diversity has caused the crop to be more prone to climate extremities. In such cases, novel gene sources are required, which confer tolerance to abiotic and biotic stresses in a better way. Traits classified as "must have" are high yield, seed size, early maturity, enhanced terminal drought tolerance, heat tolerance at flowering stage, blast and downy mildew resistance, zinc and iron biofortification while "value-added" traits comprise of lodging tolerance, improved heat tolerance, better forage yield, and reduced rancidity. Pre-breeding plays a crucial role in improvement of both trait types, which are generally not available in cultivated gene pool. On quality aspect, rancidity of pearl millet flour has been a major problem due to high fat and lipase content. Variability for this trait in available germplasm is very low making a strong case to look out for gene/alleles in wild relatives providing better shelf life and reducing rancidity in pearl millet flour.

11.4 ALLELES FROM CROP WILD RELATIVES (CWRS) IN PEARL MILLET IMPROVEMENT

Although pearl millet is a hardy crop adapted to harsh arid and semi-arid ecologies, still a variety of abiotic and biotic stresses impede in attaining high production (Satyavathi et al. 2021b,c). In a study on 529 wild pearl millet species consisted of *P. violaceum*, *P. purpureum*, *P. mollissimum*, *P. pedicellatum* and *P. polystachyon* under artificial and field conditions, 223 accessions were identified with no disease symptoms (Singh and Navi 2000). Crosses with two wild species *P. polystachion* and *P. pedicellatum* attempted to introgress downy mildew resistance but no seed set was noted in hybrids formed (Dujardin and Hanna 1987). Since last decade pearl millet farmers have been grappling with blast disease (*Magnaporthe grisea*), which was earlier believed to be a disease of minor importance in pearl millet. Since then efforts were put to introduce blast-resistant breeding material. The wild species *P. glaucum* ssp. *monodii* accession from the USA was found to be resistant to blast disease and regulated by three independent dominant genes (Hanna and Wells 1989), while single resistance gene derived from *P. glaucum* ssp. *monodii* was reported to confer blast resistance in Tift 85DB (Wilson et al. 1989). Seventeen potential sources of blast resistance were screened from *P. violaceum* in greenhouse against five pathotype-isolates Pg 53, Pg 56, Pg 45, Pg 119 and Pg 118 (Sharma et al. 2020a). *Pennisetum purpureum* can be an effective donor to improve forage quality and yield, stalk strength and also to provide restorer genes of the A_1 CMS system (Jauhar and Hanna 1998). Its hybrids with cultivated pearl millet has led to forage hybrids with high biomass and improved quality traits (Obok 2013). Genes controlling apomixis have been transferred from *P. squamulatum* and *P. orientale*, drought tolerance, pest resistance and perennial growth habit from *P. orientale*. Ornamental *Pennisetum* have been developed using *P. purpureum* and *P. squamulatum* species.

In one of the recent initiative for utilization of pearl millet wild germplasm, four advanced backcross populations were developed with two blast-resistant *P. violaceum* accessions. Stable introgression lines (ILs) and trait-specific gene pools having blast resistance were found and evaluated under terminal drought and flowering stage heat stress conditions for identification of tolerant lines (Sharma et al. 2020b). The rich wealth of genetic resources with considerable diversity has provided an abundance of natural variation for utilization in pearl millet improvement programs. Wild species and available germplasm and their utilization as a source of stress tolerant genes/alleles for pearl millet improvement are summarized in Table 11.1 and Table 11.2.

11.5 ALLELE MINING STRATEGIES IN PEARL MILLET

11.5.1 Allele Mining and Evolution

"Allele mining is a collection of techniques to dissect naturally occurring allelic variation in candidate genes that regulate important agronomic traits and has potential usage in crop improvement programs". It put forwards the vision for accelerated recovery of valuable adaptations conserved in gene banks. Allele mining experimentation helps in identifying naturally occurring variations

TABLE 11.1
Novel Allelic Sources from Wild *Pennisetums*

Trait	Species	References
Abiotic stresses		
Drought	*P. orientale* L.C. Rich, *P. ciliare* L. Mant., *P. mezianum* Leeke.	Dujardin and Hanna (1987), Rai et al. (1997)
Salinity	*P. clandestinum Hochst.* Ex Chiov., *P. purpureum Schumach*	Muscolo et al. (2013)
Freezing and winter hardiness	*P. flaccidum Griseb., P. orientale* L.C. Rich.	Dujardin and Hanna (1987), Stair et al. (1998)
Biotic stresses		
Blast	*P. setosum (Sw.) Rich., P. hordeoides* Steud., *P. pedicellatum* Trin, *P. polystachion ssp. atrichum* Stapf & C.E. Hubb.	Rai et al. (1997), Singh and Navi (2000), Shivhare and Lata (2017)
Downy mildew	*P. glaucum ssp. monodii* (Maire) Brunken, *P. pedicellatum* Trin., *P. polystachoin* L. Schult., *P. schweinfurthii*	Rai et al. (1997), Singh and Navi (2000),Wilson et al. (2004), Shivhare and Lata (2017)
Rust	*P. glaucum* ssp. *monodii* (Maire) Brunken, *P. polystachion* (L.) Schult., *P. pedicellatum* Trin., *P. violaceum (*Lam.) Rich., *P. subangustum Stapf* & C.E. Hubbard	Singh and Navi (2000), Rao et al. (2003), Shivhare and Lata (2017), Sharma et al. (2020b)
Leaf spot	*P. glaucum* ssp. *monodii* (Maire) Brunken	Hammons (1970)
Striga resistance	*P. glaucum* ssp. *Monodii, P. hordeoides* Steud., *P. glaucum ssp. Stenostachyum, P. pedicellatum* Trin.	Sy (1994), Koulengar (1995), Wilson et al. (2004)
Forage yield/ quality and related traits	*P. hordeoides* Steud., *P. cenchroides* Rich., *P. purpureum* Schum.,*P. setosum (*Sw.*)* Rich., *P. pedicellatum* Trin.,	Schmelzer (1997), Robert et al.(2011), Marshall et al. (2012)
Male sterility (MS) and fertility restoration sources	*P. schweinfurthii, P. glaucum subsp. monodii* (Maire) Brunken, *P. violaceum* (Lam.) Rich., *P. purpureum* Schumach.	Hanna (1989), Marchais and Perne's (1985), Rai et al. (1997), Jauhar and Hanna (1998),Upadhyaya et al. (2007)
Apomictic gene/ Apomixis	*P. setaceum* (Forssk.) Chiov., *P. ciliare* L. Mant., *P. orientale* L.C. Rich, *P. squamulatum* Fresen.	Goel et al. (2006), Robert et al. (2011)
Growth rate and yield traits	*P. glaucum subsp. monodii* (Maire) Brunken	Bramel-Cox et al. (1986)
Large seed size	*P. schweinfurthii* Pilg.	Serba et al. (2017)
Ornamentals		
	P. setaceum (Forssk.) Chiov., *P alopecuroides* (L.) Spreng., *P. orientale* L.C. Rich., *P. flassidum* Griseb.	Robert et al. (2011)

among alleles at loci of agronomic importance, which affect crop performance and distinctiveness. Agronomic loci can be recognized by different approaches like mutant screens, QTL analysis, association mapping and genome-wide surveys for artificial selection (Reeves et al. 2012). Novel alleles discovered at loci of agronomic importance might be incorporated into crop breeding programs via conventional or molecular tools and could be used in developing resistance against diseases like downy mildew, foliar blast and rust (Ambawat et al. 2016; Punnuri et al. 2016; Chelpuri et al. 2019) to promote yield increases (Liang et al. 2018; Patil et al. 2020), to produce better storage and nutritional properties (Anuradha et al. 2017; Yadav et al. 2021), or with better nutrient use efficiency

TABLE 11.2
Germplasm Accessions Identified for Biotic Stresses in Pearl Millet

Trait	Germplasm Accessions Identified	References
Blast	IP 22160, 21988, 21987, 21724, 21721, 21720, 21719, 21716, 21711, 21706, 21640, 21610, 21594, 21540, 21536, 21531, 21525	Sharma et al. (2020b)
Downy mildew	PS 202, IPW 407, IPW 151, 152, 153, 155	Singh and Navi, (2000),Wilson et al. (2004) Shivhare and Lata (2017)
Rust	IP 21629, 21645, 21658, 21660, 21662, 21711, 21974, 21975, 22038	Sharma et al. (2020b)
Striga Resistance	PS 755, 727, 639, 637, 622, 555, 549, 459, 428, 427, 287, 212, 208, 202, 190, 132, 64	Wilson et al. (2004)

(Gemenet et al. 2015; Pujarula et al. 2021) or in improving stress tolerance in pearl millet (Seghal et al. 2015; Tharanya et al. 2018).

Identification of right combination of alleles present among a set of germplasm lines are critical for efficient way to increase their utilization in crop improvement. Within genetic resource repositories, allelic variation of pertinent traits can be determined through allele mining. It is very useful to trace the evolution of alleles and understanding the molecular basis of unique trait variations, which aids in the discovery of novel haplotypes and the creation of allele-specific markers used in marker-assisted selection (MAS). Allelic variations usually arise through mutations, crossing over and environmental conditions selectively change the frequency of alleles within a population. Mutations typically take place in the genic regions of the genome can results in creation of new alleles which results in differential trait expressions. Mutations can affect a single nucleotide in the genome sequence, leading to single-nucleotide polymorphism (or SNP), or they can give rise to insertions or deletions, which are collectively referred to as InDels. The detection of mutations should be made even faster owing to allele mining, which also makes it possible to pinpoint and characterize allelic variants. Innovative strategies to accumulate beneficial alleles in genotypes are made possible by higher trait discovery resolution, improving genomic selection (GS) accuracy and precise genome editing technology. Identification of allelic variations affecting plant phenotype is very important to use genetic resources for plant variety development. Numerous quantitative trait nucleotides (QTNs) with varying phenotypic effects regulate agronomic characteristics. The loss of beneficial alleles brought on due to genetic drift and the reduced selection pressure on other QTNs may not be recovered by conventional breeding. The improvement of crops through the use of nucleotide sequence variants is a key objective of contemporary agricultural science.

11.5.2 Applications of Allele Mining in Pearl Millet

Allele mining has proved to be a potential tool in crop improvement programmes. The key applications of allele mining include identification of superior haplotypes, exploring and assessment of allelic diversity in gene banks, development of allele specific molecular markers, evolutionary studies etc. (Figure 11.1).

11.5.2.1 Exploring the Allelic Diversity in Gene Banks

Allele mining helps in exploring the huge diversity among plant genetic resources stored in gene banks, which eventually helps in utilizing the germplasm in crop improvement programs. It also aids in tapping the untapped diversity in crop wild relatives.

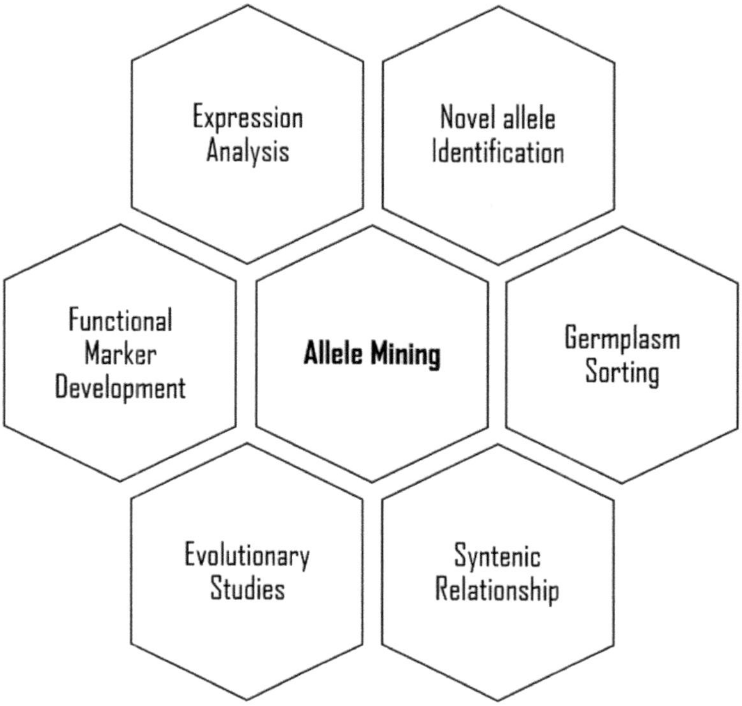

FIGURE 11.1 Applications of allele mining.

11.5.2.2 Haplotype Identification

Allele mining greatly assists in identification of superior haplotypes, which in-turn helps in improving the accuracy of genomic prediction and dissecting the functional polymorphism.

11.5.2.3 Functional Molecular Marker Development

Identifying molecular markers in the candidate gene itself helps in precise introgression of superior alleles to desired background by eliminating the linkage drag.

11.5.2.4 Deploying Evolutionary Relationship and Syntenic Relationship

The sequence information from allele mining studies may have potential application in deploying the evolutionary relationship among cultivated types and its progenitors. The sequence information may be compared across the genera to derive syntenic relationships.

11.5.3 Tools and Techniques for Allele Mining

11.5.3.1 Techniques to Identify New Alleles

Various techniques are followed for identification of allelic polymorphisms among genes. They may include non-sequencing-based methods [restriction digestion based, DNA confirmation based, chip based and targeting induced local lesions in genomes (TILLING)], genome/transcriptome sequence-based methods (whole genome shot gun method, locus-specific PCR amplification, alignment of available genomic sequences, reduced representation shot gun, overlapping of large insert cloned vector sequences by bacterial artificial chromosome (BAC) or P1-derived artificial chromosome (PAC), re-sequencing based techniques) [matrix-assisted laser desorption ionization-time of flight mass spectrometry (MALDI-TOF-MS) and next-generation sequencing (NGS) based], advanced genomic techniques, PCR-based allele priming, eco-TILLING based allele mining, *in silico* methods.

11.5.3.1.1 Advanced Genomic Techniques

A wide range of genomic tools and techniques have been developed for use in crop breeding for designing future crops. These include different approaches such as marker-assisted selection (MAS), marker-assisted recurrent selection (MARS), marker-assisted backcrossing (MABC), promotion/removal of allele through genome editing (PAGE/RAGE), genomic selection (GS), and haplotype-based breeding that can be combined with speed breeding. Modern genetic designs developed on the basis of multi-parent artificial populations have been applied for trait discovery in parallel with advances in genomic technologies. These designs have advantages of linkage analysis and association mapping like better power for quantitative trait locus (QTL) detection, more genetic diversity, controlled structure and higher mapping accuracy. Genomics-assisted breeding (GAB) techniques have aided towards thorough characterization of allelic variation for significant agronomic features and their effective integration into processes of cultivar development and germplasm augmentation (Varshney et al. 2021). Genome-wide understanding of pearl millet germplasm is required to get full information and explore the different genomic resources in this crop. Further, advanced biotechnological approaches like 'omics' are very useful and have much potential for improving pearl millet crop. 'Omics' includes several biotechnological applications like genomics, functional genomics, transcriptomics, proteomics, metabolomics, which leads to significant alterations in the plant transcriptome, proteome and metabolome (Ahuja et al. 2010; Ambawat et al. 2020).

Advent of next-generation sequencing (NGS) technologies has resulted into easy and rapid detection of genome-wide single nucleotide polymorphism (SNP) markers. Sequencing method can be fastened and used to genotype large populations by using genotyping-by-sequencing (GBS) and restriction site-associated DNA sequencing (RAD-seq) approaches, which are based on reduced representation sequencing technique (Elshire et al. 2011; Poland et al. 2012). Genotyping-by-sequencing can be used to create genome-wide SNP data sets in an effective way (Elshire et al. 2011). It has been thoroughly examined and demonstrated that GBS markers may be effective in creating high-density molecular maps for a number of crops like barley (*Hordeum vulgare* L.), oat (*Avena sativa* L.), maize (*Zea mays*) (Heet al. 2014). Utilizing a modified GBS procedure, a linkage map for pearl millet was also created using 2809 high-quality SNP markers (Moumouni et al. 2015). A new modeling research based on RAGE implies that in future multiplex genome editing can be an effective and essential technique to delete harmful alleles rather than practising selection against carriers. Breeding efforts might have decreased the fitness and genetic diversity of different present-day crops.

A contemporary method for enhancing various crops called genome editing or genome engineering via targeted nucleases can promise a significant boost in yield in the future. Recently, targeted nucleases like zinc finger nucleases (ZFNs), transcription activator-like effector nucleases (TALENs) and clustered regularly interspaced short palindromic repeat (CRISPR)/CRISPR associated protein (Cas) have been used successfully in several functional studies of genes for altering genes in plants by addition or deletion of desirable or undesirable traits for crop improvement programs (Tiwari and Lata 2019). These methods generate new possibilities for developing superior lines in pearl millet. Pearl millet (*P. glaucum* L.) was recently subjected to a genome-wide investigation of WRKY transcription factors under drought and salinity (Chanwala et al. 2020). The authors proposed that PgWRKYs is useful to boost crop productivity through genome editing techniques and can be helpful to ensure food security in the future.

RNAi silencing and genome editing tools are some of the recently developed biotechnological techniques, which can play key role in pearl millet improvement and methods for these are still being developed in pearl millet. MicroRNAs (miRNAs) act as regulators in plants, either stimulating or inhibiting gene expression. Shinde et al. (2020) used small RNA sequencing to find conserved and new miRNAs from salinity-tolerant genotype of pearl millet.

A number of transcriptomic studies were used in pearl millet to divulge the functions of some salinity stress-responsive genes like PgVDAC, PgDHN, PgNHX1, PgDREB2A (Desai et al. 2006; Verma et al. 2007; Mishra et al. 2007; Agarwal et al. 2010; Reddy et al. 2012; Singh et al. 2015; Shinde et al. 2018). Recently, comprehensive transcriptome analysis for drought stress response and Fe/Zn analysis was performed in pearl millet (Choudhary et al. 2015; Dudhate et al. 2018; Jaiswal et al. 2018; Shivhare et al. 2020 a,b; Satyavathi et al. 2022). In an another study, transcriptomic analysis was carried out in downy mildew susceptible and resistant genotypes on 454 Roche NGS platform and a total of ~685 Mb data was obtained with 1,575,290 raw reads (Kulkarni et al. 2016). RNA sequencing (RNA-Seq) has emerged as a popular technique for studying gene expression and discovering new genes and pathways among the various transcriptome analysis techniques. Unknown genes and novel transcripts can be efficiently found using RNA-Seq (Hrdlickova et al. 2017).

11.5.3.1.2 PCR-Based Allele Priming

Alleles in various genotypes are amplified by PCR in sequencing-based allele mining, which is followed by the discovery of nucleotide variation. Allele-specific PCR (AS-PCR) refers to a group of PCR variants that use a competitive reaction with allele-specific primers to amplify only specific alleles. Originally known as Kompetitive allele specific PCR (KASP), this technique is an AS-PCR version designed for fluorescence-based detection of amplified products. Both insertion-deletions (InDels) and single nucleotide polymorphisms (SNPs) may be targeted using probe sequences that are tailored in both orientations.

The technology also enables discrimination between up to four alleles at a single SNP location. Every allele-specific primer is made with penultimate base before the 3′ end placed at the SNP site in order to enhance the response specificity and differentiating capacity of SNP genotyping. The tool enables the building of unique fluorescence resonance energy transfer (FRET) cassette reporter systems for tests based on fluorescence. Allele-specific PCR (AS-PCR) (Spierings et al. 2006), PCR-amplification of specific alleles (PASA) (Sarkar et al. 1990), allele-specific amplification (ASA) and amplification refractory mutation system (ARMS) (Wu et al. 1989) are examples of closely related techniques that may be used for rapid detection of single-base changes in genomic DNA. All of these techniques make use of PCR-primer sets that were especially created and comprise allele-specific primers (ASP), whose allele specificity is dictated by nucleotide at or near the 3′ end. Practically any allelic variation can be analyzed with AS-PCR tests (Ye et al. 1992). Allele-specific PCR with SNP targeting is another technique (Kaur et al. 2020). Marker-assisted selection in plant breeding can be benefitted by the incorporation of single-plex PCRs for SNP analysis like TaqMan (Life Technologies, USA) and simple probe (Roche Applied Science, USA) (Makhoul et al., 2020; Della Coletta et al., 2021). The detection of the amplification of a particular allele using fluorescence is advancement for AS-PCR. Genotyping methods that have evolved from the initial AS-PCR strategy include PCR allele competitive extension (PACE) (Integrated DNA Technologies, Inc.) and Kompetitive allele specific PCR (KASP) (LGC Biosearch Technologies, Teddington, UK). Both of these methods enable precise bi-allelic discrimination of known SNPs and insertion-deletion polymorphisms using FRET for signal production (InDels).

11.5.3.1.3 Eco-TILLING-Based Allele Mining

The three major approaches for allele mining include modified tilling procedure called eco-tilling-based allele mining, sequencing-based allele mining and association mapping-based allele mining. Various steps involved in allele mining are described in Figure 11.2. Development of advanced genomic techniques including PCR-based allele priming and Eco-TILLING-based allele mining are being widely used now for mining superior alleles primarily underlying agronomically desirable traits including yield, quality, resistance to biotic and abiotic stresses and input use efficiency.

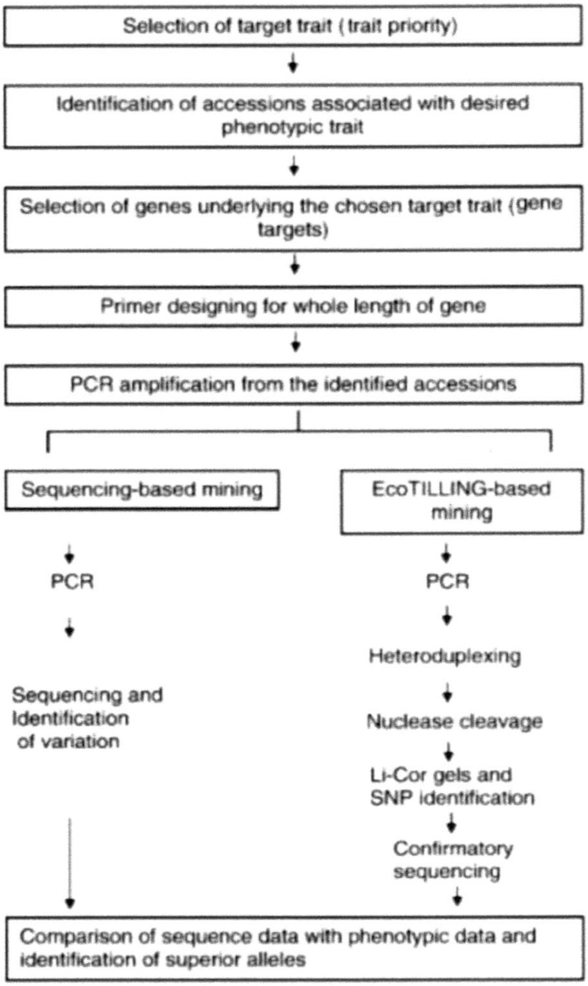

FIGURE 11.2 Various steps of allele mining.

Source: Adapted from Kumar et al. (2010).

Eco-TILLING is a molecular technique used to unravel natural genetic variation in comparison to induced mutations. Eco-Tilling technique detects natural mutation. It helps to rapidly detect variations and is cost effective since only one individual for each haplotype is to be sequenced. Eco-TILLING is termed to be one of the most powerful reverse genetic methods for research on the functional genomics of both plants and animals. Since many species cannot be chemically altered, Eco-TILLING can help identify natural variations and their potential gene functions (Gilchrist and Haughn 2005). This method enables a large number of samples to be quickly screened for a gene of interest to find naturally occurring SNPs and/or tiny INDELS and thus can reduce time and effort required for SNP discovery by eliminating identical haplotypes. The technique has demonstrated success in identifying DNA polymorphisms including changes in satellite repeat quantity (Till et al. 2006). Further, Eco-TILLING can be utilized in identifying disease-resistance variants and to assess the heterozygosity levels within a gene segment in highly heterozygous outcrossing species (Gilchrist and Haughn 2005).

Eco-TILLING can prove to be a useful tool in the case of well-established populations having systematically characterized morphological data. Various studies are available, which used Eco-TILLING to determine allelic variation in various crops like *Arabidopsis thaliana*, barley (*Hordeum vulgare*), maize (*Zea mays*), rice (*Oryza sativa*) etc. (Barkley and Wang 2008). In comparison to classical methods of detecting disease resistance alleles, Eco-TILLING offers numerous allelic variants in two resistance genes, which could be utilized to breed improved resistance cultivars (Mejlhede et al. 2006). Three different genetic backgrounds including world's reference germplasm of Tift 23DB1-P1-P5 were used by ICRISAT to generate TILLING populations or reverse genetic stocks, which can be further used for studying functional genomics of Fe and Zn as well as other significant traits (Srivastava et al. 2021).

The TILLING approach was efficiently used in different cereal crops like wheat, rice, maize, and sorghum (Slade et al. 2005; Sikora et al. 2011; Kurowska et al. 2011; Chen et al. 2014). Ethyl methane sulphate (EMS) was used to alter pearl millet [*Pennisetum typhoides* (Burn) Stapf.] Var. Co (CU)-9 seeds for morphological and chlorophyll mutation demonstrating that EMS can be an efficient chemical mutagen to cause mutation, which might be further helpful for breeding programs (Ambli and Mullainathan 2015). Use of TILLING in mutant breeding is crucial in order to employ radiation or chemical therapy to promote specific features (Szurman-Zubrzycka et al. 2018).

The M3 population of pearl millet was examined by Maryono et al. (2020) in order to assess performance and estimate genetic variability. High heritability was seen in a population exposed to various dosages of gamma rays for features like stem diameter, panicle diameter, plant number of nodes (Maryono et al. 2020). Allele mining based on association mapping is used to identify areas of genome linked to important traits through association or linkage-disequilibrium mapping. This strategy is based on the idea that alleles determining a phenotype and the markers that flank the locus are inherited as a block. It has been proposed to employ DNA markers to find valuable alleles among the enormous pools of genetic diversity.

11.5.3.1.4 In Silico Methods and Databases

Presently, there are different bioinformatics tools employed in successful determination of allelic variations associated with SNP-trait associations. These tools or software can be either open or paid and provide us information related to comparison between DNA sequences, detection of sequence polymorphism, prediction of amino acid changes, transcriptional regulatory elements, identification of transcription factor binding sites (TFBSs), and a database of plant promoter sequences. Specific pearl millet based genomic resource database such as GigaDB and TGIF-DB were recently used by many researchers for allele mining (Sankar et al. 2021, Singhal et al. 2021; Jaiswal et al. 2018). A set of bioinformatics tools to carry out allele mining assessment are listed in Table 11.3.

11.6 ALLELE MINING IN PEARL MILLET MOLECULAR BREEDING

In the last several years, enormous development has taken place in the field of genomics, which has led to the use and availability of a variety of new tools and techniques for specific and rapid breeding programs (Bollam et al. 2018; Meena et al. 2020; Srivastava et al. 2020a). Massive marker data are being used in genomics assisted breeding rather than one or a few loci linked with trait and development of a resource poor crop like pearl millet. In order to produce successful gene-based strategies that can eventually progress plant breeding programs for boosting abiotic stress tolerance in pearl millet, support for genomics and genetic transformation technologies is necessary, which will promote sustainable agriculture in dry areas. In the last few years, there have been significant advancement in the science of genomics, which has made a variety of unique techniques available and usable for breeding programs that are more precise and quick (Bollam et al. 2018; Meena et al. 2020; Srivastava et al. 2020a). Genomics-assisted breeding (GAB) is crucial in maximizing the prospective of contemporary genomic resources, identifying, utilizing and enhancing

TABLE 11.3
List of Bioinformatics Tools Used to Undertake an Investigation of Allele Mining

S.No.	Name of Webtool/ Databases	Function	Source	Reference
1.	PLACE (plant *cis*-acting regulatory DNA elements)	Database of Transcription Factor (TF) binding motifs	www.dna.affrc.go.jp/PLACE/index.html	Higo et al. (1999)
2.	plantCARE	Plant *cis* acting regulatory elements database	http://bioinformatics.psb.ugent.be/webtools/plantcare/html	Lescot et al. (2002)
3.	TRANSFAC	Database of TF and TF binding motifs	www.gene-regulation.com/pub/programs.html	Matys et al. (2003)
4.	JASPAR	Transcription Factor Binding Site Database	http://jaspar.genereg.net/	Bryne et al. (2008)
5.	W-AlignACE	Motif discovery tool	www1. spms.ntu.edu.sg/~chenxin/W-AlignACE/	Chen et al. (2008)
6.	MEME (Multiple EM for Motif Elicitation)	Motif discovery tool	http://meme.nbcr.net/meme4_1/cgi-bin/meme.cgi	Bailey et al. (2006)
7.	Plantprom DB	Plant promoter database	http://mendel.cs.rhul.ac.uk/mendel.php?topic=plantprom	Shahmuradov et al. (2003)
8.	MATinspector	To predict TFBS and of promoter analysis	www.genomatix.de/products/MatInspector/	Cartharius et al. (2005)
9.	EPD	Eukaryotic Promoter Database	www.epd.isb-sib.ch/	Schmid et al. (2004)
10.	TRRD	Regulatory regions database (description of Regulatory Elements and TFBS)	http://wwwmgs.bionet.nsc.ru/mgs/gnw/trrd/	Kolchanov et al. (2000)
11.	Gigadb	Pearl millet genomic resource database	http://gigadb.org/dataset/100192	Hunter et al., 2019
12.	TGIF-DB	Terse Genomics Interface for Developing Botany	https://webpark2116.sakura.ne.jp/rlgpr/blast.php	Tsugama and Takano (2021)
13.	PMDTDb (Pearl Millet Drought Transcriptome Database)	Transcriptomic signature of drought response database	http://webtom.cabgrid.res.in/pmdtdb/	Jaiswal et al. (2018)
14.	ooTFD	Object oriented-transcription factors database	www.ifti. org/ootfd/	Ghosh (2000)
15.	AGRIS	TF and RE databases	http:// arabidopsis.med. ohio-state.edu	Davuluri et al. (2003)
16.	ModuleFinder and CoReg	Gene co-expression and link with promoter elements	–	Holt et al. (2006)
17.	FastPCR	Nucleotide sequence analysis and primer design	–	Kalendar et al. (2017)
18.	Primer3	Primer design	http://frodo.wi. mit.edu/primer3/	Untergasser et al. (2012)
19.	BioEdit	Nucleotide sequence analysis	www.mbio.ncsu.edu/BioEdit/BioEdit.html	Hall (2004)
20.	Clustal W	Sequence alignment	www.ebi.ac.uk/Tools/clustralw/	Thompson et al. (1994)

allelic variation for cultivar and germplasm improvement. Future GAB (GAB 2.0) maintenance will depend on a number of unique strategies that expedite the focused exploitation of allelic variations in alleles for creation of novel diversity and enable their quick and effective integration in crop improvement programs. With increasing the selection efficiency and shortening time-consuming phenotypic selection process, marker-assisted breeding has generally become increasingly important in addressing these long-standing difficulties associated with plant breeding. However, molecular markers cannot entirely replace the traditional breeding techniques; they will only cover very specific elements of plant breeding. Therefore, it is recommended that modern genomic techniques be used in a systematic supplement to traditional breeding methods in order to speed up the process. Using molecular markers in conjunction with procedures like "speed breeding" and RGT (rapid generation turn over), conventional breeding may take less time (Watson et al. 2018). The adoption of double haploid (DH) lines to speed up homozygosity by reducing 6–8 generations to 2–3 generations and saving 5–6 years could be the other game-changer. Based on amino acid conservation modeling and genomic evolutionary rate profiling of sequence data of several accessions, it was clear that conventional breeding was ineffective at eliminating harmful alleles.

Since the development of the first map of pearl millet, marker system and genotyping speed have changed over time and significant advancements have been made in genome coverage, resolution and marker density (Srivastava et al. 2020a). Both distribution and density of markers have drastically risen and the coverage has more than doubled. Marker-aided breeding efforts and genetic research requires genomic resources like information on gene sequences, DNA-based molecular markers and genetic linkage maps. Development of a pearl millet linkage map using restriction fragment length polymorphism (RFLP) as a molecular tool was the first significant accomplishment (Liu et al. 1994). In the last few decades, several DNA-based molecular markers including restriction fragment length polymorphism (RFLP), amplified fragment length polymorphism (AFLP), random amplified polymorphic DNA (RAPD), conserved intron specific primers (CISP), expressed sequence tag-derived simple sequence repeats (EST-SSRs) markers, diversity arrays technology (DArT), sequence-tagged sites (STSs), genomic simple sequence repeat (gSSRs), single nucleotide polymorphisms (SNPs), single-stranded conformation polymorphism-SNP (SSCP-SNP) were developed to discriminate genetic variability, marker assisted breeding and QTL/gene identification to boost up pearl millet breeding programs (Allouis et al. 2001; Qi et al. 2004; Bertin et al. 2005; Senthilvel et al. 2010; Supriya et al. 2011; Sehgal et al. 2012; Rajaram et al. 2013; Srivastava et al. 2020a).

Mapping populations and molecular linkage maps are additional important genetic resources. Four F_2 populations generated through crosses between PT 732B x P1449-2, 81B x ICMP 451, LGD x ICMP 85410, ICMB 841 x 863B and were used to develop first integrated map (Qi et al. 2004). F_2 population of ICMB 841-P3 x 863B-P2 cross was also utilized to include recently created SSR markers into earlier maps. RILs were then employed to trace the genetic linkage of the pearl millet genome. RILs produced from the crosses Tift23DB x PI536400 (Pedraza-Garcia et al. 2010), H 77/833-2 x PRLT 2/89-33 (Supriya et al. 2011; Sehgal et al. 2012) were used for genetic linkage mapping. A consensus linkage map was created using four RIL populations derived from 81B-P6 × ICMP 451-P8, ICMB 841-P3 × 863B-P2, PT 732B-P2 × P1449-2-P1, H77/833-2 × PRLT 2/ 89-33 (Rajaram et al. 2013). F_2 population generated from a cross between wild pearl millet [11611-(PS202-14)-121] and a cultivated pearl millet (SOSAT-IBL-197) was also utilized for constructing high density map (Moumouni et al. 2015). A 96-plex ApeKI GBS library was developed from the 186 RILs and from their parents ("Tift 99D2B1" and "Tift 454") and F_1 population (Punnuri et al. 2016). Sequencing was done for these populations and data generated was used further for development of reference genetic map with 150 RILs having 333,567 sequence tags and 16,650 SNPs dispersed throughout seven pearl millet chromosomes with a genetic distance of 716.7 cM and 23.23/cM overall SNP average density. Using this map, the QTLs for blooming and resistance to *Pyricularia* leaf spot disease were also mapped in pearl millet.

DArT and SSR markers were used to develop a genetic linkage map using 317 RILs offspring from the cross between ICMS 8511-S1-17-2-1-1-B-P03×AIMP 92901-S1-183-2-2-B-08 (Kumar et al. 2018). Mahendrakar et al. (2020) discovered novel genes in seven linkage groups. Recently, Srivastava et al. (2021) used the same mapping population to study agronomic parameters on seven linkage groups. Chromosomal segment substitution lines (CSSLs) for 7 LGs were generated by introducing overlapping chromosome segments from 863B line into genetic background of elite ICMB 841 line in order to avoid masking of minor-effect alleles/QTLs and issues of segregation distortion (Kumari et al. 2014). They are beneficial genetic stocks for studying the mechanisms behind complex characteristics in pearl millet, as well as for minor-effect QTL discovery, fine mapping and trait mechanism investigations. These genetic tools were also used in comparative genetics (Gale et al. 2005), diversity assessment (Bhattacharjee et al. 2002), recombination rate studies (Liu et al. 1996), domestication syndrome analysis (Poncet et al. 2002), identifying and mapping QTLs for terminal drought (Yadav et al. 2002), reduced salt uptake (Sharma et al. 2014), grain micronutrient enrichment (Kumar et al. 2018; Singhal et al. 2021), grain and stover yield (Yadav et al. 2004), fertility restorer genes (Pucher et al. 2018), rust and blast resistance (Morgan et al. 1998; Punnuri et al. 2016), downy mildew resistance (Jones et al. 2002; Breese et al. 2002; Gulia et al. 2007). Unique genetic resources were generated in order to map Fe and Zn content of grain in pearl millet. This comprised reverse genetic stocks like TILLING populations as well as forward genetic stocks including association mapping panels and bi-parental mapping populations. Based on a previously published consensus map, Kumar et al. (2016) created a linkage map with 305 markers (96 SSRs and 208 DArTs) having 1749 cM length and 5.73 cM inter-marker distance.

NGS-based sequencing platforms like RAD-seq, GBS and others are currently gaining popularity because of their quick, higher density genome coverage and affordable platform for development of markers and genotyping (Shivhare et al. 2020 a,b). A change from bi-parental to second-generation multi-parent linkage mapping populations, like nested association mapping population (NAM) (Yu et al. 2008) and multi-parent advanced generation intercross (MAGIC) population (Kover et al. 2009) was also brought about by technological advancements in NGS (Davey et al. 2011). These populations were used to maximize the benefits of multi-parent allelic diversity for fine QTL mapping for desirable traits by means of an integrated mapping population. The benefits of both association and linkage mapping for crop development can be used because of second-generation populations and advanced computing resources (Bevan and Uauy 2013). It was suggested to use the MAGIC population for genome-wide association mapping in order to identify key genes, quantitative trait loci (QTLs) and potential novel loci linked to important traits (Kover et al. 2009; Bandillo et al. 2013). It is difficult to decipher allelic differences for stress tolerance and yield-related variables without knowledge of MAGIC and NAM populations in pearl millet. Pearl millet accessions have unique alleles and can be controlled by taking use of multi-parental populations. To capture more extensive genetic variability of characteristics, the pearl millet inbred germplasm association panel (PMiGAP) has been established as a germplasm-based population (Sehgal et al. 2015). This rigorously produced association panel from the ICRISAT global collection underwent a structural analysis, which showed six subpopulations relating primarily to pedigree or related agronomic variables (Sehgal et al. 2015). It demonstrated significance of inclusion of several morphological forms in a molecular investigation for identification of genome-wide markers, detection of QTLs or estimation of breeding value based on genomics. A thorough development of genomic tools for marker-assisted breeding and the discovery of novel allelic variations are anticipated to be aided by both within- and between-race evaluation.

Future crop designing will require the use of genomic breeding techniques that optimize crop genomes by accumulating beneficial alleles and eliminating harmful alleles. Pearl millet genetic advances have been made due to genomics and genomics-assisted breeding, which has also improved the program's accuracy and efficacy. The sequencing of 1,000 genomes has been a significant step forward for pearl millet (Varshney et al. 2017). The groundwork for mapping, trait discovery, use of alleles, QTLs and candidate genes connected with important traits was set by this effort. Additionally, it has aided in the creation and application of whole-genome prediction models for global pearl millet community (Jarquin et al. 2020; Srivastava et al. 2020a). A massive (>32 million) collection of genome-wide SNPs has been developed by mapping population parents, whole-genome resequencing of Pearl Millet Inbred Germplasm Association Panel and elite hybrid parental lines. These improvements present prospects for both rapid resequencing lines to dig and map genes of interest as well as for rapid mapping and deployment of genes of agronomic value. It can also be helpful to identify superior haplotypes for many attributes using the sequencing-based mapping approach and can lay foundation for haplotype-based breeding (Sinha et al. 2020). Several agronomically significant traits were mapped using SSR and SNP markers, such as traits associated with terminal drought, diseases, heterotic gene pools, combining ability loci, fodder and grain quality (Kumar et al. 2018; Basava et al. 2019; Gupta et al. 2020; Srivastava et al. 2020 a,b). SNP-based high-throughput platform can also be used to remap the available QTLs. This will enable incorporation into current high-throughput genotyping technologies for pearl millet breeding pipelines (Srivastava et al. 2020 a,b). Additionally, it is now possible to map breeding-related features quickly because of accessibility of cost-effective genotyping tools, reference genome sequence, resequencing data for large-scale whole-genome and accurate phenotyping platforms (Bohra 2020). The Pearl Millet Improvement Program will be accelerated in future by the use of genomic breeding technologies like forward breeding, gene editing, haplotype-based breeding and genomic prediction (Varshney et al. 2020).

11.7 ALLELE MINING AND QTL MAPPING IN PEARL MILLET

Molecular markers were utilized to create a genetic linkage map for pearl millet as well as to locate and map QTLs for other biotic and abiotic stress conditions and agronomic variables (Liu et al. 1994; Morgan et al. 1998; Breese et al. 2002; Jones et al. 2002; Yadav et al. 2004; Gulia et al. 2007; Sharma et al. 2014). These genetic techniques have boosted recombination rates, domestication syndrome studies and comparative genetics (Devos and Gale 2000; Poncet et al. 2002; Bhattacharjee et al. 2002). In a study, DArT markers and 70 SSRs were used for mapping 168 F_7 pearl millet RILs derived from the hybrid of 81B-P6 and ICMP 451-P8 to find QTLs on LG1 with a LOD score of 27 for rust resistance (Ambawat et al. 2016). A linkage map was assembled for mapping QTLs for grain zinc and iron content using a total of 106 RILs (F_6) obtained from ICMB 841 × P3 863B-P2 hybrid with 305 (96 SSRs and 208 DArT) markers (Kumar et al. 2016). Genetic linkage map was constructed utilizing 317 RIL population members obtained from (AIMP 92901-S1-183-2-2-B-08 × ICMS 8511-S1-17-2-1-1-B-P03) cross employing Zn and Fe content QTLs coupled with SSR and DArT markers (Kumar et al. 2018). RIL population of cross ICMB 841-P3×863B-P2 was employed for mapping important agronomical traits like plant height, panicle length, flowering time, grain weight, and a total of 44 QTLs were found (Kumar et al. 2017). Kumar et al. (2021) used 317 RILs population derived from cross of AIMP 92901-S1-183-2-2-B-08×ICMS 8511-S1-17-2-1-1-B-P03 hybrid for QTL mapping of different agronomic parameters to boost pearl millet's adaptation in low-input farms and increase grain yield. Chelpuri et al. (2019) identified QTLs for resistance against three different pathotype-isolates of *S. graminicola*. Recently, Singhal et al. (2021) identified 14 QTLs associated with Fe and eight for Zn. The detected QTLs and candidate genes might be suitable for pearl millet population improvement programs, seed restorer parents and MAS programs (Singhal et al. 2021). Syntenic relationships among rice and finger millet were exploited to map genes for blast resistance via association mapping approach by means of expressed sequence tag (EST) sequences of NBS-LRR, *M. grisea* and Pi genes of rice (Babu et al. 2014). QTLs of finger blast and neck blast were reported to be linked to genes such as Pi-d(t), Pi5, NBS-LRR, Pi21 and *M. grisea*. Hence, these genes may possibly be targeted for allele mining in finger millet. Similarly, syntenic relationships can be studied between pearl millet and finger millet and thus the genes and alleles of interest can be mapped further.

11.8 GENOMIC SELECTION (GS) AND GENOME-WIDE ASSOCIATION STUDIES (GWAS) FOR ALLELE DISCOVERY IN PEARL MILLET

Association mapping (AM) and genome wide association study (GWAS) are emerging as alternative approaches of QTL mapping (Gómez et al. 2011). They are based on linkage disequilibrium (LD) principle to find a significant relationship between a target trait and DNA marker (Gupta et al. 2005). It may simultaneously test and characterize many alleles in diploid and polyploid crops (Zhao et al. 2007). In comparison to linkage mapping, GWAS provides a number of advantages, including better mapping accuracy brought about by historical mutational alterations and recombination events among genetic pedigrees resulting into identification of markers in close proximity to regulatory genes (Liu et al. 2016).

The GWAS technique is helpful for discovery of genes or QTLs with high resolution (100–1000 Kb) that are firmly connected with wide variety of phenotypic traits by employing multiple historical recombination events among distinct accessions after their domestication (Mackay et al. 2009). The genetic factors causing variations in flowering time at phytochrome C (PHYC) (866 bp) locus were identified by association studies reported by Saïdou et al. (2014). Mariac et al. (2011) employed an integrated strategy of genome scanning and association mapping to discover that *PgMADS11* gene is a MADS-box gene family member that has important function in reproductive and somatic phase development under various weather circumstances. Only a few association mapping approaches have been presented owing to higher amount of diversity and heterozygozity in most pearl millet germplasm accessions and they may broaden the field of future pearl millet association mapping research (Kannan et al. 2014). They further described the relationships between the selective SSR markers and panicle length, plant height, flowering period, stover and grain yield using GWAS with 250 full-sib and progenies 34 SSR markers.

Gemenet et al. (2015) conducted the very first association study of phenotypic data with 285 DArT markers produced from 151 PMiGAP lines belonging to West Africa in six experimental sets under low and high P circumstances. *PgPb12954* marker revealed a substantial relationship with grain yield while *PgPb11603* DArT marker exhibited a steady relationship with flowering time. Association results exhibited that Xibmsp11/AP6.1, an acetyl CoA carboxylase gene SNP marker, is intensely linked with panicle yield, grain yield and grain harvest index under both treatments, while chlorophyll a/b binding protein gene associated InDel markers including Xibmcp09/AP10.1 and Xibmcp09/AP10.2 were observed to be linked with grain yield and stay-green traits. Anuradha et al. (2017) used the association mapping method to study grain iron and zinc concentration alleles. They developed marker trait associations (MTAs) between 250 SSR and 17 genic markers with grain iron and zinc content for 130 different lines in different environments. On the other hand, Xipes0180 exhibit association with LG 3, Xpsmp2261 with LG 5 and Xipes0096 with LG 7 for grain iron and zinc. Varshney et al. (2017) conducted another association analysis that yielded important results when building MTAs. A total of 3,117,056 SNPs were identified for GWAS investigation utilizing whole-genome SNP data and phenotypic information for 20 agro-morphological parameters was collected from 288 TCH (testcross hybrids) under two-stage (early and late) drought stress conditions, as well as controls. On pseudomolecules Pg1 and Pg5, there was a substantial relationship of markers with the target attribute, i.e., grain number per panicle. Expanding AM research in pearl millet will progressively pay off for other crop improvement activities since AM can provide a high-resolution power with the species demonstrating genotypic variability across the germplasm (Álvarez et al. 2014). The use of linkage mapping and genome-wide association mapping to analyze QTL effects for small QTLs is frequently skewed. As a result, scientists have been working for years to figure out how to deal with these complicated features and consequences in terms of genomic selection (GS).

Genomic (or genome-wide) selection is a promising method with a lot of potential for exploring and increasing genetic gain per selection in a breeding scheme per unit timescale and hence breeding program efficacy and speed (Spindel et al., 2015). GS has proved to be a viable and affordable alternative to MAS and phenotype selection (PS) for quantitative traits and rapid crop development projects in different crops (Muleta et al. 2019; Poland et al. 2012). The available PMiGAP lines' whole-genome resequencing (WGRS) data as well as phenotyping information for a number of characteristics are being used by ICRISAT for GWAS and GS. Varshney and his coworkers used WGRS data for genomic selection by ridge regression to make highly accurate predictions for the four scenarios of control, early stress, late stress and across habitats when predicting grain yield for test crosses (RR-BLUP) (Varshney et al. 2017). Additionally, it was revealed that the hybrid performance could be predicted by applying the GS method (additive and dominant effects) to analyze grain yield data with 302,110 SNPs. A total of 170 interesting hybrid combinations were discovered, out of which 11 had already been employed to produce hybrids with higher performance and rest 159 hybrid combinations might be candidates for the development of high-yielding hybrids. A hierarchical clustering investigation of feasible single cross combinations (167910) found two sets of lines with 8% higher hybrid performance. By creating high-yielding heterotic gene pools, these hybrids may be used as a starting point to produce pearl millet hybrids with improved yield potential (Varshney et al. 2017). Liang et al. (2018) analyzed the efficacy of genomic selection/prediction using two potential genotyping techniques, tGBS and RAD-seq, to characterize a group of ICRISAT generated inbred pearl millet lines. Twenty arbitrary rounds of five-fold cross-validation were carried out for a tested SNP set and assessed for each trait in pearl millet using projected hybrids from both (RADseq and tGBS) methodologies and four genomic prediction algorithms. Genomic prediction scheme (RR-BLUP) was claimed to have produced median prediction ranges (in parenthesis) for a variety of variables including plant height (0.72–0.73), grain yield (0.48–0.51), days to flowering (0.87–0.89), and 1,000 grain weight (0.73–0.74) using hybrid data. Other traits with fewer or no heterosis, hybrid-only schemes and hybrid/inbred schemes also executed similar results. Once inbred, inbred phenotypic data sets and hybrid trait values in relation to that population's mean trait values have also been observed to aid moderately improve hybrid genome-assisted breeding values (GEBVs). The integration of previous inbred phenotypic data into hybrid breeding operations has also been shown to decrease the predictability of traits showing heterosis. The precision of GEBVs for hybrids, however, might be increased by reducing the effects of heterosis within the inbred genotype and trait data, which would strengthen pearl millet hybrid breeding efforts.

11.9 FUNCTIONAL CHARACTERIZATION OF PEARL MILLET GENES

Transgenic technology, which aids in the investigation of molecular mechanisms of abiotic stress regulation, relies on functional analysis *in vivo*. Overexpression, gene silencing and genome editing in a plant system are all important strategies for functionally validating candidate genes for stress tolerance. The presence of potential abiotic stress tolerant genes was discovered after abiotic stress induced gene expression profiling of several pearl millet genotypes. Overexpression of pearl millet genes in heterologous systems exhibited improved stress tolerance in response to a variety of abiotic stimuli. *PgLEA* protein recombinant *E. coli* (Reddy et al. 2012) and *PgDHN* (dehydrin) protein transformed *E. coli* and yeast cells demonstrated chaperon activity and exhibited improved tolerance and growth levels under heat stress and high salt concentration. Salt stress resistance is conferred by moderate overexpression of *PgVDAC* (voltage-dependent anion channel) in transgenic rice, according to research (Desai et al. 2006). Pearl millet transcription factors are induced by abiotic stress. Overexpression of *PgNAC21* in Arabidopsis and *PgDREB2A* in tobacco lead to increase in resistance to osmotic stress (Agarwal et al. 2010; Shinde et al. 2018). Some studies have revealed that expression of pearl millet genes in other systems, such as stress-inducible small GTP-binding protein (*PgRab7*), vacuolar Na+/H+ antiporter gene (*PgNHX1*), heat shock protein (*Hsp90*), and others, is responsible for their improved performance under various abiotic conditions (Islam et al. 2010; Reddy et al. 2010). Varshney et al. (2017) reported that genes for secondary metabolites synthesis such as wax biosynthesis, terpene, phenolic compounds, flavonols and anthocyanin biosynthesis, etc. were more enriched in pearl millet genome and they might be responsible for pearl millet drought tolerance response in dry areas. In light of this, a study was done for the functional characterization of pearl millet flavone synthesis (PgF3H). PgF3H gene transformed in *Arabidopsis thaliana* and the transgenic plants have more stress tolerance capacity with high germination percentage, more amounts of flavonols and anthocyanin content compared to wild type plants under drought stress.

11.10 CHALLENGES FACED IN ALLELE MINING

Selection of genotypes, demarcation of promoter region, characterization of regulatory region, higher sequencing costs, handling genomic resources are some of the challenges that are confronted by the pearl millet researchers. Computational tools for assessing functional nucleotide diversity and predicting particular nucleotide changes responsible for changed function must be developed to keep up with the rapidly accumulating nucleotide and gene expression data. Characterizing the level of cis-acting regulatory variation in comparison to detecting variation in coding regions of genome is a big challenge as it is not unraveled even in completely sequenced genomes. Specialized computing techniques and tools must be developed in order to predict the core promoter region based on the representation or over representation of regulatory motifs. Minimizing the time and effort is needed while lowering the cost per data point is another big challenge. These difficulties may be partially overcome by using cheaper and quicker sequencing technologies for detection of high-throughput allelic variation.

11.11 FUTURE PROSPECTS AND CONCLUSION

Pearl millet possesses huge potential to fill new niches in the world and genomic tools are anticipated to be essential to develop more climate-smart crop cultivars with improved nutritional qualities in a timely and cost-effective way. Understanding breeding lines, germplasm, functional, and translational genomics is crucial to increase the genetic gains for yield and grain micronutrient contents. Pearl millet CWRs can also be used to enhance richness of alleles in cultivated germplasm by boosting diversity at the gene, haplotype, or allele level using next-generation genotyping platforms and most recent genomic resources. It is possible to quickly locate, transmit and track introgressed genes into cultivated background using genomic techniques. The precise mapping of traits and introgressions are possible with these genomic technologies. The establishment of a "super-pangenome or pangenome" of several wild species within a genus may be helpful for cataloguing full genome collection of a genus. To create a super-pan genome, diverse TGSs will be the best source of material. They will also be helpful for capturing the most amount of variety possible for a species while limiting within-accession variability found in germplasm preserved in gene banks. Transcriptome analysis and whole-genome sequencing have been revolutionized by the availability of affordable NGS-based technology. This has also prompted genome-wide genetic analyses of untapped useful alleles. Open pollinated varieties (OPVs) and elite hybrid parental lines cab be bred for high grain Fe/Zn cultivars using pan-genome marker-assisted selection approach and available 4-SNP panel. Using these techniques, promising lines with advantageous alleles may be used to create new cultivars that acquire all of the key alleles controlling grain iron and zinc metabolism.

REFERENCES

Agarwal P, Agarwal PK, Joshi AJ, Sopory SK, Reddy MK (2010) Overexpression of PgDREB2A transcription factor enhances abiotic stress tolerance and activates downstream stress responsive genes. *Mol Biol Rep* 37(2): 1125–1135.

Ahuja I, De Vos RC, Bones AM, Hall RD (2010) Plant molecular stress responses face climate change. *Trends Plant Sci* 15(12): 664–674.
Allouis S, Qi X, Lindup S, Gale MD, Devos KM (2001) Construction of a BAC library of pearl millet, *Pennisetum Glaucum*. *Theor Appl Genet* 102(8): 1200–1205.
Álvarez MF, Mosquera T, Blair MW (2014) The use of association genetics approaches in plant breeding. *Plant Breed Rev* 38: 17–68.
Ambawat S, Senthilvel S, Hash CT, Nepolean T, Rajaram V et al (2016) QTL mapping of pearl millet rust resistance using an integrated DArT and SSR based linkage map. *Euphytica* 209(2): 461–476.
Ambawat S, Singh S, Shobhit, Meena RC, Satyavathi CT (2020) Biotechnological applications for improvement of the pearl millet crop. In: Gahlawat SK, Punia S, Siroha AK, Sandhu KS, Kaur M (eds) *Pearl millet: Properties, functionality and its applications*, pp 115–138. Taylor & Francis (CRC Press), Boca Raton, Finland. doi.10.1201/9780429331732-7
Ambli K, Mullainathan L (2015) Chlorophyll and morphological mutants of Pearl millet (Pennisetum tphoides (Burn.) stapf. Var. CO(cu) 9. *Eur J Exp Biol* 5(3):72–77
Anuradha N, Satyavathi CT, Bharadwaj C, Nepolean T, Sankar SM et al (2017) Deciphering genomic regions for high grain iron and zinc content using association mapping in pearl millet. *Front Plant Sci* 8: 412.
Babu BK, Dinesh P, Agrawal PK, Sood S, Chandrashekara C, Bhatt JC, Kumar A (2014) Comparative genomics and association mapping approaches for blast resistant genes in finger millet using SSRs. *PLOS One* 9: 182.
Bailey TL, Williams N, Misleh C, Li WW (2006) MEME: Discovering and analyzing DNA and protein sequence motifs. *Nucl Acids Res* 34(2): 369–373.
Bandillo N, Raghavan C, Muyco PA, Sevilla MAL, Lobina IT, Dilla-Ermita CJ (2013) Multi-parent advanced generation inter-cross (MAGIC) populations in rice: Progress and potential for genetics research and breeding. *Rice* 6(11). doi:10.1186/1939-8433-6-11
Barkley NA, Wang ML (2008) Application of TILLING and Eco TILLING as reverse genetic approaches to elucidate the function of genes in plants and animals. *Curr Genom* 9: 212–226.
Basava RK, Hash CT Mahendrakar MD, Kishor PBK, Satyavathi CT et al (2019) Discerning combining ability loci for divergent environments using chromosome segment substitution lines (CSSLs) in pearl millet. *PLoS One* 14. https://doi.org/10.1371/journal.pone.0218916
Bertin I, Zhu JH, Gale MD (2005) SSCP-SNP in pearl millet: A new marker system for comparative genetics. *Theor Appl Genet* 8: 1467–1472.
Bevan MW, Uauy C (2013) Genomics reveals new landscapes for crop improvement. *Genome Biol* 14: 206.
Bhattacharjee R, Bramel PJ, Hash CT, Kolesnikova-Allen MA, Khairwal IS (2002) Assessment of genetic diversity within and between pearl millet landraces. *Theor Appl Genet* 105: 666–673.
Bohra A (2020) Genomic interventions for sustainable agriculture. *Plant Biotechnol J* 18: 2388–2405.
Bollam S, Pujarula V, Srivastava RK, Gupta RK (2018) Genomic approaches to enhance stress tolerance for productivity improvements in pearl millet. In: Gosal SS, Wani SH (eds) *Biotechnologies of crop improvement*, Vol 3, pp 239–264. Springer, Berlin/Heidelberg, Germany.
Bramel-Cox PJ, Andrews DJ, Frey KJ (1986) Exotic germplasm for improving grain yield and growth rate in pearl millet. *Crop Sci* 26: 687–690. https://doi.org/10.2135/cropsci1986.0011183X002600040009x
Breese WA, Hash CT, Devos KM, Howarth CJ, Leslie JF (2002) Pearl millet genomics and breeding for resistance to downy mildew. In: John F. Leslie (ed.) *Sorghum Millets Diseases2000* pp 243–246, Wiley-Blackwell.
Brunken JN, deWet JMJ, Harlan JR (1977) The morphology and domestication of pearl millet. *Econ Bot* 31: 163–174. https://doi.org/10.1007/BF02866587
Bryne JC, Valen E, Tang MHE, Marstrand T, Winther O (2008) Isabelle da Piedade, Anders Krogh, Boris Lenhard, and Albin Sandelin. JASPAR, The open access database of transcription factor-binding profiles: New content and tools. *Nucl Acids Res* 36: 102–106.
Cartharius K, Frech K, Grote K, Klocke B, Haltmeier M et al (2005) MatInspector and beyond: promoter analysis based on transcription factor binding sites. *Bioinformatics* 21(13): 2933–2942.
Chandel P, Tiwari J, Ali N, Devi S, Sharma S et al (2015) Interspecific potato somatic hybrids between *Solanum tuberosum* and *S. cardiophyllum* potential sources of late blight resistance breeding. *Plant Cell Tiss Org Cult* 123: 579–589. https://doi.org/10.1007/s11240-015-0862-8

Chanwala J, Satpati S, Dixit A, Parida A, Giri MK, Dey N (2020) Genome-wide identification and expression analysis of WRKY transcription factors in pearl millet (*Pennisetum glaucum*) under dehydration and salinity stress. *BMC Genomics* 21(1): 1–16.

Chelpuri D, Sharma R, Durga KK, Katiyar P, Mahendrakar MD et al (2019) Mapping quantitative trait loci (QTLs) associated with resistance to major pathotype-isolates of pearl millet downy mildew pathogen. *Eur J Plant Pathol* 154(4): 983–994.

Chen L, Hao L, Parry MA, Phillips AL, Hu YG (2014) Progress in TILLING as a tool for functional genomics and improvement of crops. *J Integr Plant Biol* 56: 425–443.

Chen X, Guo L, Fan Z, Jiang T (2008) W-AlignACE: An improved Gibbs sampling algorithm based on more accurate position weight matrices learned from sequence and gene expression/ChIP-chip data. *Bioinformatics* 24(9): 1121–1128.

Choudhary M, Jayanand Padaria JC (2015) Transcriptional profiling in pearl millet (*Pennisetum glaucum* L.R. Br.) for identification of differentially expressed drought responsive genes. *Physiol Mol Biol Plants* 21(2): 187–196.

Cobb JN, Juma RU, Biswas PS, Arbelaez JD, Rutkoski J et al (2019) Enhancing the rate of genetic gain in public-sector plant breeding programs: Lessons from the breeder's equation. *Theor Appl Genet* 132: 627–645. https://doi.org/10.1007/s00122-019-03317-0

Davey JW, Hohenlohe PA, Etter PD, Boone JQ, Catchen JM, Blaxter ML (2011) Genome-wide genetic marker discovery and genotyping using next-generation sequencing. *Nat Rev Genet* 12: 499–510.

Davuluri RV, Sun H, Palaniswamy SK, Matthews N, Molina C et al (2003) AGRIS: Arabidopsis gene regulatory information server, an information resource of Arabidopsis cis-regulatory elements and transcription factors. *BMC Bioinformatics* 4(1): 1–11.

Della CR, Qiu Y, Ou S, Hufford MB, Hirsch CN (2021) How the pan-genome is changing crop genomics and improvement. *Genome Biol* 22: 3. doi:10.1186/s13059-020-02224-8

Desai MK, Mishra RN, Verma D, Nair S, Sopory SK, Reddy MK (2006) Structural and functional analysis of a salt stress inducible gene encoding voltage dependent anion channel (VDAC) from pearl millet (*Pennisetum glaucum*). *Plant Physiol Biochem* 44: 483–493.

Devos KM, Gale MD (2000) Genome relationships: The grass model in current research. *Plant Cell* 12: 637–646.

Dudhate A, Shinde H, Tsugama D, Liu S, Takano T (2018) Transcriptomic analysis reveals the differentially expressed genes and pathways involved in drought tolerance in pearl millet [*Pennisetum glaucum* (L.) R. Br]. *PLoS One* 13(4): e0195908.

Dujardin M, Hanna WW (1987) Inducing male fertility in crosses between pearl millet and *Pennisetum orientale* Rich. *Crop Sci* 27: 65–68. ttps://doi.org/10.2135/cropsci1987.0011183X002700010017x

Dwivedi SL, Upadhyaya HD, Stalker HT, Blair MW, Bertioli DJ, Nielen S, Ortiz R (2008) Enhancing crop gene pools with beneficial traits using wild relatives. *Plant Breed Rev* 30: 179–230. https://doi.org/10.1002/9780470380130

Elshire RJ, Glaubitz JC, Sun Q, Poland JA, Kawamoto K, Buckler ES (2011) A robust, simple genotyping-by-sequencing (GBS) approach for high diversity species. *PLoS One* 6: 19379. doi:10.1371/journal.pone.0019379

Gale MD, Devos KM, Zhu JH, Allouis S, Couchman MS, Liu H, Pittaway TS, Qi XQ, Kolesnikova-Allen M, Hash CT (2005) New molecular marker technologies for pearl millet improvement. *SAT eJournal* 1: 1–7.

Gemenet DC, Leiser WL, Zangre RG, Angarawai II, Sanogo MD et al (2015) Association analysis of low-phosphorus tolerance in West African pearl millet using DArT markers. *Mol Breed* 35(8): 1–20.

Ghosh D (2000) Object-oriented transcription factors database (ooTFD). *Nucl Acids Res* 28(1): 308–310.

Gilchrist EJ, Haughn GW (2005) TILLING without a plough: A newmethod with applications for reverse genetics. *Curr Opinion Plant Bio* 8: 211–215.

Goel S, Chen Z, Akiyama Y, Conner JA, Basu M, Gualtieri G, Ozias-Akins P (2006) Comparative physical mapping of the apospory-specific genomic region in two apomictic grasses *Pennisetum squamulatum* and Cenchrus ciliaris. *Genetics* 173: 389–400.https://doi.org/10.1534/genetics.105.054429

Gómez G, Álvarez MF, Mosquera T (2011) Association mapping, a method to detect quantitative trait loci: Statistical bases. *Agron Colomb* 29: 367–376.

Govindaraj M, Rai KN, Cherian B, Pfeiffer WH, Kanatti A, Shivade H (2019) Breeding biofortified pearl millet varieties and hybrids to enhance millet markets for human nutrition. *Agriculture* 9(5): 106. doi:10.3390/agriculture9050106

Govindaraj M, Rai KN, Kanatti A, Upadhyaya HD, Shivade H, Rao AS (2020) Exploring the genetic variability and diversity of pearl millet core collection germplasm for grain nutritional traits improvement. *Sci Rep* 10(1): 1–13.

Gulia SK, Hash CT, Thakur RP, Breese WA, Sangwan RS (2007) Mapping new QTLs for downy mildew [*Sclerospora graminicola* (Sacc.)J. Schroet.] resistance in pearl millet [*Pennisetum glaucum* (L.) R. Br.]. In: Singh DS, Tomar VS, Behl RK, Upadhyaya SD, Bhale MS, Khare D (eds) *Crop production in stress environments – Genetic and management options*, pp 373–386. Agrobios, Jodhpur.

Gupta PK, Rustgi S, Kulwal PL (2005) Linkage disequilibrium and association studies in higher plants: Present status and future prospects. *Plant Mol Biol* 57: 461–485.

Gupta SK, Patil KS, Rathore A, Yadav DV, Sharma LD, Mungra KD (2020) Identification of heterotic groups in South-Asianbred hybrid parents of pearl millet. *Theor Appl Genet* 133: 873–888. doi: 10.1007/s00122-019-03512-z

Hajjar R, Hodgkin T (2007) The use of wild relatives in crop improvement: A survey of developments over the last 20 years. *Euphytica* 156: 1–1. https://doi.org/10.1007/s10681-007-9363-0

Hall T (2004) *BioEdit version 7.0. 0.* website: www.mbio.ncsu.edu/BioEdit/bioedit.html

Hammons RO (1970) Registration of Spancross peanuts (Reg. no.3). *Crop Sci* 10: 459–460. https://doi.org/10.2135/cropsci1970.0011183X001000040056x

Hanna WW (1989) Characteristics and stability of a new cytoplasmic- nuclear male sterile source in pearl millet. *Crop Sci* 29: 1457–1459. https://doi.org/10.2135/cropsci1989.0011183X002900060026x

Hanna WW, Wells HD (1989) Inheritance of *Pyricularia* leaf spot resistance in pearl millet. *J Hered* 80: 145–147. https://doi.org/10.1093/oxfordjournals.jhered.a110814

Heet J, Zhao X, Laroche A, Lu ZX, Liu H, Li Z (2014) Genotyping-by-sequencing (GBS), an ultimate marker-assisted selection (MAS) tool to accelerate plant breeding. *Front Plant Sci* 5: 484.

Higo K, Ugawa Y, Iwamoto M, Korenaga T (1999) Plant cis-acting regulatory DNA elements (PLACE) database. *Nucl Acids Res* 27(1): 297–300.

Holt KE, Harvey Millar A, Whelan J (2006) Module finder and co reg: Alternative tools for linking gene expression modules with promoter sequences motifs to uncover gene regulation mechanisms in plants. *Plant Methods* 2(1): 1–15.

Hrdlickova R, Toloue M, Tian B (2017) RNA-Seq methods for transcriptome analysis. *Wiley Interdiscip Rev RNA* 8(1): 1364.

Hunter CI, Zhe XS, Armit CJ, Edmunds SC, Li P, Tuli MA, Goodman L (2019) *Publishing Genomes in Gigascience Database (GigaDB)*. In: PAG ASIA Conference, PAG

Islam SMT, Tammi RS, Singla-Pareek SL, Seraj ZI (2010) Enhanced salinity tolerance and improved yield properties in Bangladeshi rice Binnatoa through Agrobacterium-mediated transformation of *PgNHX1* from *Pennisetum glaucum*. *Acta Physiol Planta* 32: 657–663.

Jaiswal S, Antala TJ, Mandavia MK, Chopra M, Jasrotia RS et al (2018) Transcriptomic signature of drought response in pearl millet [*Pennisetum glaucum* (L.)] and development of web-genomic resources. *Sci Rep* 8: 3382.

Jarquin D, Howard R, Liang Z, Gupta SK, Schnable JC, Crossa J (2020) Enhancing hybrid prediction in pearl millet using genomic and/or multi-environment phenotypic information of inbreds. *Front Genet* 10: 1294.

Jauhar PP, Hanna WW (1998) Cytogenetics and genetics of pearl millet. *Adva Agron* 64: 1–26.

Jones ES, Breese WA, Liu CJ, Singh SD, Shaw DS, Witcombe JR (2002) Mapping quantitative trait loci for resistance to downy mildew in pearl millet: Field and glass house detect the same QTL. *Crop Sci* 42: 1316–1323.

Kalendar R, Khassenov B, Ramankulov Y, Samuilova O, Ivanov KI (2017) Fast PCR: An in silico tool for fast primer and probe design and advanced sequence analysis. *Genomics* 109(3–4): 312–319.

Kannan B, Senapathy S, Raj B, Gajaraj A, Chandra S, Muthiah A (2014) Association analysis of SSR markers with phenology, grain, and stover yield related traits in pearl millet [*Pennisetum glaucum* (L.) R. Br.]. *Scientific World J* 2014(562327): 1–14. http://dx.doi.org/10.1155/2014/562327

Kaur A, Kaur P, Ahuja S (2020) Fluorescence resonance energy transfer (FRET) and applications thereof. *Anal Methods* 12: 5532–5550. doi:10.1039/ d0ay01961e

Kilian B, Mammen K, Millet E, Sharma R, Graner A, Salamini F, Ozkan H (2011) Aegilops. In: Kole C (ed) *Wild crop relatives: Genomic and breeding resources*. Springer, Berlin, Heidelberg. https: //doi.org/10.1007/978-3-642-14228-4_1

Kolchanov NA, Podkolodnaya OA, Ananko EA, Ignatieva EV, Stepanenko IL et al (2000) Transcription regulatory regions database (TRRD): Its status. *Nucl Acids Res* 28(1): 298–301.

Koulengar D (1995) Le sensible de Pennisetum polystachion (L.) Schult. In: *De différentes provenances de Sahel pour le Striga hermonthica (Del.) Benth*, pp 1–13. Département de Formationen Protection des Végétaux (DFPV), Centre AGRHYMET, Niamey, Niger.

Kover PX, Valdar W, Trakalo J, Scarcelli N, Ehrenreich IM et al (2009) A multiparent advanced generation intercross to fine-map quantitative traits in *Arabidopsis thaliana*. *PLoS Genet* 5(7): e1000551. doi:10.1371/journal.pgen.1000551

Kulkarni KS, Zala HN, Bosamia TC, Shukla YM, Kumar S et al (2016) De novo transcriptome sequencing to dissect candidate genes associated with pearl millet-downy mildew (*Sclerospora graminicola* Sacc.) interaction. *Front Plant Sci* 7: 847. https://doi.org/10.3389/fpls.2016.00847

Kumar GR, Sakthivel K, Sundaram RM, Neeraja CN, Balachandran SM, Rani NS, Viraktamath BC, Madhav MS (2010) Allele mining in crops: Prospects and potentials. *Biotech Adv* 28: 451–461.

Kumar S, Hash C, Nepolean T, Mahendrakar M, Satyavathi C, Singh G (2018) Mapping grain iron and zinc content quantitative trait loci in an iniadi-derived immortal population of pearl millet. *Genes* 9: 248.

Kumar S, Hash CT, Nepolean T, Satyavathi CT, Singh G et al (2017) Mapping QTLs controlling flowering time and important agronomic traits in pearl millet. *Front Plant Sci* 8: 1731.

Kumar S, Hash CT, Singh G, Nepolean T, Srivastava RK (2021) Mapping QTLs for important agronomic traits in an Iniadi-derived immortal population of pearl millet. *Biotechnol Notes* 2: 26–32.

Kumar S, Hash CT, Thirunavukkarasu N, Singh G, Rajaram V, Rathore A (2016) Mapping quantitave trait loci controlling high iron and zinc in self and open pollinated grains of pearl millet [*Pennisetum glaucum* (L) R. Br.]. *Front Plant Sci* 7: 1636.

Kumari BR, Kolesnikova-Allen MA, Hash CT, Senthilvel S, Nepolean T et al (2014) Development of a set of chromosome segment substitution lines in pearl millet [*Pennisetum glaucum* (L.) Br.]. *Crop Sci* 54: 2175–2182. https://doi.org/10.2135/cropsci2013.09.0589

Kumari R, Kumar P, Sharma VK, Kumar H (2018) Allele mining for crop improvement. *Int J Pure App Biosci* 6(1): 1456–1465 doi: http://dx.doi.org/10.18782/2320-7051.6073

Kurowska M, Daszkowska-Golec A, Gruszka D, Marzec M, Szurman M et al (2011) TILLING-a shortcutin functional genomics. *J Appl Genet* 52: 371–390.

Lescot M, De hais P, Thijs G, Marchal K, Moreau Y, Van de Peer Y (2002) Plant CARE, a database of plant cis-acting regulatory elements and a portal to tools for in silico analysis of promoter sequences. *Nucl Acids Res* 30(1): 325–327.

Liang Z, Gupta SK, Yeh CT, Zhang Y, Ngu DW, Kumar R, Schnable JC (2018) Phenotypic data from inbred parents can improve genomic prediction in pearl millet hybrids. *G3: Genes Genomes Genet* 8(7): 2513–2522.

Liu C, Devos K, Witcombe J, Pittaway T, Gale M (1996). The effect of genome and sex on recombination rates in *Pennisetum* species. *Theor Appl Genet* 93(5–6): 902–908.

Liu CJ, Witcombe JR, Pittaway TS, Nash M, Hash CT, Busso CS (1994) An RFLP-based genetic map of pearl millet (*Pennisetum glaucum*). *Theor Appl Genet* 89: 481–487.

Liu N, Xue Y, Guo Z, Li W, Tang J (2016) Genome-wide association study identifies candidate genes for starch content regulation in maize kernels. *Front Plant Sci* 7: 1046.

Mackay TF, Stone EA, Ayroles JF (2009) The genetics of quantitative traits: Challenges and prospects. *Nat Rev Genet* 10: 565–577.

Mahendrakar MD, Parveda M, Kishor PK, Srivastava RK (2020) Discovery and validation of candidate genes for grain iron and zinc metabolism in pearl millet [*Pennisetum glaucum* (L.) R. Br.]. *Sci Rep* 10(1): 1–16.

Makhoul M, Rambla C, Voss-Fels KP, Hickey LT, Snowdon RJ, Obermeier C (2020) Overcoming polyploidy pitfalls: A user guide for effective snp conversion into kasp markers in wheat. *Theor Appl Genet* 133: 2413–2430. doi:10.1007/s00122-020-03608-x

Marchais L, Pernes J (1985) Genetic divergence between wild and cultivated pearl millets (*Pennisetum typhoides*). *1. Male sterility. Z Pflanzenzücht* 95: 103–112.

Mariac C, Jehin L, Saïdou AA, Thuillet AC, Couderc M, Sire P (2011) Genetic basis of pearl millet adaptation along an environmental gradient investigated by a combination of genome scan and association mapping. *Mol Ecol* 20: 80–91.

Mariac C, Luong V, Kapran I, Mamadou A, Sagnard F, Deu M, Vigouroux Y (2006) Diversity of wild and cultivated pearl millet accessions (*Pennisetum glaucum* [L.] Br.) in Niger assessed by microsatellites. *Theor Appl Genet* 114: 49–58. https://doi.org/10.1007/s00122-006-0409-9

Marshall VM, Lewis MM, Ostendorf B (2012) Buffel grass (*Cenchrus ciliaris*) as an invader and threat to biodiversity in arid environments: A review. *J Arid Environ* 78: 1–12. https://doi.org/10.1016/j.jaridenv.2011.11.005

Martel E, De Nay D, Siljak-Yakovlev S, Brown S, Sarr A (1997) Genome size variation and basic chromosome number in pearl millet and fourteen related Pennisetum species. *J Hered* 88: 139–143. https://doi.org/10.1093/oxfordjournals.jhered.a023072

Maryono MY, Indriatama WM, Human S (2020) Performance and estimation genetic variability of M3 pearl millet (*Pennisetum glaucum*) populations. In: *IOP Conference Series: Earth and Environmental Science, IOP Publishing*, 484(1): 12–21.

Matys V, Fricke E, Geffers R, Gobling E, Haubrock M, Hehl R, Wingender E (2003) TRANSFAC®: Transcriptional regulation, from patterns to profiles. *Nucl Acids Res* 31(1): 374–378.

Meena RP, Vishwakarma H, Ghosh G, Gaikwad K, Chellapilla TS, Singh MP, Padaria JC (2020) Novel ASR isolated from drought stress responsive SSH library in pearl millet confers multiple abiotic stress tolerance in PgASR3 transgenic Arabidopsis. *Plant Physiol Biochem* 156: 7–19.

Mejlhede N, Kyjovska Z, Backes G, Burhenne K, Rasmussen SK, Jahoor A (2006) Eco TILLING for the identification of allelic variation in the powdery mildew resistance genes mlo and Mla of barley. *Plant Breed* 125: 461–467.

Mishra R, Reddy P, Nair S, Markandeya G, Reddy A, Sopory S, Reddy M (2007) Isolation and characterization of expressed sequence tags (ESTs) from subtracted cDNA libraries of *Pennisetum glaucum* seedlings. *Plant Mol Biol* 64:713–732.

Morgan RN, Wilson JP, Hanna WW, Ozias-Akins P (1998) Molecular markers for rust and pyricularia leaf spot disease resistance in pearl millet. *Theor Appl Genet* 96: 413–420.

Moumouni KH, Kountche BA, Jean M, Hash CT, Vigouroux Y, Haussmann BI, Belzile F (2015) Construction of a genetic map for pearl millet [*Pennisetum glaucum* (L.) R. Br.] using a genotyping-by- sequencing (GBS) approach. *Mol Breed* 35(1): 5–10.

Muleta KT, Pressoir G, Morris GP (2019) Optimizing genomic selection for a sorghum breeding program in Haiti: a simulation study. *G3. Genes Genomes Genet* 9: 391–401.

Muscolo A, Panuccio MR, Eshel A (2013) Ecophysiology of *Pennisetum clandestinum*: A valuable salt tolerant grass. *Environ Exp Bot* 92: 55–63. https://doi.org/10.1016/j.envexpbot.2012.07.009

Nazeer W, Ahmad S, Mahmood K, Tipu AL, Mahmood A, Zhou B (2014) Introgression of genes for cotton leaf curl virus resistance and increased fiber strength from *Gossypium stocksii* into upland cotton (*G. hirsutum*). *Genet Mol Res* 13:1133–1143. https://doi.org/10.4238/2014.February.21.2

Obok EE (2013) Mineral contents of selected pearl millet (*Pennisetum glaucum* (L.) R. Br.) × elephant grass (*Pennisetum purpureum*(Schum.)) interspecific hybrids of Nigerian origin. *J Plants Studies* 2: 22–27.

Ogbonnaya FC, Abdalla O, Mujeeb-Kazi A, Kazi AG, Xu SS, Gosman N, Lagudah Evans S, Bonnett D, Sorrells Mark E, Tsujimoto H (2013) Synthetic hexaploids: Harnessing species of the primary gene pool for wheat improvement. *Plant Breed Rev* 37: 35–122.

Oumar I, Mariac C, Pham JL, Vigouroux Y (2008) Phylogeny and origin of pearl millet (*Pennisetum glaucum* [L.] R. Br) as reveal ed by microsatellite loci. *Theor Appl Genet* 117: 489–497. https://doi.org/10.1007/s00122-008-0793-4

Patil KS, Gupta SK, Marathi B, Danam S, Thatikunta R, Rathore A, Yadav OP (2020) African and Asian origin pearl millet populations: Genetic diversity pattern and its association with yield heterosis. *Crop Sci* 60(6): 3035–3048.

Pedraza-Garcia F, Specht JE, Dweikat I (2010) A new PCR-based linkage map in pearl millet. *Crop Sci* 50(5): 1754–1760.

Poland J, Endelman J, Dawson J, Rutkoski J, Wu S, Manes Y (2012) Genomic selection in wheat breeding using genotyping-by-sequencing. *Plant Genome* 5: 103–113.

Poncet V, Martel E, Allouis S, Devos KM, Lamy F, Sarr A (2002) Comparative analysis of QTLs affecting domestication traits between two domesticated x wild pearl millet (*Pennisetum glaucum* L., Poaceae) crosses. *Theor Appl Genet* 104: 965–975.

Prohens J, Gramazio P, Plazas M, Dempewolf H, Kilian B et al (2017) Introgressiomics: A new approach for using crop wild relatives in breeding for adaptation to climate change. *Euphytica* 213: 158 https://doi.org/10.1007/s10681-017-1938-9

Pucher A, Hash CT, Wallace JG, Han S, Leiser WL, Haussmann BIG (2018) Mapping a male-fertility restoration locus for the A4 cytoplasmic-genic male-sterility system in pearl millet using a genotyping-by-sequencing-based linkage map. *BMC Plant* 18: 65.

Pujarula V, Pusuluri M, Bollam S, Das RR, Ratnala R, Adapala G, Gupta R (2021) Genetic variation for nitrogen use efficiency traits in global diversity panel and parents of mapping populations in pearl millet. *Front Plant Sci* 12: 625915.

Punnuri SM, Wallace JG, Knoll JE, Hyma KE, Mitchell SE et al (2016) Development of a high-density linkage map and tagging leaf spot resistance in pearl millet using genotyping-by-sequencing markers. *Plant Genome* 9(2): 2010–2015.

Qi X, Pittaway TS, Lindup S, Liu H, Waterman E et al (2004) An integrated genetic map and a new set of simple sequence repeat markers for pearl millet (*Pennisetum glaucum*). *Theor Appl Genet* 109(7): 1485–1493.

Rai KN, Appa Rao S, Reddy KN (1997) Pearl millet. In: Fuciillo D, Sears L, Stapleton P (eds) *Biodiversity in trust: Conservationand use of plant genetic resources in CGIAR centers*, pp 243–258. Cambridge University Press, Cambridge.

Rajaram V, Nepolean T, Senthilvel S, Varshney RK, Vadez V et al (2013) Pearl millet [*Pennisetum glaucum* (L.) R. Br.] consensus linkage map constructed using four RIL mapping populations and newly developed EST-SSRs. *BMC Genom* 14: 159.

Rao NK, Reddy LJ, Bramel PJ (2003) Potential of wild species for genetic enhancement of some semi-arid food crops. *Genet Resour Crop Evol* 50: 707–721. https://doi.org/10.1023/A:1025055018954

Reddy PK, Reddy GM, Pandey P, Chandrasekhar K, Reddy MK (2012) Cloning and molecular characterization of a gene encoding late embryogenesis abundant protein from *Pennisetum glaucum*: Protection against abiotic stresses. *Mol Biol Rep* 39: 7163–7174.

Reddy PS, Mallikarjuna G, Kaul T, Chakradhar T, Sopory SK, Reddy MK (2010) Molecular cloning and characterization of gene encoding for cytoplasmic Hsc70 from *Pennisetum glaucum* may play a protective role against abiotic stresses. *Mol Genet Genom* 283: 243–254. doi: 10.1007/s00438-010-0518-7

Reeves PA, Panella LW, Richards CM (2012) Retention of agronomically important variation in germplasm core collections: Implications for allele mining. *Theor Appl Genet* 124(6): 1155–1171.

Robert T, Khalfallah N, Martel E, Lamy F, Poncet V, Allinne C, Sarr A (2011) *Pennisetum*. In: Kole C (ed) *Wild crop relatives: Genomic and breeding resources*, pp 217–255. Springer, Berlin, Heidelberg.

Ross-Ibarra J, Morrell PL, Gaut BS (2007) Plant domestication, a unique opportunity to identify the genetic basis of adaptation. *Proc Natl Acad Sci USA* 104: 8641–8648. https://doi.org/10.1073/pnas.0700643104

Saïdou AA, Clotault J, Couderc M, Mariac C, Devos KM, Thuillet AC (2014) Association mapping, patterns of linkage disequilibrium and selection in the vicinity of the PHYTOCHROME C gene in pearl millet. *Theor Appl Genet* 127: 19–32. doi: 10.1534/genetics.109.102756

Sankar MS, Tara Satyavathi C, Barthakur S, Singh SP, Bharadwaj C, Soumya SL (2021) Differential modulation of heat-inducible genes across diverse genotypes and molecular cloning of a sHSP from pearl millet [*Pennisetum glaucum* (L.) R. Br.]. *Front Plant Sci* 12: 659893.

Sarika GU, Amruta N (2014) Allele mining and its importance in seed quality. *Int J Dev Res* 4: 819–824.

Sarkar G, Cassady J, Bottema CDK, Sommer SS (1990) Characterization of polymerase chain reaction amplification of specific alleles. *Anal Biochem* 186: 64–68. doi:10.1016/0003-2697(90)90573-r

Satyavathi CT, Ambawat S, Khandelwal V, Govindaraj M, Neeraja CN (2021a) *Micronutrient rich pearl millet for nutritionally secure India ICAR-All India coordinated research project on pearl millet.* Jodhpur, Rajasthan, India: 18P.

Satyavathi CT, Ambawat S, Sehgal D, Lata C, Tiwari S et al (2021b) Genomic designing for abiotic stress tolerance in pearl millet [*Pennisetum glaucum* (L.) R. Br.]. In: Kole C (ed) *Genomic designing for abiotic stress resistant cereal crops*, pp 223–253, Springer, Cham. https://doi.org/10.1007/978-3-030-75875-2_6

Satyavathi CT, Ambawat S, Singh S, Lata C, Tiwari S, Siddaiah CN (2021c) Genomic designing for biotic stress resistance in pearl millet [*Pennisetum glaucum* (L.) R. Br.]. In: Kole C (ed) *Genomic designing for biotic stress resistant cereal crops*, pp 257–294. Springer, Cham. https://doi.org/10.1007/978-3-030-75879-0_6 (ISBN: 978-3-030-75878-3)

Satyavathi CT, Ambawat S, Khandelwal V, Srivastava RK (2021d) Pearl millet: A climate-resilient nutricereal for mitigating hidden hunger and provide nutritional security. *Front Plant Sci* 12: 659938. doi: 10.3389/fpls.2021.659938

Satyavathi CT, Tiwari S, Bharadwaj C, Rao AR, Bhat J, Singh SP (2013) Genetic diversity analysis in a novel set of restorer lines of pearl millet [*Pennisetum glaucum* (L.) R. Br] using SSR markers. *Vegetos* 26: 72–82. https://doi.org/10.5958/j.2229-4473.26.1.011

Satyavathi CT, Tomar RS, Ambawat S, Kheni J, Padhiyar SM, Desai H, Khandelwal V (2022) Stage specific comparative transcriptomic analysis to reveal gene networks regulating iron and zinc content in pearl millet [*Pennisetum glaucum* (L.) R. Br.]. *Sci Rep* 12(1): 1–13.

Schmelzer GH (1997) Review of *Pennisetum* section *Brevivalvula* (Poaceae). *Euphytica* 97: 1–20. https://doi.org/10.1023/A:1003077906928

Schmid CD, Praz V, Delorenzi M, Périer R, Bucher P (2004) The Eukaryotic Promoter Database EPD: The impact of in silico primer extension. *Nucl Acids Res* 32(1): 82–85.

Sehgal D, Rajaram V, Armstead IP, Vadez V, Yadav YP, Hash CT (2012) Integration of gene-based markers in a pearl millet genetic map for identification of candidate genes underlying drought tolerance quantitative trait loci. *BMC Plant Biol* 12(1): 9.

Sehgal D, Skot L, Singh R, Srivastava RK, Das SP, Taunk J, Yadav RS (2015) Exploring potential of pearl millet germplasm association panel for association mapping of drought tolerance traits. *PLoS One* 10(5): 0122165.

Senthilvel S, Nepolean T, Supriya A, Rajaram V, Kumar S, Hash CT (2010) Development of a molecular linkage map of pearl millet integrating DArT and SSR markers. In: *Proceedings of the plant and animal genome 18 Conference*, pp 9–13. San Diego, CA.

Serba DD, Perumal R, Tesso TT, Min D (2017) Status of global pearl millet breeding programs and the way forward. *Crop Sci* 57: 2891–2905. https://doi.org/10.2135/cropsci2016.11.0936

Serba DD, Yadav RS, Varshney RK, Gupta SK, Mahalingam G et al (2020) Genomic designing of pearl millet: A resilient crop for arid and semi-arid nvironments. In: Kole C (ed) *Genomic designing of climate-smart cereal crops*, pp 221–286. Springer, Cham.

Shahmuradov IA, Gammerman AJ, Hancock JM, Bramley PM, Solovyev VV (2003) PlantProm: A database of plant promoter sequences. *Nucl Acids Res* 31(1): 114–117.

Sharma PC, Singh D, Sehgal D, Singh G, Hash CT, Yadav RS (2014) Further evidence that terminal drought tolerance QTL of pearl millet is associated with reduced salt uptake. *Environ Exp Bot* 102: 48–57. doi: 10.1016/j.envexpbot.2014.01.013

Sharma R, Sharma S, Gate VL (2020a) Tapping *Pennisetum violaceum*, a wild relative of pearl millet (*Pennisetum glaucum*), for resistance to blast (caused by *Magnaporthe grisea*) and rust (caused by *Puccinia substriata* var. *indica*). *Plant Dis* 104: 1487–1491. https://doi.org/10.1094/PDIS-08-19-1602-RE

Sharma S, Pandey MK, Sudini HK, Upadhyaya HD, Varshney RK (2017) Harnessing genetic diversity of wild species for genetic enhancement of cultivated peanut. *Crop Sci* 57: 1121–1131. https://doi.org/10.2135/cropsci2016.10.0871

Sharma S, Sharma R, Pujar M, Yadav D, Yadav Y et al (2020b) Use of wild *Pennisetum* species for improving biotic and abiotic stress tolerance in pearl millet. *Crop Sci* 61(1): 289–304. https://doi.org/10.1002/csc2.20408

Sharma S, Upadhyaya HD, Varshney RK, Gowda CLL (2013) Pre-breeding for diversification of primary gene pool and genetic enhancement of grain legumes. *Front Plant Sci* 4: 309. https://doi.org/10.3389/fpls.2013.00309

Shinde H, Dudhate A, Anand L, Tsugama D, Gupta SK, Liu S, Takano T (2020) Small RNA sequencing reveals the role of pearl millet miRNAs and their targets in salinity stress responses. *S Afr J Bot* 132: 395–402.

Shinde H, Tanaka Dudhate A, Tsugama D, Mine Y, Kamiya T et al (2018) Comparative de novo transcriptomic profiling of the salinity stress responsiveness in contrasting pearl millet lines. *Environ Exp Bot* 155: 619–627.

Shivhare R, Asif MH, Lata C (2020a) Comparative transcriptome analysis reveals the genes and pathways involved in terminal drought tolerance in pearl millet. *Plant Mol Biol* 103(6): 639–652. https://doi.org/10.1007/s11103-020-01015-w

Shivhare R, Lakhwani D, Asif MH, Chauhan PS (2020b) De novo assembly and comparative transcriptome analysis of contrasting pearl millet (*Pennisetum glaucum* L.) genotypes under terminal drought stress using illumina sequencing. *Nucleus* 63: 34–352.

Shivhare R, Lata C (2017) Exploration of genetic and genomic resources for abiotic and biotic stress tolerance in pearl millet. *Front Plant Sci* 7: 2069. https://doi.org/10.3389/fpls.2016.02069

Sikora P, Chawade A, Larsson M, Olsson J, Olsson O (2011) Mutagenesis as a tool in plant genetics, functional genomics, and breeding. *Int J Plant Genom* 2011(314829), pages 13. doi:10.1155/2011/314829

Singh AK, Rana MK, Singh S, Kumar S, Durgesh K, Arya L (2013) Assessment of genetic diversity among pearl millet [*Pennisetum glaucum* (L.) R Br.] cultivars using SSR markers. *Range Manag Agrofores* 34: 77–81.

Singh J, Reddy PS, Reddy S, Reddy M (2015) Molecular cloning and characterization of salt inducible dehydrin gene from the C_4 plant *Pennisetum glaucum*. *Plant Gene* 4: 55–63. 10.1016/j.plgene.2015.08.002

Singh M, Nara U (2022) Genetic insights in pearl millet breeding in the genomic era: Challenges and prospects *Plant Biotechnol Rep* 17(1): 15–37. https://doi.org/10.1007/s11816-022-00767-9

Singh SD, Navi SS (2000) Genetic resistance to pearl millet downy mildew II. Resistance in wild relatives *Mycol Plant Pathol* 30: 167–171.

Singhal T, Satyavathi CT, Singh SP, Kumar A, Sankar SM et al (2021) Multi-environment quantitative trait loci mapping for grain iron and zinc content using bi-parental recombinant inbred line mapping population in pearl millet. *Front Plant Sci* 12: 659789.

Sinha P, Singh VK, Saxena RK, Khan AW, Abbai R, Chitikineni A (2020) Superior haplotypes for haplotype-based breeding for drought tolerance in pigeonpea (*Cajanus cajan* L.). *Plant Biotechnol J* 1: 1. doi: 10.1111/pbi.13422

Slade AJ, Fuerstenberg SI, Loeffler D, Steine MN, Facciotti D (2005) Areverse genetic, nontransgenic approach to wheat crop improvementby TILLING. *Nat Biotechnol* 23: 75–81.

Spierings E, Drabbels J, Hendriks M, Pool J, Spruyt-Gerritse M, Claas F (2006) A uniform genomic minor histocompatibility antigen typing methodology and database designed to facilitate clinical applications. *PLoS One* 1: 42. doi:10.1371/journal.pone.0000042

Spindel J, Begum H, Akdemir D, Virk P, Collard B, Redona E (2015) Genomic selection and association mapping in rice (*Oryza sativa*): effect of trait genetic architecture, training population composition, marker number and statistical model on accuracy of rice genomic selection in elite, tropical rice breeding lines. *PloS Genet* 11: 1004982.

Srivastava RK, Singh RB, Pujarula VL, Bollam S, Pusuluri M et al (2020a) Genome-wide association studies and genomic selection in Pearl Millet: Advances and prospects. *Front Genet* 10: 1389. doi:10.3389/fgene.2019.01389

Srivastava RK, Bollam S, Pujarula V, Pusuluri M, Singh RB, Potupureddi G (2020b) Exploitation of heterosis in pearl millet: A review. *Plants* 9: 807. doi: 10.3390/plants9070807

Srivastava RK, Satyavathi CT, Mahendrakar MD, Singh RB, Kumar S, Govindaraj M, Ghazi IA (2021) Addressing iron and zinc micronutrient malnutrition through nutrigenomics in pearl millet: Advances and prospects. *Front Genet* 12: 723472.doi: 10.3389/fgene.2021.723472

Stair DW, Dahmer ML, Bashaw EC, Hussey MA (1998) Freezing tolerance of selected *Pennisetum* species. *Int J Plant Sci* 159: 599–605. https://doi.org/10.1086/297578

Stapf O, Hubbard CE (1934) Pennisetum. In: Prain D (ed) *Flora of tropical Africa* 19: pp. 954–1070, L. Reeves, Ashford.

Supriya A, Senthilvel S, Nepolean T, Eshwar K, Rajaram V, Shaw R (2011) Development of a molecular linkage map of pearl millet integrating DArT and SSR markers. *Theor Appl Genet* 123: 239–250.

Suszkiw J (2005) Hessian fly-resistant wheat germplasm available. *USDA–ARS*. Retrieved from www.ars.usda.gov/newsevents/news/research-news/2005/hessian-fly-resistant-wheat germplasm- available

Sy M (1994) La *sensihilite pour le Striga hermonthica (Del.) Benth.des clones de Pennesitum pedicellatum de differentes provenances duSahel*. Département de Formation en Protection desVégétaux (DFPV) Centre AGRHYMET, Niamey, Niger.

Szurman-Zubrzycka ME, Zbieszczyk J, Marzec M, Jelonek J, Chmielewska B et al (2018) HorTILLUS-a rich and renewable source of induced mutations for forward/reverse genetics and pre-breeding programs in barley (*Hordeum vulgare* L.). *Front Plant Sci* 9: 216.

Tanksley SD, Nelson JC (1996) Advanced backcross QTL analysis: A method for the simultaneous discovery and transfer of valuable QTLs from unadapted germplasm into elite breeding lines. *Theor Appl Genet* 92: 191–203. https://doi.org/10.1007/BF00223376

Tarwacka J, Polkowska-Kowalczyk L, Bozena K, Jadwiga S, Bernard W (2013) Interspecific somatic hybrids Solanum villosum (+) *S. tuberosum*, resistant to *Phytophthora infestans*. *J Plant Physiol* 170: 1541–1548. https://doi.org/10.1016/j.jplph.2013.06. 013

Tharanya M, Kholova J, Sivasakthi K, Seghal D, Hash CT, Raj B, Vadez V (2018) Quantitative trait loci (QTLs) for water use and crop production traits co-locate with major QTL for tolerance to water deficit in a fine-mapping population of pearl millet [*Pennisetum glaucum* (L.) R. Br.]. *Theor Appl Genet* 131(7): 1509–1529.

Thompson JD, Higgins DG, Gibson TJ (1994) CLUSTAL W: Improving the sensitivity of progressive multiple sequence alignment through sequence weighting, position-specific gap penalties and weight matrix choice. *Nucl Acids Res* 22(22): 4673–4680.

Till BJ, Zerr T, Comai L, Henikoff S (2006) A protocol for TILL-ING and Eco-Tilling in plants and animals. *Nat Protocols* 1: 2465–2477.

Tiwari S, Lata C (2019) Genome engineering in rice: Applications, advancements and future perspectives. In: Singh SP, Upadhyay SK, Pandey A, Kumar S (eds) *Molecular approaches in plant biology and environmental challenges*, pp 323–337. Springer, Singapore.

Tsugama D, Takano T (2021) TGIF-DB: Terse genomics interface for developing botany. *BMC Res Notes* 14(1): 1–3.

Untergasser A, Cutcutache I, Koressaar T, Ye J, Faircloth BC, Remm M, Rozen SG (2012) Primer3-new capabilities and interfaces. *Nucl Acids Res* 40(15): 15–115.

Upadhyaya HD (2011) Development of pearl millet minicore collection for enhanced utilization of germplasm. *Crop Sci* 51: 217–223.

Upadhyaya HD, Reddy KN, Gowda CLL (2007) Pearl millet germplasm at ICRISAT genebank-Status and impact. *J SAT Agric Res* 3: 1–6.

van Heerwaarden J, Doebley J, Briggs WH, Glaubitz JC, Goodman MM et al (2011) Genetic signals of origin, spread, and introgression in a large sample of maize landraces. *Proc Natl Acad Sci USA* 108: 1088–1092. https://doi.org/10.1073/pnas.1013011108

Varshney RK, Shi C, Thudi M, Mariac C, Wallace J, Qi P, Xu X (2017) Pearl millet genome sequence provides a resource to improve agronomic traits in arid environments. *Nat Biotechnol* 35: 969–976. https://doi.org/10.1038/nbt.3943

Varshney RK, Bohra A, Yu J, Graner A, Zhang Q, Sorrells ME (2021) Designing future crops: Genomics-assisted breeding comes of age. *Trends Plant Sci* 26(6): 631–649.

Varshney RK, Sinha P, Singh VK, Kumar A, Zhang Q, Bennetzen JL (2020) 5Gs for crop genetic improvement. *Curr Opin Plant Biol* 56: 190–196. doi: 10.1016/j.pbi.2019.12.004

Verma D, Singla-Pareek SL, Rajagopal D, Reddy MK, Sopory SK (2007) Functional validation of a novel isoform of Na+/H+ antiporter from *Pennisetum glaucum* for enhancing salinity tolerance in rice. *J Biosci* 32(3): 621–628.

Watson A, Ghosh S, Williams MJ, Cuddy WS, Simmonds J et al (2018) Speed breeding is a powerful tool to accelerate crop research and breeding. *Nat Plants* 4: 23–29.

Wilson JP, Hess DE, Hanna WW, Kumar KA, Gupta SC (2004) *Pennisetum glaucum* subsp. *monodii* accessions with *Striga* resistance in West Africa. *Crop Protec* 23: 865–870. https://doi.org/10.1016/j.cropro.2004.01.006

Wilson JP, Wells HD, Burton GW (1989) Inheritance of resistanceto *Pyricularia grisea* in pearl millet accessions from Burkino Faso and inbred Tift 85DB. *J Hered* 80: 499–501. https://doi.org/10.1093/oxfordjournals.jhered.a110912

Wu DY, Ugozzoli L, Pal BK, Wallace RB (1989) Allele-specific enzymatic amplification of beta-globin genomic dna for diagnosis of sickle cell anemia. *Proc Natl Acad Sci USA* 86: 2757–2760. doi:10.1073/pnas.86. 8.2757

Yadav OP, Gupta SK, Govindaraj M, Sharma R, Varshney RK et al (2021) Genetic gains in Pearl Millet in India: In sights into historic breeding strategies and future perspective. *Front Plant Sci* 12: 396.

Yadav RS, Hash CT, Bidinger FR, Cavan GP, Howarth CJ (2002) Quantitative trait loci associated with traits determining grain and stover yield in pearl millet under terminal drought stress conditions. *Theor Appl Genet* 104: 67–83.

Yadav RS, Hash CT, Bidinger FR, Devos KM, Howarth CJ (2004) Genomic regions associated with grain yield and aspects of post flowering drought tolerance in pearl millet across stress environments and tester background. *Euphytica* 136: 265–277.

Ye S, Humphries S, Green F (1992) Allele specific amplification by tetra-primer PCR. *Nucl Acids Res* 20: 1152. doi:10.1093/nar/20.5.115

Yu J, Holland JB, McMullen MD, Buckler ES (2008) Genetic design and statistical power of nested association mapping in maize. *Genetics* 178: 539–551.

12 Allele Mining for Key Traits in Finger Millet

*Salej Sood, Kalyana Babu, Dinesh C. Joshi, and Arunava Pattanyak**
*Corresponding author: apat.icarneh@gmail.com

12.1 INTRODUCTION

Crop plants have wide genetic diversity for traits of economic importance for utilization in breeding programs. Traditionally, the variants were picked based on phenotype data and used as donors to develop improved varieties. The exponential increase in the availability of sequence information in most crop plants directed the efforts towards novel allelic variants present in wild species and landraces. However, much of the genetic diversity for agronomic, biotic, and nutritional traits is untapped in crops like finger millet, where the genomics information is limited.

In the last two decades, systematic efforts have been made to generate genomic resources of finger millet and its wild relatives using comparative genomics, synteny, genotyping by sequencing, transcriptomics, and candidate gene sequencing approaches (Sood et al. 2016; Sood et al. 2019). Recently, the finger millet genome has been sequenced by two groups (Hittalmani et al. 2017; Hatakeyama et al. 2018); however, chromosome-wise assembly and annotation is still lacking. The availability of genomic sequences provides the opportunities to isolate useful alleles of genes using PCR based strategies. This enables direct access to important alleles for biotic, abiotic stresses, nutritional quality and grain yield enhancement. This dissection of naturally occurring variation in the germplasm for a given trait at a locus in the genome is called allele mining. Here in this chapter we will be discussing the allele mining efforts and opportunities in finger millet for better future varieties.

12.2 ALLELIC VARIATIONS IN THE GENOME

The use of nucleotide sequence variation to improve crops is an important goal. The variations could be naturally occurring or may be induced mutations. Mutations may have desirable positive or undesirable negative effects. The mutations with positive effects are selected while negative mutants are discarded. Mutations may be in the form of insertions and deletions, which are called InDels or may alter a single nucleotide in the genome resulting in single nucleotide polymorphisms (SNPs). The changes in coding region have major phenotypic effect due to change in protein. Allele mining helps in capturing the allelic variants for a candidate gene in the genome and their characterization to produce new genotypes. True allele mining, on the other hand, includes analysis of non-coding and regulatory regions in addition to the sequence variations in the coding region.

The allele mining success depends on the availability of diverse germplasm resources in a crop. Most of the allelic variations at any given locus occur in wild relatives and not in the cultivated species due to directional selection during domestication (Reeves et al. 2012). It shows that allele mining efforts should have more focus on wild species to find allelic variants. However, lack of

DOI: 10.1201/9781003385004-12

phenotyping for the target trait complicates the efficient characterization and exploitation of genetic variation. Therefore, core collection, which represents the complete genetic diversity in a subset, is the best option for allele mining. Allele mining helps in tracing the evolution of alleles in a genome, identification of new haplotypes and allele-specific markers for various traits to select the best genotypes using marker-assisted selection (MAS). Allele mining also provides insights into the molecular basis of novel allelic variations and identifies superior alleles.

12.3 APPROACHES FOR ALLELE MINING

12.3.1 TILLING-Based Allele Mining

It is a powerful reverse genetics tool for functional genomics, identification and characterization of variants. Most Targeting Induced Local Lesions in Genomes (TILLING) studies used chemical mutagen, ethyl methane sulfonate (EMS) for point mutations and random mutations. Use of TILLING in natural population to identify natural allelic variants is called Eco-TILLING. Both TILLING and Eco-TILLING has not been used in finger millet for allele mining but hold great promise in the future to identify new variations (Figure 12.1).

12.3.2 Sequencing Based Allele Mining

Another important approach is PCR-based amplification of alleles of a gene in diverse genotypes to identify nucleotide variation by DNA sequencing. This technique has been widely used in crop plants, including finger millet for allelic variants identification, characterization and utilization in breeding programs. Although a finger millet annotated genome is unavailable, many important genes for biotic stresses, abiotic stresses, nutritional quality, and agronomic traits have been targeted to identify superior alleles through comparative genomics. Most of the studies have been carried out using gene sequences from related species particularly, rice, maize, wheat, and foxtail millet. Besides, association mapping studies using core collection of finger millet have resulted in the identification of several candidate genes for blast resistance, protein content, calcium content, grain yield, and its component traits (Sharma et al. 2018, 2022; Tiwari et al. 2020).

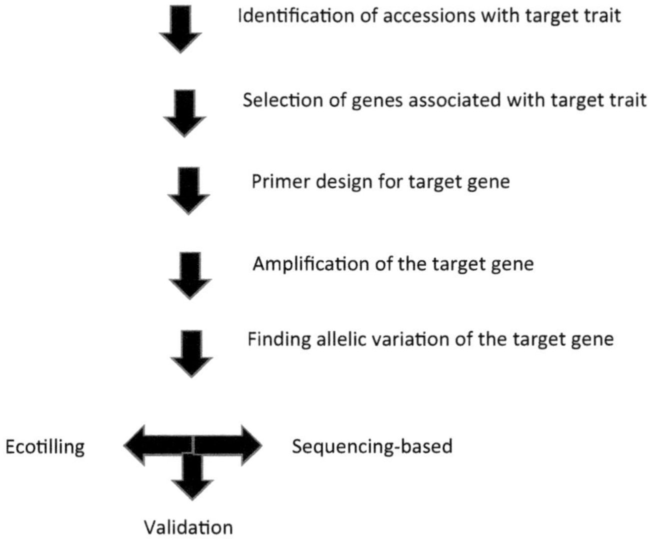

FIGURE 12.1 Protocol of allele mining in crop plants.

12.4 GENE POOL AND GENETIC DIVERSITY FOR ALLELE MINING IN FINGER MILLET

The cultivated FM (*Eleusine coracana* subsp. *coracana;* $2n = 4x = 36$) is evolved from its wild progenitor *E. coracana* subsp. *africana* in the highlands of Africa and domesticated in the semiarid regions and dryland of Africa and Asia (Sood et al. 2022). The continuous existence of landraces together with wild and weedy relatives in distinct adaptive environments resulted in the development of two separate gene pools represented by *E. coracana* spp. *coracana* (Indian gene pool) and *E. coracana* spp. *africana* (African gene pool) for the crop. These two distinct yet cross compatible subspecies constitute the primary gene pool of finger millet and provide ample scope for allele mining of yield components, nutritional quality, and stress tolerance. The secondary gene pool is represented by the diploid species viz., *E. indica* ($2n = 2x = 18$; *AA* genome), *E. floccifolia* ($2n = 18$; *BB* genome), *E. tristachya* ($2n = 18$; *AA* genome), and *E. intermedia* ($2n = 18$; *AB* genome) (Kumar et al. 2021b). Likewise, diploid species *viz*., *E. jaegeri* ($2n = 2x = 20$; *DD* genome), *E. multiflora* ($2n = 2x = 16$; *CC* genome), and *E. verticillata* ($2n = 2x = 18$) constitutes the tertiary gene pool of finger millet. These diverse species could be the reservoir of novel genes and alleles that can be mined for developing climate-resilient finger millet cultivars.

At present more than 35,000 germplasm accessions are being conserved in various national and international gene banks of which the gene bank at the International Crops Research Institute for the Semi Arid Tropics (ICRISAT), Hyderabad, India is the largest repository of finger millet germplasm (Meena et al. 2021). However, their widespread utilization in genetic studies is often restricted due to the cost associated with phenotyping of such large collections in replicated trials across multienvironments (Joshi et al. 2018). Therefore, to maximize the use of this existing genetic diversity, core collection represented by 622 accessions (Upadhyaya et al. 2006) and minicore collection represented by 80 accessions (Upadhyaya et al. 2010) have been developed. The multilocational evaluation of these collections have identified genotypes for early maturity, salinity tolerance and blast resistance. The accessions of diverse geographic origin form these collections have also been utilized in genome-wide association studies (GWAS) for identifying genomic regions governing agronomic traits (Sharma et al. 2018) and nutritional quality (Puranik et al. 2020) in finger millet.

12.5 ALLELE MINING FOR BLAST RESISTANCE IN FINGER MILLET

Allele mining involves effective utilization of variations occurred at genetic level among the naturally available germplasm for special traits like blast resistance, quality parameters, biotic and abiotic stress characters. Eco-TILLING and sequence-based allele mining are most commonly used mining approaches for mining alleles of interest. The model cereal crop, rice is well studied for allele mining for various traits like blast resistance due to availability of large germplasm. Finger millet and rice have a high amount of homogeneity at the genetic level and share more than 85% similarity of the whole genome sequence. Hence, allele mining in finger millet was carried out by comparing the sequence of rice genome. In rice, most of the genes for a trait like blast were cloned and sequenced and hence was taken as the bench mark for allele mining for finger millet.

Blast is a major and well-investigated disease in rice and a lot of reviews are available on molecular biology of the blast fungus (Caracuel-Rios and Talbot 2007; Dida et al. 2007) and blast resistance. After the availability of rice and blast genome sequences, this disease has become a model for homologous genome species for plant–disease interactions, especially in monocots. In severe cases, blast disease becomes a major obstacle to the rice yields, resulting in 50 to 90% yield loss. Molecular breeding programs like gene pyramiding (Bonman et al. 1992) were mostly able to obstruct the evolution of blast disease. Tackling the resistance at host-plant level can be a major approach to increase the degree of resistance. Nearly 48 R-genes were sequenced from different crops like rice, wheat and *Arabidopsis* etc. using transposon-based and map-based approaches of cloning. The alleles like *PiK* and *Pita* were cloned from wild and rice landraces for their control against blast disease. Similarly, *Pi54* orthologous allele cloned and sequenced from *Oryza rhizomatis* belongs to one of the nucleotide binding site-Leucine-rich repeat (NBS-LRR) family. Qu et al. (2006) cloned *Pi9* gene, which is presently being used in marker-assisted selection programs for rice blast disease.

Comparative genomics plays crucial role in allele mining for blast resistance in finger millet (Babu et al. 2014a) by utilizing the large genetic data available at public domain. Finger millet draft genome sequence has been published recently (Hittalmani et al. 2017), allele mining in finger millet is mostly dependent on rice genome and expressed sequence tags (ESTs). Similarly, for allele mining in rye, homologous alleles for aluminum tolerance were taken from wheat crop since it is well-explored crop and homologous to rye crop. Identification of core germplasm and their development plays an important role in any crop and also in finger millet. Genes like *PiKh*, *Pita*, and *Pi9* can be more explored for blast disease resistance in finger millet. Babu et al. (2014a) by using comparative genomics, identified several significant quantitative loci linked for three types of blast (leaf neck, and finger blast) resistance using genome wide association study and comparative genomics (CG). They developed 58 microsatellite markers from intron regions, 5' and 3' untranslated region (UTR). This is the first report of allele mining for blast disease resistance in finger millet using the CG approach. They also sequenced the alleles, which were found significant and showed that NBS-LRR regions in finger millet consisted of *kinase2* and *kinase 3a* domains. These may be further cloned and validated for resistance against blast in finger millet. The results found that *Pi-d(t)* and *P5* homologous genes might be playing a significant role for finger blast disease resistance, which were located on chromosome 2A and 6B, respectively. They also found *Pi21* homologous gene located on chromosome 4B of finger millet confer resistance for leaf blast in finger millet. Though the researchers used markers from several genes for association mapping, only three genes viz., *Pi-d(t)*, *Pi5* and *Pi21* were found to be linked to blast disease resistance.

12.6 ALLELE MINING FOR ESSENTIAL AMINO ACID IMPROVEMENT IN FINGER MILLET

Essential amino acids are important source of dietary proteins, which are mostly supplied by cereals; however, they are present in an imbalanced manner (Kumar et al. 2016b). Lysine and tryptophan are the major essential amino acids required for human metabolism, which are limiting in most of the cereals. In maize crop, it was reported that *opauqe2* mutation could increase the amount of lysine and tryptophan to double the content which is notably known as quality protein maize (Vasal et al. 1980). Along with *opaque2*, its modifiers also play a significant role in achieving the maximum level of essential amino acids. Several candidate genes *viz.*, *aspartate kinase-homoserine dehydrogenase* (*AK-HSDH*) and *aspartate kinase2* (*AK2*), have been reported in maize, which are influencing the amount of essential amino acids (Wang et al. 2007). In this connection, comparative genomics is a vital tool in exploiting the large genetic database of maize to go for mining the alleles responsible for above said essential amino acids in finger millet also.

Though finger millet has high nutritional value parameters, it also lacks essential amino acids. Through comparative genomics, it is possible to mine and use the alleles responsible for lysine and tryptophan to improve the total content of proteins. Babu et al. (2014b), reported for the first time, identification of allelic loci responsible for *opaque2* modifiers in finger millet using primers derived from expressed sequence tag (EST) sequences of candidate genes of maize *opaque2*. Their results found five genetic loci on chromosome 5, 7, and 9 of finger millet affecting the essential amino acids. A high amount of phenotypic variance was explained by one simple sequence repeat (SSR) loci on chromosome 9, which was found linked to *Waxy1* gene (Gibbon et al. 2003) of maize. The homolog of the same gene loci also might be significantly influencing the amino acid levels in finger millet, which is a positive sign for further improvement. The loci found on the chromosome was found to be associated with *Zea mays aspartate kinase 3* gene (*ZmASK3*), which is responsible for transforming from aspartate to aspartate semi aldehyde. Thus, close association of homologous markers in finger millet genome for different candidate genes influencing kinase linked protein phosphatise, aspartate kinase, and waxy2 gene might be playing an important role in enhancing the essential amino acid levels in finger millet. Nirgude et al. (2014) designed EST-based SSRs for *opaque2* modifiers and found primers FMO2E33 and FMO2E27 were associated to opaque gene.

12.7 ALLELE MINING FOR CALCIUM AND PROTEIN IMPROVEMENT IN FINGER MILLET

Finger millet is perceived as super cereal crop due to its high nutritious values and well adapted to harsh climatic conditions. It is a rich source of minerals, proteins, anti-oxidants among the global major cereal crops. It also has high linoleic and alpha linoleic acid content, which plays a significant role in the central nervous system of human beings. Diabetes is a rapidly growing disease, which can be controlled by finger millet-based products, which drastically brings down the blood glucose levels (Pradhan et al. 2010). The genomics approaches allows the researchers to find and trace the gene or loci influencing several important nutritional parameters in finger millet and their further use in cloning of the genes. Allele mining approaches like Eco-TILLING and sequence-based tilling helps in discovering significant loci or alleles of importance. For this purpose, utilization of landraces and germplasm with high diversity helps in unraveling the mechanisms involved in nutritional pathway genes. Putative genetic loci for calcium trait in finger millet have been mapped by several workers, which are the basis for further molecular studies (Yadav et al. 2014; Kumar et al. 2015). Nirgude et al. (2014) designed and developed EST-based microsatellite markers for finger millet calcium and protein content. These were designed from the calmodulin and calcium transporter genes. They observed a large amount of allelic variation with polymorphic primers and estimated the genetic diversity among the landraces. A more recent study by Sharma et al. (2022) described the genetics of candidate genes involved for calcium content using association mapping methodologies using SNP. They used 2977 high quality SNP loci for marker trait analysis. These genotypes also included core germplasm, which is a basic requirement of allele mining studies to unravel the genetic loci. Several putative genes like *EcCBP* and *EcCIPK7* were identified.

12.8 ALLELE MINING FOR AGRONOMIC TRAITS

Genetic architecture of the yield contributing traits like plant height, flowering, maturity, panicle length, and tillering plays a pivotal role in designing a suitable breeding strategy for their improvement (Paschapur et al. 2021). Therefore, deciphering genes and loci governing these traits is of paramount importance for realizing rapid genetic gains in finger millet. Adequate phenotypic variation for the various agronomic traits has been documented in global (Upadhyaya et al. 2006) and regional (Manyasa et al. 2015) germplasm collections of finger millet. Integration of this phenotypic variation with next generation molecular markers like SNPs described the allelic variation and genetics of flowering, maturity, plant height, and panicle length in finger millet (Sharma et al. 2018). Further, *in silico* comparative genomics of the identified genomic regions in finger millet were found to be associated with several cross genome orthologs that were proximal to several candidate genes known to regulate the respective traits in other grass species. For instance, the identified SNPs for agronomic traits were found to be associated with genes that are known to encode *ATP synthase, Photosystem I and Photosystem II*, and CO_2 assimilation, which may play an important role in enhancing finger millet grain yield (Sharma et al. 2018). In a subsequent GWAS study conducted on a diversity panel of 128 finger millet genotypes of Indian and exotic origin, seven QTLs were identified for various agronomic traits in finger millet (Ramakrishnan et al. 2016). To aid in the understanding of genetic architecture of agronomic traits, Pendergast et al. (2021) constructed a 18-linkage-group genetic map comprising 5,422 markers and compared it with rice, foxtail millet and sorghum, which identified 16 putative candidate genes that may regulate the trait variation in finger millet. In a more recent study, 13 important agronomic traits were characterized using a set of more than 8,000 SNPs derived from genome assembly of finger millet (Bančič et al. 2021). The GWAS panel comprising 423 landraces of Asian an African origin revealed that in terms of genetic complexity, threshing percentage and days to flowering were simple traits, whereas finger length and grain yield were found to be governed by a more complex genetic architecture.

12.9 ALLELE MINING FOR TRAITS GOVERNING ABIOTIC STRESS TOLERANCE

Information regarding the genomic resources and allelic variation for traits conferring tolerance to drought, salinity, water logging, heat stress, and aluminum toxicity is very scanty in finger millet (Kumar et al. 2021a). A survey of the available literature revealed that there are considerably fewer (< 2000) publically available ESTs, which are associated with drought, salinity, and other stress-related traits (Ceasar et al. 2018). Apart from sequence-based variation, mutant database for the above-mentioned traits is also lacking in finger millet. The genome sequence of finger millet (Hittalmani et al. 2017) revealed 2,866 drought-responsive genes, 146 C_4 pathway genes, 56 families of transcription factors for public use through National Centre for Biotechnology Information, USA (NCBI) data base. These resources will be a great platform for mining robust markers like SSRs and SNPs linked with various kinds of genes governing abiotic stress tolerance in finger millet. Many researchers have reported cross transferability of genic SSRs linked with abiotic stress tolerance and agronomic trails from closely related grass genera to finger millet (Yadav et al. 2014; Ramakrishnan et al. 2016). Such studies supported with extensive *in silico* comparative genomics studies will be highly useful to elaborate allelic variation for abiotic stress tolerance in finger millet. To investigate the genetic mechanism of drought tolerance in finger millet, Li et al. (2021) integrated transcriptome and proteome analysis and identified a total of 80,602 differentially expressed genes, and 3,009 differentially expressed proteins.

Root system architecture (RSA) plays a very important role in determining the response of a plant towards various abiotic stresses including drought (Joshi et al. 2016). In general, very little has been done for understanding the allelic variation for RSA in core, minicore, diversity panels, landraces, and wild species of finger millet. Therefore, adopting the high-throughput root phenotyping platforms developed for closely related species like sorghum (Joshi et al. 2017) and their use in large-scale phenomics and genomics studies will decipher the complex genetic architecture and allelic variation in finger millet RSA.

12.10 GAPS IN FINGER MILLET ALLELE MINING

Numerous crops, including rice, maize, and wheat, have benefited from the DNA-aided breeding methods like marker-aided selection, marker-assisted backcross breeding, pyramiding of desirable genes, speed breeding technology. So far, no finger millet variation has been improved and released using the MAS technique, despite the fact that it has the potential to improve disease resistance as well as nutritional values of finger millet (Mabinda and Masaki 2021). Neck and finger blast being the most important disease of finger millet, needs immediate attention in this direction. Utilizing these modern techniques will ultimately speed up efforts to develop finger millet to combat blast.

Expressed sequences of NBS-LRR regions of finger millet have synteny with *Pi-kh* and *Pi21* genes of rice, indicating that rice blast R-gene orthologs probably play a crucial role in finger millet blast resistance (Kumar et al. 2016a). Association mapping has been used in some cases to identify QTLs (Ramakrishnan et al. 2016). However, more extensive use of genomic resources of finger millet, now available in the public database, is required. Similarly, for foot rot and shoot fly resistance, exploration of new alleles through comparative analysis of the genome information is required. To understand the molecular mechanisms involved in the expression of key agronomic traits like yield, grain quality, stress tolerance, and tolerance/resistance to diseases and pests' new alleles/markers are required to be developed and in this regard sequencing of other species of the genera will be an important step forward. According to studies on genome collinearity, finger millet and rice, followed by foxtail millet and maize, exhibit substantial genomic synteny (Hittalmani et al. 2017; Pandian and Ramesh 2019), which is required to be used more vigorously. Mutation breeding is an efficient approach to generate new alleles as well as identify and functionally validate existing allelic variations. There is currently no literature describing the use of mutation breeding to create new kinds of finger millet. Nevertheless, given the synteny observed with rice and foxtail millet (Hittalmani et al. 2017) and the availability of a large number of mutants in both the crops, the use of information from rice and foxtail millet mutants for discovery of new alleles need serious consideration.

A number of calcium transporters have been reported in finger millet (Singh et al. 2015; Chinchole et al. 2017). All those were reported before the availability of whole genome sequence information. Although Hitalmani et al. (2017) reported some analysis of those transporters, they need to be analyzed in more detail for detecting allelic variation especially for homoeostasis. Moisture stress is also a well-studied trait in finger millet. Four phosphorous transporters of the *PHT1* family have also been reported (Pudake et al. 2017); they need further study. Finger millet is an orphan crop and like many other such crops, it has been selected for centuries in marginal soil ecologies. Such crops exhibit tolerance by being able to restrict absorption (Lambers et al. 2006) as well as increasing homoeostasis. Expression profiling of stress adaptive genes and search for new alleles in stress-tolerant cereals like finger millet will help improve many other cereals.

12.11.1 Allele Mining in the Era of NGS in Finger Millet

Since the availability of next-generation sequencing (NGS) platforms for commercial use in 2005, the technology has been extensively used for whole genome sequencing and resequencing, transcriptome sequencing (RNA-Seq), *de novo* assembly, etc. Among the various NGS technologies, sequencing by synthesis is most commonly used nowadays. The power of NGS in identifying SSRs and SNPs and subsequent allele mining has been proved beyond doubt in various crops. Although the benefit is yet to be fully realized in finger millet (Taheri et al. 2018), the technology has helped identify new SSRS, SNPs, and the two whole genome studies. In the absence of a well-annotated finger millet genome sequence, the NGS data were generally filtered/compared using rice sequences.

A genome-wide transcriptional analysis of two finger millet genotypes differing in their level of salinity tolerance was done by Rahman et al. (2014). Using the IonProton technology, they sequenced the salinity responsive leaf transcriptome of two finger millet genotypes and produced 27.91 million reads. Finger millet transcripts were mapped and annotated against rice gene models, resulting in the discovery of genes that respond to salinity and genotype-specific responses. It was discovered that the tolerant genotype had much higher levels of several functional categories of genes, including transporters, transcription factors, genes involved in cell signaling, osmotic equilibrium, and the manufacture of compatible solutes. Salinity stress inhibited photosynthetic capacity and photosynthesis-related genes in the susceptible genotype. Several genes involved in cell growth and differentiation were found to be up-regulated in both the genotypes but more specifically in the tolerant genotype. However, genes involved in the flavonoid biosynthetic pathway were down-regulated in the tolerant genotype.

Gimode et al. (2016) used NGS technology to assemble a 207 Mb region representing about 13% of the finger millet genome. They identified 10,327 SSRs, 23,285 non-homeologous SNPs, and tested 101 (each) of them. The study identified 49 and 80 polymorphic SSRs and SNPs, respectively.

They also identified many homeologous SNPs suggesting a recent origin of the allotetraploidy in finger millet. Hitalmani et al. (2017), using NGS technology, assembled 1196 Mb finger millet genome covering about 82% of the estimated genome size of the species. The genome showed high collinearity with rice followed by foxtail millet and contained 85,243genes (half of which were repeats) and 114,083 SSRs. Transcriptome analysis revealed an abundance of drought tolerance-associated genes, several calcium transporters, and C_4 pathway genes.

Using a unique polyploidy genome assembly workflow, Hatakeyama et al. (2018) reported whole genome sequencing and assembly of the finger millet genotype PR-202 (IC: 479099). Their coverage was 1189 Mb (78.2% of the estimated whole genome size). The genome size for genotype PR-202 was 2,387 scaffolds, with a maximum sequence length of 5 Mb and an N50 value of 905.318 Kb. This study identified 62,348 genes, of which 96.5 percent were single-copy genes, and roughly 91 percent had functional annotations. In the genome of PR-202, a total of 57,913 genes were duplicated with more than two copies, according to the NCBI BLAST analysis (Hatakeyama et al. 2018). The availability of whole genome sequences of ML-365 and PR-202 can be utilized for further genomic studies like SNP identification, NGS-based allele discovery, linkage and association map construction, identification of candidate genes for important agronomic traits, functional characterization of candidate genes, and marker-assisted breeding programs (Hatakeyama et al. 2018).

Genome assembly and annotation for polyploid species have proven to be particularly problematic since homeologous genes or duplicated genes resulting from genome duplication are very similar and difficult to differentiate. The scaffold length is usually of inferior quality in polyploid construction (Hatakeyama et al. 2018). The allotetraploid *Brassica juncea* genome was successfully assembled using a combination strategy of PacBio long reads and Bionano Genomics optical genome mapping. The N50 scaffold for this genome was 1.5 Mb in length. The hybrid *de novo* assembly workflow (Hatakeyama et al. 2018) involved contig assembly, and only paired-end reads were used. The contigs of >1 kb and long reads were used for hybrid assembly, followed by polishing the assembled contigs with paired-end reads. Two independent steps were followed – first scaffolding of the polished contigs with mate-pair reads and optical genome mapping. The optical genome mapping results were used for *de novo* assembly and fed for hybrid scaffolding along with the NGS scaffolds. After gap closing of the scaffolds with the contigs from pair-end reads, super scaffolds were generated.

12.12 PROSPECTS OF ALLELE MINING IN FINGER MILLET

Due to a lack of funding and scientific interest, finger millet is still an orphan crop, and its genetic and genomic status lags behind that of major cereal crops and even other millets. NGS data on transcriptomes and whole genome data has opened new possibilities for mining new alleles for various traits. Combining data from related plants like rice, pearl millet, barley, and others with syntenic research may be possible to find novel alleles. Blast resistance, moisture stress tolerance, nutritional quality (especially homeostasis), tolerance to low phosphorous and nitrogen, and florate size are most important from an allele mining point of view. The root system architecture is an important trait controlled by a complex system of signaling and hormonal pathways involving both underground and above-ground parts of the plant and environment. It is not clearly known whether root biomass increase or root turnover increase plays a crucial role in finger millet conferring tolerance to moisture and immobile nutrients like phosphorous. Finger millet will immensely benefit from allelic variation studies on these aspects using the information available in other cereals, especially rice, maize, and foxtail millet.

Similarly, studying allelic variations in genes involved in high calcium content in grains without seriously affecting nitrogen assimilation will benefit major cereals like rice, wheat, and maize. One of the critical elements of all genomic studies is a well-annotated genome, which is lacking in cultivated finger millet as the second progenitor is unknown. Available whole genome data, sequencing of suggested progenitors, and study of syntenic genomic information need immediate attention to identify the progenitor. Storability of grains under normal storage conditions has conferred the status of famine reserve to most of the millets, and it is valid for finger millet also. However, storage of finger millet flour is an issue, and genes involved in causing rancidity need to be identified and modified by creating new variations or adding new alleles.

Innovative genome editing technologies that produce altered alleles without creating transgenic plants are now being investigated to avoid the controversy surrounding genetically modified organisms (Luo et al. 2015; Miroshnichenko et al. 2019). However, the technology requires a robust tissue culture system, which is lacking in finger millet. A robust tissue culture technique combined with the use of recent DNA delivery systems like cell-penetrating peptides (Wu et al. 2021), lyposome encapsulated delivery will not only create new alleles in finger millet but also help in functional validations of existing alleles. Only a few biotechnology methods have been tested for crop improvement in finger millet (Sood et al. 2019). But as the cost of technologies falls, the cost of producing new information falls exponentially, and this is a reality now. Numerous institutes worldwide have access to a significant collection of more than 28,041 finger millet germplasm for genetic and breeding research (Ceasar et al. 2018).

12.13 CONCLUSION

A nutrient-rich and drought-resistant crop, finger millet is mainly grown and eaten by resource-constrained farmers in developing nations in Asia and Africa. Although recent studies revealed that finger millet is also susceptible to drought, salinity, and low nutritional stressors in addition to fungal blast, this crop has traditionally been thought of as a climate-resilient crop for the impoverished world. Before the introduction of whole genome sequencing (WGS), only a small number of research on the characterization of finger millet's functionally significant genes had been conducted. While most big genomes are made up of species-specific repetitive DNA sequences, plant species contain highly conserved areas at the DNA sequence level. By examining tiny, less complicated genomes, comparative genomics has created new opportunities for the map-based positional cloning of genes encoding significant features on large, convoluted genomes. To effectively compare sequences between different species of interest, comparative genomics studies on finger millet will mostly rely on low-cost sequencing methods and computer tools that can handle vast amounts of sequences. The strong proof of the finger millet genome's collinearity with the rice and foxtail millet genomes will accelerate progress and produce valuable databases and datasets for real-world breeding.

Additionally, thanks to enhanced sequencing capacity, researchers can utilize sorghum and some of the less studied cereals as models for sequencing the genes and alleles responsible for distinctive features, such as broad adaptation to stress-prone settings. To manage and evaluate the massive datasets that these technological breakthroughs will produce, it is necessary to invest in cutting-edge computational and bioinformatics tools. In the field of finger millet genomics, the most recent innovation in the hybrid *de novo* assembly technique will be a game changer.

REFERENCES

Babu BK, Dinesh P, Agrawal PK, Sood S, Chandrashekara C, Bhatt JC, et al. (2014a) Comparative genomics and association mapping approaches for blast resistant genes in finger millet using SSRs. *PLoS One* 9: e99182. doi: 10.1371/journal.pone.0099182

Babu BK, Pandey D, Agrawal PK, Sood S, Kumar A (2014b) *In silico* mining, type and frequency analysis of genic microsatellites of finger millet (*Eleusine coracana* (L.) Gaertn.): A comparative genomic analysis of NBS–LRR regions of finger millet with rice. *Mol Biol Rep* 41: 3081–3090. doi: 10.1007/s11033-014-3168-8

Bančič J, Odeny DA, Ojulong HF, Josiah SM, Buntjer Jaap, Gaynor RC, Hoad SP, Gorjanc G, Dawson IK (2021) Genomic and phenotypic characterization of finger millet indicates a complex diversification history. https://doi.org/10.1101/2021.04.06.438618

Bonman JM, Khush GS, Nelson RJ (1992) Breeding rice for resistance to pests. *Annu Rev Phytopathol* 30: 507–528.

Caracuel-Rios Z, Talbot NJ (2007) Cellular differentiation and host invasion by the rice blast fungus *Magnaporthe grisea. Curr Opin Microbiol* 10: 339–345.

Ceasar SA, Maharajan T, Ajeesh Krishna TP, Ramakrishnan M, Victor Roch G, Satish L, Ignacimuthu S (2018) Finger millet [*Eleusine Coracana* (L.) Gaertn] improvement: Current status and future interventions of whole genome sequence. *Front Plant Sci* 9: 1–16.

Chinchole M, Pathak RK, Singh UM, Kumar A (2017) Molecular characterization of EcCIPK24 gene of finger millet (*Eleusine coracana* (L.) Gaertn) for investigating its regulatory role in calcium transport. 3 *Biotech* 7: 1–10. doi:10.1007/s13205-017-0874-7

Dida MM, Srinivasachary, Ramakrishnan S, Bennetzen JL, Gale MD, Devos KM (2007) The genetic map of finger millet, *Eleusine coracana. Theor Appl Genet* 114: 321–333

Gibbon BC, Wang X, Larkins BA (2003) Altered starch structure is associated with endosperm modification in Quality Protein Maize. *Proc Natl Acad Sci USA* 100: 15329–15334.

Gimode D, Odeny DA, De Villiers EP, Wanyonyi S, Dida MM, Mneney EE, Muchugi A, Machuka J, De Villiers SM (2016) Identification of SNP and SSR markers in finger millet using next generation sequencing technologies. *PLoS One* 11: 1–23.

Hatakeyama M, Aluri S, Balachadran MT, Sivarajan SR, Patrignani A, Grüter S, Poveda L, Shimizu-Inatsugi R, Baeten J, Francoijs KJ, Nataraja KN, Reddy YAN, Phadnis S, Ravikumar RL, Schlapbach R, Sreeman SM, Shimizu KK (2018) Multiple hybrid *de-novo* genome assembly of finger millet, an orphan allotetraploid crop. *DNA Res* 25(1): 39–47. doi: 10.1093/dnares/dsx036

Hittalmani S, Mahesh HB, Deepak Shirke M, Biradar H, Uday G, Aruna YR, Lohithaswa HC, Mohanrao A (2017) Genome and transcriptome sequence of finger millet (*Eleusine coracana* (L.) Gaertn.) provides insights into drought tolerance and nutraceutical properties. *BMC Genom* 18: 465. DOI 10.1186/s12864-017-3850-z

Joshi D, Singh V, van Oosterom E, Mace E, Jordan D, Hammer G (2016) Genetic manipulation of root system architecture to improve drought adaptation in sorghum. In: Rakshit S, Wang YH (eds) *The Sorghum Genome. Compendium of Plant Genomes*. Springer, Cham. https://doi.org/10.1007/978-3-319-47789-3_11

Joshi DC, Singh V, Hunt C, Maee E, Oosterom E, Sulman R, Jordan D, Hammer G (2017) Development of a phenotyping platform for high throughput screening of nodal root angle in sorghum. *Plant Methods* 13: 56. https://doi.org/10.1186/s13007-017-0206-2

Joshi DC, Sood S, Lakshmi Hosahatti R, Kant L, Pattanayak A, Kumar A, Yadav D, Stetter MG (2018) From zero to hero: The past, present and future of grain amaranth breeding. *Theor Appl Genet* 131: 1807–1823.

Kumar A, Babu BK, Yadav S, Agrawal PK (2016a) Allele mining for resistance gene analogs (RGAs) in crop plants: A special emphasis on blast resistance in finger millet *(Eleusine coracana* (L.)). *Indian J Genet* 76: 1–9. doi: 10.5958/0975-6906.2016.00001.8

Kumar A, Metwal M, Kaur S, Gupta AK, Puranik S, Singh S, et al. (2016b) Nutraceutical value of finger millet [*Eleusine coracana* (L.) Gaertn.], and their improvement using omics approaches. *Front Plant Sci* 7: 00934. doi: 10.3389/fpls.2016.00934

Kumar A, Mirza N, Charan T, Sharma N, Gaur VS (2015) Isolation, characterization and immunolocalization of a seed dominant CaM from finger millet (*Eleusinecoracana* (L.) Gartn.) for studying its functional role in differential accumulation of calcium in developing grains. *Appl Biochem Biotechnol* 172: 2955–2973. doi: 10.1007/s12010-013-0714-0

Kumar A, Tomar RSS, Chandra A, Joshi D, Tiwari S, Singh P, Choudhary RK, Kumar V (2021b) Genomics-Assisted improvement of grain quality and Nutraceutical properties in Millets. In: Kumar A, Tripathi MK, Joshi D, Kumar V (eds) *Millets and Millet Technology*. Springer, Singapore. https://doi.org/10.1007/978-981-16-0676-2_17

Kumar A, Tripathi MK, Joshi D, Kumar V (2021a) *Millets and Millet Technology*. Springer, Singapore, p 438.

Lambers H, Michael W, Shane W, Cramer MD, Pearse SJ, Veneklaas EJ (2006) Root structure and functioning for efficient acquisition of phosphorous: Matching morphological and physiological traits. *Ann Bot* 98: 693–713.

Li J, Wang Y, Wang L, Zhu J, Deng J, Tang R, et al. (2021) Integration of transcriptomic and proteomic analyses for finger millet [*Eleusine coracana* (L.) Gaertn.] in response to drought stress. *PLoS One* 16: e0247181. https://doi.org/10.1371/journal.pone.0247181

Luo S, Li J, Stoddard TJ, Baltes NJ, Demorest ZL, Clasen BM, et al. (2015) Non-transgenic plant genome editing using purified sequence-specific nucleases. *Mol Plant* 8: 1425–1427. doi: 10.1016/j.molp.2015.05.012

Manyasa EO, Tongoona P, Shanahan P, Mgonja MA, de Villiers S (2015) Genetic diversity in East African finger millet (Eleusine coracana (L.) Gaertn) landraces based on SSR markers and some qualitative traits. *Plant Genet Resour* 13: 45–55. doi:10.1017/S1479262114000628

Mbinda W, Masaki H (2021) Breeding strategies and challenges in the improvement of blast disease resistance in finger millet. A current review. *Front Plant Sci* 11: 602882. doi: 10.3389/fpls.2020.602882

Meena RP, Joshi D, Bisht JK, Kant L (2021) Global scenario of millets cultivation. In: Kumar A, Tripathi MK, Joshi D, Kumar V (eds) *Millets and Millet Technology*. Springer, Singapore. https://doi.org/10.1007/978-981-16-0676-2_2

Miroshnichenko DN, Shulga OA, Timerbaev VR, Dolgov SV (2019) Generation of non-transgenic genome-edited plants: Achievements, challenges and prospects. *Biotekhnologiya* 35, 3–26. doi: 10.21519 / 0234-2758-2019-35-1-3-26

Nirgude M, Babu BK, Shambhavi Y, Singh UM, Upadhyaya HD, Kumar A (2014) Development and molecular characterization of genic molecular markers for grain protein and calcium content in finger millet (*Eleusine coracana* (L.) Gaertn.). *Mol Biol Rep* 41: 1189–1200. doi: 10.1007/s11033-013-2825-7

Pandian S, Ramesh M (2019) Decoding of finger millet genome: A milestone of millet genomics. *Signal Transduct Insights* 8: 1–3. doi: 10.1177/1178643418820541

Paschapur AU, Joshi D, Mishra KK, Kant L, Kumar V, Kumar A (2021) Millets for life: A brief introduction. In: Kumar A, Tripathi MK, Joshi D, Kumar V (eds) *Millets and Millet Technology*. Springer, Singapore. https://doi.org/10.1007/978-981-16-0676-2_1

Pendergast TH, Qi P, Odeny DA, Dida MM, Devos KM (2021) A high-density linkage map of finger millet provides QTL for blast resistance and other agronomic traits. *Plant Genome* 15(1): e20175. doi: 10.1002/tpg2.20175. Epub 2021 Dec 13. DOI: 10.1002/tpg2.20175

Pradhan A, Nag SK, Patil SK (2010) Dietary mamagement of finger millet (*Elusine coracana* L. Gaertn) controls diabetes. *Curr Sci* 98: 763–765.

Pudake RN, Mehta CM, Mohanta TK, Sharma S, Varma A, Sharma AK (2017) Expression of four phosphate transporter genes from finger millet (*Eleusine coracana* L.) in response to mycorrhizal colonization and Pi stress. 3 *Biotech* 7: 1–13. https://doi.org/10.1007/s13205-017-0609–9

Puranik S, Sahu PP, Beynon S, Srivastava RK, Sehgal D, Ojulong H Yadav R (2020) Genome-wide association mapping and comparative genomics identifies genomic regions governing grain nutritional traits in finger millet (*Eleusine coracana* L. Gaertn.). *Plants People Planet* 2: 649–662 http://dx.doi.org/10.1002/ppp3.10120

Qu S, Liu G, Zhou B, Bellizzi M, Zeng L, Dai L, Han B, Wang GL (2006) The broad-spectrum blast resistance gene Pi9 encodes a nucleotide-binding site-leucine-rich repeat protein and is a member of a multigene family in rice. *Genetics* 172: 1901–1914.

Rahman H, Jagadeeshselvam N, Valarmathi R, Sachin B, Sasikala R, Senthil ND, Sudhakar D, Robin S, Raveendran M (2014) Transcriptome analysis of salinity responsiveness in contrasting genotypes of finger millet (*Eleusine coracana* L.) through RNA-sequencing. *Plant Mol Biol* 85: 485–503.

Ramakrishnan MSA, Ceasar V, Duraipandiyan KK, Vinod K, Kalpana NA, Al-Dhabi, Ignacimuthu S (2016) Tracing QTLs for leaf blast resistance and agronomic performance of finger millet (*Eleusine Coracana* (L.) Gaertn.) genotypes through association mapping and *in Silico* comparative genomics analyses. *PLoS One* 11: 1–23.

Reeves PA, Panella LW, Richards CM (2012) Retention of agronomically important variation in germplasm core collections: Implications for allele mining. *Theor Appl Genet* 124: 1155–1171.

Sharma D, Tiwari A, Sood S, Jamra G, Singh NK, Meher PK, Kumar A (2018) Genome wide association mapping of agro-morphological traits among a diverse collection of finger millet (*Eleusine coracana* L.) genotypes using SNP markers. *PLoS One* 13(8): e0199444. https://doi.org/10.1371/journal.pone.0199444

Sharma D, Tiwari A, Sood S, Meher PK, Kumar A (2022) Identification and validation of candidate genes for high calcium content in finger millet [*Eleusine coracana* (L.) Gaertn.] through genome-wide association study. *J Cereal Sci* 107: 103517.

Singh UM, Metwal M, Singh M, Taj G, Kumar A (2015) Identification and characterization of calcium transporter gene family in finger millet in relation to grain calcium content. *Gene* 566: 37–46.

Sood S, Babu BK, Joshi D (2022) History, Botanical and Taxonomic Description, Domestication, and Spread. In: Kumar A, Sood S, Babu BK, Gupta SM, Rao BD (eds) *The Finger Millet Genome. Compendium of Plant Genomes*. Springer, Cham. https://doi.org/10.1007/978-3-031-00868-9_1

Sood S, Joshi DC, Chandra A, Kumar A (2019) Phenomics and genomics of finger millet: Current status and future prospects. *Planta* DOI: 10.1007/s00425-019-03159-6

Sood S, Kumar A, Babu BK, Gaur VS, Pandey D, Kant L, et al. (2016) Gene Discovery and advances in finger millet [*Eleusine coracana* (L.) Gaertn.] genomics-an important nutri-cereal of future. *Front Plant Sci* 7: 1634. doi: 10.3389/fpls.2016.01634

Taheri S, Abdullah TL, Yusop MR, Hanafi MM, Sahebi M, Azizi P, Shamshiri RR (2018) Mining and development of novel SSR markers using next generation sequencing (NGS) data in plants. *Molecules* 23: 399. doi:10.3390/molecules23020399

Tiwari A, Sharma D, Sood S, Jaiswal JP, Pachauri SP, Ramteke PW, Kumar A (2020) Genome-wide association mapping for seed protein content in finger millet (*Eleusine coracana*) global collection through genotyping by sequencing. *J Cereal Sci* 91: https://doi.org/10.1016/j.jcs.2019.102888

Upadhyaya HD, Albrecht T, Sube S, Dwivedi SL (2010) Developing a mini-core collection in finger millet *Crop Sci* 50: 1924–1931.

Upadhyaya HD, Gowda CLL, Pundir RPS, Reddy VG, Singh S (2006) Development of core subset of finger millet germplasm using geographical origin and data on 14 quantitative traits. *Genet Resour Crop Evol* 53: 679–685.

Vasal SK, Villegas E, Bjarnason M, Gelaw B, Goertz P (1980) Genetic modifiers and breeding strategies in developing hard endosperm opaque2 materials. In: Pollmer WG, Phillips RH (eds) *Quality traits for maize for grain and silage use*. MartinusNijhoff, London.

Wang X, Jose A, Bryan CG, Lopez-Valenzuela JA, Gakiere B, Galili G, Larkins BA (2007) Characterization of monofunctional aspartate kinase genes in maize and their relationship with free amino acid content in the endosperm. *J Exp Bot* 58(10): 2653–2660.

Wu H, Zhang Z, Ke F, Jia P, Zhu K, Zhou J (2021) Cell penetrating peptides and DNA-free genome editing in plants. *Int J Plant Anim Environ Sci* 11: 474–484.

Yadav S, Gaur VS, Jaiswal JP, Kumar A (2014) Simple sequence repeat (SSR) analysis in relation to calcium transport and signaling genes reveals transferability among grasses and a conserved behaviour within finger millet genotypes. *Plant Syst Evol* 300: 1561–1568. doi: 10.1007/s00606-014-0982-3